LECTURES ON
Chemical Bonding
and
Quantum Chemistry

LECTURES ON
Chemical Bonding and Quantum Chemistry

S. N. Datta
Indian Institute of Technology, Bombay, India

World Scientific
NEW JERSEY • LONDON • SINGAPORE • BEIJING • SHANGHAI • HONG KONG • TAIPEI • CHENNAI • TOKYO

Published by

World Scientific Publishing Co. Pte. Ltd.
5 Toh Tuck Link, Singapore 596224
USA office: 27 Warren Street, Suite 401-402, Hackensack, NJ 07601
UK office: 57 Shelton Street, Covent Garden, London WC2H 9HE

British Library Cataloguing-in-Publication Data
A catalogue record for this book is available from the British Library.

LECTURES ON CHEMICAL BONDING AND QUANTUM CHEMISTRY

ISBN 978-981-120-000-7

For any available supplementary material, please visit
https://www.worldscientific.com/worldscibooks/10.1142/11268#t=suppl

Typeset by Stallion Press
Email: enquiries@stallionpress.com

PREFACE

The first edition was published in 1998. It found an enthusiastic response and acceptance as a text book in many universities in India. Perhaps the Lecture Note style made all the difference.

The subject matter of the text, chemical bonding and quantum chemistry, covers two one-semester core courses for the post-graduate (M.Sc.) programmes in chemistry which are routinely offered by Indian universities. It was found that this subject matter is more or less a complete novelty for many chemistry graduates. Though some aspects are supposed to be taught at the high school level and graduate (B.Sc.) level, the material actually taught is often repetitive and rudimentary. Only a few exceptions are known. For instance, the integrated M. Sc. programme in Indian Institute of Technology - Bombay exposes students to this subject during their formative years. But these instances are truly rare. It takes an enormous effort to consume the newly learnt materials of a novel subject and then to assimilate the knowledge for advanced learning and making applications such as those involved in M.Sc. project work in a restricted period of time. Orthodoxy in presentation fails here, and many of the students become completely bewildered by the accumulated information in most of the available texts. In reality the breadth of information is to be imparted to them stepwise in courses at different levels. This leads to a conflict between the time allotted for study and the requirement for the absorption of knowledge. A new approach was deemed necessary.

The present text was written in the student-friendly lecture note style. Besides, the fundamental and conceptual aspects were separated from the detailed derivations and applications – a move to benefit students of all types: some need to have only a general concept of the chemical bond, while some others are interested in the detailed theory though learning it at first at the qualitative level becomes helpful. A secondary benefit accrued from the lecture note style. The division of the whole text into a certain number of notes for one-hour lectures was found useful for teaching and learning. Depending on the course curriculum of the university, or because of other constraints such as limitation of time, a teacher or a student may choose to skip a few topics. The set of lectures presented herein more or less remains valid for post-graduate studies in almost all Indian universities. So the text may be found useful for preparing a teaching schedule suitable for each individual. The third distinctive feature of this text is that applications have been presented immediately after the involved principles in order to maintain the natural flow of discussion in its pristine simplicity. This strategy always works in the classroom. For instance, the electronic structure of atoms and the molecular orbital theory have been discussed right after the presentation of the variation principle. Perturbation theory has been followed by the traditional examples of Zeeman and Stark effects, and deviating from tradition, by further examples of crystal field theory, ligand field theory and the anharmonic molecular oscillator. Separation of the same topic has been accomplished at different levels of learning and associated rigors. Angular momenta, Hartree-Fock treatment and CI techniques have been discussed separately in both parts, at first qualitatively in chemical bonding and then quantitatively in quantum chemistry. Barring a few unavoidable expressions, the presentation in the second step is not really a repetition. In actual teaching practice too, these aspects are discussed twice in almost all schools of thought, once qualitatively and then quantitatively. This separation is in striking contrast with the exclusively qualitative style adopted by the English genius Coulson and the exclusively quantitative approaches made by the great American theoretical chemists like Eyring and Pauling. It is generally difficult to develop a text that may combine the two steps together in practice. This task has been done here. The fifth distinctive aspect of this book is that certain not-so-conventional topics have been included in quantum chemistry. Photons have been the first understood quantum mechanical systems. They are involved in different quantum chemical processes and of course in spectroscopy. There is absolutely no reason why a chemistry student would not learn about the quantum states of a photon. A discourse on photon has been rightfully accommodated. Electron and proton transfer rates have been discussed. Vibronic coupling and molecular non-rigidity have been explained. A short exposition of many-body techniques has been provided. These fundamentals may guide an advanced student towards developing an appreciation and understanding for more complicated topics.

The subjects, chemical bonding and quantum chemistry on the one hand and group theory on the other hand, cannot be mixed up at the level of knowledge of postgraduate students in Indian universities, (undergraduates in I. I. T.'s). Therefore, a short discussion on the aspects of symmetry has been incorporated as Appendix III. The outline given in this Appendix only serves the purpose of reference. Lists of character tables are given in many books including books on inorganic chemistry, physical chemistry and general physics. Hence a list is omitted here.

After almost two decades, certain additions and deletions become necessary. Also, some errors are normally found in almost every book and corrections become due. This has given rise to the need for bringing out the second edition.

A final note is on acknowledgement and appreciation. First thanks go to all students who have attended the courses and persuaded me to prepare the second edition. Carl Trindle from Virginia has been a solid pillar of inspiration. D. S. Ray and S. Adhikari from Indian Association for Cultivation of Sciences, Kolkata, D. K. Maity from Homi Bhabha National Instiute, Mumbai, Anindya Datta from IIT Bombay, and S. Pati from Jawaharlal Nehru Centre for Advanced Scientific Research, Bangalore, have constantly encouraged me to bring out the second edition. This text would not have materialized without their encouragement.

The maximum help has come from A. K. Pal who has painstakingly remade all Figures and Illusttrations using the previously computed data and various softwares, mostly ChemBioDraw but sometimes Adobe illustrator, Windows paint and Excel 2016. I have also received full support from my family members, Anasuya and daughter Gargi. I am grateful to them.

Sambhu N. Datta
Mumbai, July 2018

CONTENTS

Part II. Elemenatry Quantum Chemistry

PART I
CHEMICAL BOND AND MOLECULAR GEOMETRY

CHAPTER 1

INTRODUCTION

Lecture 1: Review of Classical Mechanics

Classical mechanics that is so successful in explaining the dynamics of bodies of macroscopic dimensions is based on Newton's postulate

$$\mathbf{F}_i = \dot{\mathbf{p}}_i \tag{1.1}$$

where $\mathbf{F}_i$ and $\mathbf{p}_i$ ($\mathbf{p}_i = m_i \dot{\mathbf{r}}_i$) are the force acting on the ith particle (body) and its linear momentum, respectively. The single dot over a dynamical variable denotes the rate of change of the same variable with time. Thus Newton's equation is stated as "force is the rate of change of linear momentum with time". This is also known as Newton's second law of motion, but it is in reality a postulate and not a law. There are actually three equations corresponding to the three Cartesian components of force in a three-dimensional space. To specify the dynamics completely, one needs to describe the force as a function of the spatial coordinates. Newton completed this task by defining the force to be derivable from a function $V(\{\mathbf{r}\})$ called the potential. For a *conservative system* the force is defined as the gradient of the potential:

$$\mathbf{F}_i = -\nabla_i V. \tag{1.2}$$

Equations (1.1) and (1.2) together construct a complete dynamical representation of the conservative system:

$$\boxed{\dot{\mathbf{p}}_i = -\nabla_i V.} \tag{1.3}$$

When the position vector for the ith particle is known at a specific instant of time, the position vector at a different instant can be precisely determined by this equation provided that the functional form of V is known. Examples of a few simple mechanical problems are given below.

Example 1: $\mathbf{F} = 0 \Rightarrow \mathbf{p}$ is independent of time. (1.4a)

This is valid for stationary bodies moving with uniform velocities when no force acts on them. This result is also known as Newton's first law of motion.

Example 2: $\mathbf{F}_1 + \mathbf{F}_2 = 0 \Rightarrow d\mathbf{p}_1 \cdot d\mathbf{s} = -d\mathbf{p}_2 \cdot d\mathbf{s}$ (in time interval dt). (1.4b)

Every action has an equal and opposite reaction. This result is known as Newton's third law..

Example 3: $V = mgh \Rightarrow h(t) = h(0) + ut - \frac{1}{2}gt^2$ (1.4c)

where $u = \dot{h}(0)$. This describes the movement of a projectile thrown with the initial velocity u in the vertical direction; h(t) is the height attained by the projectile at time instant t.

Example 4: $V = \frac{1}{2}k(x-x_0)^2 \Rightarrow \ddot{x} = -\omega^2(x-x_0),\ \omega = \sqrt{(m^{-1}k)}$ (1.4d)

$\Rightarrow x = x_0 + a \sin(\omega t + \delta)$.

This is the solution of a simple harmonic oscillation along the x axis around the mean point x_0. The circular frequency is $\nu = \omega/2\pi$ and δ is the phase.

Although Newton's equations are very simple in form, a much more abstract version of classical mechanics was developed later by Lagrange and subsequently by Hamilton. The abstract forms eventually paved the way for the formulation of a totally new system of mechanics, namely, quantum mechanics.

Lagrange's equations are obtained as follows. The total kinetic energy of a collection of N particles is

$$T = \frac{1}{2}\sum_{i=1}^{N} m_i \dot{\mathbf{r}}_i \cdot \dot{\mathbf{r}}_i \tag{1.5}$$

so that

$$m_i \ddot{x}_i = \frac{d}{dt}\left(\frac{\partial T}{\partial \dot{x}_i}\right) \tag{1.6}$$

etc. But $\dot{p}_{ix} = m_i \ddot{x}_i$, and from Eq. (1.3) one obtains

$$\frac{d}{dt}\left(\frac{\partial T}{\partial \dot{x}_i}\right) + \left(\frac{\partial V}{\partial x_i}\right) = 0. \tag{1.7}$$

Two similar equations can be written for the variation of y and z coordinates. The *Lagrangian* is defined by

$$L(x, y, z; \dot{x}, \dot{y}, \dot{z}; t) = T(\dot{x}, \dot{y}, \dot{z}) - V(x, y, z; t). \tag{1.8}$$

It is easy to show

$$\frac{d}{dt}\left(\frac{\partial L}{\partial \dot{x}_i}\right) - \left(\frac{\partial L}{\partial x_i}\right) = 0 \tag{1.9}$$

and two similar equations for the Cartesian coordinates. Equation (1.9) is in reality a transformed version of Eq. (1.3), that is, Newton's equation for a conservative system for the movement along the x-axis.

To see the advantage of Lagrange's formulation, let us carry out a transformation by defining the cartesian coordinates of N particles in terms of a set of $3N$ generalised coordinates $q_1, q_2, \ldots, q_{3N}$ and time t as

$$x_i = x_i(q_1, \ldots, q_{3N}; t) \tag{1.10}$$

etc. We write

$$\frac{\partial \dot{x}_i}{\partial \dot{q}_j} = \frac{\partial x_i}{\partial q_j} \tag{1.11a}$$

and

$$\frac{d}{dt}\left(\frac{\partial x_i}{\partial q_j}\right) = \frac{\partial \dot{x}_i}{\partial q_j}, \tag{1.11b}$$

and incorporate these results in the Lagrangian equation of motion, Eq. (1.9). When the Lagrangian L has no explicit time dependence we get the Lagrangian equation in the generalized coordinates

$$\boxed{\frac{d}{dt}\left(\frac{\partial \mathrm{L}}{\partial \dot{q}_j}\right) - \left(\frac{\partial \mathrm{L}}{\partial q_j}\right) = 0} \tag{1.12}$$

for $j = 1, 2, \ldots, 3N$. Thus the Lagrangian equation of motion remains invariant under a general transformation of coordinates so long as the system is conservative (no friction, infinite life, etc.) and the Lagrangian is explicitly independent of time. This invariance is missing from Newton's equations, Eq. (1.3). One may easily check this point by converting the Cartesian coordinates into spherical polar coordinates and studying how Newton's and Lagrange's equations appear in the latter coordinate system.

Another procedure for obtaining Lagrange's equations is to use the classical action integral

$$A = \int_{t_1}^{t_2} dt\ \mathrm{L}(q_1, \ldots, q_{3N}; \dot{q}_1, \ldots, \dot{q}_{3N}; t) \tag{1.13}$$

and to postulate that the classical trajectory is obtained by minimizing the action (*Principle of least action*). Consider the small changes of the variables $q_i(t)$ and $\dot{q}_i(t)$, write them as $\delta q_i(t)$ and $\delta \dot{q}_i(t)$, respectively, which create the small change in the action integral

$$\delta A = \int_{t_1}^{t_2} dt \left(\frac{\partial \mathrm{L}}{\partial q_i}\delta q_i + \frac{\partial \mathrm{L}}{\partial \dot{q}_i}\delta \dot{q}_i\right). \tag{1.14}$$

If δq_i vanishes at $t = t_1$ and $t = t_2$, and the Lagrangian has no explicit time dependence, this integral may be written as

$$\delta A = -\int_{t_1}^{t_2} dt \left[\frac{d}{dt}\left(\frac{\partial L}{\partial \dot{q}_i} \right) - \left(\frac{\partial L}{\partial q_i} \right) \right] \delta q_i \,. \tag{1.15}$$

For least action, $\delta A = 0$ which leads to Lagrange's equations, Eq. (1.12), as the change δq_i is as such completely arbitrary. The path of least action is depicted in Figure 1.1.

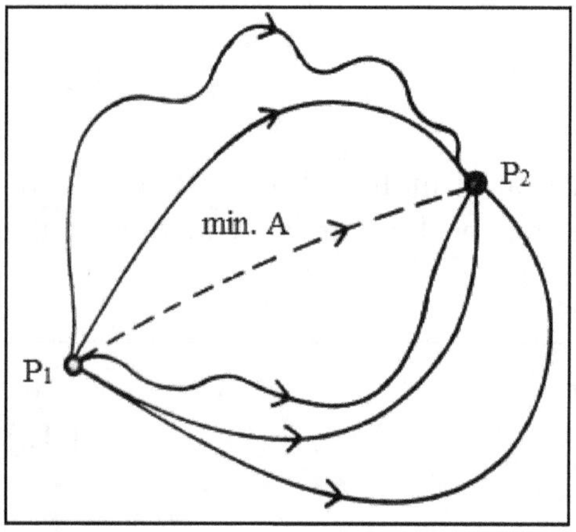

Figure 1.1 Migration starting from the phase point P_1 at time instant t_1 and reaching the phase point P_2 at instant t_2 by various possible paths in phase space. The classical path is indicated by dashes.

We have seen that Lagrange's equation of motion not only remains invariant under a coordinate transformation but also it ascribes uniqueness to the classical path. Nevertheless, there is a major difficulty with the Lagrangian formulation as the Lagrangian L is not conserved in time, that is, it is not a constant of motion.

If we compare Eq. (1.3) with Eq. (1.12) we see that $(\partial L/\partial \dot{q}_i)$ behaves as the momentum corresponding to the coordinate q_i. We then write

$$p_i = \left(\frac{\partial L}{\partial \dot{q}_i} \right) \tag{1.16}$$

and call it the *conjugate momentum*. In Hamilton's development of classical mechanics a new function H is defined by

$$H(q_1, \ldots, q_{3N}; p_1, \ldots, p_{3N}; t) = \sum_{j=1}^{3N} \dot{q}_j p_j - L(\{q\},\{\dot{q}\};t) \,. \tag{1.17}$$

Using Lagrange's equation (1.12), and the definition of the conjugate momentum as given by Eq. (1.16), the exact differential dH can be found as

$$dH = \sum_{j=1}^{3N} (\dot{q}_j dp_j - \dot{p}_j dq_j) - \frac{\partial L}{\partial t} dt. \quad (1.18)$$

This result immediately leads to Hamilton's equations

$$\dot{q}_j = \frac{dH}{dp_j}; \; \dot{p}_j = -\frac{dH}{dq_j}, \text{ and } -\frac{\partial L}{\partial t} = \frac{\partial H}{\partial t} \quad (1.19)$$

for $j = 1, \ldots, 3N$. The function H is called the classical Hamiltonian.

Like Lagrange's equations, Hamilton's equations are equivalent to Newton's equations. By appearance they are invariant under transformations of coordinates. As they are equivalent to Lagrange's equations they also follow the principle of least action. The Hamiltonian has a nice physical interpretation. It equals the total energy of the system if the coordinate transformations do not explicitly depend on time:

$$H = \sum_{j=1}^{3N} (\dot{x}_j p_{xj} + \dot{y}_j p_{yj} + \dot{z}_j p_{zj}) - L = T + V = E. \quad (1.20)$$

There is one important advantage. If the Lagrangian L does not explicitly depend on time, the Hamiltonian function becomes independent of time,

$$\frac{dH}{dt} = 0. \quad (1.21)$$

In other words, H is a *constant of motion*.

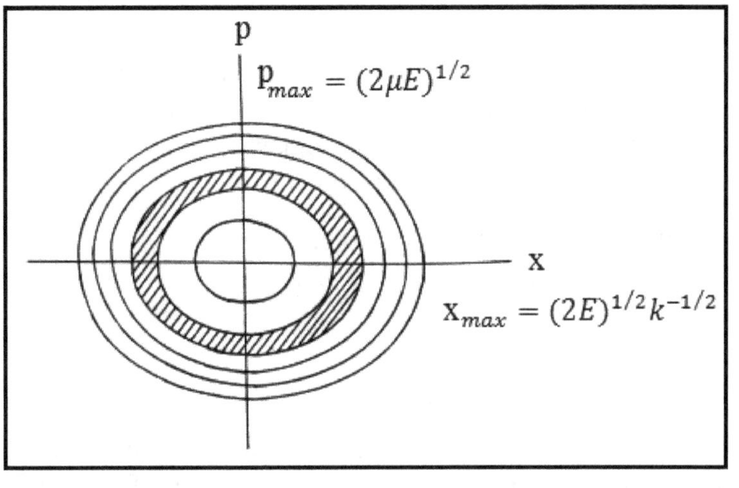

Figure 1.2 The classical path of a one-dimensional simple harmonic oscillator in the two-dimensional phase space.

The classical path of a one-dimensional simple harmonic oscillator in the two-dimensional phase space is shown in Figure 1.2. The Lagrangian equals $(\frac{1}{2}\mu\dot{x}^2 - \frac{1}{2}kx^2)$

and the Hamiltonian is given by $H = \frac{p^2}{2\mu} + \frac{1}{2}kx^2$ where $k = \mu\omega^2$ and $\omega = 2\pi\nu$. Action for one complete oscillation is $\oint pdx$, the area enclosed by the ellipse. Planck's atomic oscillators correspond to $\oint pdx = nh + c$ where $n = 0, 1, 2, \ldots$, h is Planck's constant, and c is an arbitrary constant. Bohr's orbits correspond to $\oint pdq = nh$, dq being the infinitesimally small change in the position vector in two dimension and n being the positive integers 1, 2, etc.

The time-dependence of any function of the coordinates, conjugate momenta and time along a moving point in the 6N-dimensional phase space is written as

$$\frac{d}{dt}\mathrm{f}\left(q_1, \ldots, q_{3N}; p_1, \ldots, p_{3N}; t\right) = \frac{\partial \mathrm{f}}{\partial \mathrm{t}} + \sum_{j=1}^{3N}\left[\frac{\partial \mathrm{f}}{\partial q_j}\dot{q}_j + \frac{\partial \mathrm{f}}{\partial p_j}\dot{p}_j\right] \tag{1.22}$$

which, by virtue of Hamilton's equations, takes the form

$$\boxed{\frac{d\mathrm{f}}{dt} = \frac{\partial \mathrm{f}}{\partial t} + \{\mathrm{f}, H\}_{PB}.} \tag{1.23}$$

In the above, the *Poisson bracket* $\{A,B\}_{PB}$ is defined as

$$\{A, B\}_{PB} = \sum_{j=1}^{3N}\left[\frac{\partial A}{\partial q_j}\frac{\partial B}{\partial p_j} - \frac{\partial B}{\partial q_j}\frac{\partial A}{\partial p_j}\right]. \tag{1.24}$$

Equations (1.23) and (1.24) show that the rate of change of the function f with time arises from the explicit time dependence of the same function as well as from the motion of the phase points which change with time.

The Hamiltonian function $H = (T+V)$ appears in quantum mechanics in the form of an operator. In Schrödinger's wave equation the Hamiltonian is essentially equated to an energy operator. In Heisenberg's formulation of quantum mechanics, the Poisson bracket in Eq. (1.23) is replaced by a commutator bracket with an appropriate multiplicative factor $(-\frac{i}{\hbar})$ that keeps up the dimensional equivalence.

Lecture 2: The Need for a Wave Equation

Although classical mechanics was very successful in accounting for the movement of macroscopic bodies or of an ordinary macroscopic collection of molecules, it is unable to explain the motion of electrons in atoms and molecules, and the effects arising out of this motion such as the interatomic and intermolecular forces, binding energies, frequencies and intensities of spectroscopic transitions, etc. The inadequacy of classical mechanics and classical electrodynamics became evident from a series of experiments. The distribution of the intensity of radiation emitted by a perfectly black body, the photoelectric effect, the electron-photon scattering and the frequencies of the hydrogen atomic spectra could be explained by Planck (1900), Einstein (1905), Compton(1923), and Bohr (1913) respectively only by using the hypothesis of quantized radiation or photon. Bohr's atomic model also required the quantization of the angular momentum of the electron in the hydrogen atom. Wilson (1915) and Sommerfeld (1916) proposed a more general quantization rule to extend Bohr's treatment to many more atoms. The explanation of the specific heat of solids by Einstein (1907) and Debye (1912), the Franck-Hertz experiment (1914) and the Stern-Gerlach experiment all required the idea of discrete values of physical quantities. In an attempt to resolve the problem of the nature of matter and radiation de Broglie (1924) suggested that every entity has a dual character — wave-like and particle-like. For photons the wavelength λ is related to its momentum p by

$$\lambda = \frac{h}{p}. \tag{2.1}$$

De Broglie hypothesized that this relationship holds for every entity. The hypothesis was soon confirmed by Davisson and Germer (1927) and by Thomson (1927, 1928) who observed the diffraction of electrons by a crystal lattice.

There is a conceptual difficulty in the old quantum theory. If a diaphragm with two narrow slits in it is placed in front of a source of light a diffraction pattern appears on a screen placed beyond it as illustrated in Figure 1.3. The pattern is easily explained by classical electromagnetic waves. One may also view the pattern to emerge as a large number of photons accumulate over a long period of time. However, if the intensity of the light is diminished, the intensity of the pattern decreases but the pattern itself remains unchanged with no observed deviation. This leads to the conclusion that even when the intensity of the light is decreased to such an extent that only one photon is in transit between the source and the screen, the diffraction and interference pattern would persist. Thus the pattern is not thought of being the statistical property of a large collection of photons. It is to be considered as the statistical property of a single photon.

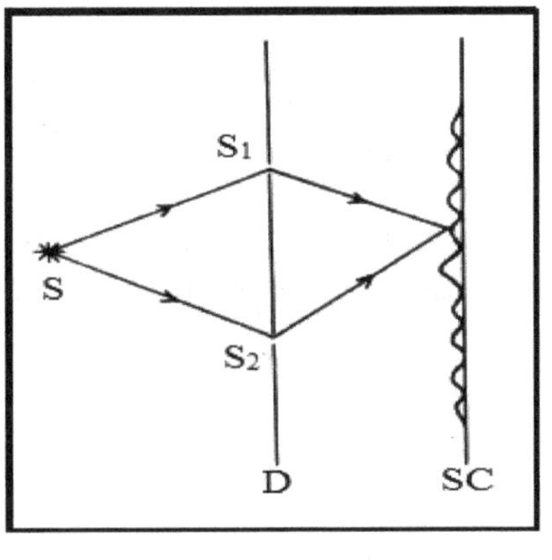

Figure 1.3 Diffraction of light: S is a light source and D is a diaphragm with slits S1 and S2. An interference pattern is produced on a photosensitive screen SC. The intensity distribution of the pattern is indicated by the curly line.

One may also cite the ad-hoc assumption in Bohr's model of hydrogen atom that the orbital angular momentum is quantized. The stability of Bohr orbits remains unexplained. Also, a detailed mechanism of the interaction of the electron with radiation is missing from Bohr's treatment. There are a few more objections but the ones already cited are enough to indicate the inadequacy of the old quantum theory and the need for a detailed mechanics.

From a chemist's viewpoint the situation can be best illustrated as follows. An electron moving along the n-th Bohr orbit of a hydrogen-like atom or ion with nuclear charge $Z|e|$ has the linear momentum

$$p_n = \frac{Z\hbar}{na_0} \tag{2.2a}$$

where $\hbar = h/2\pi$ and a_0 is the *Bohr radius*, the radius of the first Bohr orbit: $a_0 = \hbar^2/me^2$. De Broglie's hypothesis yields the wavelength

$$\lambda_n = \frac{2\pi n}{Z} a_0 \tag{2.2b}$$

which for small n is of the order of the radius of the orbit

$$r_n = \frac{n^2}{Z} a_0 . \tag{2.2c}$$

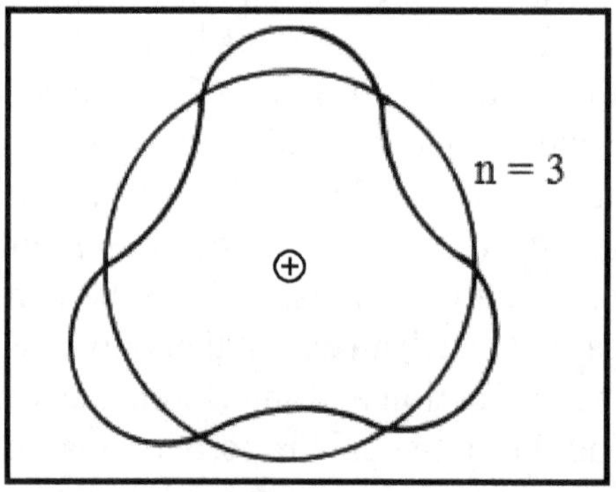

Figure 1.4 The de Broglie wavelength in Bohr's model of hydrogen atom.

Hence for small n the de Broglie wavelength is comparable to the dimension of the atom: $\lambda_n = 2\pi r_n/n$. See Figure 1.4 for illustration. Although this result lends credence to the consideration of Bohr orbits as unique orbits (closed orbits), it can in no way explain why the orbits would be stable. Instead it strongly indicates that the atomic electron should be strongly scattered by the nucleus as the distance involved is comparable to the wavelength. The electron should behave less like a particle and more like a wave, and one realizes that *a wave equation is needed* to describe the movement of electrons in atoms and molecules.

Lecture 3: Review of the Chemical Bond

The discovery of the laws of definite and multiple proportions was the earliest advancement made for a quantitative understanding of the chemical bond. These laws were complemented by Kekule's proposal of a tetravalent carbon atom and in particular by his visualization of the alternating carbon-carbon single and double bonds in the benzene molecule (1857, 1858). Later, van't Hoff (1874, 1875) and Le Bel (1874) proposed a tetrahedral bonding for the carbon atom so that they could successfully explain the optical isomerism of tartaric acid. These progresses opened up the custom of depicting a molecule as an aggregate of atoms joined by 'bonds' in three dimension in a regular fashion. The physical picture became complete with Werner's proposal of symmetric coordination in metal complexes (1893) and Sidgwick's idea of the "directed valence" (1927). To know the nature of bonding of atoms in a molecule and the molecular geometry one must know the "number" of bonds, the "type" of bonds, the bond lengths and the bond angles. Some molecules are known to have no rigid geometrical structure under normal conditions of temperature and pressure. These are called fluxional or non-rigid molecules. Most of the molecules are described by rigid "equilibrium" structures, and significant deviations from these geometries occur only at a high temperature or under extreme agitation.

Bond types are determined from the range of bond energies. The enthalpy of bond formation or the heat of bonding (ΔH) is measured from thermochemical experiments. The Gibbs free energy of bonding (ΔG) is given in terms of the heat of bonding and the entropy of bonding (ΔS) as

$$\Delta G = \Delta H - T\Delta S. \tag{3.1}$$

Spontaneous bonding takes place when ΔG is negative. The enthalpy change is in turn written in terms of the change in the internal energy E as

$$\Delta H = \Delta E + P\Delta V. \tag{3.2}$$

The internal energy E can be decomposed into contributions arising from the translational, rotational and vibrational motions of the reactants and products,

$$E = E_{\text{translation}} + E_{\text{rotation}} + E_{\text{vibration}} + E_{\text{intrinsic}}. \tag{3.3}$$

The translational and rotational contributions, and at a sufficiently high temperature, the vibrational contribution, may be estimated from the Equipartition theorem. The term $E_{\text{intrinsic}}$ is the difference between the intrinsic internal energy of the products and that of the reactants while both the reactants and the products remain in their respective equilibrium configurations. The intrinsic internal energy of an atom or a molecule owes its origin to the electromagnetic interaction between the constituent particles, electrons and nuclei, if we choose to neglect aspects of the internal structure of each nucleus.

Let us think of the formation of a molecule when two atoms (or two fragments) which are otherwise at rest and have intrinsic internal energies E_1 and E_2 are brought close to a distance R from an infinite separation ($R = \infty$). We will write the intrinsic energy of the molecule as

$$E_{\text{intrinsic}}(R) = E_1 + E_2 + E_{12}(R) \tag{3.4}$$

where $E_{12}(R)$ is the potential energy of interaction of the two fragments. That $E_{12}(R)$ is nonzero is well-known. For instance, we know of (i) the van der Waals interaction (van der Waals 1899-1900) (Examples: real gases, and liquefaction of noble gases), and (ii) a change of phase (gas to liquid, liquid to solid, and gas to solid). We would like to use $E_{12}(R)$ to understand (iii) the equilibrium molecular structure which determines the molecular properties and hence the spectral characteristics, (iv) the properties of an ensemble of molecules in thermal equilibrium, and (v) the change of molecular geometry due to chemical reactions.

The dynamical bond

The theoretical concept of the chemical bond is built up from the study of the intrinsic internal energy or the *electronic* energy of a diatomic molecule as a function of the inter-nuclear distance. The plot of the electronic energy against the inter-nuclear distance is often called the *potential energy curve* since it approximately serves as the potential energy for the relative movement of the two nuclei, (or simply the *energy curve*). A polyatomic molecule has more than one bond. For two independent inter-nuclear separations the electronic energy plot looks like a surface and so it is called a *potential energy surface*. Let us go back to the situation of a diatomic molecule, and consider an analogy favoured by H. Eyring. If one nucleus is kept fixed at $R = 0$, the other nucleus behaves just like a steel ball let loose on a curved surface. Under the influence of gravity the ball is seen to cross the deepest point and go over to the other side until it reaches a maximum height, then it turns back and comes across the depth to the original side, thus undergoing a to and fro motion. Similarly the second nucleus oscillates around the point of minimum energy. Hence the bond length corresponding to the minimum point is called the *mean bond length*. For reasons to be discussed in a moment, the mean inter-nuclear separation is also called the *equilibrium bond length* (R_{eq}).

The molecular oscillation is best understood in terms of the Hellman (1937) - Feynman (1939) theorem that in essence states that the force acting on the second nucleus is equal to the negative gradient of the energy curve:

$$F = -\frac{dE}{dR}. \tag{3.5}$$

Thus force is always in the direction of the minimum point where it vanishes. This leads to an oscillation around the minimum and the corresponding inter-nuclear distance (R_{eq}) is in turn interpreted as the mean point of oscillation of R.

A molecule not only translates and rotates, but also oscillates. What we have is truly a "dancing molecule" that remains perpetually in motion – pulsating, twisting, gyrating, freely rotating, moving and even bumping into each other! We will see later that even at absolute zero, where the translational and rotational movements freeze, the molecular oscillation continues. The *zero-point vibration* implies that under no condition the bond angles and bond lengths may remain static. One can at best talk about the mean values or the values in thermal equilibrium for these geometrical features.

Various physical properties like the electronic dissociation energy, the equilibrium bond length, vibrational frequency, the experimental dissociation energy and the anharmonic constant can be calculated from the potential energy surface. In Figure 1.5 we show that the energy curve can be approximated to a harmonic oscillator potential energy curve given by $V = \frac{1}{2}k(R-R_{eq})^2$ around the equilibrium point. The force constant k is the curvature at the minimum

$$k = \left(\frac{d^2 E}{dR^2} \right)_{R=R_{eq}} . \tag{3.6}$$

The classical vibrational frequency is given by

$$\nu = \frac{1}{2\pi} \sqrt{\frac{k}{\mu}} \tag{3.7}$$

where μ is the reduced mass of the molecular system. The harmonic oscillator energy levels are spaced uniformly by energy hν. Because of the deviation from harmonicity the successive energy levels of the molecular oscillator become progressively close to each other as the vibrational quantum number increases. It is instructive to expand $E(R)$ in a Taylor series and identify the anharmonic constant x_e of vibrational spectroscopy. Note that in Figure 1.5 we have conveniently placed the total electronic energy of the isolated atoms as zero of the scale ($E_1 + E_2 = 0$) such that $E_{12}(R)$ equals the electronic energy of the combined species. The electronic dissociation energy has been indicated as D_e in Figure 1.5. We may fairly assume that for an ensemble of molecules in thermal equilibrium the vibrational energy levels are populated according to the Boltzmann distribution equation. At an ordinary temperature $k_B T << h\nu$, almost all the molecules would remain at the zero-point level and the observed dissociation energy would be given by

$$D_o = D_e - \frac{1}{2} h\nu . \tag{3.8}$$

At a considerably high temperature, the zero-point energy in Eq. (3.8) would be replaced by $k_B T$, the average energy for each vibrational degree of freedom.

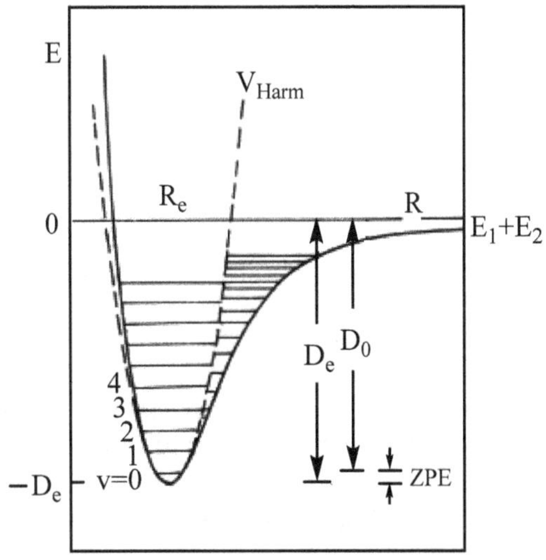

Figure 1.5 Potential energy curve $E(R)$ for a diatomic molecule. The dashed line indicates the harmonic oscillator potential $V_{Harm}(R-R_{eq})$. Two sets of vibrational energy levels have been shown by horizontal lines. The equidistant lines are for a simple harmonic oscillator. Vibrational levels for the anharmonic oscillator potential $E(R)$ come progressively close to each other.

From the discussion above there emerges a dynamical view of the chemical bond that can be defined as *the phenomenon of an optimum separation existing between two nuclei in a molecule, with the inter-nuclear distance oscillating around this optimum value*. A polyatomic molecule also has interatomic angles or bond angles, and these angles oscillate around their optimum values. This outlook gives rise to a query: at what separation can we consider the bond to be dissociated? The answer may be found from qualitative reasoning. The bond can be viewed as dissociated when the inter-nuclear separation is greater than the mean free path of the dissociated fragments in the gaseous phase, but the exact choice of the critical length for dissociation seems to be up one's own sleeve. Actually when the fragments behave more like infinitely separated species and less like associated ones, the bond may be considered ruptured. The latter logic can be very useful for bond dissociations in condensed phases where, until diffusion occurs or another reaction takes place, the fragmented species may remain in close proximity for a considerable length of time and may even recombine.

Three empirical potential energy surfaces are popular in literature and hence an outline is presented for each of them. The reader should try to work out the dissociation energy and the force constant in every case. See Figure 1.6.

Example 5: *The Lennard-Jones (6, n) potential* (Lennard-Jones 1924)

$$V_{LJ} = AR^{-n} - BR^{-6} \qquad n = 12, 13... \tag{3.9}$$

has certain attractive features. It becomes repulsive as R approaches zero, tends to zero as R increases indefinitely and has a minimum at an intermediate value of R. It is good for representing the dipole-induced dipole interaction and the induced dipole-induced dipole interaction.

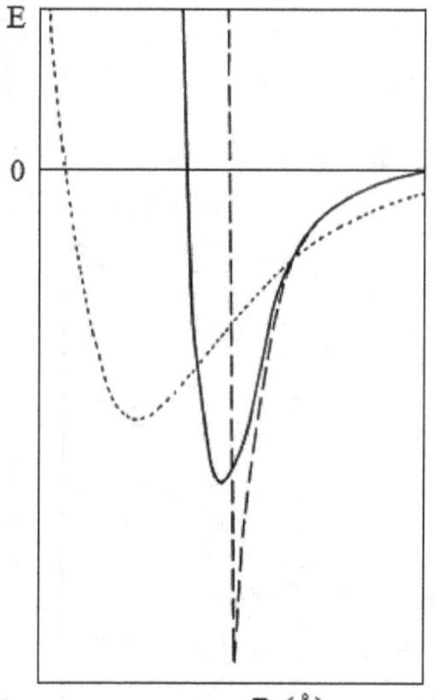

Figure 1.6 The Lennard-Jones potential (the continuous line), the Morse potential (the dotted line) and the rigid-sphere potential (the dashed line).

Example 6: *The Morse curve* (Morse 1929)

$$V_M = D[1-\exp\{-\alpha(R-R_{eq})\}]^2 - D \tag{3.10}$$

has the drawback that it is finite at $R = 0$, thereby indicating that the two nuclei can coalesce with relative ease. It gives a good representation of the right hand side of the potential energy curve.

Example 7: *The rigid sphere model,* often applied to the studies of nonpolar liquids, considers two rigid spheres each of radius R_0 and attracting each other with a force inversely proportional to the square of the distance R between their centers for $R \geq 2R_0$. For $R < 2R_0$, the potential becomes infinitely large. Thus,

$$V_{RS} = -\frac{C}{R} \qquad \text{for } R \geq 2R_0$$

where C is a constant, and (3.11)

$$V_{RS} = \infty \qquad \text{for } R < 2R_0.$$

Potential energy curves are of tremendous importance in chemistry. They play a fundamental role in the development of the equilibrium structure and the shape of a molecule, and they dictate the course and the speed of chemical reactions. Structure and reactivity cover almost all aspects of chemistry. Therefore, in a course of chemical bonding, the basic task is to know how the electronic energy $E(R)$ may be calculated as a function of the inter-nuclear distance. Our principal objectives in these lectures are to discuss the theories and methodologies for a systematic and quantitative determination of $E(R)$ and to build up a conceptual framework in which all the different aspects of molecular structure and reactivity are accommodated so that qualitative explanations can be given for a wide variety of chemical phenomena. The dynamics of electrons in atoms, molecules and condensed phases are determined by the electromagnetic interaction for which quantum physics has been proved to be entirely suitable. Discussions presented in the following lectures will be necessarily held in the language of quantum mechanics.

Texts Suggested for Further Reading

For classical mechanics:

H. Goldstein, "Classical Mechanics", (Addison-Wesley, Reading, Mass., 1950).

A. Katz, "Classical Mechanics, Quantum Mechanics, Field Theory", Chap. 1, (Academic, New York, 1965).

For historical developments:

M. Jammer, "The Conceptual Development of Quantum Mechanics", (McGraw-Hill, New York, 1966).

L. Pauling and E. B. Wilson, Jr., "Introduction to Quantum Mechanics with Applications to Chemistry", Chap.2, (McGraw-Hill, New York, 1935).

For atomic physics:

F. K. Richtmyer, E. H. Kennard and T. Lauritsen, "Introduction to Modern Physics", (McGraw-Hill, New York, 1955).

For overall conceptual understanding:

R. McWeeny, "Coulson's Valence", 3rd Edition, Chap. 1, (Oxford University Press, Oxford, 1979).

L. I. Schiff, "Quantum Mechanics", 3rd Edition, Chap. 1, (McGraw-Hill, New York, 1968).

For intermolecular forces:

P. W. Atkins, "Physical Chemistry", 4th Edition, (ELBS with Oxford University Press, 1990).

J. O. Hirschfelder, C. F. Curtiss and R. B. Bird, "Molecular Theory of Gases and Liquids", (Wiley, New York, 1964).

PROBLEMS

1. Show that the total energy of a system is conserved when the system moves under a conservative force.
2. Find the Lagrangian equations of motion of a particle in spherical polar coordinates.
3. Consider a two-particle system with potential energy as a function of only the inter-particle distance $|\mathbf{r}_2 - \mathbf{r}_1|$. Transform the Cartesian coordinates into the center of mass coordinates $X = \frac{m_1 x_1 + m_2 x_2}{M}$, $(M = m_1 + m_2)$, etc., and the relative coordinates $x = x_2 - x_1$, etc. Write the Lagrangian equations in the transformed coordinates and interpret these equations.
4. Prove that the kinetic energy of a system of N bodies can be written as

$$\mathrm{T} = \frac{1}{2}\sum_{i=1}^{3N}\sum_{j=1}^{3N} t_{ij}\,\dot{q}_i\,\dot{q}_j$$

where q_i etc. are generalized coordinates. Use the definition of conjugate momentum to show that T is given by

$$\mathrm{T} = \frac{1}{2}\sum_{i=1}^{3N} p_i\,\dot{q}_i\,.$$

5. The Kroenecker delta function is defined by

$$\delta_{i,j} = \begin{cases} 1 & \text{for } i = j, \\ 0 & \text{otherwise.} \end{cases}$$

Evaluate the following sums which will be useful in many quantum chemical calculations:

$$\sum_{j=1}^{3N}\delta_{i,j}\,;\ \sum_{i=1}^{3N}\sum_{j=1}^{3N}\delta_{i,j}\,;\ \sum_{j=1}^{3N} f_j\delta_{i,j}\,;\ \sum_{i=1}^{3N}\sum_{j=1}^{3N} f_j\delta_{i,j}\,;\ \sum_{j=1}^{3N} g_{ij}\delta_{i,j}\,;\ \sum_{i=1}^{3N}\sum_{j=1}^{3N} g_{ij}\delta_{i,j}\,.$$

6. Find the condition, in terms of Poisson bracket, for a physical quantity to be a constant of motion, (that is, conserved in time). What is the quantum mechanical correspondence of this conservation law?
7. Find the equilibrium bond length and force constant for the Lennard-Jones energy curve and the force constant associated with the Morse potential.
8. Can you guess the shape of the energy curve for an antibonding state? (Hint: There is no equilibrium bond length – hence the absence of a minimum). If you succeed, rationalize the antibonding nature of the state by using the Hellmann-Feynman theorem.

CHAPTER 2

FUNDAMENTAL BACKGROUND

Lecture 4: Postulates of Quantum Mechanics

Every system of mechanics starts from a set of axioms or postulates. These postulates cannot be verified directly, but using them a mechanics can be developed in order to explain the movement of a wide variety of dynamical systems of a particular class. Results derived from the mechanics confirm the experimental findings on such systems. For example, classical mechanics grew up from the Newtonian postulate Eq. (1.1) and Newton's definition of work or of the potential function. Classical mechanics has been successfully applied to the problems of our daily life which are mostly concerned with the movement of bodies of macroscopic dimensions and mass, and a macroscopically large collection of particles. Applications involve, to name a few, planetary motion, kinetic theory of gases, all sorts of known mechanical vehicles and almost all branches of engineering. Classical electrodynamics found its application in electrical engineering. In contrast to these applications, computer science and engineering, electronics, materials science and modern technologies, and nuclear engineering grew up mainly as applications of quantum mechanics. Our interest in quantum mechanics lies in its ability to explain the phenomenon of chemical bonding. To learn quantum mechanics we must begin with its postulates.

Therefore, a physical theory is merely a mathematical model. The chief purpose of it is to *confirm* the results of experiments. Theory also has its predictive value, but it cannot be stretched too far. Every model has its limitations. Like every model every theory fails when applied to the study of radically different entities. Quantum theory is highly successful in explaining or confirming and sometimes even predicting electromagnetic phenomena which control the behavior of atoms, molecules and solids. It has also found reasonable success in the case of weak interaction, but it is less successful in cases involving strong or very weak (gravitational) interactions. This introduction serves to summarize the applicability and limitations of quantum theory.

We now proceed to discuss the postulates of quantum mechanics. We will see later how the same postulates may be used to determine the dynamical properties of a quantum mechanical system such as a collection of electrons in an atom or in a molecule.

Postulate 1. Every observable (physical variable, measurable quantity) is represented by a mathematical operator.

A list of observables and corresponding operators is given in Table 1. Note that we have given two operators for the total energy E. One of them is the operator $i\hbar\frac{d}{dt}$ that is primarily a time derivative, and the other is the Hamiltonian operator **H**. Equivalence of these two operators gives the wave equation (Postulate 4).

An operator (**O**) operates on a function f(***r***) of coordinates, yielding another function g(***r***) as result:

$$\mathbf{O}f(\boldsymbol{r}) = g(\boldsymbol{r}). \tag{4.1}$$

Table 2.1 Observables and operators.[1]

Observable	Classical variable	Operator
Position coordinates	x, y, z	x, y, z
	r, θ, φ	r, θ, φ
	$\boldsymbol{r} = \mathbf{i}x + \mathbf{j}y + \mathbf{k}z$	$\boldsymbol{r} = \mathbf{i}x + \mathbf{j}y + \mathbf{k}z$
Linear momentum	$p_x = m\dot{x}$	$\mathbf{p_x} = -i\hbar\frac{d}{dx}$
	$\mathbf{p} = m\dot{\mathbf{r}}$	$\mathbf{p} = -i\hbar\,\nabla$
Orbital angular momentum	$\mathbf{L} = \mathbf{r} \times \mathbf{p}$	$\mathbf{L} = -i\hbar\,\mathbf{r} \times \nabla$
Kinetic energy	$\mathrm{T} = \frac{1}{2\mathrm{m}}\mathbf{p} \cdot \mathbf{p}$	$\mathbf{T} = -\frac{\hbar^2}{2m}\nabla \cdot \nabla = -\frac{\hbar^2}{2m}\nabla^2$
Potential energy	$\mathrm{V}(\mathbf{r})$	$\mathbf{V} = \mathrm{V}(\mathbf{r})$
Total energy	$E = \mathrm{T} + \mathrm{V}$	$\mathbf{H} = \mathbf{T} + \mathbf{V}$
	E	$\mathbf{E} = i\hbar\frac{d}{dt}$

[1]We have used the notation $\hbar$ for $h/2\pi$ where h is Planck's constant.

In quantum mechanics we deal with *linear operators* ($\mathbf{O}_1$, $\mathbf{O}_2$, etc.) which satisfy the properties

$$(\mathbf{O}_1+\mathbf{O}_2)f = \mathbf{O}_1 f+\mathbf{O}_2 f \tag{4.2a}$$

$$\mathbf{O}cf = c\mathbf{O}f \tag{4.2b}$$

$$(\mathbf{O}_1\mathbf{O}_2)f = \mathbf{O}_1(\mathbf{O}_2 f) \tag{4.2c}$$

where c is a constant. The function f is called a *state function* since we demand that it must uniquely characterize the dynamical state of the system. To be a state function it should be well-behaved, that is, it must be

(i) *finite*, as the measurable characteristics must be finite;
(ii) *single-valued*, as the state is to be characterized in a unique way;
(iii) *continuous*, as the elementary properties must not have any discontinuity; and
(iv) *smoothly varying*, as the first derivative must exist and must be continuous.

Consider the operator of the x-component of linear momentum $p_x = -i\hbar\frac{d}{dx}$. It operates on the two functions, $f_1(x) = \sin ax$ and $f_2(x) = e^{ikx}$, and yields different types of results:

$p_x \sin ax = -i\hbar a \cos ax = g_1(x)$;

$p_x e^{ikx} = \hbar k\, e^{ikx} = g_2(x)$.

The first result $g_1(x)$ is different from $f_1(x)$, but the second one $g_2(x)$ is merely a constant $\hbar k$ times $f_2(x)$. The second equation represents an *eigenvalue problem.* An eigenvalue equation is of the form

$$\mathbf{O}f = of \tag{4.3}$$

where o is a constant; (more precisely, o is independent of the variable involved in the operator **O**). The function f and the constant o are called an *eigenfunction* and the corresponding *eigenvalue* of the operator **O**.

In general, an operator has an infinite number of eigenfunctions and eigenvalues:

$$\mathbf{O}f_i = o_i f_i \qquad (i = 1, 2, 3, \ldots). \tag{4.4}$$

It may turn out that a set of eigenfunctions have the same eigenvalue. Functions belonging to this set will be called degenerate. We now prove a very important theorem on degenerate eigenfunction.

Theorem 1. Any arbitrary linear combination of the members of a set of degenerate functions is another degenerate function.

Proof: Let f_1, f_2, etc. be degenerate eigenfunctions of the operator **O** with eigenvalue o,

$\mathbf{O}f_1 = of_1$,

$\mathbf{O}f_2 = of_2$,

etc. By the linearity of **O**,

$$\mathbf{O}(c_1f_1 + c_2f_2 + \ldots) = c_1\mathbf{O}f_1 + c_2\mathbf{O}f_2 + \ldots = o(c_1f_1 + c_2f_2 + \ldots) \tag{4.5}$$

where c_1, c_2, etc. are arbitrary constants. Thus the arbitrary linear combination $(c_1f_1 + c_2f_2 + \ldots)$ is an eigenfunction of the operator **O** with the same eigenvalue o. **QED**

Linear combinations of the members of a degenerate set form another set of degenerate eigenfunctions. This property will be used later.

Hermitean operators form an important class of operators. An operator **O** is hermitean if it satisfies the *Turn over rule:*

$$\int dv\, f^*\mathbf{O}g = \left[\int dv\, g^*\mathbf{O}f\right]^* = \int dv\,(\mathbf{O}f)^*g \tag{4.6}$$

where the asterisk denotes complex conjugation and dv is the appropriate volume element. The integration is carried out over the concerned region, normally over all space. The equality must be valid for arbitrary well-behaved functions f and g. These functions need not necessarily be the eigenfunctions of the operator **O**.

The importance of hermitean operators stems from the following theorem.

Theorem 2. Hermitean operators have real eigenvalues and orthogonal eigenfunctions.

Proof: Let **O** be hermitean, and f_1 and f_2 be two of its eigenfunctions corresponding to the eigenvalues o_1 and o_2 respectively:

$\mathbf{O}f_1 = o_1 f_1,$
$\mathbf{O}f_2 = o_2 f_2.$

If we take complex conjugates of both sides of the first eigenvalue equation, multiply both the sides by f_2 from left and the integrate over all space, we get

$$\int dv\, f_2 \mathbf{O}^* f_1^* = o_1^* \int dv\, f_2 f_1^*. \tag{4.6a}$$

Multiplying both the sides of the second equation by f_1^* from left and integrating one obtains

$$\int dv\, f_1^* \mathbf{O} f_2 = o_2 \int dv\, f_1^* f_2 . \tag{4.6b}$$

As **O** is hermitean the left sides of the equations (4.6a) and (4.6b) are by definition equal. Hence

$$(o_1^* - o_2) \int dv\, f_1^* f_2 = 0$$

which implies that either $o_1^* = o_2$, or $\int dv\, f_1^* f_2 = 0$.

Case (i). Let f_1 and f_2 be the same function. Then $f_1^* f_2 = |f_1|^2 \geq 0$ for all points in space so that

$$\int dv\, f_1^* f_2 \geq 0,$$

the equality being valid if and only if $f_1 = 0$ at every point in space. For a nontrivial f_1 the integral is nonzero and so $o_1^* = o_2$. But $o_2 = o_1$ as the two functions are the same. Thus every eigenvalue of **O** is real.

Case (ii). If f_1 and f_2 are different nondegenerate functions, $(o_1^* - o_2) = (o_1 - o_2)$ is not equal to zero. Then

$$\int dv\, f_1^* f_2 = 0,$$

that is, nondegenerate eigenfunctions are orthogonal to each other.

Case (iii). If f_1 and f_2 are different functions belonging to a degenerate set, $(o_1^* - o_2) = 0$. Then the integral

$$\int dv\, f_1^* f_2$$

may or may not be zero. Nevertheless one can always make two linear combinations of f_1 and f_2, $g_1 = c_{11} f_1 + c_{21} f_2$ and $g_2 = c_{12} f_1 + c_{22} f_2$, such that

$$\int dv\, g_1^* g_2 = 0$$

is maintained. By Theorem 1, g_1 and g_2 are degenerate eigenfunctions of operator **O**. **QED**

You should try to prove the converse theorem,

Theorem 3. If all the eigenvalues of an operator are real and the eigenfunctions are mutually orthogonal then the operator is hermitean.

Hint: Start from the relation $(o_1^* - o_2) \int dv\, f_1^* f_2 = 0$ and work backwards.

Some operators like x, t, etc. are obviously hermitean. But when an operator involves the derivative of a variable of integration, the hermitean nature evolves from the boundary conditions. To see this, consider example 8.

Example 8: The operator $\mathbf{p}_x$ is hermitean:

$$\int_a^b dx\, f^* \mathbf{p}_x g = -i\hbar \int_a^b dx\, f^* \frac{d}{dx} g = -i\hbar \left[f^* g\right]_a^b + i\hbar \int_a^b dx\, g \frac{d}{dx} f^*.$$

If the functions f and g vanish at the boundaries $x = a$ and $x = b$ then the right hand side can be written as

$$\text{R.H.S.} = \left(\int_a^b dx\, g^* \mathbf{p}_x f \right)^*$$

and hence $\mathbf{p}_x$ is hermitean. Note that $\mathbf{p}_x$ is hermitean as it is used along with well-behaved functions that become zero at the boundaries. Also see that the multiplicative factor $-i$ makes the operator hermitean. The $\frac{d}{dx}$ operator satisfies

$$\int_a^b dx\, f^* \frac{d}{dx} g = -\left(\int_a^b dx\, g^* \frac{d}{dx} f \right)^*$$

when f and g are well-behaved, and it is an *antihermitean* operator. One can similarly check that $\frac{d}{dt}$ is antihermitean while the energy operator $i\hbar \frac{d}{dt}$ is hermitean.

Lecture 5: More Postulates

Quantum mechanics directly addresses the question: what is the result of a specific type of measurement? An experiment may be performed to measure a particular physical quantity. One usually makes more than one measurement while carrying out the experiment. The first postulate has been necessary to prepare the background in this context. The second and the third postulates of quantum mechanics specify the result of a single measurement and the average result from multiple measurements respectively.

Postulate 2. The observed value in a single measurement is equal to one of the eigenvalues of the operator.

Since the measured quantities are real, eigenvalues of observables must be real. Thus operators corresponding to the observables must be hermitean. It is also envisaged that a single measurement yields information on one of the "pure" states of the physical system. Eigenfunctions are functions representing the pure states for the dynamical variables involved.

A physical system need not necessarily be found in a pure state of the physical quantity being observed. An impure state is a mixture of the pure states. So each measurement need not necessarily yield the same observed value. In general, different values are obtained from a sequence of measurement. The average value measured from a sequence is prescribed by the third postulate of quantum mechanics.

Postulate 3. The average value measured ($\bar{o}$) equals the expectation value $< \mathbf{O} >$ that is given by

$$< \mathbf{O} > = \frac{\int dv\, f^* \mathbf{O} f}{\int dv\, f^* f} \,. \tag{5.1}$$

Note that if the system is in a pure state, that is, if its state is represented by an eigenfunction f_i, the average value equals the eigenvalue:

$$\int dv\, f_i^* \mathbf{O} f_i = \int dv\, f_i^* o_i f_i = o_i \int dv\, f_i^* f_i$$

which implies $< \mathbf{O} > = o_i$ so that $\bar{o} = o_i$. Since this would hold for any number of measurements, o_i is the observed value in every measurement when the system is in the *i*th pure state.

Let us assign the probability density $P(\mathbf{r})$ in space. Then for the ordinary operators (those not involving derivatives),

$$\bar{o} = \int dv\, \mathbf{O} P(\mathbf{r}). \tag{5.2}$$

Comparing Eq. (5.1) with Eq. (5.2) we find

$$P(\mathbf{r}) = \frac{|\, f(\mathbf{r}) \,|^2}{\int dv' \,|\, f(\mathbf{r}') \,|^2} \,. \tag{5.3}$$

The state function is said to be normalized to unity if the following equality holds:

$$\int dv\, |f(\mathbf{r})||^2 = 1. \tag{5.4}$$

The probability density corresponding to the normalized function f is easily found by substituting Eq. (5.4) in Eq. (5.3),

$$P(\mathbf{r}) = |f(\mathbf{r})|^2. \tag{5.5}$$

This leads to a *physical interpretation* of the state function or the wave function $f(\mathbf{r})$. We can regard $f(\mathbf{r})$ as the *probability amplitude* in space such that the square modulus $|f(\mathbf{r})|^2$ equals the *probability density* for *finding* the system (rather, a particle) at point **r** (rather, in the neighborhood of the point represented by the position vector **r**). That the probability refers to the finding is due to an interpretation put forward by Born (1926, 1927) on the results of scattering experiments.

The description of the process of measurement remains incomplete until one specifies how the state function develops in time. At this point we adopt the more common notation $\Psi(\mathbf{r}, t)$ instead of $f(\mathbf{r})$ for the wave function. The evolution of Ψ in time is given by the general wave equation proposed by Schrödinger (1926):

Postulate 4. Schrödinger equation: $i\hbar \dfrac{d}{dt}\Psi(\{\mathbf{r}\},t) = \mathbf{H}\Psi(\{\mathbf{r}\},t)$. (5.6)

The wave function $\Psi(\{\mathbf{r}\},t)$ is a function of the coordinates of all the particles in the system and also of time. The Hamiltonian operator **H** is for the total system.

A special situation arises when the coordinate-dependences and the time-dependence of Ψ are multiplicatively separable. For the sake of convenience we consider only one particle in the system and write

$$\Psi(\mathbf{r},t) = \psi(\mathbf{r})\, \eta(t) \tag{5.7}$$

where ψ is independent of t and η is independent of **r**. After substituting Eq. (5.7) in Eq. (5.6) and simplifying the resulting expression we get

$$i\hbar\frac{d\ln\eta(t)}{dt}=\frac{\mathbf{H}\psi(\mathbf{r})}{\psi(\mathbf{r})}. \tag{5.8}$$

This equality holds for all spatial points **r** at all instants of time *t*, and the only way it can be so is that each of the terms must be equal to a constant quantity independent of both **r** and *t*. Since **H** is the operator for total energy, this constant must have the dimension of energy. So it is written as *E*. Subsequently, from the right side of Eq. (5.8) it is found as an eigenvalue of **H**,

$$\boxed{\mathbf{H}\psi = E\psi} \tag{5.9}$$

and hence it is interpreted as the total energy of the system. We also get

$$d\ln\eta(t) = -\frac{i}{\hbar}Edt \tag{5.10}$$

with the solution

$$\eta(t) = \eta(0)e^{-iEt/\hbar}. \tag{5.11}$$

Thus the total wave function Ψ is found as

$$\Psi(\mathbf{r},t) = \psi(\mathbf{r})\eta(0)e^{-iEt/\hbar}.$$

For this wave function to be normalized,

$$|\eta(0)|^2\int dv\,|\psi(\mathbf{r})|^2 = 1.$$

Without any loss of generality we can put $\eta(0) = 1$ such that

$$\eta(t) = e^{-iEt/\hbar}, \tag{5.12a}$$

$$\Psi(\mathbf{r},t) = \psi(\mathbf{r})e^{-iEt/\hbar}, \tag{5.12b}$$

and

$$\int \mathrm{dv}\,|\psi(\mathbf{r})|^2 = 1. \tag{5.12c}$$

Equation (5.12a) gives the normalized time dependent part of the total wave function in Eq. (5.12b). The spatial dependence has to be found out from Eq. (5.9) and $\psi(\mathbf{r})$ must be normalized as shown in Eq. (5.12c). Observe that $|\eta(t)|^2=1$ such that $\rho(\mathbf{r},t) = |\Psi(\mathbf{r},t)|^2 = |\psi(\mathbf{r})|^2$, is independent of time. As the probability density remains unchanged, average values do not vary with time. Such a state is called a *stationary state*, because its properties remain unchanged with time. Thus Eq. (5.12b) shows a stationary state wave function that is necessarily associated with periodic time dependence. The spatial dependence of the stationary state is determined from Eq. (5.9) which is referred to as the *time-independent Schrödinger equation.*

Our main task in quantum mechanics is two-fold. First, we need to solve the time-independent Schrödinger equation and obtain the stationary state wave functions and the corresponding energy eigenvalues, so that we may calculate the average value of any observable in a specific stationary state. Second, we want to solve the general Schrödinger equation, Eq. (5.6), and to account for the changes in physical properties with time. A standard approach in quantum mechanics is as follows. The system is taken to be in a specific stationary state (for Hamiltonian $\mathbf{H}^0$) before an experiment begins. It is perturbed during the experiment, such as by switching on an external electric or magnetic field, whence the Hamiltonian becomes $\mathbf{H}^0+\mathbf{H}'$. The perturbation $\mathbf{H}'$ leads to a mixing of the stationary states of $\mathbf{H}^0$ during the experiment. Once the perturbation is switched off, the experiment is over and the system again has the Hamiltonian $\mathbf{H}^0$ and it continues to remain in the final mixture of stationary states. Now, Eq. (5.6) is fundamentally a first order differential equation. It gives the rate of change of the state function with time. So, if one knows the initial wave function, one can find the wave function at a later time. By solving Eq. (5.6) one can find the probability of transition from the initial stationary state to another stationary state in a given time, that is, during the course of the experiment. If these transition probabilities are known, the final state $\Psi(\mathbf{r}, t)$ can be easily written down.

Lecture 6: Commuting Operators, Complete Set and Closure

Two operators $\mathbf{O}_1$ and $\mathbf{O}_2$ are said to commute with each other if

$$[\mathbf{O}_1, \mathbf{O}_2]\,\mathrm{f} \equiv \mathbf{O}_1\mathbf{O}_2\mathrm{f} - \mathbf{O}_2\mathbf{O}_1\mathrm{f} = 0 \tag{6.1}$$

is satisfied for every arbitrary function f. We say that the commutator is equal to zero,

$$[\mathbf{O}_1, \mathbf{O}_2] = 0. \tag{6.2}$$

For example, x commutes with y as $xy\mathrm{f} = yx\mathrm{f}$, but x does not commute with $\mathbf{p}_x$ as

$$[x, \mathbf{p}_x]\mathrm{f} = -i\hbar x\frac{d\mathrm{f}}{dx} + i\hbar\frac{dx\mathrm{f}}{dx} \mathrm{f} = i\hbar\,\mathrm{f}.$$

So we write $[x, \mathbf{p}_x] = i\hbar$. One may show that

$$[\varphi, \boldsymbol{L}_z] = i\hbar$$

where $\boldsymbol{L}_z$ is the operator for the z-component of the orbital angular momentum, $\boldsymbol{L}_z = -i\hbar(d/d\varphi)$, and

$$[\mathbf{E}, \mathrm{t}] = i\hbar$$

where $\mathbf{E}$ is the energy operator, $\mathbf{E} = i\hbar(d/dt)$. We now prove a very important theorem in quantum mechanics.

Theorem 4. If two operators commute with each other they *can* have common eigen-functions.

Proof: Let ψ_i be an eigenfunction of the operator $\mathbf{O}$ with the eigenvalue o_i, and let $\mathbf{O}$ commute with $\mathbf{O}'$. Then

$$\mathbf{OO}'\psi_i = \mathbf{O}'\mathbf{O}\psi_i = \mathbf{O}'o_i\psi_i = o_i\mathbf{O}'\psi_i \tag{6.3}$$

which indicates that $\mathbf{O}'\psi_i$ is an eigenfunction of the operator $\mathbf{O}$ with the same eigenvalue o_i. If ψ_i is a nondegenerate eigenfunction then $\mathbf{O}'\psi_i$ can differ from it at most by a multiplicative factor o'_i,

$$\mathbf{O}'\psi_i = o'_i\,\psi_i. \tag{6.4a}$$

This implies that ψ_i is also an eigenfunction of the operator $\mathbf{O}'$ with eigenvalue o'_i which proves the theorem in the nondegenerate case.

In case ψ_i is degenerate with eigenfunctions ϕ_i, η_i, etc., then $\mathbf{O}'\psi_i$ must be a linear combination of these degenerate eigenfunctions,

$$\mathbf{O}'\psi_i = c_1\psi_i + c_2\phi_i + c_3\eta_i + \ldots$$

Similar equations can be written for $\mathbf{O}'\phi_i$, $\mathbf{O}'\eta_i$, etc. Then it will be possible find a particular linear combination $f_i = d_1\psi_i + d_2\phi_i + d_3\eta_i + \ldots$ such that

$$\mathbf{O}'f_i = o'_i f_i. \tag{6.4b}$$

Because f_i is an eigenfunction of $\mathbf{O}$ (by Theorem 1), the theorem is proved for the degenerate case. **QED**

Note the stress on the word "*can*" in the statement of the theorem. That all the eigenfunctions may not be common can be understood in the following way. Let $\mathbf{O}$ involve operators which are functions of r_1 or its derivatives. Let $\mathbf{O}'$ involve functions or derivatives of one more variable r_2. Also let ψ_i be a function of r_1 and f an arbitrary function of r_2. Then the product $\psi_i(r_1)f(r_2)$ is certainly an eigenfunction of the operator $\mathbf{O}$,

$$\mathbf{O}[r_1]\ \psi_i(r_1)f(r_2) = f(r_2)\ \mathbf{O}\psi_i(r_1) = o_i\psi_i(r_1)f(r_2),$$

but it need not necessarily be an eigenfunction of $\mathbf{O}'(r_1,r_2)$ as f is arbitrary. We will have an occasion to discuss this point while considering the forms of atomic orbitals.

We now define a *complete set*. A set of *orthonormal* (mutually orthogonal and normalized) functions $\{\psi\}$ is complete if any arbitrary function ϕ can be completely described by a linear expansion in the set,

$$\phi = \sum_i c_i\psi_i\,. \tag{6.5}$$

This expansion gives

$$\int dv\,\psi_j^*(\mathbf{r})\phi(\mathbf{r}) = \sum_i c_i \int dv\,\psi_j^*(\mathbf{r})\psi_i(\mathbf{r}) = \sum_i c_i\delta_{j,i}$$

so that from the property of the Kroenecker delta,

$$c_j = \int dv\,\psi_j^*(\mathbf{r})\phi(\mathbf{r}). \tag{6.6}$$

When we insert this expression in Eq. (6.5), we get

$$\phi(\mathbf{r}) = \int dv' \sum_i \psi_i(\mathbf{r})\psi_i^*(\mathbf{r}')\phi(\mathbf{r}') \tag{6.7}$$

where the variable $\mathbf{r}'$ is distinct from $\mathbf{r}$. Since ϕ is an arbitrary continuous function of the spatial coordinates, we must have

$$\text{(i)}\ \sum_i \psi_i(\mathbf{r})\psi_i^*(\mathbf{r}') = 0 \qquad (\text{for } \mathbf{r} \neq \mathbf{r}'), \tag{6.8a}$$

and

$$\text{(ii)}\ \int dv' \sum_i \psi_i(\mathbf{r})\psi_i^*(\mathbf{r}') = 1 \tag{6.8b}$$

(when the point indicated by the position vector $\mathbf{r}$ is included in the range of integration).

Equations (6.8) describe the *closure* property of the orthonormal set of functions. They can be written in a more compact form as

$$\sum_i \psi_i(\mathbf{r})\psi_i^*(\mathbf{r}') = \delta^3(\mathbf{r}-\mathbf{r}') \tag{6.9}$$

where $\delta^3(\mathbf{r}-\mathbf{r}')$ is the Dirac delta function in three dimensions.

The Dirac delta function $\delta(x - x_0)$ is defined by

$$\delta(x - x_0) = 0 \qquad (\text{for } x \neq x_0), \tag{6.10a}$$

And

$$\int_a^b dx\, \delta(x - x_0)\, f(x) = f(x_0) \qquad (\text{for } a < x_0 < b), \tag{6.10b}$$

$$= 0 \qquad (\text{when } x \notin (a, b)).$$

In particular,

$$\int_a^b dx\, \delta(x - x_0) = 1 \qquad (\text{for } a < x_0 < b). \tag{6.10c}$$

Two useful expressions are

$$\delta(x) = \lim_{g\to\infty} \frac{\sin gx}{\pi x} \tag{6.11}$$

and

$$\delta(k) = \frac{1}{2\pi}\int_{-\infty}^{+\infty} dx\, e^{ikx}. \tag{6.12}$$

The function $\delta^3(\mathbf{r}-\mathbf{r}')$ is the product of the three one-dimensional delta functions:

$$\delta^3(\mathbf{r}-\mathbf{r}') = \delta(x - x')\,\delta(y - y')\,\delta(z - z').$$

We now prove the converse of Theorem 4.

Theorem 5. If two operators **O** and **O'** have a set of common eigenfunctions and the set is complete, the operators commute with each other.

Proof: Let $\{\psi\}$ be the set of common eigenfunctions. Then for every ψ_i,

$$\mathbf{OO'}\psi_i = \mathbf{O}o'_i\psi_i = o_i o'_i\psi_i$$

and

$$\mathbf{O'O}\psi_i = \mathbf{O'}o_i\psi_i = o'_i o_i\psi_i.$$

As o_i and o'_i are the eigenvalues, they commute with each other and the right hand sides of the two equations are equal. Hence $[\mathbf{O}, \mathbf{O'}]\psi_i(\mathbf{r}) = 0$ for every ψ_i. Therefore,

$$[\mathbf{O}, \mathbf{O'}] \sum_i \int dv'\, \psi_i(\mathbf{r})\psi_i^*(\mathbf{r}') = 0.$$

The closure relation Eq. (6.9) and the property of Dirac delta function (6.10c) yields the desired result, $[\mathbf{O}, \mathbf{O'}] = 0$. **QED**

Lecture 7: The Uncertainty Principle

Deviations

Suppose the two observables a and b are measured by the same experiment. The corresponding operators are **A** and **B** and the average values are $\bar{a} = \langle\mathbf{A}\rangle$ and $\bar{b} = \langle\mathbf{B}\rangle$, the system being in a general state represented by the normalized function ψ. The operators for *deviation* are written as $(\mathbf{A}-\bar{a})$ and $(\mathbf{B}-\bar{b})$, respectively. The average deviations are always equal to zero,

$$\langle \mathbf{A}-\bar{a} \rangle = \langle\mathbf{A}\rangle - \bar{a} = \bar{a} - \bar{a} = 0,$$

$$\langle \mathbf{B}-\bar{b} \rangle = \langle\mathbf{B}\rangle - \bar{b} = \bar{b} - \bar{b} = 0. \tag{7.1}$$

The *square deviation* operators are $(\mathbf{A}-\bar{a})^2$ and $(\mathbf{B}-\bar{b})^2$, and the *mean square deviations* are

$$\sigma_a^2 = \langle (\mathbf{A}-\bar{a})^2 \rangle = \langle \mathbf{A}^2 \rangle - 2\langle \mathbf{A} \rangle \bar{a} + \bar{a}^2 = \langle \mathbf{A}^2 \rangle - \bar{a}^2,$$

$$\sigma_b^2 = \langle (\mathbf{B}-\bar{b})^2 \rangle = \langle \mathbf{B}^2 \rangle - 2\langle \mathbf{B} \rangle \bar{b} + \bar{b}^2 = \langle \mathbf{B}^2 \rangle - \bar{b}^2. \tag{7.2}$$

The quantities σ_a and σ_b are the *root mean square* (*rms*) *deviations* or the *standard deviations*.

We now successively prove three important theorems.

Theorem 6. For a well-behaved function f,

$$\int dv\,|f|^2 \geq 0, \tag{7.3}$$

the equality being valid only in the trivial case that f = 0 at every point in space.

Proof: This theorem can be proved by writing an integral as a sum over points. In the case of one dimension,

$$\int dx\,|f|^2 = \lim_{N\to\infty} \sum_{i=1}^{N} |f(x_i)|^2\,\Delta x_i$$

As $\Delta x_i \geq 0$ and $|f(x_i)|^2 \geq 0$, the right hand side is always greater than zero except when $f(x)$ vanishes at every point in space. This leads to the desired result in one dimension. Equation (7.3) is merely a three-dimensional version that may be derived by carrying out integration over three variables. **QED**

Theorem 7. Schwarz inequality: For well-behaved functions u_1 and u_2,

$$\int dv\,|u_1|^2 \int dv\,|u_2|^2 \geq \frac{1}{4}\left[\int dv\,(u_1^* u_2 + u_2^* u_1)\right]^2. \tag{7.4}$$

Proof: Write $f = \mu u_1 + u_2$ where μ is a chosen constant and real. Since u_1 and u_2 are well-behaved, f is also well-behaved so that

$$\int dv\,|f|^2 = C \geq 0$$

where C is a constant and real. But

$$\int dv\,|f|^2 = \mu^2 R + \mu(S + S^*) + T$$

where $R = \int dv\,|u_1|^2$, $S = \int dv\, u_1{}^* u_2$ and $T = \int dv\,|u_2|^2$. Therefore,

$$\mu^2 R + \mu(S + S^*) + (T - C) = 0.$$

For μ to be *real* and *unique* one must have

$$(S + S^*)^2 - 4R(T - C) = 0$$

which gives

$$RT \geq \frac{1}{4}(S + S^*)^2$$

as $R > 0$ and $T > 0$ for nontrivial functions u_1 and u_2, respectively. Equation (7.4) immediately follows. **QED**

Theorem 8. The Uncertainty Theorem: The product of standard deviations for two observables A and B that are simultaneously measured is bounded from below by the inequality

$$\sigma_a \sigma_b \geq \frac{1}{2}\left| i < [\mathbf{A}, \mathbf{B}] > \right|. \tag{7.5}$$

Proof: Let us write $u_1 = (\mathbf{A} - \bar{a})\psi$ and $u_2 = i(\mathbf{B} - \bar{b})\psi$. As **A** and **B** are hermitean operators, $\bar{a}$ and $\bar{b}$ are real by the turnover rule, Eq.(4.6). Hence the operators for deviations $(\mathbf{A} - \bar{a})$ and $(\mathbf{B} - \bar{b})$ are hermitean. Therefore,

$$\int dv\,|u_1|^2 = \sigma_a{}^2,$$

$$\int dv\,|u_2|^2 = \sigma_b{}^2,$$

and

$$\int dv\,(u_1{}^* u_2 + u_2{}^* u_1) = i < [\mathbf{A}, \mathbf{B}] >.$$

Using Schwarz inequality Eq.(7.4) we get

$$\sigma_a{}^2 \sigma_b{}^2 \geq \frac{1}{4}[i < [\mathbf{A}, \mathbf{B}] >]^2$$

which leads to Eq.(7.5). **QED**

The uncertainty theorem is a quantitative statement of Heisenberg's **Uncertainty Principle:** The minimum product of standard deviations of two simultaneously measured properties equals half of the absolute magnitude of the expectation value of the commutator of the two operators.

If the two operators commute with each other like **r** and t do, then the minimum product is zero. In such cases it is possible to simultaneously measure the two quantities with complete accuracy. Of course some deviations are unavoidable in each measurement process: they arise from the limitations of the instrument and from human sources (subjective error). By improving the instrument and by taking care to reduce the individual errors committed by the experimentalist, the error in the measurement of the two quantities from the same experiment can be made progressively smaller. However, for quantities like x and p_x or E and t, the minimum product is not zero as the commutators are non-vanishing: $[x, p_x] = i\hbar$ and $[E, t] = i\hbar$. In fact, $\sigma_x\sigma_{px} \geq \frac{\hbar}{2}$ and $\sigma_E\sigma_t \geq \frac{\hbar}{2}$. Even if one improves the instrument and becomes careful enough to reduce the error one may make while taking a reading, one cannot accurately determine x and p_x, or E and t, from a simultaneous measurement. The inability owes its origin to the way experiments are designed and is referred to as the *natural uncertainty*. In order to measure linear momentum accurately one must design the experiment in a specific way and the measurement of position will be increasingly uncertain. One just cannot avoid it – this is how measurements are allowed by nature.

Let us try to understand the physical basis of this uncertainty. Suppose we are observing an "electron" through a microscope, as illustrated in Figure 2.1. The maximum accuracy obtainable for the position of the electron is found as

$$|\Delta x| \simeq \frac{\lambda}{\sin\delta} \tag{7.6}$$

where δ is the half angle subtended at the electron by the lens, and λ is the wave length of light. The photon momentum is $p_{photon} = \frac{h}{\lambda}$, and its x-component varies from $-\frac{h}{\lambda}\sin\delta$ to $\frac{h}{\lambda}\sin\delta$. As the initial momentum of the electron and that of the photon can be exactly known, the momentum uncertainty for the electron can be estimated from the conservation of total linear momentum. This uncertainty is

$$|\Delta p_x| \simeq \frac{h}{\lambda}\sin\delta \tag{7.7}$$

and it leads to the product of deviations

$$|\Delta x|\ |\Delta p_x| \sim h \tag{7.8}$$

which is unavoidable because of the known properties of radiation and matter. What is shown in Figure 2.1 is Compton effect: photon and electron scatter each other while the combined linear momentum and energy remain conserved. Thus, nature does not allow the simultaneous measurement of x and p_x with a greater accuracy.

For a "macroscopic" entity such as a bird in flight, a stone hurled or a running train, the mass is so great that for $|\Delta x| \cong 1$ bohr, $\Delta p_x/m$ is too small to be detected. For the electron, however, $|\Delta p_x|$ and m would be roughly of the same order in atomic unit ($|e| = 1$, $\hbar = 1$, $m_e = 1$). For a ball of steel the impact of a photon with frequency in the visible range is like the impact of a snowflake on a mountain. But the electron is a tiny entity, and for it the photon is quite a monster. Both the photon and the electron violently push each other, resulting in the change of momentum for each of them.

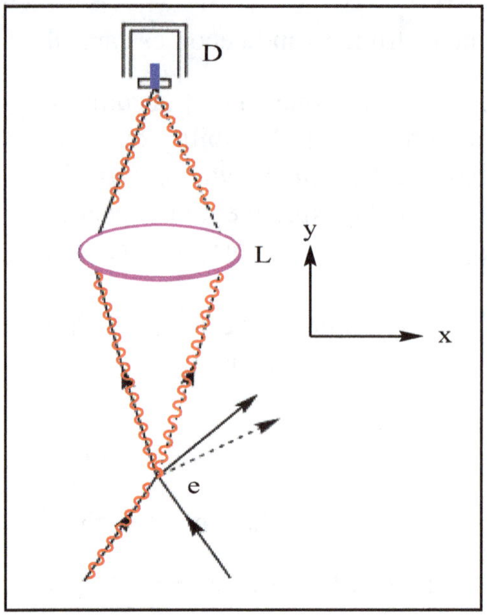

Figure 2.1 A thought experiment for the measurement of momentum and position of the electron: the Compton scattering. The trajectory of the electron (e) is shown by straight lines. The path of the photon is depicted by wavy lines. The scattered photon passes through the lens L and is received by the detector D. Alternative paths are shown for the scattered electron and the scattered photon.

If the electron is in an eigenstate of linear momentum, its position is infinitely un-certain ($\sigma_X \to \infty$ as $\sigma_p \to 0$). If it is in an eigenstate of $\mathbf{L}_z$, its orientation φ is infinitely uncertain. If it is in an eigenstate of energy it has an infinite lifetime. This is why while considering a measurement process one may take the quantum mechanical system initially in a stationary state (from time $t = -\infty$ to $t = 0$) and finally in another stationary state (from time $t = \tau$ to $t = \infty$), solve the time-dependent Schrödinger equation (5.6) for the finite period $0 \le t \le \tau$, and determine the probability of transition from the initial stationary state to the final state in time τ. During the period $t = 0$ to $t = \tau$ the energy remains uncertain by an amount of $\sigma_E \ge \dfrac{\hbar}{2\tau}$, and the system remains in an impure state.

The energy-time uncertainty is best exemplified by a spectroscopic transition. See Figure 2.2. The finite lifetime of the excited state, Δt, causes the uncertainty

$$h\Delta\nu = \Delta E = \frac{\hbar}{2\Delta t} \tag{7.9}$$

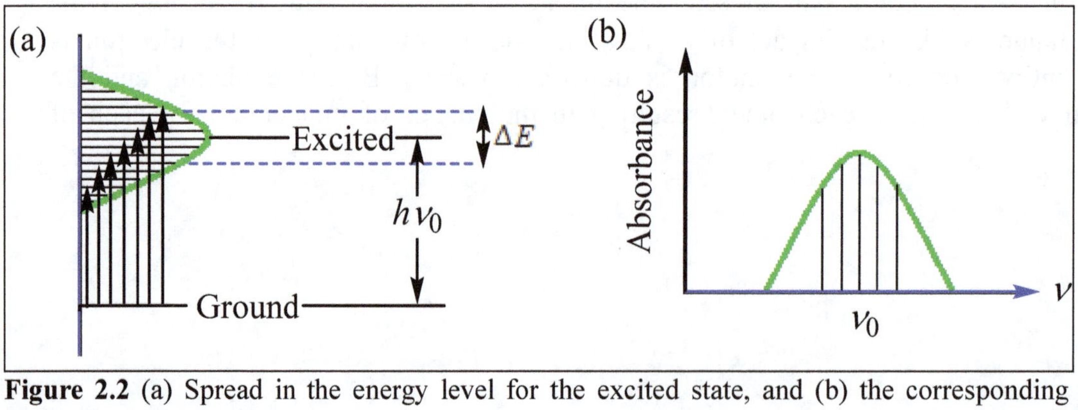

Figure 2.2 (a) Spread in the energy level for the excited state, and (b) the corresponding absorption spectrum.

in the excited state energy, and it imparts a line width Δν to the absorption spectrum. The line width Δν acquires contributions from various sources such as (i) the natural lifetime of the excited state, (ii) collision of molecules, (iii) vibronic interactions, etc.

The first law of thermodynamics may also be viewed as an outcome of the energy-time uncertainty relationship while the whole universe is considered as a quantum mechanical object. The infinite uncertainty associated with time leads to the statement that energy can neither be created nor be destroyed, but it can be converted from one form to another. The same uncertainty at the microscopic level gives rise to the *principle of conservation of energy* which is obeyed by classical-mechanical systems (bodies) as well as quantum-mechanical entities (particles).

Lecture 8: How to Write the Hamiltonian?

The first objective in any quantum mechanical treatment is to solve the stationary state problem, that is, the time-independent Schrödinger equation

$$\mathbf{H}\psi = \mathrm{E}\psi. \tag{5.9}$$

The wave functions for stationary states are of the form $\psi(\mathbf{r})e^{-iEt/\hbar}$. One can express a time-dependent wave function $\Psi(\mathbf{r}, t)$ as a suitable linear combination of the stationary state wave functions. It should be noted that once one knows $\Psi(\mathbf{r}, t)$ one gets a complete description of the dynamical state of the system since the average value of any observable can be calculated as the expectation value of the corresponding operator over Ψ. It is, therefore, crucial to know how to write the Hamiltonian operator for each atom or each molecule, or for each specific quantum mechanical system.

The Hamiltonian consists of the kinetic and the potential energy operators. The kinetic energy operator for a particle of mass m and moving in one dimension (along the x-axis) is

$$\mathbf{T} = \frac{p_x^2}{2m} = -\frac{\hbar^2}{2m}\frac{d^2}{dx^2}$$

while for the movement in three dimension one writes

$$\mathbf{T} = \frac{1}{2m}(p_x^2 + p_y^2 + p_z^2) = -\frac{\hbar^2}{2m}\left(\frac{\partial^2}{\partial x^2} + \frac{\partial^2}{\partial y^2} + \frac{\partial^2}{\partial z^2}\right)$$

or

$$\mathbf{T} = \frac{1}{2m}\mathbf{p}\bullet\mathbf{p} = -\frac{\hbar^2}{2m}\nabla\bullet\nabla = -\frac{\hbar^2}{2m}\nabla^2.$$

Total kinetic energy of N particles of mass $m_1, m_2, \ldots, m_N$ is given by the sum of the individual kinetic energies,

$$\mathbf{T} = \mathbf{T} = \sum_{i=1}^{N}\mathbf{T}(i) = -\hbar^2\sum_{i=1}^{N}\frac{1}{2m_i}\nabla_i^2.$$

For instance, the kinetic energy of N electrons each of mass m_e is

$$\mathbf{T} = -\frac{\hbar^2}{2m_e}\sum_{i=1}^{N}\nabla_i^2 .$$

Two different systems may have the same number of electrons (such as He atom and Li^+ ion). Then they have the same electronic kinetic energy operators if we neglect the difference in the reduced mass. These systems are characterized by different potential energy operators. It is often the potential energy (operator) which makes a (quantum mechanical) system distinct from any other (quantum mechanical) system.

A free particle is not under any force. This implies that the potential is independent of spatial coordinates, that is, $\mathbf{V} = \mathrm{c}$ where c is a constant. See that by virtue of Eq. (5.9),

$$(\mathbf{H} + \mathrm{c})\psi = (\mathrm{E} + \mathrm{c})\psi \tag{8.1}$$

holds. Experiment always yields energy differences and never an absolute value of energy. As the zero of the energy can be arbitrary, one may take $\mathrm{c} = 0$ without any loss of generality. Thus we put $\mathbf{V} = 0$ for a free particle. For a quasi-free particle such as an electron confined to a one-dimensional box of length L, the potential is described by

$$\mathrm{V}(x) = 0 \qquad \text{for } 0 \leq x \leq L,$$

and

$$\mathrm{V}(x) = \infty \qquad \text{for } x < 0 \text{ and } x > L.$$

For a particle undergoing a simple harmonic oscillation along the x axis,

$$\mathrm{V}(x) = \frac{1}{2}k(x - x_0)^2$$

where k is the force constant and x_0 is the mean point of vibration.

The electron in a hydrogen-like atom interacts electrostatically with the nucleus of charge $Z|e|$ where Z is the atomic number, and so

$$\mathbf{V}_{en}(\mathbf{r}) = -\frac{Ze^2}{r}$$

where r is the distance of the electron from the nucleus. Note that $\mathbf{V}_{en}(\mathbf{r})$ is spherically symmetric – it depends only on the distance and is independent of the orientation: $\mathbf{V}_{en}(\mathbf{r}) = \mathbf{V}_{en}(r)$. For the movement of electrons in a many-electron atom we write

$$\mathbf{V} = \mathbf{V}_{en} + \mathbf{V}_{ee}$$

where $\mathbf{V}_{en}$ is the sum of the electrostatic electron-nucleus potential energy terms for all electrons and $\mathbf{V}_{ee}$ is the net potential energy of the instantaneous electron-electron repulsion. These components are explicitly written as

$$\mathbf{V}_{en} = -\sum_{\substack{i=1 \\ (electron)}}^{N}\frac{Ze^2}{r_i}$$

and (8.2)

$$\mathbf{V}_{ee} = \sum_{\substack{i<j \\ (electron)}}\frac{e^2}{r_{ij}} = \frac{1}{2}\sum_{\substack{i\neq j \\ 1}}^{N}\frac{e^2}{r_{ij}},$$

where N is the total number of electrons. We have written r_{ij} for the interelectronic distance $|\mathbf{r}_i - \mathbf{r}_j|$. The factor $\frac{1}{2}$ in the second expression is required as the interaction between every pair of electrons is counted twice.

A molecule has an additional type of interaction, viz., the nucleus-nucleus interaction. The molecular potential energy operator is

$$\mathbf{V} = \mathbf{V}_{en} + \mathbf{V}_{ee} + \mathbf{V}_{nn}$$

where $\mathbf{V}_{nn}$ is the potential energy of the electrostatic repulsion between the nuclei,

$$\mathbf{V}_{nn} = \sum_{\substack{m<n \\ (nucleus)}}^{M} \frac{Z_m Z_n e^2}{R_{mn}} = \frac{1}{2} \sum_{\substack{m \neq n \\ 1}}^{M} \frac{Z_m Z_n e^2}{R_{mn}}. \tag{8.3}$$

In the above R_{mn} denotes the distance $|\mathbf{R}_m - \mathbf{R}_n|$, the vector $\mathbf{R}_p$ signifying the position vector for the pth nucleus ($p = m, n$, etc.), and M is the number of nuclei in the molecule. In this case the form of $\mathbf{V}_{ee}$ remains unchanged from that given in Eqs. (8.2), but $\mathbf{V}_{en}$ becomes more elaborate because there are more than one nucleus in a molecule. We write $\mathbf{V}_{en}$ as

$$\mathbf{V}_{en} = -\sum_{i=1}^{N} \sum_{m=1}^{M} \frac{Z_m e^2}{|\mathbf{r}_i - \mathbf{R}_m|}, \tag{8.4}$$

where i is an electron index and m is the index for the nuclei.

We are now ready to write the Hamiltonian operators for different systems, and the corresponding time-independent Schrödinger equations. Notice that the wave function ψ is always a function of the coordinates involved in writing the operator $\mathbf{H}$. Electrostatic interactions between different charged particles in a few atomic or molecular systems are illustrated in Figure 2.3.

Example 9: Free particle of mass m

The Hamiltonian is written as

$$\mathbf{H} = -\frac{\hbar^2}{2m}\frac{d^2}{dx^2} \qquad \text{(for movement in one dimension)},$$

and

$$\mathbf{H} = -\frac{\hbar^2}{2m}\nabla^2 \qquad \text{(for the three-dimensional problem)}.$$

The Schrödinger equation is

$$-\frac{\hbar^2}{2m}\frac{d^2\psi(x)}{dx^2} = E\psi(x) \qquad \text{(1-dimension),}$$

and

$$-\frac{\hbar^2}{2m}\nabla^2\psi(x,y,z) = E\psi(x,y,z) \qquad \text{(3-dimension).}$$

Example 10: Particle in a one-dimensional potential box

The Hamiltonian is written as

$$\mathbf{H} = -\frac{\hbar^2}{2m}\frac{d^2}{dx^2} + \mathrm{V}(x)$$

where

$\mathrm{V}(x) = 0$ for $0 < x < L$,

$\mathrm{V}(x) = \infty$ otherwise.

The Schrödinger equation is

$$-\frac{\hbar^2}{2m}\frac{d^2\psi(x)}{dx^2} + V(x)\psi(x) = E\psi(x).$$

Example 11: One-dimensional simple harmonic oscillator of reduced mass m

The Hamiltonian is written as

$$\mathbf{H} = -\frac{\hbar^2}{2m}\frac{d^2}{dx^2} + \frac{1}{2}kx^2.$$

The Schrödinger equation is

$$-\frac{\hbar^2}{2\mathrm{m}}\frac{d^2\psi(x)}{dx^2} + \frac{1}{2}kx^2\psi(x) = E\psi(x)$$

Let us consider the *internal motion of the electrons* in atoms and molecules. We assume that the coordinates of the center of mass have been separated, that is, the Hamiltonian operator for the translational motion of the atom as a whole aggregate of the electrons and the nucleus has been subtracted from the total Hamiltonian operator, leaving aside only the Hamiltonian for the movement of the electrons in the internal coordinates. In the case of molecules both the translational and rotational Hamiltonian operators have been subtracted. These assumptions are justified as the coupling between the translational motion (and the rotational motion in the case of molecules) with the internal motion (the movement of the electrons relative to the nucleus or nuclei and in the case of molecules, vibrational motion of the nuclear frame which is generally coupled to the electronic motion) is exceedingly weak and negligible. In the following the Hamiltonian operators and Schrödinger equations have been written in terms of relative coordinates (coordinates of the electrons relative to those of one of the nuclei).

Figure 2.3 Instantaneous electrostatic interaction in (i) the hydrogen atom, (ii) the helium atom, (iii) the H_2^+ molecule ion and (iv) the H_2 molecule.

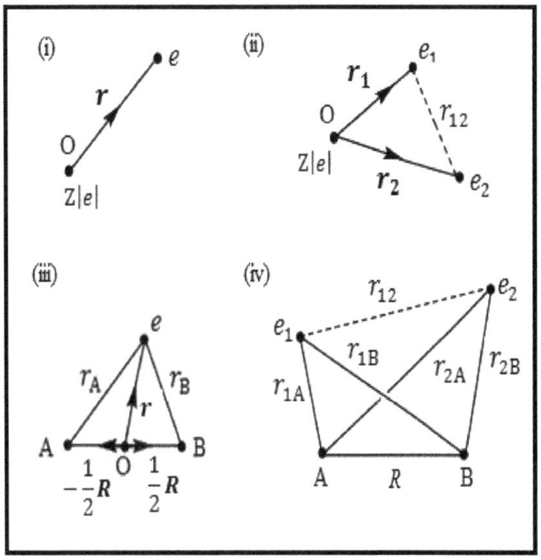

Example 12: Electron in a hydrogen-like atom

The Hamiltonian is written as

$$\mathbf{H} = -\frac{\hbar^2}{2m}\nabla^2 - \frac{Ze^2}{r}.$$

The Schrödinger equation is

$$-\frac{\hbar^2}{2m}\nabla^2\psi(\mathbf{r}) - \frac{Ze^2}{\mathrm{r}}\psi(\mathbf{r}) = E\psi(\mathbf{r}).$$

Here m is the reduced mass of the electron-nucleus system.

Example 13: The helium atom (N= 2, Z= 2)

The Hamiltonian is written as

$$\mathbf{H} = -\frac{\hbar^2}{2m}(\nabla_1^2 + \nabla_2^2) - (\frac{Ze^2}{r_1} + \frac{Ze^2}{r_2}) + \frac{e^2}{r_{12}}.$$

The Schrödinger equation is

$$-\frac{\hbar^2}{2m}(\nabla_1^2 + \nabla_2^2)\Psi(\mathbf{r}_1, \mathbf{r}_2) - (\frac{Ze^2}{r_1} + \frac{Ze^2}{r_2})\Psi(\mathbf{r}_1, \mathbf{r}_2) + \frac{e^2}{r_{12}}\Psi(\mathbf{r}_1, \mathbf{r}_2) = E\Psi(\mathbf{r}_1, \mathbf{r}_2).$$

Example 14: The hydrogen molecule ion, H_2^+ (M= 2, N= 1; Z_A= 1, Z_B= 1)

The Hamiltonian is written as

$$H = -\frac{\hbar^2}{2m}\nabla^2 - (\frac{e^2}{r_A} + \frac{e^2}{r_B}) + \frac{e^2}{R_{AB}}$$

where $r_A = |\mathbf{r} - \mathbf{R}_A|$, $r_B = |\mathbf{r} - \mathbf{R}_B|$ and $R_{AB} = |\mathbf{R}_A - \mathbf{R}_B|$. The Schrödinger equation is

$$-\frac{\hbar^2}{2m}\nabla^2\psi(\mathbf{r}, R_{AB}) - \left(\frac{e^2}{r_A}+\frac{e^2}{r_B}\right)\psi(\mathbf{r}, R_{AB}) + \frac{e^2}{R_{AB}}\psi(\mathbf{r}, R_{AB}) = E\psi(\mathbf{r}, R_{AB}).$$

Example 15: The hydrogen molecule , H_2 (M= 2, N= 2; Z_A= 1, Z_B= 1)

The Hamiltonian is written as

$$\boldsymbol{H} = -\frac{\hbar^2}{2m}(\nabla_1^2 + \nabla_2^2) - \left(\frac{e^2}{r_{1A}}+\frac{e^2}{r_{1B}}+\frac{e^2}{r_{2A}}+\frac{e^2}{r_{2B}}\right)+\frac{e^2}{r_{12}}+\frac{e^2}{R_{AB}}.$$

The Schrödinger equation is

$$-\frac{\hbar^2}{2m}(\nabla_1^2 + \nabla_2^2)\,\Psi(\mathbf{r}_1, \mathbf{r}_2; R_{AB}) - \left(\frac{e^2}{r_{1A}}+\frac{e^2}{r_{1B}}+\frac{e^2}{r_{2A}}+\frac{e^2}{r_{2B}}\right)\Psi(\mathbf{r}_1, \mathbf{r}_2; R_{AB})$$

$$+\frac{e^2}{r_{12}}\Psi(\mathbf{r}_1, \mathbf{r}_2; R_{AB}) + \frac{e^2}{R_{AB}}\Psi(\mathbf{r}_1, \mathbf{r}_2; \mathrm{R}_{AB}) = E\Psi(\mathbf{r}_1, \mathbf{r}_2; R_{AB}).$$

In Examples 14 and 15, R_{AB} is the inter-nuclear distance.

We now consider a general atom or molecule with N electrons and M nuclei. For a general atom M = 1. The electronic coordinates are still given by the set $\{\mathbf{r}\}$, and the nuclear coordinates by $\{\mathbf{R}\}$. We first write the total Hamiltonian operator **H** that accounts for all types of motion – translational, rotational, vibrational and electronic – and the corresponding Schrödinger equation. Next we point out the Hamiltonian in internal coordinates, ***H***, and the Schrödinger equation in these coordinates.

Example 16: The general atom or molecule having N electrons and M nuclei (M=1 for atoms)

A general atom (or a general molecule) has the Hamiltonian

$$\mathbf{H} = (\mathbf{T}_e + \mathbf{T}_n) + (\mathbf{V}_{en} + \mathbf{V}_{ee} + \mathbf{V}_{nn}). \tag{8.5}$$

It accounts for the movement of all the electrons and the nucleus (nuclei). The operators $\mathbf{T}_e$ and $\mathbf{T}_n$ stand for the kinetic energies of all the particles of the respective kinds. Thus we write

$$\mathbf{T}_e = -\frac{\hbar^2}{2m_e}\sum_{i=1}^{N}\nabla_i^2, \tag{8.6}$$

and

$$\mathbf{T}_n = -\sum_{m=1}^{M}\frac{\hbar^2}{2M_m}\nabla_m^2 \tag{8.7}$$

where M_m is the mass of the mth nucleus and ∇_m^2 refers to the second partial derivatives with respect to the nuclear coordinates. The terms $\mathbf{V}_{ee}$, $\mathbf{V}_{nn}$ and $\mathbf{V}_{en}$ are given by Eqs. (8.2), (8.3) and (8.4), respectively. The wave function is now a function of the coordinates of all electrons and nuclei. The Schrödinger equation is

$$[(\mathbf{T}_e+\mathbf{T}_n) + (\mathbf{V}_{en}+\mathbf{V}_{ee}+\mathbf{V}_{nn})]\Psi(\{\mathbf{r}\}, \{\mathbf{R}\}) = E\Psi(\{\mathbf{r}\}, \{\mathbf{R}\}). \tag{8.8}$$

Equations (8.5) through (8.8) define the general problem in quantum chemistry. Therefore, these are normally referred to in all the textbooks on the subject. What is usually solved in quantum chemistry is somewhat different. Normally one solves the Schrödinger equation in relative coordinates, like those in examples 12-15.

Suppose that somehow we have separated the translations and the bodily rotations of the general molecule; (there is no bodily rotation for an atom). The very weak coupling of these motions with the electronic movement and the vibration of the molecular frame is neglected. Furthermore, assume all the electrons to have the same reduced mass m and that the internal coordinates for molecular vibration, Q_v's, are associated with the reduced masses μ_v's.

Example 17: The electronic-vibrational motion
The concerned Hamiltonian is

$$\boldsymbol{H} = (\boldsymbol{T}_e + \boldsymbol{T}_n) + (\mathbf{V}_{\mathrm{en}} + \mathbf{V}_{\mathrm{ee}} + \mathbf{V}_{\mathrm{nn}}) \tag{8.9}$$

where

$$\boldsymbol{T}_e = -\frac{\hbar^2}{2m}\sum_{i=1}^{N}\nabla_i^2, \tag{8.10}$$

and

$$\boldsymbol{T}_n = -\sum_{v=1}^{N_v}\frac{\hbar^2}{2\mu_v}\nabla_v^2. \tag{8.11}$$

The quantity m is a sort of *average* reduced mass for all the electrons. It is in effect very close to the rest mass m_e. The terms $\mathbf{V}_{en}$, $\mathbf{V}_{ee}$ and $\mathbf{V}_{nn}$ remain unaltered. The Schrödinger equation is now written as

$$\boldsymbol{H}\Psi(\{\mathbf{r}\}, \{\mathbf{Q}\}) = E\Psi(\{\mathbf{r}\}, \{\mathbf{Q}\}). \tag{8.12}$$

The usual practice in quantum chemistry is to solve Eq. (8.12) in lieu of Eq. (8.8). A good starting point for solving Eq. (8.12) was suggested by Born and Oppenheimer (1927). This will be discussed later in Chapter 6.

Lecture 9: Separability of the Wave Equation; Role of Symmetry

Separability

A very important aspect of the Hamiltonian is whether it is separable into a sum of two or more operators without any common variable. If the Hamiltonian satisfies this criterion we call it separable. For example, the operator $\mathbf{H}(x,y)$ is additively separable if it can be written as

$$\mathbf{H}(x,y) = \mathbf{H}_1(x) + \mathbf{H}_2(y) \tag{9.1}$$

where $\mathbf{H}_1$ is independent of the variable y and derivatives with respect to it, and $\mathbf{H}_2$ involves neither x nor derivatives with respect to x. Let us take the eigenfunctions $\psi_i(x)$ and $\phi_j(y)$ such that

$$\mathbf{H}_1(x)\psi_i(x) = \lambda_i\psi_i(x)$$

and (9.2)

$$\mathbf{H}_2(y)\phi_j(y) = \mu_j\phi_j(y).$$

One can easily verify that the product

$$\Psi_{ij}(x,y) = \psi_i(x)\phi_j(y) \tag{9.3}$$

is an eigenfunction of $\mathbf{H}(x,y)$:

$$\mathbf{H}(x,y)\Psi_{ij}(x,y) = E_{ij}\Psi_{ij}(x,y). \tag{9.4}$$

The eigenvalue of the total Hamiltonian is the sum of the eigenvalues of the parts $\mathbf{H}_1$ and $\mathbf{H}_2$:

$$E_{ij} = \lambda_i + \mu_j. \tag{9.5}$$

Thus we can choose a set of wave functions $\Psi_{ij}(x,y)$ such that the x- and the y-dependences of each member of the set can be separated into factors. In other words, the wave functions are multiplicatively separable. We also say that the wave equation (9.4) is separable into the two wave equations given in Eqs.(9.2).

The Schrödiger equation can be exactly solved only in a few cases. The merit of the separability of the wave equation lies herein. Suppose that the Hamiltonian is reasonably complex. If by some method we can convert it into a separable form, we can determine the exact solutions provided that the separate wave equations are exactly solvable. Elementary quantum chemical problems can be classified into two categories,

a) Problems where an exact solution of the Schrödinger equation is possible (Chapter 3), and

b) Problems in which exact analytical solutions are not derivable and hence approximate solutions to the Schrödinger equation are found by following reliable methodologies (most of the remaining chapters of this book).

The approximate methods are further classified into two broad subdivisions, viz., the variation method (Chapter 4) and the perturbation theory (Chapter 7). Often the most useful tactics is to employ complex methodologies such as the variation-perturbation technique, transformations followed by or even preceded by either variation or perturbation treatments, etc.

Symmetry

Some systems are symmetric, that is, by applying certain (linear) transformations on the coordinates of all the constituent particles, the (atomic or the molecular) system can be converted from a specific instantaneous configuration into an equivalent instantaneous configuration. Coordinate transformations (or geometrical operations such as rotations around an axis, reflections in a plane or through a center, and even rotations followed by reflections) which generate the equivalent configurations are called symmetry operations.

Let **S** be the operator for a symmetry operation. Since it generates an equivalent configuration we must have

$$\mathbf{S}^{-1}\mathbf{H}\mathbf{S} = \mathbf{H} \tag{9.6}$$

where $\mathbf{S}^{-1}$ is the operator for the reverse of the (essentially geometrical) operation **S**. Equation (9.6) states that the Hamiltonian remains invariant under a *similarity transformation.* Symmetry is in fact the symmetry of the Hamiltonian. Note that $\mathbf{S}^{-1}$ itself is a symmetry operation. Multiplying both sides of Eq. (9.6) by **S** from the left, one obtains

$$[\mathbf{H}, \mathbf{S}] = 0. \tag{9.7}$$

The Hamiltonian commutes with the symmetry operator.

Any set of linear transformations of coordinates ($\mathtt{L}$) must leave the derivative operators and hence the kinetic energy operator **T** unaltered, that is, $\mathtt{L}^{-1}\mathbf{T}\mathtt{L} = \mathbf{T}$, such that

$$[\mathbf{T}, \mathtt{L}] = 0. \tag{9.8}$$

Since **S** is a linear transformation, it satisfies this criterion. But it also commutes with the Hamiltonian. Therefore, a set of coordinate transformations (or a geometrical operation) becomes a symmetry operation if and only if it commutes with the potential energy operator,

$$[\mathbf{V}, \mathbf{S}] = 0. \tag{9.9}$$

We say that the symmetry of the potential is the symmetry of the Hamiltonian operator. In the earlier lecture we have already discussed that it is the potential that characterizes a physical system. Systems with the same potential energy operator but differing in the masses must have the same symmetry in their dynamics. They have similarity in dynamical behavior.

The importance of symmetry becomes evident from Eq. (9.7). By Theorem 4, **H** and **S** can have common eigenfunctions, that is, we can choose a set of eigenfunctions ψ_i of **H** such that the members of this set are also eigenfunctions of **S**:

$$\mathbf{H}\psi_i = E_i\psi_i \Longrightarrow \mathbf{S}\psi_i = s_i\psi_i. \tag{9.10}$$

In other words, the eigenfunctions of **H** are symmetrized. A consideration of the symmetry of the wave functions allows us to write the form for the solutions of the Schrödinger equation in a systematic way. The mathematical theory that deals with the subject is known as the theory of groups. A brief account of Group theory is given in Appendix III.

Texts Suggested for Further Reading

H. Eyring, J. Walter and G. E. Kimball, "Quantum Chemistry", Chap. 3, (J. Wiley, New York, 1944).

PROBLEMS

1. Find which of the operators are hermitean:

$$\frac{d}{dx}, -i\frac{d}{dx}, \ \nabla, \ -i\nabla, \ \nabla^2, \ i\nabla^2, \ \frac{d}{dt}, \ i\frac{d}{dt}.$$

2. The function $\phi(\mathbf{r})$ is not normalized to unity. Normalize it.
   Hint: Set $\psi(\mathbf{r}) = c\phi(\mathbf{r})$ and demand that $\psi(\mathbf{r})$ is normalized. Find c.
3. Calculate the following commutators explicitly:

   $[x, p_x]$, $[\mathbf{r}, \mathbf{p}]$, $[E, t]$ and $[\varphi, \mathbf{L}_z]$.

4. Consider the wave functions $\psi(x)$ and $\psi(K)$ shown in the figures below. The coordinate x is Cartesian while the variable K is a wave vector in one-dimension. Identify the regions of high and low kinetic energy.

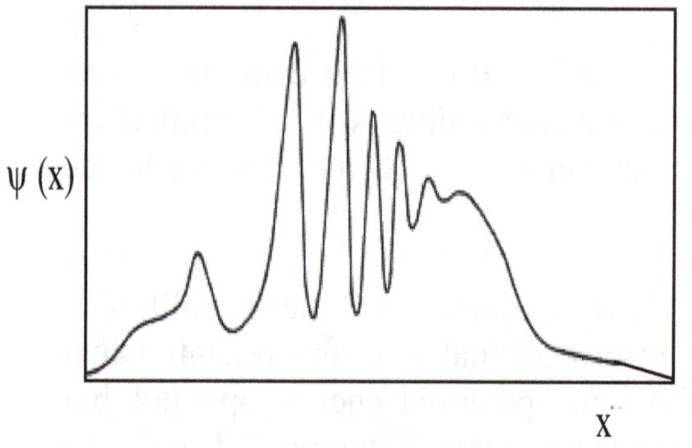

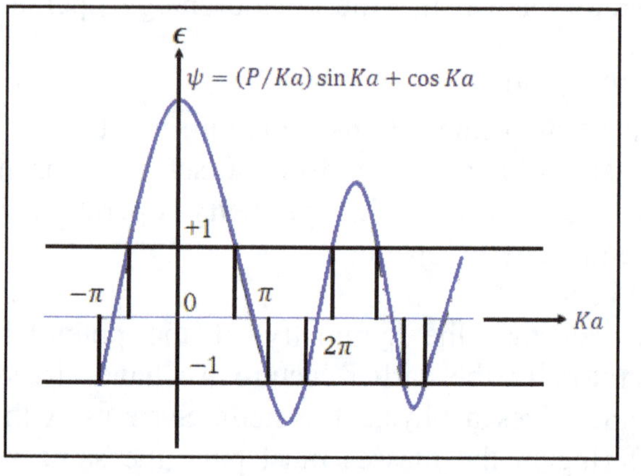

5. Write the electronic Hamiltonian operators for (i) the lithium atom and (ii) the H_3 molecule.
6. Calculate the de Broglie wavelength of an electron that has been accelarated practically from rest through a potential difference of 10 kilovolt.
7. What is the minimum uncertainty in the speed of an electron in the Bohr model of hydrogen atom?

CHAPTER 3

EXACTLY SOLVABLE PROBLEMS

Lecture 10: Exact Solutions; Free Particles

The time-independent Schrödinger equation is basically a second order differential equation which can be solved by analytical methods when the potential energy operator has an exceedingly simple dependence on coordinates. In more complicated cases the wave equation can be analytically solved when it is separable into several exactly solvable wave equations.

Free particle

The eigenvalue equation for the linear momentum is

$$\mathbf{p}_x\psi(x) \equiv -i\hbar\frac{d\psi(x)}{dx} = p_x\psi(x). \tag{10.1}$$

This is a first order differential equation with solutions

$$\psi(x) = e^{ik_x x} \tag{10.2}$$

where $\hbar k_x = p_x$. The Schrödinger equation for a particle of mass m moving along the x-axis and subject to no force (the so-called free particle) is

$$\mathbf{H}^x_{f.p.}\psi(x) \equiv -\frac{\hbar^2}{2m}\frac{d^2\psi(x)}{dx^2} = E_x\psi(x). \tag{10.3}$$

The wave function $\psi(x)$ that is given by Eq.(10.2) also satisfies the above equation with eigenvalue

$$E_{k_x} = \frac{p_x^2}{2m} = \frac{\hbar^2 k_x^2}{2m}. \tag{10.4}$$

Both $e^{ik_x x}$ and $e^{-ik_x x}$ are eigenfunctions of $\mathbf{H}^{x}_{\text{f.p.}}$ with the same eigenvalue E_{k_x}, but they correspond to different momentum eigenvalues, p_x and $-p_x$.

Each of the functions $e^{ik_x x}$ and $e^{-ik_x x}$ are of modulus unity, and hence they are not normalizable. Therefore we apply the so-called box normalization condition

$$\int_{-L_x/2}^{L_x/2} dx\,|\,\psi(x)\,|^2 = 1 \tag{10.5}$$

such that the normalized eigenfunctions are

$$\psi(x) = L_x^{-1/2} e^{\pm ik_x x} . \tag{10.6}$$

Next we calculate the expectation values, and then let L_x tend to infinity to obtain the final results for infinite length.

The Hamiltonian for a free particle moving in three dimensions is separable into three Hamiltonian operators each valid for an individual dimension:

$$\mathbf{H}_{\text{f.p.}} = -\frac{\hbar^2}{2\text{m}}\nabla^2 = -\frac{\hbar^2}{2\text{m}}\left(\frac{\partial^2}{\partial x^2} + \frac{\partial^2}{\partial y^2} + \frac{\partial^2}{\partial z^2}\right)$$

$$= \mathbf{H}^{\text{x}}_{\text{f.p.}} + \mathbf{H}^{\text{y}}_{\text{f.p.}} + \mathbf{H}^{\text{z}}_{\text{f.p.}} \tag{10.7}$$

Hence the eigenfunction is given by the product

$$\psi_{\mathbf{k}}(\mathbf{r}) = L_x^{-1/2} e^{\pm ik_x x} L_y^{-1/2} e^{\pm ik_y y} L_z^{-1/2} e^{\pm ik_z z} = V^{-1/2} e^{\pm i\mathbf{k}\cdot\mathbf{r}} \tag{10.8}$$

where V is the volume, $V = L_x L_y L_z$. The eigenvalue $E_{\mathbf{k}}$ is given by the sum

$$E_{\mathbf{k}} = E_{k_x} + E_{k_x} + E_{k_x} = \frac{\hbar^2 k^2}{2m} = \frac{p^2}{2m} . \tag{10.9}$$

Thus the stationary state wave function is

$$\Psi_{\mathbf{k}}(\mathbf{r},t) = V^{-1/2} e^{i(\mathbf{k}\cdot\mathbf{r} - \omega_k t)} \tag{10.10}$$

where $\hbar\omega_{\mathbf{k}} = E_{\mathbf{k}}$. Again we may use $\psi_{\mathbf{k}}(\mathbf{r})$ in the quantum mechanical treatment, calculate the expectation values and finally let V tend to infinity to obtain the desired results for a free particle moving in a three-dimensional space.

The free particle wave function is called a plane wave. At any instant of time the amplitude remains constant at every point on any plane perpendicular to the direction of the propagation vector **k**, and the constant amplitude planes move with the constant speed $\frac{\omega_{\text{k}}}{\text{k}}$. The probability density is constant everywhere within the volume V,

$$\rho_{\mathbf{k}}(\mathbf{r},t) = |\Psi_{\mathbf{k}}(\mathbf{r},t)|^2 = \frac{1}{\text{V}} . \tag{10.11}$$

It is also independent of time t and the wave vector **k**. Thus the free particle is completely delocalized in space. This happens since the state is an eigenstate of linear momentum with $\sigma_{\mathbf{p}}=0$ and consequently $\sigma_{\mathbf{r}}$ is infinitely large. To describe a particle localized in space yet with an overall average velocity we write

$$\Phi(\mathbf{r},t) = \int d^3k \; \text{c}(\mathbf{k})\Psi_{\mathbf{k}}(\mathbf{r},t) \tag{10.12}$$

where $\text{c}(\mathbf{k})$ is a function with a peak somewhere, say, at $\mathbf{k} = \mathbf{k}_0$. The function $\Phi(\mathbf{r},t)$ is called a wave packet. It represents a superposition of plane waves. It is normalized when

$$\int d^3k \, |\text{c}(\mathbf{k})|^2 = 1. \tag{10.13}$$

This condition is a consequence of the orthogonality of the $\psi_{\mathbf{k}}$'s. One finds

$$\mathbf{p}\Phi(\mathbf{r},t) = \hbar\mathbf{k}_0\Phi(\mathbf{r},t) + \hbar\int d^3k \; \text{c}(\mathbf{k})(\mathbf{k}-\mathbf{k}_0)\Psi_{\mathbf{k}}(\mathbf{r},t)$$

$$\cong \hbar\mathbf{k}_0\Phi(\mathbf{r},t).$$

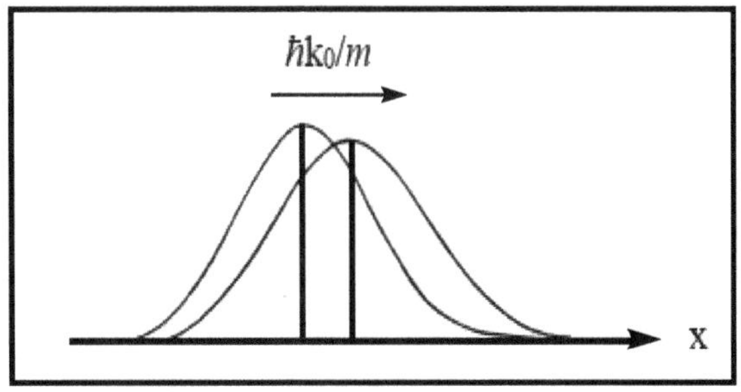

Figure 3.1 Schematic representation of the shift and the spread of the wave packet in one dimension.

This approximation works as the integral on the right hand side of the first step is close to zero for a carefully chosen coefficient c(**k**). A very popular choice for $\Phi(\mathbf{r},t)$ is a Gaussian wave packet defined in the terms of the Gaussian distribution of coefficients

$$c(\mathbf{k}) = N\alpha e^{-\alpha|\mathbf{k}-\mathbf{k}_0|^2}. \qquad (10.14)$$

The constant $N\alpha$ may be calculated from the normalization condition given by Eq.(10.13). We get

$$N\alpha = \left(\frac{2\alpha}{\pi}\right)^{\frac{3}{4}}. \qquad (10.15)$$

Such descriptions have found usage in quantum trajectory analysis of unimolecular decompositions where one considers that two molecular fragments interact with each other while they are separated by a short distance whereas at a larger separation, beyond a critical distance, they behave as effectively free particles.

Lecture 11: Quasi-Free Particle; Particle in a One-Dimensional Potential Box

While the free particle is of immense importance in physics, say, in solid-state physics or in the scattering theory, a particle in a potential box is of more interest in chemistry. The Schrödinger equation for a spin-less particle moving in a one-dimensional potential box of length L is

$$-\frac{\hbar^2}{2m}\frac{d^2\psi(x)}{dx^2} + V(x)\psi(x) = E\psi(x) \qquad (11.1)$$

where

$$V(x) = 0 \qquad \text{for } -\frac{L}{2} \le x \le \frac{L}{2}$$

and (11.2)

$$V(x) = \infty \qquad \text{for } x < -\frac{L}{2} \text{ and } x > \frac{L}{2}.$$

See Figure 3.2. Since the kinetic energy operator is essentially the square of a Hermitian operator, its expectation value is always positive. In the regions $x<-\frac{L}{2}$ and $x>\frac{L}{2}$, the potential V is infinitely large and positive so that the kinetic energy must be infinitely large and negative for a finite value of total energy E, and there is only one possible well-behaved solution. This is

$$\psi(x) = 0 \qquad \text{for } x < -\frac{L}{2} \text{ and } x > \frac{L}{2}. \tag{11.3}$$

In the range $-\frac{L}{2} \leq x \leq \frac{L}{2}$, the Schrödinger equation reduces to the free particle equation, Eq. (10.3), and the general solution is

$$\psi(x) = \mathrm{a}e^{ik(x+L/2)} + \mathrm{b}e^{-ik(x+L/2)} \tag{11.4}$$

with

$$E = \frac{\hbar^2 k^2}{2m}. \tag{11.5}$$

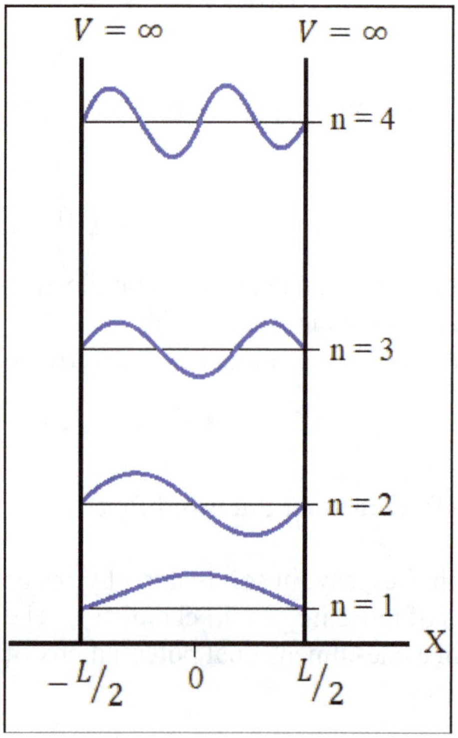

Figure 3.2 Energy levels for the particle in a one-dimensional potential box model. These levels are indicated with energy along the y-axis. The wave functions $\psi_n(x)$ for various n are plotted as blue lines while the solid horizontal black lines show E_n.

The wave function must satisfy the boundary conditions

$$\text{(i) } \psi(x) = 0 \text{ at } x = -\frac{L}{2}. \tag{11.6}$$

This implies

$$\mathrm{a} + \mathrm{b} = 0 \tag{11.7}$$

such that

$$\psi(x) = 2i\mathrm{a} \sin k(x + \frac{L}{2}); \tag{11.8}$$

$$\text{(ii) } \psi(x) = 0 \text{ at } x = \frac{L}{2} \tag{11.9}$$

which yields

$$k = \frac{n\pi}{L}, \text{ where } n = 0, \pm 1, \pm 2, \ldots \text{ (positive or negative integer).} \tag{11.10}$$

Therefore, n and hence k do not vary continuously, but they have discrete values. Because of the integral values, n is to be called a quantum number. Consequently the eigenvalue E in Eq. (11.5) can be written as

$$E_n = \frac{n^2 h^2}{8mL^2}. \tag{11.11}$$

The eigenvalues are discrete and depend on n. Quantization has been brought in by boundary conditions. The same effect will be repeatedly witnessed later.

In case we normalize the wave function $\psi(x)$ in Eq. (11.8), we find

$$\psi_n(x) = \sqrt{\frac{2}{L}} \sin\left(\frac{n\pi x}{L} + \frac{n\pi}{2}\right). \tag{11.12}$$

For $n = |n|$ and $n = -|n|$ we have the same $|\psi(x)|^2$. Hence the states represented by ψ_n and ψ_{-n} cannot be distinguished from each other, that is, these two functions represent the same state. The two functions merely differ by a factor of -1. Therefore one takes only zero and the positive values for the quantum number,

$n = 0, 1, 2, 3,$ etc.

As $\psi_0(x) = 0$ for all x, the total probability of finding the particle in state ψ_0 is zero. Hence the value n = 0 is discarded. Thus one is left with the quantum number

$$n = 1, 2, 3, \ldots \text{ (only the positive integers).} \tag{11.13}$$

With this result we find only two possibilities under inversion of coordinates $x \rightarrow -x$,

$$\psi(x) = \pm\psi(-x),$$

so that (11.14)

$$\rho(x) = |\psi(x)|^2 = |\psi(-x)|^2 = \rho(-x).$$

This result is a consequence of the symmetry of the potential around $x = 0$.

Points where the wave function becomes zero are called *nodes*. From Eq. (11.12) and Figure 3.2 one observes that the number of nodes of $\psi_n(x)$ is $(n-1)$ excluding the boundary points $x = \pm L/2$. The number of nodes increases with the quantum number n, and simultaneously the energy increases. This is another standard feature in quantum mechanics – energy increases with the number of nodes of the wave function.

The stationary state wave function $\Psi_n(x,t) = \psi_n(x) e^{-iE_n t/\hbar}$ is a standing wave, not the propagating wave of the free particle. The momentum expectation value is zero: $\langle p_x \rangle = 0$. The zero momentum is a consequence of the fact that $\psi_n(x)$ is a superposition of two equally weighted contributions from the propagating waves of linear momenta $\hbar k$ and $-\hbar k$. See (11.4) and (11.7). The stationary state wave function is just like the standing wave in a vibrating string.

Conjugated Polyenes

This simple model can be applied to the treatment of π-electrons in conjugated polyenes. The π-electrons (electrons residing in the p_z orbitals of the carbon atoms) in these molecules are relatively free to move around the framework of the σ bonds. The π-electron treatment reduces to the particle in a one-dimensional box model when the following assumptions are made:

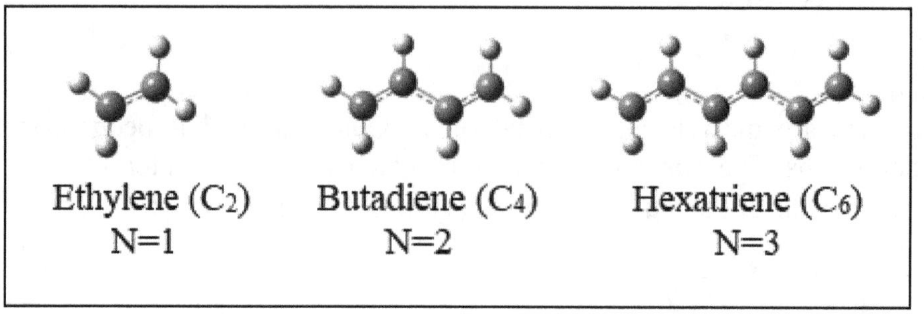

Figure 3.3 Some linear conjugated polyenes.

(1) the potential energy of the instantaneous electron-electron repulsion can be neglected; (2) the potential energy of the interaction of the π electrons with the σ electrons and the positively charged cores of the atoms belonging to the conjugated system is uniform over the entire length of the chain of carbon atoms; and (3) only the movement along the carbon skeleton matters, that is, the zigzag nature of the chain does not affect the smooth flow of the π electrons along the length of the molecule.

Under these assumptions, the π-electron Hamiltonian reduces to the sum of the Hamiltonians for individual π electrons with each electron behaving as a particle moving in a one-dimensional potential box. The total energy of the π electrons in this model is

$$E_\pi = \sum_{n=1}^{\infty} f_n E_n \tag{11.15}$$

where E_n is given in terms of L, the length of the carbon chain in the polyene, and f_n is the population of the π electrons in the nth energy level.

The energy levels are filled in accordance with Pauli's *exclusion principle* which states that *no more than two electrons can be in the same single particle state*. If we allow each electron to be in either of the two spin states α and β, then there can be at most two electrons in each energy level of the model problem – one in each spin state. The ground state is defined as the state of minimum energy. The model π-electron ground state is the one in which the lowermost energy levels are fully occupied.

Take the case of the C_{2N} conjugated polyene (in Figure 3.3). The most obvious canonical structure gives the length of the carbon chain $(N-1)\,l_{C\text{-}C} + N\,l_{C=C}$ where $l_{C\text{-}C}$ and $l_{C=C}$ are the carbon-carbon single bond length and the carbon-carbon double bond length, respectively. In real molecules, the potential does not suddenly jump to infinity. Hence one may consider a sharp but smooth rise at the two ends such that these rises contribute in average $\frac{1}{2}l_{C\text{-}C}$ each to the length of the potential box. This is illustrated in Figure 3.4. Thus we get

$$L = N\,(l_{C\text{-}C} + l_{C=C}). \tag{11.16}$$

The number of π electrons in the C_{2N} polyene is $2N$, one from each carbon atom. The ground state configuration involves $f_n = 2$ for $n = 1, 2, \ldots, N$ and $f_n = 0$ for $n > N$. The highest occupied molecular orbital (HOMO) and the lowest unoccupied molecular orbital (LUMO) correspond to $n=N$ and $n=N+1$, respectively. The minimum-energy spectroscopic transition corresponds to the excitation of one electron from the HOMO to the LUMO, indicated by $N \leftrightarrow N+1$. Since the total energy is conserved, the energy difference must equal the energy of the radiation absorbed, $h\nu = \dfrac{hc}{\lambda}$. Thus for the longest wavelength transition,

$$\frac{hc}{\lambda_{longest}} = E_{N+1} - E_N. \tag{11.17}$$

Using Eq. (11.11) and Eq. (11.16), one obtains

$$\lambda_{longest} = \frac{8mc}{h} \frac{(l_{C\text{-}C} + l_{C=C})^2}{(2N^{-1} + N^{-2})}. \tag{11.18}$$

Figure 3.4 The one-dimensional π-electron model for the C_6 conjugated polyene (N=3).

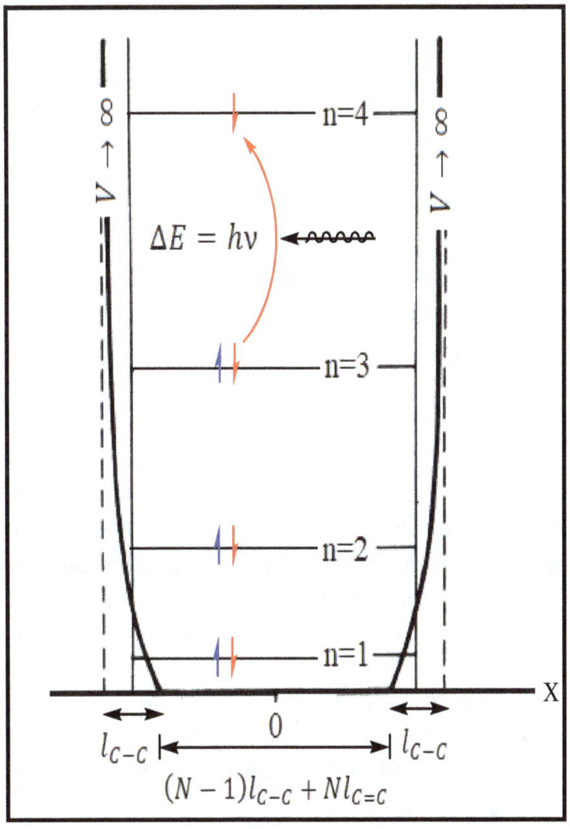

This relationship qualitatively agrees with the longest wavelength of absorption of conjugated C_{2N} polyenes. The longest wavelength of transition $\lambda_{longest}$ increases with increase of N. Using the values $m = 9.1 \times 10^{-31}$ Kg, $c = 3 \times 10^{8}$ m s^{-1}, $l_{C\text{-}C} = 1.53$ Å, $l_{C=C} = 1.34$ Å, and $h = 6.626 \times 10^{-34}$ J s, one may check that for a small N the longest wavelength transition is in the ultraviolet range. The longest wavelength enters the visible range ($\lambda_{longest} \sim 4000$ Å) at $N \sim 3.38$ when the violet end of the visible light is absorbed, and at this point the linear conjugated polyene starts appearing colored. The color of any substance is complementary to the color of the light absorbed. However, N must be an integer, and so the polyene starts being colored at $N = 4$, that is, from C_8 onwards.

An investigation of the electronic structure of a molecule essentially requires the investigation of *bound states*. As a general trend bound state investigations are more popular in chemistry, whereas physicists often engage themselves with studies on scattering phenomena. The latter mainly involve the continuum states, although in some cases the bound states are known to exert a remarkable influence on the scattering amplitude.

Lecture 12: The Simple Harmonic Oscillator

The classical simple harmonic oscillator in one dimension is described by F = $-kx$ which is equivalent to writing the potential as a quadratic function of the coordinate, V(x) = $\frac{1}{2}kx^2$ + C, C being a constant. The constant C can be chosen as zero. Comparing with Newton's equation (1.1), one finds the solution $x(t) = a\cos(\omega t+\delta)$ where ω is the angular frequency given by

$$\omega = 2\pi\nu = \sqrt{\frac{k}{m}}, \tag{12.1}$$

a is the maximum displacement or the amplitude of the oscillating particle or body with reduced mass m, and δ is called the phase. The time period is the time for one complete vibration, $T = \nu^{-1}$. The total energy is independent of the coordinate but depends on the amplitude, $E = \frac{1}{2}m\omega^2a^2$.

At $x=\pm a$, (i) $-$F = $\pm ka$, that is, the magnitude of force is maximum, and (ii) $\dot{x} = 0$ such that the kinetic energy is also zero and the potential energy equals the total energy. At these two points the particle turns and goes back towards the mean point x=0. Therefore, these two points are called as the classical turning points.

At x=0, (i) F = 0, (ii) V = 0 so that the total energy is entirely kinetic in form, and (iii) $|\dot{x}| = \omega a$, that is, the speed is maximum. One can easily verify that for a complete cycle of vibration the average kinetic energy equals the average potential energy.

The Schrödinger equation for the one-dimensional harmonic oscillator is

$$-\frac{\hbar^2}{2m}\frac{d^2\psi(x)}{dx^2} + \frac{1}{2}kx^2\psi(x) = E\psi(x). \tag{12.2}$$

The standard way of solving this equation is to first convert it into the dimensionless form

$$-\frac{d^2\psi(\xi)}{d\xi^2} + \xi^2\psi(\xi) = \varepsilon\psi(\xi) \tag{12.3}$$

where $\xi = \sqrt{\frac{m\omega}{h}}x$ and $\varepsilon = \frac{2E}{h\omega}$. As $x\to\infty$, $\xi\to\infty$ as well so that for a finite ε, Eq. (12.3) can be approximated by

$$\frac{d^2\psi(\xi)}{d\xi^2} = \xi^2\psi(\xi) \tag{12.4}$$

that has the asymptotic solutions $\psi(\xi) = e^{\pm\xi^2/2}$. The asymptotic solution $\psi(\xi) = e^{\xi^2/2}$ is discarded because it becomes infinitely large as ξ tends towards infinity. Therefore, the exact solution of Eq.(12.3) for any ξ can be written as

$$\psi(\xi) = H(\xi)e^{-\xi^2/2}. \tag{12.5}$$

Next, the function $H(\xi)$ is to be determined. If one inserts Eq.(12.5) into Eq.(12.3) one obtains the differential equation satisfied by $H(\xi)$,

$$\frac{d^2H(\xi)}{d\xi^2} - 2\xi\frac{dH(\xi)}{d\xi} + (\varepsilon - 1)H(\xi) = 0. \tag{12.6}$$

To solve this equation one expands $H(\xi)$ into a power series as

$$H(\xi) = \sum_i b_i\, \xi^i. \tag{12.7}$$

When the series expansion is incorporated into Eq.(12.6) one finds the recursion formula for the coefficients b_i's:

$$b_{i+2} = -\frac{(\varepsilon - 1 - 2i)}{(i+1)(i+2)}b_i. \tag{12.8}$$

This relationship shows that there are two choices for the coefficients. We can have either

even solutions with $b_0 \neq 0$ but $b_1 = 0$,

or (12.9)

odd solutions starting from $b_0 = 0$ and $b_1 \neq 0$.

As $i\to\infty$, the ratio b_{i+2}/b_i tends to $2/i$. The same trend is observed for the ratio of successive coefficients in the power series expansion for e^{ξ^2}. This indicates that in case the series expansion in Eq. (12.7) is not terminated, $H(\xi)$ would be expressible as

$$H(\xi) = H'(\xi) + c\,e^{\xi^2}$$

and $\psi(\xi)$ as

$$\psi(\xi) = H'(\xi)\,e^{-\xi^2/2} + c\,e^{\xi^2/2}.$$

But then $\psi(\xi)$ does not satisfy the boundary condition – it diverges as ξ tends to infinity.

Therefore, for a well-behaved solution, the series expansion in Eq. (12.7) must terminate. Suppose it does so after the nth term. Hence $b_{n+2} = 0$ though $b_n \neq 0$. From Eq. (12.8), this happens when

$$\varepsilon = 2n + 1$$

which in turn implies

$$E_n = h\nu(n + \tfrac{1}{2}) \qquad \text{(where } n = 0, 1, 2, 3, \ldots\text{).} \tag{12.10}$$

The integer n is the vibrational quantum number. Again we witness that quantization has arisen from the application of boundary conditions. The differential equation satisfied by the function $H_n(\xi)$ is found by substituting the value of ε in Eq. (12.6). We get

$$\frac{d^2H_n(\xi)}{d\xi^2} - 2\xi\frac{dH_n(\xi)}{d\xi} + 2nH_n(\xi) = 0 \qquad (12.11)$$

which is known as Hermite's differential equation. The function $H_n(\xi)$ is called the Hermite polynomial of the nth degree. Properties of these polynomials are given in the Table 3.1.

If one normalizes the wave function in Eq.(12.5) one finds

$$\psi_n(\xi) = (\pi^{1/2}2^n n!)^{-1/2} H_n(\xi)\, e^{-\xi^2/2}. \qquad (12.12)$$

A few of these wave functions, those with n = 0, 1, 2, and 3, have been plotted in Figure 3.5.

Table 3.1 Properties of Hermite Polynomials.

Generating function	$G(\xi,p) = e^{\xi^2-(p-\xi)^2}$ $= \sum_{n=0}^{\infty}(n!)^{-1}H_n(\xi)p^n$
	$H_n(\xi) = \left[\frac{d^nG(\xi,p)}{dp^n}\right]_{p=0}$
Rodrigue's formula	$H_n(\xi) = (-1)^n e^{\xi^2}\frac{d^n e^{-\xi^2}}{d\xi^n}$
Hermite's differential equation	$\frac{d^2H_n(\xi)}{d\xi^2} - 2\xi\frac{dH_n(\xi)}{d\xi} + 2nH_n(\xi) = 0$
Recurrence relation	$H_{n+1} - 2\xi H_n + 2nH_{n-1} = 0$
Differential relation	$\frac{dH_n}{d\xi} - 2nH_{n-1} = 0$
Orthonormality	$\int_{-\infty}^{\infty} d\xi H_m(\xi)H_n(\xi)\, e^{-\xi^2} = 2^n n!\, \pi^{1/2}\delta_{m,n}$
The lowest degree polynomials	
$H_0 = 1$	$H_1 = 2\xi$
$H_2 = 4\xi^2-2$	$H_3 = 8\xi^3-12\xi$
$H_4 = 16\xi^4-48\xi^2+12$	$H_5 = 32\xi^5-160\xi^3+120\xi$

The symmetry of the Hamiltonian, $\mathbf{H}(x) = \mathbf{H}(-x)$, indicates that the solutions $\psi_n(x)$ should be either symmetric or antisymmetric around the mean point $x = 0$. In fact,

$\psi_n(-x) = \psi_n(x)$ (for $n = 0, 2, 4, \ldots$ even solutions); (12.13)

$\psi_n(-x) = -\psi_n(x)$ (for $n = 1, 3, 5, \ldots$ odd solutions).

This is illustrated in Figure 3.5.

Figure 3.5 Quantized energy levels (parallel lines) and wave functions (oscillating lines) of the one-dimensional harmonic oscillator.

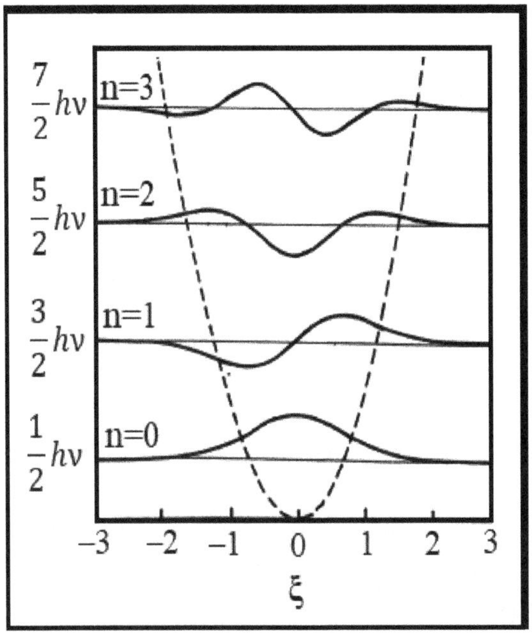

From the antisymmetry of the integrands, the expectation values of x and p_x are found to be zeroes,

$$\langle x\rangle_n = \int_{-\infty}^{\infty} dx\, |\psi_n(x)|^2\, x = 0, \tag{12.14}$$

$$\langle p_x\rangle = -i\hbar \int_{-\infty}^{\infty} dx\, \psi_n^*(x)\frac{d\psi_n(x)}{dx} = 0.$$

Using the recurrence relations of the Hermite polynomials, their normalization and orthogonality relationships, (see Table 3.1) one can show

$$\langle x^2\rangle_n = \frac{E_n}{m\omega^2}$$

and (12.15)

$$\langle p_x^2\rangle_n = mE_n.$$

In other words, the energy of the harmonic oscillator in *any* eigenstate is purely the energy of uncertainty:

$$E_n = \frac{1}{2m}\sigma_p^2 + \frac{1}{2}m\omega^2\sigma_x^2\,. \tag{12.16}$$

For every eigenstate, one obtains the uncertainty product

$$\sigma_x\sigma_p = (n+\frac{1}{2})\hbar = E_n/\omega. \tag{12.17}$$

The zero point wave function can be written as a function of the dimensionless variable ξ,

$$\psi_0(\xi) = \pi^{-1/4} e^{-\xi^2/2}, \tag{12.18}$$

or as a function of the Cartesian coordinate x,

$$\psi_0(x) = \left(\frac{m\omega}{\pi\hbar}\right)^{1/4} e^{-\frac{m\omega}{2\hbar}x^2}. \tag{12.19}$$

It is the minimum uncertainty wave as $n = 0$ and correspondingly, $\sigma_x\sigma_p = \hbar/2$.It is a pure Gaussian function.

The quantum mechanical harmonic oscillator treatment has found wide application in different branches of physics, solid state chemistry and spectroscopy. In bonding theory, the molecular oscillator is often very nicely described as an approximate harmonic oscillator with only a slight anharmonicity. Harmonic oscillator models have also been used in the discussion of normal modes and chemical reactivity.

Lecture 13: Orbital Angular Momentum

The orbital angular momentum is defined by

$$\mathbf{L} = \mathbf{r} \times \mathbf{p} = \mathbf{i}L_x + \mathbf{j}L_y + \mathbf{k}L_z. \tag{13.1}$$

It has the cartesian components

$$L_x = yp_z - zp_y,$$
$$L_y = zp_x - xp_z, \tag{13.2}$$
$$L_z = xp_y - yp_x.$$

The square of the orbital angular momentum is

$$L^2 = \mathbf{L}\cdot\mathbf{L} = L_x^2 + L_y^2 + L_z^2. \tag{13.3}$$

The corresponding quantum mechanical operators are

$$\boldsymbol{L_x} = -i\hbar\left(y\frac{\partial}{\partial z} - z\frac{\partial}{\partial y}\right) \qquad (x, y, z \text{ cyclically varying}) \tag{13.4}$$

where (x, y, z cyclically varying) means that two similar relations hold with (x, y and z) replaced by (y, z, and x) and (z, x, and y) respectively. We also have

$$\boldsymbol{L}^2 = \boldsymbol{L}\cdot\boldsymbol{L} = \boldsymbol{L_x}^2 + \boldsymbol{L_y}^2 + \boldsymbol{L_z}^2. \tag{13.5}$$

With these definitions we find

$$[\boldsymbol{L_x}, \boldsymbol{L_y}] = i\hbar\boldsymbol{L_z} \qquad (x, y, z \text{ cyclically varying}) \tag{13.6}$$

so that $\boldsymbol{L_x}$, $\boldsymbol{L_y}$, and $\boldsymbol{L_z}$ cannot have common eigenfunctions. Also,

$$[\boldsymbol{L_u}, \boldsymbol{L}^2] = 0 \qquad (u = x, y, z) \tag{13.7}$$

so that each of the operators $\boldsymbol{L_x}$, $\boldsymbol{L_y}$, and $\boldsymbol{L_z}$ can have common eigenfunctions with $\boldsymbol{L}^2$. One may seek common eigenfunctions either of $\boldsymbol{L_x}$ and $\boldsymbol{L}^2$, or of $\boldsymbol{L_y}$ and $\boldsymbol{L}^2$, or even of $\boldsymbol{L_z}$ and $\boldsymbol{L}^2$. It is conventional to use the common eigenfunctions Y of the last pair. These satisfy

$$\boldsymbol{L_z}\mathrm{Y} = \lambda\mathrm{Y} \tag{13.8a}$$

and

$$L^2\mathrm{Y} = \Lambda \mathrm{Y}, \tag{13.8b}$$

In these lectures we will stick to the conventional choice.

Since the orbital angular motion is in fact rotation in three-dimensional space while maintaining a constant distance from the center, it will be useful to employ the spherical polar coordinates for studying it. Figure 3.6 shows the spherical polar coordinates r, θ, and φ, and the corresponding Cartesian coordinates x, y, and z of a point. The following relations are found between the two sets of coordinates:

(Cartesian coordinates in terms of the spherical polar coordinates)

$$\begin{aligned} x &= r\sin\theta\cos\varphi \quad (-\infty \leq x \leq \infty), \\ y &= r\sin\theta\sin\varphi \quad (-\infty \leq y \leq \infty), \\ z &= r\cos\theta \quad (-\infty \leq z \leq \infty); \end{aligned} \tag{13.9}$$

and

(Spherical polar coordinates in terms of the Cartesian ones)

$$\begin{aligned} r &= \sqrt{(x^2 + y^2 + z^2)} \quad (0 \leq r \leq \infty), \\ \theta &= \tan^{-1}\frac{\sqrt{x^2 + y^2}}{z} \quad (0 \leq \theta \leq \pi), \\ \varphi &= \tan^{-1}\frac{y}{x} \quad (0 \leq \varphi \leq 2\pi) \end{aligned} \tag{13.10}$$

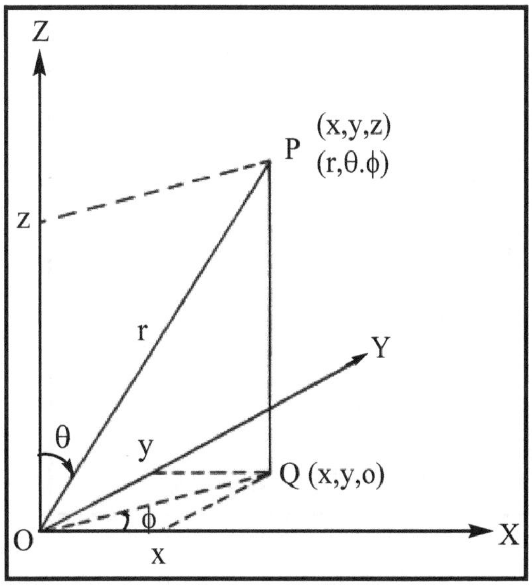

Figure 3.6 Relations between the spherical polar coordinates r, θ and φ and Cartesian coordinates x, y and z. OQ is the projection of OP onto the xy plane.

The infinitesimally small volume element is written in Cartesian coordinates as $dv = dx\, dy\, dz$. As illustrated in Figure 3.7, it is given in the spherical polar coordinates by

$$dv = r^2 dr d\Omega = r^2 dr \sin\theta d\theta d\varphi \tag{13.11}$$

where $d\Omega$ is the infinitesimally small element of solid angle

$$d\Omega = \sin\theta d\theta d\varphi. \tag{13.12}$$

One can easily derive the following expressions:

$$L_X = i\hbar\left(\cot\theta\cos\varphi\frac{\partial}{\partial\varphi} + \sin\varphi\frac{\partial}{\partial\theta}\right),$$

$$L_y = i\hbar\left(\cot\theta\sin\varphi\frac{\partial}{\partial\varphi} - \cos\varphi\frac{\partial}{\partial\theta}\right), \tag{13.13}$$

$$L_z = -i\hbar\frac{\partial}{\partial\varphi},$$

and

$$L^2 = -\hbar^2\left[\frac{1}{\sin\theta}\frac{\partial}{\partial\theta}\left(\sin\theta\frac{\partial}{\partial\theta}\right) + \frac{1}{\sin^2\theta}\frac{\partial^2}{\partial\varphi^2}\right]. \tag{13.14}$$

Because all these operators contain trigonometric functions and derivatives of only the angles θ and φ and are independent of r and its derivatives, the common eigenfunction of L_z and L^2 in equations (13.8a) and (13.8b) must be a function of the same angles and will be denoted by $Y(\theta, \varphi)$. These solutions are called the *spherical harmonics*.

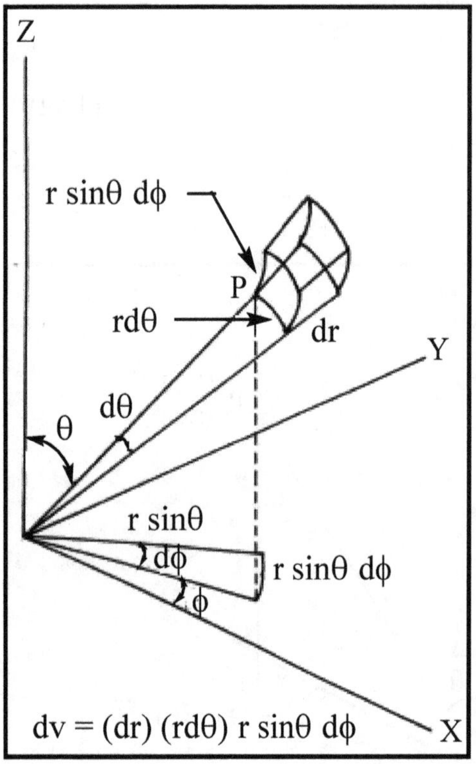

Figure 3.7 The infinitesimally small volume element in the spherical polar coordinates.

Let us find the eigenfunctions of the operator L_z. We write

$$L_z\Phi(\varphi) = -i\hbar\frac{d\Phi(\varphi)}{d\varphi} = \lambda\Phi(\varphi) \tag{13.15}$$

which is merely a first order differential equation, with solutions of the form

$$\Phi(\varphi) = Ne^{im\varphi}$$

corresponding to (13.16)

$$\lambda = m\hbar.$$

The eigenvalues are not yet quantized. However, the moment the boundary condition

$$\Phi(\varphi + 2\pi) = \Phi(\varphi) \tag{13.17}$$

is imposed, we get $e^{2\pi im} = 1$ which implies that the m value is quantized,

$$m = 0, \pm1, \pm2, \pm3, \ldots \quad \text{(zero, and positive or negative integers).} \tag{13.18}$$

These values (m) are called the *magnetic quantum numbers*. The reason for this nomenclature will become clear later when we discuss the Zeeman effect that involves the interaction of an atomic electron with a uniform magnetic field. The function $\Phi(\varphi)$ should be characterized by this quantum number and written as $\Phi_m(\varphi)$. For a normalized $\Phi_m(\varphi)$,

$$\int_0^{2\pi} d\varphi |\Phi_m(\varphi)|^2 = 1$$

which gives $N = \dfrac{1}{\sqrt{2\pi}}$. Therefore, the eigenfunctions of the operator $\boldsymbol{L}_Z$ are

$$\Phi_m(\varphi) = \frac{1}{\sqrt{2\pi}} e^{im\varphi} \qquad (m = 0, \pm1, \pm2, \pm3, \ldots) \tag{13.19}$$

with eigenvalues

$$\lambda_m = \hbar m. \tag{13.20}$$

We now separate the spherical harmonics as

$$Y(\theta,\varphi) = \Theta(\theta)\Phi_m(\varphi). \tag{13.21}$$

Introducing this solution to Eq. (13.8b) and using the expressions (13.14) for $\boldsymbol{L}^2$ and (13.19) for function Φ_m, we get

$$\frac{1}{\sin\theta}\frac{d}{d\theta}\left(\sin\theta\frac{d\Theta}{d\theta}\right) + \left(\frac{\Lambda}{\hbar^2} - \frac{m^2}{\sin^2\theta}\right)\Theta = 0. \tag{13.22}$$

This is a standard differential equation (Associated Legendre equation) with solutions characterized by the two quantum numbers l and m when $\Lambda/\hbar^2$ is replaced by $l(l+1)$. The number l varies as

$$l = 0, 1, 2, \ldots \tag{13.23}$$

subject to the condition $l \geq |m|$ such that the allowed values of m are

$$m = -l, -l+1, \ldots, -1, 0, 1, \ldots, l-1, l. \tag{13.24}$$

Actually the function $\Theta(\theta)$ depends only on $|m|$ as m is involved in Eq. (13.22) only through m^2. The solution $\Theta_{l,|m|}(\theta)$ is the normalized form of an associated Legendre polynomial $P_{l,m}(\cos\theta)$. The normalization condition is written as

$$\int_0^{\pi} d\theta \sin\theta \, \Theta_{l,|m|}^2(\theta) = 1. \tag{13.25}$$

Of course, the eigenvalue Λ is characterized by the quantum number l,

$$\Lambda = \hbar^2 l(l+1) \qquad (l = 0, 1, 2, 3, \ldots). \tag{13.26}$$

The quantum number l is called the *azimuthal quantum number* or the *orbital angular momentum quantum number*. Solution of Eq. (13.22) is explicitly discussed in Part II of these lectures.

Thus the conventional spherical harmonics are given by

$$Y_{l,m}(\theta,\varphi) = \Theta_{l,|m|}(\theta)\Phi_m(\varphi). \tag{13.27}$$

They obey the eigenvalue equations

$$\boldsymbol{L}_z Y_{l,m}(\theta,\varphi) = \hbar m Y_{l,m}(\theta,\varphi) \tag{13.28}$$

and

$$\boldsymbol{L}^2 Y_{l,m}(\theta,\varphi) = \hbar^2 l(l+1) Y_{l,m}(\theta,\varphi). \tag{13.29}$$

The spherical harmonics form an orthonormal set, as

$$\int_{4\pi} d\Omega\, Y_{l,m}^*(\theta,\varphi) Y_{l',m'}(\theta,\varphi) = \delta_{l,l'}\delta_{m,m'}. \tag{13.30}$$

The first few spherical harmonics are given in Table 3.2. These are known as s, p, d, f, g, h, etc., angular functions as l equals 0, 1, 2, 3, 4, 5, and so on. This nomenclature originated from atomic spectroscopy where lines were called sharp (s), principal (p), diffuse (d), and fundamental (f). The alphabetic nomenclature has been adopted for l values starting from $l = 3$ (f) onwards.

Table 3.2 The first few Spherical Harmonics.

$$Y_{l,m}(\theta,\varphi) = \Theta_{l,|m|}(\theta)\Phi_m(\varphi)$$

$$\Phi_m(\varphi) = \frac{1}{\sqrt{2\pi}} e^{im\varphi} \qquad (m = 0, \pm 1, \pm 2, \pm 3, \ldots)$$

$$\Theta_{l,m}(\theta) = \left[\frac{2l+1}{2}\frac{(l-m)!}{(l+m)!}\right]^{\frac{1}{2}} P_{l,m}(\cos\theta)$$

where $P_{l,m}(\cos\theta)$ is an associated Legendre polynomial.

Name	l	m	Spherical Harmonics
s	0	0	$Y_{0,0} = \left(\frac{1}{4\pi}\right)^{\frac{1}{2}}$
p	1	0	$Y_{1,0} = \left(\frac{3}{4\pi}\right)^{\frac{1}{2}} \cos\theta$
		±1	$-\,Y_{1,\pm 1} = \pm\left(\frac{3}{8\pi}\right)^{\frac{1}{2}} \sin\theta\, e^{\pm i\varphi}$
d	2	0	$Y_{2,0} = \left(\frac{5}{16\pi}\right)^{\frac{1}{2}} (3\cos^2\theta - 1)$
		±1	$-\,Y_{2,\pm 1} = \pm\left(\frac{15}{8\pi}\right)^{\frac{1}{2}} \sin\theta\cos\theta\, e^{\pm i\varphi}$
		±2	$Y_{2,\pm 2} = \left(\frac{15}{32\pi}\right)^{\frac{1}{2}} \sin^2\theta\, e^{\pm 2i\varphi}$

Table 3.2 The first few Spherical Harmonics — continued

Name	l	m	Spherical Harmonics
f	3	0	$Y_{3,0} = \left(\frac{7}{16\pi}\right)^{\frac{1}{2}} \cos\theta(5\cos^2\theta - 3)$
		±1	$-Y_{3,\pm 1} = \pm\left(\frac{21}{64\pi}\right)^{\frac{1}{2}} \sin\theta(5\cos^2\theta - 1)\, e^{\pm i\varphi}$
		±2	$Y_{3,\pm 2} = \left(\frac{105}{32\pi}\right)^{\frac{1}{2}} \sin^2\theta\cos\theta e^{\pm 2i\varphi}$
		±3	$-Y_{3,\pm 3} = \pm\left(\frac{35}{64\pi}\right)^{\frac{1}{2}} \sin^3\theta e^{\pm 3i\varphi}$
g	4	0	$Y_{4,0} = \left(\frac{9}{256\pi}\right)^{\frac{1}{2}} (35\cos^4\theta - 30\cos^2\theta + 3)$
		±1	$-Y_{4,\pm 1} = \pm\left(\frac{45}{64\pi}\right)^{\frac{1}{2}} \sin\theta(7\cos^3\theta - 3\cos\theta)\, e^{\pm i\varphi}$
		±2	$Y_{4,\pm 2} = \left(\frac{45}{128\pi}\right)^{\frac{1}{2}} \sin^2\theta(7\cos^2\theta - 1)\, e^{\pm 2i\varphi}$
		±3	$-Y_{4,\pm 3} = \pm\left(\frac{315}{64\pi}\right)^{\frac{1}{2}} \sin^3\theta\cos\theta e^{\pm 3i\varphi}$
		±4	$Y_{4,\pm 4} = \left(\frac{315}{512\pi}\right)^{\frac{1}{2}} \sin^4\theta e^{\pm 4i\varphi}$

Lecture 14: Angular Functions

Although the spherical harmonics are angular wave functions – eigenfunctions of $\boldsymbol{L}^2$ and $\boldsymbol{L}_z$ – they suffer from the difficulty that they are in general complex functions of coordinates. To obtain real functions of θ and φ one can make suitable linear combinations of $Y_{l,m}(\theta, \varphi)$ and $Y_{l,-m}(\theta, \varphi)$ such that the new functions are orthogonal to each other. These new functions are functions of the cosine and sine series,

$$Y_{l,|m|,c}(\theta, \varphi) = \pi^{-1/2}\Theta_{l,|m|}(\theta) \cos|m|\varphi,$$

and $\qquad (m \neq 0) \qquad (14.1)$

$$Y_{l,|m|,s}(\theta, \varphi) = \pi^{-1/2}\Theta_{l,|m|}(\theta) \sin|m|\varphi.$$

The harmonics $Y_{l,0}$ are always real functions. The new angular functions belonging to the sine and the cosine series are not eigenfunctions of $\boldsymbol{L}_z$ but they are still eigenfunctions of $\boldsymbol{L}^2$. Table 3.3 gives a list of these real functions. Note the name given to each function.

The functions p_x, p_y, and p_z are proportional to $\frac{x}{r}$, $\frac{y}{r}$ and $\frac{z}{r}$, respectively. The function $d_{x^2-y^2}$ is proportional to $\frac{x^2-y^2}{r^2}$. Thus the name itself indicates the angular dependence of the wave function. The remaining part of each function is in reality the normalization constant.

Table 3.3 Angular functions – real form.

Name	l $\lvert m\rvert$	$Y_{l,m=0}(\theta,\varphi)$ or $Y_{l,\lvert m\rvert,\text{c or s}}(\theta,\varphi)$
s	0 0	$Y_{0,0} = (1/4\pi)^{1/2}$
p_z	1 0	$Y_{1,0} = (3/4\pi)^{1/2}\cos\theta = (3/4\pi)^{1/2} z/r$
p_x	1 1	$Y_{1,1,c} = (3/4\pi)^{1/2}\sin\theta\cos\varphi = (3/4\pi)^{1/2} x/r$
p_y	1 1	$Y_{1,1,s} = (3/4\pi)^{1/2}\sin\theta\sin\varphi = (3/4\pi)^{1/2} y/r$
d_{z^2}	2 0	$Y_{2,0} = (5/16\pi)^{1/2}(3\cos^2\theta - 1) = (5/16\pi)^{1/2}(2z^2 - x^2 - y^2)/r^2$
d_{xz}	2 1	$Y_{2,1,c} = (15/4\pi)^{1/2}\sin\theta\cos\theta\cos\varphi = (15/4\pi)^{1/2} xz/r^2$
d_{yz}	2 1	$Y_{2,1,s} = (15/4\pi)^{1/2}\sin\theta\cos\theta\sin\varphi = (15/4\pi)^{1/2} yz/r^2$
$d_{x^2-y^2}$	2 2	$Y_{2,2,c} = (15/16\pi)^{1/2}\sin^2\theta\cos 2\varphi = (15/16\pi)^{1/2}(x^2-y^2)/r^2$
d_{xy}	2 2	$Y_{2,2,s} = (15/16\pi)^{1/2}\sin^2\theta\sin 2\varphi = (15/4\pi)^{1/2} xy/r^2$
f_{z^3}	3 0	$Y_{3,0} = (7/16\pi)^{1/2}\cos\theta(5\cos^2\theta - 3) = (7/16\pi)^{1/2} z(5z^2-3r^2)/r^3$
f_{xz^2}	3 1	$Y_{3,1,c} = (21/32\pi)^{1/2}\sin\theta(5\cos^2\theta-1)\cos\varphi = (21/32\pi)^{1/2} x(5z^2-r^2)/r^3$
f_{yz^2}	3 1	$Y_{3,1,s} = (21/32\pi)^{1/2}\sin\theta(5\cos^2\theta-1)\sin\varphi = (21/32\pi)^{1/2} y(5z^2-r^2)/r^3$
$f_{(x^2-y^2)z}$	3 2	$Y_{3,2,c} = (105/16\pi)^{1/2}\sin^2\theta\cos\theta\cos 2\varphi = (105/16\pi)^{1/2}(x^2-y^2)z/r^3$
f_{xyz}	3 2	$Y_{3,2,s} = (105/16\pi)^{1/2}\sin^2\theta\cos\theta\sin 2\varphi = (105/4\pi)^{1/2} xyz/r^3$
$f_{x^3-3xy^2}$	3 3	$Y_{3,3,c} = (35/32\pi)^{1/2}\sin^3\theta\cos 3\varphi = (35/32\pi)^{1/2} x(x^2-3y^2)/r^3$
$f_{y^3-3x^2y}$	3 3	$Y_{3,3,s} = (35/32\pi)^{1/2}\sin^3\theta\sin 3\varphi = (35/32\pi)^{1/2} y(3x^2-y^2)/r^3$

Since $Y_{l,m}(\theta,\varphi)$ is a normalized angular wave function, $|Y_{l,m}(\theta,\varphi)|^2$ gives the angular probability density at the orientation (θ,φ). The angular probability density is defined as the angular probability per unit solid angle at an orientational point, that is, the total angular probability over an infinitesimally small element of solid angle around the point divided by the same small element. In normal practice the angular probability distribution is represented by *nodal* or *lobe* diagrams as shown in Figures 3.8(a) and 3.8(b). In these plots, the distance of any point from the origin equals the weighted angular probability density $|Y_{l,m}(\theta, \varphi)|^2/(2l + 1)$ at the specific orientation (θ,φ). The weighting factor $(2l+1)^{-1}$ is chosen to bring all the diagrams to a comparable scale with the maximum distance of the order of $1/4\pi$.

The s function is spherically symmetric, that is, independent of orientation. So it is represented by a sphere in three dimension and a circle as a two-dimensional projection. The radius of the sphere is $1/4\pi$. The real p functions are axially symmetric. The p_x function is symmetric around the x-axis. The nodal diagram for the p_z function is drawn by tracing points in the xz-plane or the yz-plane in such a way that the distance of each point from the origin equals

$$\frac{1}{3}p_z^2 = \frac{1}{4\pi}\frac{z^2}{r^2} = \frac{1}{4\pi}\cos^2\theta \,. \tag{14.2}$$

The maximum distance occurs at $\theta = 0$ and $\theta = \pi$. The minimum distance zero occurs with $\theta = \frac{\pi}{2}$ for points in the xy-plane. The xy-plane is the nodal surface for the p_z function. The reader is urged to use the real functions in Table 3.3 in order to draw all the nodal diagrams shown in Figures 3.8.

The d orbital has two nodal surfaces. This is obvious from the diagrams for d_{xy}, d_{yz}, d_{xz} and $d_{x^2-y^2}$. Identifying the nodal surfaces of the d_{z^2} function is left as an exercise.

Notice the relative arrangement of the signs and the nodal surfaces. The sign changes while crossing a node. The *sign* of the angular function is shown in the nodal diagrams. A positive sign is indicated by a clear area, and a negative sign by a shaded region. The sign of the s function is positive everywhere. The two lobes of the p functions have opposite signs.

One caveat: the functions discussed here are angular functions, and not the atomic orbitals. Atomic orbitals are wave functions for one electron in an atom. Because of the spherically symmetric potential, an atomic orbital is written as the product of the radial wave function for one electron and the angular function – in real or complex form. These are called as s, p, d, etc. orbitals, depending on the angular function involved.

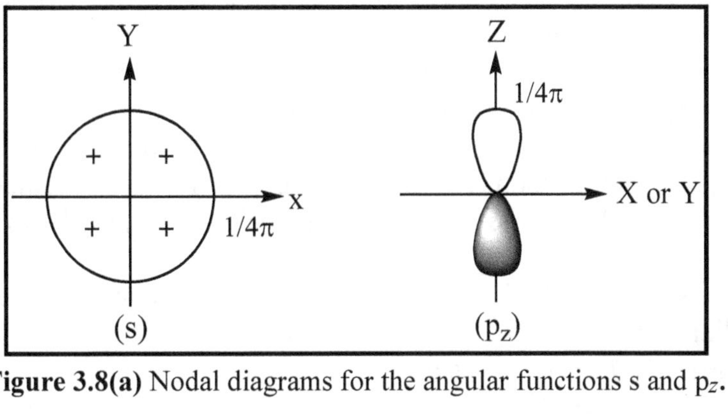

Figure 3.8(a) Nodal diagrams for the angular functions s and p_z.

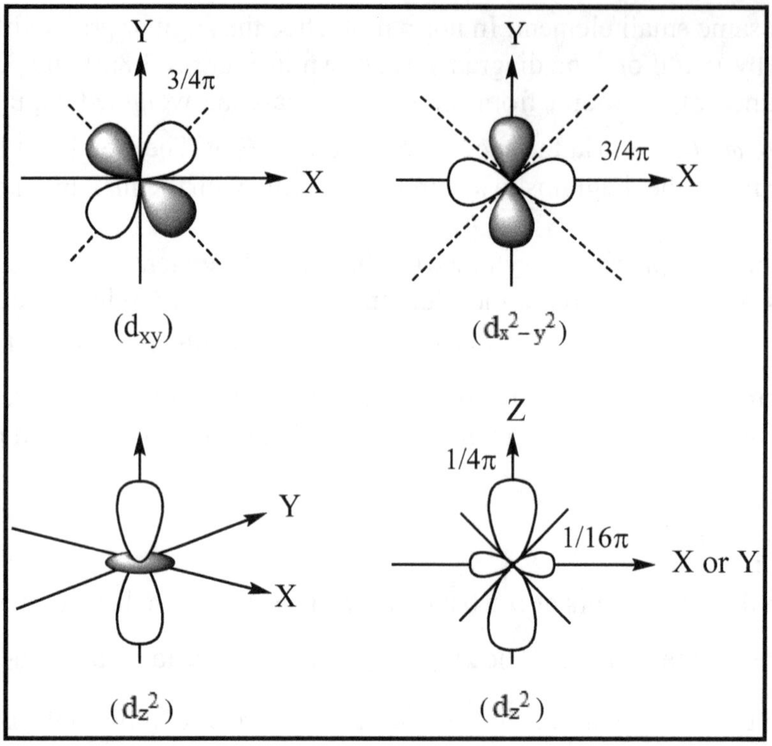

Figure 3.8(b) Nodal diagrams for the angular functions d_{xy}, $d_{x^2-y^2}$ and d_{z^2}.

Lectures 15 and 16: The Hydrogen Atom

Lecture 15: Schrödinger Equation for the Electron in Hydrogen Atom and Its Solution

Hydrogen atom is the simplest atom. There are many things known about it, and being the simplest atom the corresponding equations are the simplest to solve. Therefore, hydrogen atom constitutes one of the most fundamental problems in quantum theory. The electronic Hamiltonian for the hydrogenic system

$$\mathbf{H} = -\frac{\hbar^2}{2m}\nabla^2 - \frac{Ze^2}{r} \quad (Z = 1 \text{ for hydrogen atom}) \tag{15.1}$$

involves a centrally symmetric potential, (that is, a potential that depends only upon the radial distance r from the nucleus). Hence a treatment based on the spherical polar coordinates would be the most fruitful approach to the study of the electronic Schrödinger equation. For the hydrogen-like systems such as He^+ ion, Li^{+2} ion, etc., Z has different integral values: $Z = 2, 3, \ldots$, etc. The mass m in the Hamiltonian is the reduced mass of the electron-nucleus system which is the appropriate mass for the movement of the electron relative to the nucleus. The nucleus resides at the origin (r=0) of the relative coordinate system.

The unit vectors in the spherical polar coordinates are related to the unit vectors in Cartesian coordinates by

$$\hat{\mathbf{r}} = \sin\theta\cos\varphi\,\mathbf{i} + \sin\theta\sin\varphi\,\mathbf{j} + \cos\theta\,\mathbf{k}$$
$$\hat{\boldsymbol{\theta}} = \cos\theta\cos\varphi\,\mathbf{i} + \cos\theta\sin\varphi\,\mathbf{j} - \sin\theta\,\mathbf{k}$$

$$\hat{\boldsymbol{\varphi}} = -\sin\varphi\,\mathbf{i} + \cos\varphi\,\mathbf{j}$$

so that the vector operator ∇ is written as

$$\nabla = \hat{\mathbf{r}}\frac{\partial}{\partial r} + \hat{\boldsymbol{\theta}}\frac{1}{r}\frac{\partial}{\partial \theta} + \hat{\boldsymbol{\varphi}}\frac{1}{r\sin\theta}\frac{\partial}{\partial \varphi}.$$

Hence the Laplacian operator has the form

$$\nabla^2 = \frac{1}{r^2}\frac{\partial}{\partial r}\left(r^2\frac{\partial}{\partial r}\right) + \frac{1}{r^2\sin\theta}\frac{\partial}{\partial\theta}\left(\sin\theta\frac{\partial}{\partial\theta}\right) + \frac{1}{r^2\sin^2\theta}\frac{\partial^2}{\partial\varphi^2}. \tag{15.2}$$

We rewrite the Laplacian as

$$\nabla^2 = \frac{1}{r^2}\frac{\partial}{\partial r}\left(r^2\frac{\partial}{\partial r}\right) - \frac{1}{\hbar^2 r^2}\boldsymbol{L}^2 \tag{15.3}$$

Where $\boldsymbol{L}^2$ is the squared orbital angular momentum operator given by Eq.(13.14). Therefore, the Hamiltonian in Eq. (15.1) can be written as

$$\mathbf{H} = -\frac{\hbar^2}{2mr^2}\frac{\partial}{\partial r}\left(r^2\frac{\partial}{\partial r}\right) + \frac{1}{2mr^2}\boldsymbol{L}^2 - \frac{Ze^2}{r}. \tag{15.4}$$

The first term represents the kinetic energy for the radial motion whereas the second one stands for the kinetic energy for the angular movement, $\frac{\boldsymbol{L}^2}{2I}$ where I ($=mr^2$) is the moment of inertia. Obviously $\mathbf{H}$ commutes with $\boldsymbol{L}^2$. As $[\boldsymbol{L}^2, \boldsymbol{L}_Z] = 0$, the operator $\mathbf{H}$ also commutes with $\boldsymbol{L}_Z$. This means that $\mathbf{H}$, $\boldsymbol{L}^2$ and $\boldsymbol{L}_Z$ *can* have a set of common eigenfunctions. Eigenfunctions of $\mathbf{H}$ are functions of r, θ, and φ, whereas those of $\boldsymbol{L}^2$ need to be functions of only θ and φ and the eigenfunctions of $\boldsymbol{L}_z$ need be functions of only φ. This indicates that the angular parts of the eigenfunctions of $\mathbf{H}$ can be separated. One may take either the spherical harmonics or the real functions prepared from the spherical harmonics of a specific l value as these angular parts.

Separation

The electronic Schrödinger equation for a hydrogen-like system is

$$-\frac{\hbar^2}{2m}\nabla^2\psi - \frac{Ze^2}{r}\psi = E\psi. \tag{15.5}$$

Because of Eq. (15.3) the present equation converts into

$$-\frac{\hbar^2}{2mr^2}\frac{\partial}{\partial r}\left(r^2\frac{\partial\psi}{\partial r}\right) + \frac{1}{2mr^2}\boldsymbol{L}^2\psi - \frac{Ze^2}{r}\psi = E\psi. \tag{15.6}$$

Consider that the wave function $\psi(r,\theta,\varphi)$ is separable as

$$\psi(r,\theta,\varphi) = \mathrm{R}(r)\,\mathrm{Y}(\theta,\varphi). \tag{15.7}$$

Substituting this product for ψ in Eq.(15.6) and dividing both the sides by RY we get

$$\frac{\hbar^2}{\mathrm{R}(r)}\frac{\partial}{\partial r}\left(\mathrm{r}^2\frac{\partial \mathrm{R}}{\partial \mathrm{r}}\right) + 2mr^2\left(E + \frac{Ze^2}{r}\right) = \frac{\boldsymbol{L}^2\,\mathrm{Y}(\theta,\varphi)}{\mathrm{Y}(\theta,\varphi)} \tag{15.8}$$

except when ψ=0. The left side is dependent only on the variable r whereas the right side depends only on the angular variables θ and φ. The above equality holds for all possible r and (θ,φ), indicating that each side must be equal to a constant term independent of these variables r, θ and φ. The right hand side then identifies the function $Y(\theta,\varphi)$ as one of the spherical harmonics $Y_{l,m}(\theta,\varphi)$ or a real angular function developed from the linear combination of the spherical harmonics for a particular azimuthal quantum number l. The constant term equals $\hbar^2 l(l+1)$, the eigenvalue of $\boldsymbol{L}^2$, leaving aside the radial wave equation

$$\frac{1}{r^2}\frac{d}{dr}\left(r^2\frac{d\mathrm{R}}{dr}\right)+\left[\frac{2m}{\hbar^2}\left(E+\frac{Ze^2}{r}\right)-\frac{l(l+1)}{r^2}\right]\mathrm{R}(r)=0. \tag{15.9}$$

This equation will be explicitly solved in Part II of these lectures. For reasons of the boundary conditions, (similar to those encountered in the treatment of the harmonic oscillator), one finds that there must be a quantum number n,

$$n = 1, 2, 3, \ldots \quad \text{(positive integers)}, \tag{15.10}$$

such that $n>l$. The l quantum number is thereby restricted,

$$l = 0, 1, 2, \ldots, (n-1). \tag{15.11}$$

The quantum number n is crucial for the determination of a well-behaved solution of the form

$$\mathrm{R}_{nl}(r) = \left[\left(\frac{2Z}{na_0}\right)^3\frac{(n-l-1)!}{2n(n+l)!}\right]^{\frac{1}{2}}\rho^l e^{-\frac{1}{2}\rho}L_{n-l-1}^{2l+1}(\rho) \tag{15.12}$$

where ρ is the dimensionless quantity

$$\rho = \frac{2Z}{na_0}r. \tag{15.13}$$

The distance a_0 is the Bohr radius, given by $a_0 = \frac{\hbar^2}{me^2}$. The function $L_{n-l-1}^{2l+1}(\rho)$ is an associated Laguerre polynomial. The radial part of the wave function is thus characterized by the two quantum numbers n and l. Consequently the whole wave function is characterized by the three quantum numbers n, l and m:

$$\psi_{nlm}(r,\theta,\varphi) = \mathrm{R}_{nl}(r)Y_{l,m}(\theta, \varphi). \tag{15.14}$$

The total energy is characterized by only the *principal quantum number n*,

$$E_n = -\frac{mZ^2e^4}{2n^2\hbar^2}. \tag{15.15}$$

This is precisely the formula derived by Bohr from his celebrated model of old quantum theory. It was very successful in explaining the frequencies of the lines appearing in the hydrogen atomic spectrum. Here, however, the result is arrived at for a three-dimensional atom with no arbitrary postulate required for the angular momentum being quantized. Quantization always appears here as a result of the application of boundary conditions. Some of the radial wave functions for the electron in the hydrogen atom have been given in Table 3.4, and illustrated in Figures 3.9(a) and 3.9(b).

Table 3.4 Radial wave functions for hydrogen-like atoms.

$$\rho = \frac{2Z}{na_0} r\,; \quad a_0 = \frac{\hbar^2}{me^2} = 0.529\ \text{Å} = 52.9\ \text{pm.}$$

Shell	n	l	Orbital	Radial function $R_{nl}(r)$
K	1	0	1s	$R_{10} = 2\,(Z/a_0)^{3/2}\exp(-\rho/2)$
L	2	0	2s	$R_{20} = 2^{-3/2}\,(Z/a_0)^{3/2}\,(2-\rho)\exp(-\rho/2)$
		1	2p	$R_{21} = 2^{-1}6^{-1/2}\,(Z/a_0)^{3/2}\rho\exp(-\rho/2)$
M	3	0	3s	$R_{30} = 9^{-1}3^{-1/2}\,(Z/a_0)^{3/2}\,(6-6\rho+\rho^2)\exp(-\rho/2)$
		1	3p	$R_{31} = 9^{-1}6^{-1/2}\,(Z/a_0)^{3/2}\,(4-\rho)\rho\exp(-\rho/2)$
		2	3d	$R_{32} = 9^{-1}30^{-1/2}\,(Z/a_0)^{3/2}\rho^2\exp(-\rho/2)$
N	4	0	4s	$R_{40} = 96^{-1}\,(Z/a_0)^{3/2}\,(24-36\rho+12\rho^2-\rho^3)\exp(-\rho/2)$
		1	4p	$R_{41} = 32^{-1}15^{-1/2}\,(Z/a_0)^{3/2}\,(20-10\rho+\rho^2)\rho\exp(-\rho/2)$
		2	4d	$R_{42} = 96^{-1}5^{-1/2}\,(Z/a_0)^{3/2}\,(6-\rho)\rho^2\exp(-\rho/2)$
		3	4f	$R_{43} = 96^{-1}35^{-1/2}\,(Z/a_0)^{3/2}\rho^3\exp(-\rho/2)$
O	5	0	5s	$R_{50} = 300^{-1}5^{-1/2}\,(Z/a_0)^{3/2}$ $\times\,(120-240\rho+120\rho^2-20\rho^3+\rho^4)\exp(-\rho/2)$
		1	5p	$R_{51} = 150^{-1}30^{-1/2}\,(Z/a_0)^{3/2}$ $\times\,(120-90\rho+18\rho^2-\rho^3)\rho\exp(-\rho/2)$
		2	5d	$R_{52} = 150^{-1}70^{-1/2}\,(Z/a_0)^{3/2}\,(42-14\rho+\rho^2)\rho^2\exp(-\rho/2)$
		3	5f	$R_{53} = 300^{-1}70^{-1/2}\,(Z/a_0)^{3/2}\,(8-\rho)\rho^3\exp(-\rho/2)$
		4	5g	$R_{54} = 900^{-1}70^{-1/2}\,(Z/a_0)^{3/2}\rho^4\exp(-\rho/2)$
P	6	0	6s	$R_{60} = 2160^{-1}6^{-1/2}\,(Z/a_0)^{3/2}$ $\times\,(720-1800\rho+1200\rho^2-300\rho^3+30\rho^4-\rho^5)\exp(-\rho/2)$
		1	6p	$R_{61} = 432^{-1}210^{-1/2}\,(Z/a_0)^{3/2}$ $\times\,(840-840\rho+252\rho^2-28\rho^3+\rho^4)\rho\exp(-\rho/2)$
		2	6d	$R_{62} = 864^{-1}105^{-1/2}\,(Z/a_0)^{3/2}$ $\times\,(336-168\rho+24\rho^2-\rho^3)\rho^2\exp(-\rho/2)$
		3	6f	$R_{63} = 2592^{-1}35^{-1/2}\,(Z/a_0)^{3/2}\,(72-18\rho+\rho^2)\rho^3\exp(-\rho/2)$
		4	6g	$R_{64} = 12960^{-1}7^{-1/2}\,(Z/a_0)^{3/2}\,(10-\rho)\rho^4\exp(-\rho/2)$
		5	6h	$R_{65} = 12960^{-1}77^{-1/2}\,(Z/a_0)^{3/2}\rho^5\exp(-\rho/2)$

At this point we introduce the *atomic units*. In these units, $m_e=1$, $|e| = 1$, $\hbar = 1$ and $\kappa_0 = 1$ where m_e is the rest mass of the electron and κ_0 is the permittivity of vacuum. If we use these units, the electronic Hamiltonian in Eq. (15.1) acquires an exceedingly simple appearance,

$$\mathbf{H} = -\frac{1}{2}\nabla^2 - \frac{Z}{r} \qquad (Z = 1 \text{ for the hydrogen atom}) \qquad (15.14)$$

where we have taken the reduced mass m as the unit of mass, that is, an infinite mass has been ascribed to the nucleus. Also, the eigenvalue has a simple expression,

$$E_n = -\frac{Z^2}{2n^2} \qquad \text{(in atomic unit).} \qquad (15.14a)$$

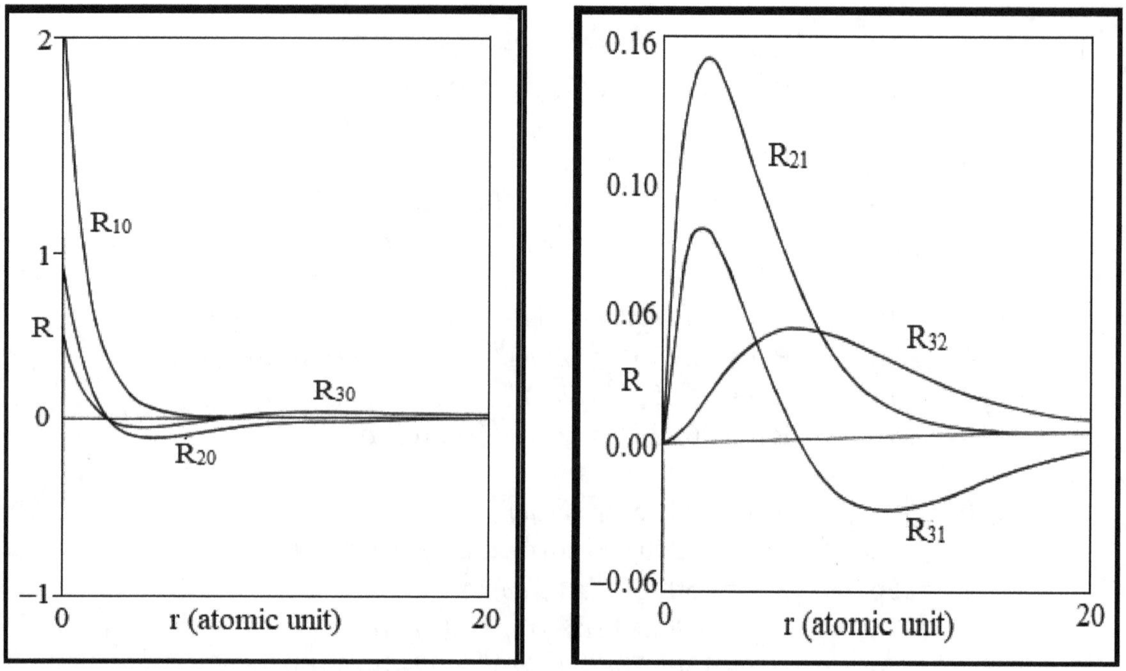

(a) **(b)**

Figure 3.9 (a)Radial wave functions R_{n0} as functions of variable r, and (b) a few other radial wave functions for the electron in a hydrogen atom.

As the atom is not infinitely massive, these expressions require a reduced mass correction. The correct expression for the electronic Hamiltonian in atomic unit is given by

$$\mathbf{H} = -\frac{m_e}{2m}\nabla^2 - \frac{Z}{r} \qquad (Z = 1 \text{ for the hydrogen atom}) \qquad (15.1b)$$

and the correct energy by

$$E_n = -\frac{mZ^2}{2m_e n^2} \qquad \text{(in atomic unit),} \qquad (15.14b)$$

but it is usual to adopt expressions like (15.1a) and (15.14a) in most of the cases with more than one electron or more than one nucleus so as to avoid the complication of accounting for the reduced mass effects. This is why some authors prefer the unit hartree for the energy in Eq. (15.14a). One atomic unit of energy is equal to 27.21 electron Volts (eV) or 627.5 kcal mol^{-1}. In atomic unit, the Bohr radius (radius of the first Bohr orbit) equals unity. Again this length requires a reduced mass correction in the form of the multiplicative factor m_e/m.

Lecture 16: Atomic orbitals

Atomic orbitals are the pure states (energy eigenstates) of one electron bound to the nucleus in an atom. Since the wave functions ψ_{nlm} represent these states they are also called atomic orbitals. The state and the state function are synonymous in our usage. Atomic orbitals are different from classical orbits or trajectories. The electron in a specific orbital has a distinct distribution of the probability of its being found at different regions of space. This distribution is characterized by the probability density

$$\rho_{nlm}(\mathbf{r}) = |\psi_{nlm}(r,\theta,\varphi)|^2. \tag{16.1}$$

The atomic orbital is not "the region of space where the electron may be found" as claimed erroneously in certain books in organic, inorganic and even physical chemistry. The orbital is a bound *single-particle state* and even in a tiny region of space located far away from the nucleus, there is a finite probability of finding the electron in this state. The atomic orbital is not even "the region of space in which the probability of finding the electron is maximum". This often quoted notion does not make it clear how such a region may be defined or found, whether it is a point or a line or a surface or even a small or not too small volume.

The function ψ_{nlm} may be viewed as a vector in an infinite-dimensional function space – the so-called infinite-dimensional Hilbert space. This is perhaps the reason for the appearance of the two wrongly conceived statements mentioned above. Function spaces are abstract mathematical spaces which should not be confused with the coordinate space, although a specific mathematical relationship can be established between the coordinate space and a function space.

The hydrogen atomic orbitals are called *n*s, *n*p, *n*d, etc. orbitals corresponding to the principal quantum number *n* and the orbital angular momentum for l = 0, 1, 2, etc. respectively. A list of the radial functions is given in Table 3.4 so that one may easily calculate the orbitals by using the spherical harmonics from Table 3.2 or the real angular functions from Table 3.3. Use of Table 3.2 yields p orbitals like $np_{m=0}$, np_1 and np_{-1} which are in general complex functions of the coordinates whereas Table 3.3 gives orbitals which are real functions like np_z, np_x, np_y, and so on.

An atomic orbital can be pictorially represented in several ways. The first and perhaps the best method is to plot the *radial distribution function*

$$P_{nl}(r) = \int_{4\pi} d\Omega\, \rho_{nl}(\mathbf{r}) = r^2 R_{nl}^2(r). \tag{16.2}$$

Several plots of $P_{nl}(r)$ are given in Figure 3.10.

The *most probable radius* is defined as the distance at which the radial probability density $P_{nl}(r)$ is maximum (or the largest maximum). For example, for the 1s orbital,

$$P(r) = 4\left(\frac{Z}{a_0}\right)^3 r^2 e^{-2Zr/a_0} \tag{16.3}$$

$$= 4Z^3 r^2 e^{-2Zr} \text{ in atomic units}$$

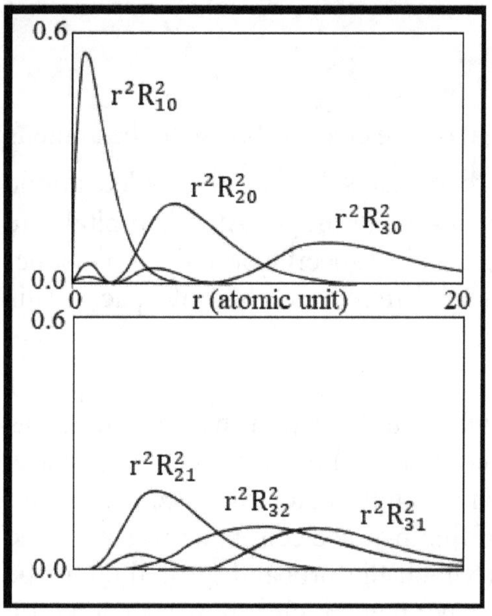

Figure 3.10 Radial probability distribution $r^2R_{nl}{}^2$ in hydrogen atom.

and

$$\frac{dP}{dr}=8\left(\frac{Z}{a_0}\right)^3 r\left(1-\frac{Zr}{a_0}\right)e^{-2Zr/a_0} \tag{16.4}$$

$$=8Z^3r(1-Zr)e^{-2Zr} \text{ in atomic units}$$

so that the first derivative vanishes at $r=0$, $r=a_0/Z$, and as $r\to\infty$. The second derivative

$$\frac{d^2P}{dr^2}=8\left(\frac{Z}{a_0}\right)^3 r\left(1-\frac{4Zr}{a_0}+\frac{2Z^2r^2}{a_0}\right)e^{-2Zr/a_0}\ \mathrm{e}^{-2Zr/a_0} \tag{16.5}$$

$$=8Z^3(1-4Zr+2Z^2r^2)e^{-2Zr} \text{ in atomic units}$$

is positive at $r=0$ and negative at $r=a_0/Z$, and it approaches zero as $r\to\infty$ since the function e^{-2Zr/a_0} decreases faster than the r-dependent term in the first bracket. Hence $r=0$ is a minimum, $r=a_0/Z$ is the most probable radius, and P(r) asymptotically approaches the r-axis as r increases to infinity. For the spherically symmetric orbital ns (with $l=0$) the most probable radius is $n^2\frac{a_0}{Z}$, the radius of the nth Bohr orbit.

The average radial distance can be easily found out. One writes

$$<r>_{nl} = \int dv\, \psi_{nlm}{}^* \, r\psi_{nlm} \tag{16.6}$$
$$= \int_0^\infty dr\, r\mathrm{P}_{n\,l}(r).$$

After carrying out an elaborate calculation in the general case (n,l), one obtains

$$<r>_{nl} = n^2 \frac{a_0}{Z}\left[1+\frac{1}{2}\left(1-\frac{l(l+1)}{n^2}\right)\right]. \tag{16.7}$$

The average distance for the spherically symmetric orbitals ($l = 0$) is always 1.5 times the most probable distance.

The s orbitals have nonzero values at r=0. Atomic orbitals of all other types vanish at this point. This feature will be important in the case of many-electron atoms, and for molecules. The valence s electrons penetrate the inner shells. Since the nucleus has a small but finite volume, a non-vanishing fraction of the valence s electron remains in contact with the nucleus. This gives rise to the contact interaction that plays a crucial role in magnetic resonance spectroscopies.

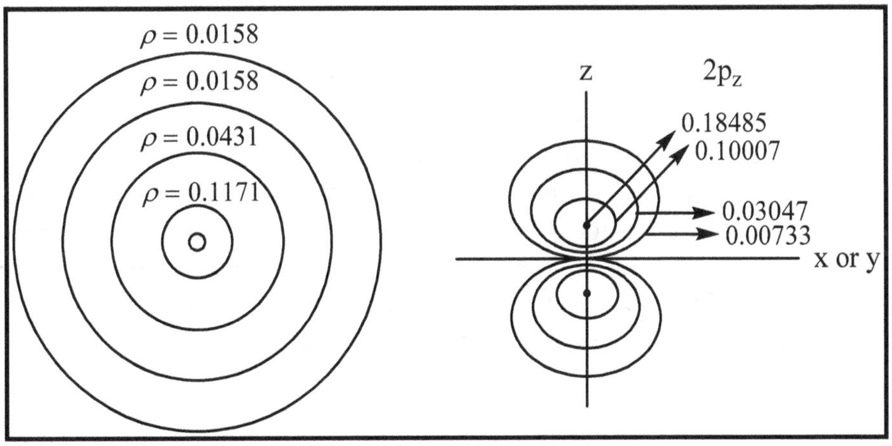

Figure 3.11 Contours of constant probability density surfaces.

There are other ways of pictorially describing an orbital. These rely on illustrating the probability distribution in various ways. One may draw contours of constant probability densities. The result is a *contour diagram.* A typical contour plot is shown in Figure 3.11. One can also show boundary surfaces with specific total probabilities inside them. For instance, the probability of finding the 1s electron within a sphere of radius r_0 with the nucleus at center is given by

$$p(r_0) = \int_0^{r_0} dr\, r^2 \int_{4\pi} d\Omega\, \rho_{1s}(r,\theta,\varphi) = \int_0^{r_0} dr\, \mathrm{P}_{1s}(r)\mathrm{P}_{1s}(r)$$

$$= 1 - \left[1 + 2\left(\frac{Zr_0}{a_0}\right) + 2\left(\frac{Zr_0}{a_0}\right)^2\right] e^{-2Zr_0/a_0}. \tag{16.8}$$

Boundary surfaces are shown in Figure 3.12.

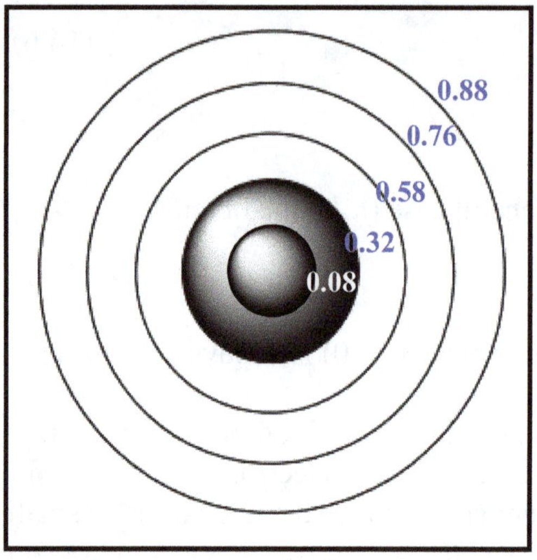

Figure 3.12 Boundary surface each with a specific probability of finding the electron within the enclosed space.

An extremely popular way of visualizing the orbital is to generate a picture of the *probabilty cloud* or the *charge cloud.* The charge density is obtained by multiplying the probability density with the charge of the electron,

$$q(\mathbf{r}) = e\rho(\mathbf{r}). \tag{16.9}$$

The probability cloud or the charge cloud diagram is shown in Figure 3.13. It is drawn by putting a large number of dots in such a way that the number of dots within every small unit region is proportional to the probability density at the center of the same region. This creates an overall impression of a misty distribution of the electronic charge in coordinate space. It is often possible to identify the region of the greatest concentration of the electronic charge in an atom or a molecule from a charge cloud picture. This is used to generate a rough idea of the size of the species. Sometimes it also helps in identifying the possible centers for an electrophilic or a nucleophilic attack.

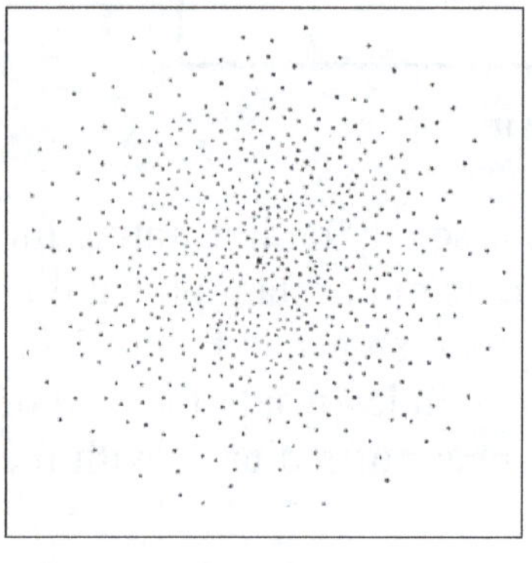

Figure 3.13 The charge cloud diagram for the distribution of the charge of the electron in a specific orbital.

Texts Suggested for Further Reading

L. I. Schiff, "Quatum Mechanics –An Introduction", (McGraw-Hill, New York, 1968) Chap. 2-4.

H. Eyring, J. Walter and G. E. Kimball, "Quantum Chemistry", (Wiley, New York, 1944) Chap. 3-6.

L. Pauling and E. B. Wilson, Jr., "Introduction to Quantum Mechanics", (McGraw-Hill, New York, 1935) Chap. 4-5.

PROBLEMS

1. Show that in the nth stationary state of a particle moving in a one-dimensional box of uniform potential $\langle x \rangle = 0$ and $\langle p_x \rangle = 0$. Find $\langle x^2 \rangle$ and $\langle p_x^2 \rangle$, and then calculate the standard deviations σ_x and σ_{px} and the uncertainty product $\sigma_x \sigma_{px}$.
2. Using the uncertainty product in the previous problem, decide whether $n = 0$ can be a valid quantum number for the particle moving in a box or not.
3. Derive an expression for the energy levels for a particle moving in a three-dimensional potential box. The three sides are of length L_x, L_y and L_z. Find the degeneracy of the eigenstates when the three sides are equally long. (Hint: The Schrödinger equation is separable into three one-dimensional wave equations).
4. Prove the uncertainty result (12.17) for an eigenstate of a harmonic oscillator.
5. Find the energy eigenvalues and the corresponding eigenfunctions of the Hamiltonian for a three-dimensional harmonic oscillator. (Hint: Use separability).
6. Normalize the Gaussian wave packet $e^{-\alpha x^2}$. Show that this is a minimum uncertainty wave packet.
7. Prove that the minimum uncertainty wave packet must be a pure Gaussian.
8. Calculate all real eigenfunctions of $\boldsymbol{L}^2$ for $l = 3$. Name these functions.
9. What is the functional expression for f_{xyz}? Hence normalize this function.
10. Find the nodal surfaces for the angular functions d_{z^2} and f_{xyz}.
11. Calculate the most probable distance of the electron from the nucleus in the 2s state of the hydrogen atom.
12. Calculate the average radial distance of the hydrogenic electron in the 2s state.
13. What is the probability of finding (i) a 2s electron and (ii) a 2p electron within a sphere of radius 2 atomic unit?
14. Calculate the mean value of $\frac{1}{r}$ for one electron in a few low-energy orbitals of a hydrogen-like atom with atomic number Z, and deduce the mean potential energy of the electron in these states. The total energy is $-\frac{Z^2}{2n^2}$ a. u. Hence prove **Theorem 9:** $\langle V \rangle / \langle T \rangle = -2$ (the virial theorem).

CHAPTER 4

THE VARIATION METHOD

Lecture 17: The Variation Theorem and Principle

Approximate methods

In the previous lectures we have seen that the Schrödinger equation can be exactly solved for a few ideal systems. Most of the real systems are not so simple in nature. They are more complicated and their Hamiltonian operators defy separation. Most physical oscillators are damped. Most atoms contain more than one electron, and because of the inter-electronic interaction represented by the r_{12}^{-1} term in the Hamiltonian, an exact solution of the many-electron Schrödinger equation cannot be found by any analytical method. Molecules have more than one nucleus, and this feature makes the molecular treatment much more complicated. The inability of explicitly solving the Schrödinger equation gives rise to the need for methods by which one can find at least some approximate solutions of the same equation.

Methods for approximately solving the wave equation should have three desirable characteristics:

1) They must yield initial solutions of reasonably high accuracy.
2) They must yield superior results while guided by physical intuition.
3) It should be possible to subsequently improve the accuracy of the solutions in a systematic way.

The variation method and the perturbation theory are two highly successful methodologies for finding approximate solutions of the Schrödinger equation. More than half of our knowledge of the electronic structures of atoms and molecules and the basic concepts of chemical bonding simply arise from the application of the variation principle, whereas the systematics of the physical interaction among different particles can be understood from a perturbative treatment. Different forms of the perturbation theory have evolved. Some are more successful and some are less so. The perturbative approach often starts from a set of the variation principle based solutions, and progressively leads to the correct energy values and wave functions.

Theorem 10. The Variation Theorem: For any arbitrary wave function ψ the energy expectation value E_ψ is greater than or equal to the ground state energy E_0. The equality holds good if and only if ψ is identical with the ground state wave function ψ_0. That is,

$$E_\psi = \int dv\, \psi^* \mathbf{H}\psi \;/\; \int dv\, \psi^* \psi \geq E_0$$

$$E_\psi = E_0 \text{ iff } \psi \equiv \psi_0 \tag{17.1}$$

The theorem is proved by using the principle of linear superposition. An arbitrary function ψ can be expanded as

$$\psi = \sum_i c_i \psi_i \tag{6.5}$$

where the ψ_i's are the eigenfunctions of the Hamiltonian operator, $\mathbf{H}\psi_i = E_i\psi_i$. Because $\mathbf{H}$ is hermitean, the eigenfunctions ψ_i's form an orthonormal complete set. We assume that the eigenvalues are ordered as $E_0 < E_1 < E_2 < \ldots$ such that E_0 is the ground state energy and correspondingly, ψ_0 is the ground state wave function. Using Eq. (6.5) in the expression for E_ψ and further using the orthonormality of the ψ_i's one obtains

$$E_\psi = \frac{\sum_i |c_i|^2 E_i}{\sum_i |c_i|^2} = E_0 + \frac{\sum_i |c_i|^2 (E_i - E_0)}{\sum_i |c_i|^2}. \tag{17.2}$$

Note that for a nontrivial wave function

$$\int dv\, \psi^* \psi = \sum_i |c_i|^2 > 0 \tag{17.3}$$

so that the denominator is a nonzero finite quantity and a singularity does not arise. The quantity $|c_i|^2$ is always greater than or equal to zero, and $(E_i - E_0)$ is greater than zero for $i \neq 0$. Hence from Eq. (17.2) we directly reach the first part of the theorem, $E_\psi \geq E_0$.

The equality holds if and only if $c_i = 0$ for all i, $i \neq 0$, in which case $\psi = c_0\psi_0$ and the normalized ψ becomes identical with ψ_0. This proves the second part.

The Variation Principle

When ψ is a function of one or more parameter (a *trial* function) E_ψ becomes a function of the same parameter(s). One may vary the wave function ψ by changing these parameters, and as a result vary E_ψ. The variation theorem guarantees that E_ψ will generally remain above E_0 and one can at most reduce it to E_0, but it can never be brought below the ground state energy. A minimum can be found by varying the parameters of the wave function, and by the variation theorem, the minimum itself must lie at or above E_0:

$$E_{\min} = \min_\psi E_\psi \geq E_0 \tag{17.4}$$

The minimum energy is called the *optimized energy* or the *optimum value* and the wave function at the energy minimum ψ_{min} is similarly called the *optimized* or the *optimum function.* The latter is the energetically closest function of the particular type of ψ and therefore the best approximation to the ground state wave function ψ_0. The ground state function ψ_0 is itself called the *optimal trial function.* The variation principle is stated as follows: *the best approximation to the ground state wave function is obtained by minimizing the energy expectation value with respect to the trial function.* The principle is illustrated in Figure 4.1 where it is shown that the optimized energy is greater than the ground state energy. From a careful consideration of the trial function one may even obtain E_0 as the optimum value, but such happy choices can be rarely made.

The ground state energy is always less than or equal to E_{min}. Hence the variation technique yields an *upper bound* to the energy of the ground state.

The key to a successful variation technique lies in the choice of a good trial function. This choice is often guided by physical intuition. Knowledge and understanding of similar but simpler systems help in building the power of intuition.

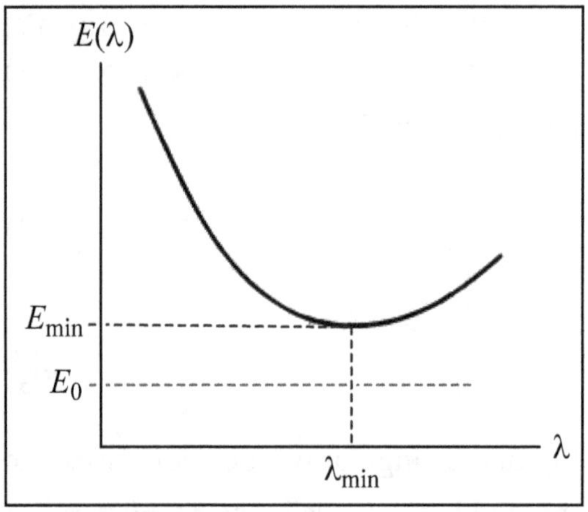

Figure 4.1 Illustration of the variation principle: the trial function is a function of the parameter λ. The optimized function is $\psi(\lambda_{min})$.

Generalization

The variation theorem can be stated for the expectation value of any hermitean operator that has a least eigenvalue, that is, when the eigenvalue spectrum is bounded from below. A hermitean operator has a least eigenvalue when it contains a squared operator (such as the kinetic energy operator, and the squared orbital angular momentum operator). In most cases, however, one is interested in the stationary state wave functions which are constructed from the solutions of the time-independent Schrödinger equation. This is why the optimization of total energy is invariably chosen, and it leads to the best approximation to the energetic ground state.

If the system has some symmetry elements in it, then the potential energy and the Hamiltonian have the same features of symmetry. The wave functions are accordingly symmetrized, and there are often a few different kinds of symmetry of the wave function. The hydrogen atomic potential is spherically symmetric. Consequently, the electronic Hamiltonian has central symmetry. The hydrogen atomic orbitals have the symmetries of orbital angular momentum which are specified by the azimuthal and the magnetic quantum numbers l and m, and these orbitals are classified as s orbitals, p orbitals, d orbitals, etc. There is a specific ground state of each symmetry, just as 1s is the ground state of the hydrogenic s electron and 2p is the triply degenerate ground state of the p electron. The variation theorem separately holds for each class or symmetry of the trial functions,

$$
\begin{aligned}
&E_{\psi(l=0)} \geq E_{n=1,l=0} \\
&E_{\psi(l=1)} \geq E_{n=2,l=1} \\
&E_{\psi(l=2)} \geq E_{n=3,l=2}
\end{aligned}
\tag{17.5}
$$

and so on.

Functions of two different symmetries are always orthogonal to each other,

$$\int dv\, \psi_1{}^*\psi_2 = 0,$$

and they do not interact with each other,

$$\int dv\, \psi_1{}^*\mathbf{H}\psi_2 = 0.$$

Therefore, they do not mix, that is, they do not linearly combine with each other. Thus ψ_l will be given by a linear combination of only the ψ_{il}'s, eigenfunctions of the Hamiltonian with angular momentum quantum number l, and so on. Hence the variation theorems indicated for different classes of functions in Eq. (17.5) can be proved one after another for each class separately.

Lecture 18: Simple Examples — Hydrogen, Helium, and Screening Constants

Example 18: The hydrogen-like system – hydrogenic trial function

Consider a normalized trial function of the form

$$\phi(\mathbf{r}) = (\zeta^3/\pi)^{1/2} \exp(-\zeta r) \tag{18.1}$$

in atomic units. The function ϕ is spherically symmetrical and therefore, an eigenfunction of $\boldsymbol{L}^2$ and $\boldsymbol{L}_z$ with eigenvalues 0 and 0, respectively. The kinetic energy expectation value can be easily calculated by using Eq. (15.3) for ∇^2 and the standard integral

$$\int_0^\infty dr\, r^m e^{-\alpha r} = \frac{m!}{\alpha^{m+1}}. \tag{18.2}$$

One finds

$$-\frac{1}{2}\langle\nabla^2\rangle = \frac{\zeta^2}{2},$$

$$-\langle r^{-1}\rangle = -\zeta,$$

and the energy in atomic units (18.3)

$$E_\phi \equiv E(\zeta) = \frac{\zeta^2}{2} - \zeta Z.$$

If we vary the parameter ζ, the function ϕ changes. The energy expectation value changes correspondingly, with $E(\zeta)$ minimum at $\zeta = Z$. Hence the optimized function and the optimum value are

$$\phi_{min} = (Z^3/\pi)^{1/2}\exp(-Zr),\quad E_{min} = -\frac{Z^2}{2}, \tag{18.4}$$

both in atomic units. In this example, we have recovered the exact ground state energy and wave function for a hydrogen-like system. For the hydrogen atom, Z=1. The next two examples would show that one is not so fortunate in all cases.

Example 19: The hydrogen atom – Gaussian trial function

Take the trial Gaussian

$$\phi(\mathbf{r}) = (2\zeta/\pi)^{3/4}\exp(-\zeta r^2) \tag{18.5}$$

which is normalized in three-dimension. Using the standard results

$$\int_0^\infty dr\, r^{2m} e^{-\alpha r^2} = \frac{(2m-1)!!}{2^{m+1}\alpha^m}\left(\frac{\pi}{\alpha}\right)^{1/2}$$

and (18.6)

$$\int_0^\infty dr\, r^{2m+1} e^{-\alpha r^2} = \frac{m!}{2\alpha^{m+1}}$$

where

$$(2m+1)!! = 1.3.5.7...(2m-1) = \frac{(2m)!}{2^m(m!)},$$

one may obtain the expectation values

$$-\frac{1}{2} < \nabla^2 > = \frac{3\zeta}{2}$$

and

$$-Z < r^{-1} > = -2\left(\frac{2\zeta}{\pi}\right)^{1/2} Z.$$

Therefore,

$$E_\phi \equiv E(\zeta) = \frac{3\zeta}{2} - 2\left(\frac{2\zeta}{\pi}\right)^{1/2} Z$$

which yields

$$\zeta_{min} = 8Z^2/9\pi,$$

$$\phi_{min} = (4Z/3\pi)^{3/2}\exp(-8Z^2 r/9\pi), \tag{18.7}$$

$$E_{min} = -4Z^2/3\pi > -\frac{1}{2}Z^2.$$

This is the best result one can obtain with a single Gaussian function. The calculated E_{min} is an upper bound for the ground state energy of hydrogen atom.

We now look for a slightly more complicated case, namely, a two-electron atom.

Example 20: The helium atom

The Hamiltonian in atomic unit is

$$\mathbf{H} = -\frac{1}{2}(\nabla_1^2+\nabla_2^2) - Z(r_1^{-1} + r_2^{-1}) + r_{12}^{-1} \tag{18.8}$$

with Z=2 and $r_{12} = |\mathbf{r}_1-\mathbf{r}_2|$. Let us think that the two electrons are in the same hydrogen-like orbital but with opposite spins, and write the spin-independent trial function as the product of two hydrogenic 1s orbitals,

$$\Psi(\mathbf{r}_1,\mathbf{r}_2) = (\zeta^3/\pi) \exp[-\zeta(r_1+r_2)]. \tag{18.9}$$

Later we will make it clear that this is the simplest electronic wave function for helium atom in the so-called Hartree model. The trial function given above is known as the Eckart-Kellner wave function.

Calculation of the expectation values in this case involves the calculation of integrals over the coordinates of the two electrons. One gets

$$\langle\nabla_1^2\rangle = (\zeta^6/\pi^2) \int dv_1 \int dv_2 \ \exp[-\zeta(r_1+r_2)] \ \nabla_1^2 \exp[-\zeta(r_1+r_2)]$$

$$= (\zeta^3/\pi) \int dv_1 \exp[-\zeta r_1] \ \nabla_1^2 \exp[-\zeta r_1] = -\zeta^2 ,$$

$$\langle r_1^{-1}\rangle = (\zeta^6/\pi^2) \int dv_1 \int dv_2 \ \exp[-\zeta(r_1+r_2)] \ r_1^{-1} \exp[-\zeta(r_1+r_2)]$$

$$= (\zeta^3/\pi) \int dv_1 \exp[-\zeta r_1] \ r_1^{-1} \exp[-\zeta r_1] = \zeta;$$

and similarly,

$$\langle\nabla_2^2\rangle = -\zeta^2$$

and

$$\langle r_2^{-1}\rangle = \zeta.$$

The calculation of the integral of r_{12}^{-1} is as follows, but one may skip the derivation and take the final result for granted if one so desires. The term r_{12}^{-1} can be expanded as

$$r_{12}^{-1} = \sum_{l=0}^{\infty} \sum_{m=-l}^{l} \frac{r_<^l}{r_>^{l+1}} \frac{(l-m)!}{(l+m)!} P_{lm}(\cos\theta_1) \, P_{lm}(\cos\theta_2) \, e^{im(\varphi_1-\varphi_2)} \tag{18.10}$$

where $r_>$ is defined as the larger of the two radii, r_1 and r_2, and $r_<$ is the smaller one. Now,

$$\langle r_{12}^{-1}\rangle = (\zeta^6/\pi^2) \int dv_1 \int dv_2 \ \exp[-\zeta(r_1+r_2)] \ r_{12}^{-1} \exp[-\zeta(r_1+r_2)] \tag{18.11}$$

which, by using expansion (18.10), yields

$$\langle r_{12}^{-1}\rangle = 32\zeta^6 \int_0^{\infty} dr_1 \, r_1 e^{-2\zeta r_1} \int_0^{r_1} dr_2 \, r_2^2 e^{-2\zeta r_2} . \tag{18.12}$$

It is rather easy to calculate the definite integrals in the expression above. Once this calculation is done one obtains

$$\langle r_{12}^{-1}\rangle = \frac{5\zeta}{8}. \tag{18.13}$$

Therefore, the energy expectation value is

$$E(\zeta) = \zeta^2 - 2Z\zeta + \frac{5\zeta}{8} \tag{18.14}$$

that gives

$$\zeta_{min} = \left(Z - \frac{5}{16}\right) \equiv Z_{eff}, \quad (18.15)$$

$$\Psi_{min}(\mathbf{r}_1, \mathbf{r}_2) = (Z_{eff}^3/\pi)\ \exp[-Z_{eff}(r_1 + r_2)],$$

and (18.16)

$$E_{min} = -Z_{eff}^2.$$

The optimized energy value for the helium atom is $-[2-5/16]^2 = -2.847656$ a.u. which accounts for more than 98% of the total electronic energy (−2.90372 a.u.).

Screening constant

The quantity Z_{eff} is called the effective nuclear charge since it is the optimized exponent of the hydrogenic 1s orbital in helium atom. The ground state electronic configuration is $1s^2$. As the electrons in the atom repel each other, each is held by the nucleus a little less strongly. That is, each electron screens the nuclear charge so that the other electron faces a reduced charge of the nucleus. One may write

$$Z_{eff} = Z - \sigma \quad (18.17)$$

Where σ is the screening (or shielding) constant. As each orbital has a probabilistic description, each electron penetrates the inside of the atom. Thus the screening is never fully effective, and the constant σ for each orbital lies between 0 and 1. For the helium atom, the calculated σ_{1s} is $5/16 = 0.3125$. This corresponds to an effective nuclear charge $Z_{eff} = 1.6875$. It varies from one orbital to another. For other atoms too we encounter the screening constants and different orbitals appear to have different screening constants. The parameter Z_{eff} is generally decided by comparing the optimized exponent of a specific orbital with the exponent expected for a hydrogenic orbital of same n, l and m quantum numbers. The screening constant is to be considered in greater detail in Chapter 5 where a general discourse on the electronic structure of many-electron atoms will be presented.

Lecture 19: Ritz Variation

The linear variation method is a special but very useful variation technique. It was developed by Rayleigh (in the context of classical oscillator chains) and later adopted by Ritz, and so is referred to as the Rayleigh-Ritz, or simply, the Ritz variation technique. The Ritz variation method is based on the axiom that quantum mechanics deals with a linear space such that the principle of linear superposition remains operative. By this principle any arbitrary function ϕ can be described by a linear combination of the functions belonging to a complete set.

In practice in the linear variation method the trial function ϕ is expanded as a linear combination of N known functions $\psi_1, \psi_2, \ldots, \psi_N$ where N is finite:

$$\phi = \sum_{i=1}^{N} c_i \psi_i. \quad (19.1)$$

The known functions are called the *basis functions* or the *reference functions*. These may not be eigenfunctions of the Hamiltonian **H**. These need not be necessarily orthogonal to each other, that is, they may not be independent of each other. The *basis set* is not necessarily complete. All that is required is that the basis functions are distinct functions, and preferably of the same type. The linear combination coefficients c's are the parameters of variation.

The linear or Ritz variation involves the variation of the expectation value E_ϕ with the coefficients c's so as to find a stationary value. This yields a specific relationship that has been applied to the study of many problems of interest. Of course, as the set $\{\psi\}$ is not complete the optimized function may differ from an optimal trial function (eigenfunction). But this indicates two avenues for the improvement of the optimized function and the optimized value. One way is to choose basis functions of a superior quality. The other approach is to include more functions to the basis set, that is, to increase the *basis size*.

Let us first derive the working equation for the Ritz variation technique. We calculate the energy expectation value to get

$$E_\phi = \frac{\int dv\ \phi^* \mathbf{H}\phi}{\int dv\ \phi^*\phi} = \frac{\sum_{i=1}^{N}\sum_{j=1}^{N} c_i^* H_{ij} c_j}{\sum_{i=1}^{N}\sum_{j=1}^{N} c_i^* S_{ij} c_j} \tag{19.2}$$

where

$$H_{ij} = \int dv\ \psi_i^* \mathbf{H}\psi_j$$

and (19.3)

$$S_{ij} = \int dv\ \psi_i^* \psi_j .$$

The integral H_{ii} is called the *energy* or the *coulomb integral* since it indicates the coulombic energy of the *ith* state represented by ψ_i. The integral H_{ij} with different indices i and j is called the *interaction* or *resonance integral* as it indicates the energy of interaction between the states ψ_i and ψ_j. The integral S_{ij} represents the extent to which the two states ψ_i and ψ_j overlap each other; hence it is called the *overlap integral*. The term S_{ii} is the self-overlap that equals 1 for a normalized function ψ_i. For orthogonal functions ψ_i and ψ_j, $S_{ij} = 0$. After carrying out a cross multiplication and a little bit of simplification in Eq. (19.2) one finds

$$\sum_{i=1}^{N}\sum_{j=1}^{N} c_i^*(H_{ij} - ES_{ij})c_j = 0 . \tag{19.4}$$

Consider now the small changes

$$c_m \to c_m + \delta c_m$$

and

$$c_m^* \to c_m^* + \delta c_m^*$$

for $m = 1, 2, \ldots, N$ whereby the energy changes as

$$E \to E + \delta E + \delta^2 E . \tag{19.5}$$

When we write Eq. (19.4) in terms of the changed coefficients and energy, we get

$$\sum_{i=1}^{N}\sum_{j=1}^{N}(\mathrm{c}_i{}^*+\delta\mathrm{c}_i{}^*)[H_{ij}-(E+\delta E+\delta^2 E)S_{ij}](\mathrm{c}_j+\delta\mathrm{c}_j)=0. \qquad (19.6)$$

If we subtract Eq. (19.4) from the equation above and in the resulting expression equate the sum of terms linear in the variations with zero, we obtain

$$\sum_{i=1}^{N}\sum_{j=1}^{N}\delta\mathrm{c}_i{}^*[H_{ij}-ES_{ij}]\mathrm{c}_j+\sum_{i=1}^{N}\sum_{j=1}^{N}\mathrm{c}_i{}^*[H_{ij}-ES_{ij}]\delta\mathrm{c}_j=\delta E\sum_{i=1}^{N}\sum_{j=1}^{N}\mathrm{c}_i{}^*S_{ij}\mathrm{c}_j. \qquad (19.7)$$

A somewhat similar equation is obtained from the terms that are quadratic in changes:

$$\sum_{i=1}^{N}\sum_{j=1}^{N}\delta\mathrm{c}_i{}^*[H_{ij}-ES_{ij}]\delta\mathrm{c}_j-\delta E\sum_{i=1}^{N}\sum_{j=1}^{N}(\delta\mathrm{c}_i{}^*S_{ij}\mathrm{c}_j+\mathrm{c}_i{}^*S_{ij}\delta\mathrm{c}_j)=\delta^2 E\sum_{i=1}^{N}\sum_{j=1}^{N}\mathrm{c}_i{}^*S_{ij}\mathrm{c}_j.$$

It can be used to determine the nature of the stationary point — a minimum, a maximum or a saddle point, etc.

When the energy is stationary,

$$\delta E = 0. \qquad (19.8)$$

Introducing stationarity to Eq. (19.7) and considering that all the coefficients except c_i (and c_i^*) remain unchanged one obtains

$$\sum_{j=1}^{N}\delta c_i^*(H_{ij}-ES_{ij})c_j+\sum_{j=1}^{N}c_j^*(H_{ji}-ES_{ji})\,\delta c_i=0. \qquad (19.9)$$

The variations δc_i and δc_i^* are independent of each other. For an arbitrary δc_i^* while keeping δc_i as zero, one obtains

$$\sum_{j=1}^{N}(H_{ij}-ES_{ij})c_j=0, \qquad (19.10)$$

and for an arbitrary δc_i (while $\delta c_i^*=0$) we have a similar equation which is really the complex conjugate of the above equation.

Equation (19.10) has been obtained by considering the partial variation of E with respect to c_i for a specific i while keeping all the other coefficients c_j (with $j \neq i$) constant. A similar equation will result for another i value, so that we obtain in total N equations of the form of Eq. (19.10):

$$\sum_{j=1}^{N}(H_{ij}-ES_{ij})c_j=0 \qquad \text{for } i=1, 2, \ldots, N. \qquad (19.11)$$

These N equations are linear and homogeneous in coefficients and they have to be solved simultaneously. They have the trivial solution

$$c_1 = c_2 = \ldots = c_N = 0. \qquad (19.12)$$

By a theorem in Algebra the N linear homogeneous equations will have a non-trivial solution if and only if the corresponding secular determinant vanishes, that is,

$$\|H_{ij} - ES_{ij}\| \equiv \begin{vmatrix} H_{11} - ES_{11} & H_{12} - ES_{12} & \cdots & H_{1N} - ES_{1N} \\ H_{21} - ES_{21} & H_{22} - ES_{22} & \cdots & H_{2N} - ES_{2N} \\ \cdots & \cdots & \cdots & \cdots \\ H_{N1} - ES_{N1} & H_{N2} - ES_{N2} & \cdots & H_{NN} - ES_{NN} \end{vmatrix} = 0. \tag{19.13}$$

As the determinant is of order N, Eq. (19.13) is really an algebraic equation involving the N*th* power of E, and it has N possible roots.

From N linear equations one can determine N unknown variables. A particular root E is one of the unknowns. Therefore, if we place one particular root in the N equations, the N unknown coefficients cannot be completely determined, but we will get the $(N-1)$ ratios of coefficients like $c_2{:}c_1$, $c_3{:}c_1, \ldots, c_N{:}c_1$. The coefficients are uniquely determined by applying an additional constraint, viz., the normalization condition:

$$\int dv\,|\phi|^2 = \sum_{i=1}^{N}\sum_{j=1}^{N} c_i^* S_{ij} c_j = 1. \tag{19.14}$$

This completes the derivation involved in the linear variation treatment.

The Ritz variation technique involves the four steps:

1. Write the N linear homogeneous equations of coefficients, form the secular determinant and set it equal to zero.
2. Solve the algebraic equation to obtain the roots, that is, the *matrix* eigenvalues (optimized energies).
3. Incorporate the roots in the linear homogeneous equations and solve to obtain the $(N-1)$ ratios of the coefficients.
4. Write the normalization condition and determine the coefficients of basis functions in the expressions for the optimized linear combinations.

The lowest (matrix) eigenvalue found from the linear or Ritz variation is truly a minimum for the specific basis set involved in the treatment, and the highest (matrix) eigenvalue is a true maximum. All other (matrix) eigenvalues are minimum for some of the variational degrees of freedom and maximum for the rest of the variational coordinates. So these are in general saddle points. The (matrix) eigenvalues change when one or more functions are added to or subtracted from the basis set. In case one more basis function is added to the N-component basis, the number of (matrix) eigenvalues increases to $N+1$. The new (matrix) eigenvalues $E^{(N+1)}$'s generally satisfy a set of *separation theorems* like $E_{i-1}^{(N)} \leq E_i^{(N+1)} \leq E_i^{(N)}$ where the $E^{(N)}$'s are the (matrix) eigenvalues calculated from the N-component basis.

A very important application of the method is found in the molecular orbital formalism. Molecular orbitals (m.o.s) are the amplitudes of the probability waves of individual electrons in a molecule, and they can be written as a vector sum, that is, a linear superposition, of the probability amplitudes in atoms. This is the scheme of *linear combination of atomic orbitals as molecular orbital* (LCAO-MO). This scheme is in principle correct and not an approximation. In practice, one works with a finite number of atomic orbitals and so what one obtains as the best LCAOs are in reality approximations to the exact molecular orbitals. The calculated m.o.s may be made more correct by increasing the size of the atomic bases. Often, a change in the nature of basis functions leads to a rapid improvement of the calculated results.

Lecture 20: The Non-Crossing Rule

To discuss this simple but elegant rule we first work out the simplest example of Ritz variation.

Example 21: Two state problem – basis functions

The trial function is

$$\phi = c_1\psi_1 + c_2\psi_2. \tag{20.1}$$

By Ritz variation one obtains the linear homogeneous equations

$$c_1(H_{11} - ES_{11}) + c_2(H_{12} - ES_{12}) = 0, \tag{20.2}$$
$$c_1(H_{21} - ES_{21}) + c_2(H_{22} - ES_{22}) = 0.$$

For normalized ψ_1 and ψ_2, both S_{11} and S_{22} equal unity. If ψ_1 and ψ_2 are real functions then $S_{12} = S_{21} = S$ and $H_{12} = H_{21}$. We also write $E_1 = H_{11}$ and $E_2 = H_{22}$ so that the linear homogeneous equations appear as

$$c_1(E_1 - E) + c_2(H_{12} - ES) = 0, \tag{20.3}$$
$$c_1(H_{12} - ES) + c_2(E_2 - E) = 0.$$

For nontrivial solutions,

$$\begin{vmatrix} E_1 - E & H_{12} - ES \\ H_{12} - ES & E_2 - E \end{vmatrix} = 0 \tag{20.4}$$

that yields the roots

$$E_\pm = [2(1 - S^2)]^{-1}[(E_1 + E_2 - 2H_{12}S)$$
$$\pm \{(E_1 + E_2 - 2H_{12}S)^2 - 4(1 - S^2)(E_1E_2 - H_{12}^2)\}^{1/2}]. \tag{20.5}$$

In the event ψ_1 and ψ_2 are equivalent functions such that $E_1 = E_2$ one obtains

$$E_\pm = (1 \pm S)^{-1}[E_1 \pm H_{12}] \tag{20.6}$$

which, when inserted into Eq. (20.3), yields

$$c_1{:}c_2 = \pm 1. \tag{20.7}$$

Applying the normalization condition

$$c_1^2 + 2c_1c_2S + c_2^2 = 1 \tag{20.8}$$

one obtains, corresponding to the two rooots $E = E_+$ and $E = E_-$, the coefficients

$$c_1 = c_2 = [2(1 + S)]^{-1/2}$$

and (20.9)

$$c_1 = -\,c_2 = [2(1 - S)]^{-1/2}.$$

Thus two solutions have been found,

$$\phi_+ = [2(1 + S)]^{-1/2}[\psi_1 + \psi_2] \Leftrightarrow E_+ = (1 + S)^{-1}(E_1 + H_{12}); \tag{20.10}$$
$$\phi_- = [2(1 - S)]^{-1/2}[\psi_1 - \psi_2] \Leftrightarrow E_- = (1 - S)^{-1}(E_1 - H_{12}).$$

Furthermore, H_{12} is usually negative and S is normally a positive quantity so that $E_+ < E_1$ and $E_- > E_1$.

The non-crossing rule

The functions ψ_1 and ψ_2 are generally non-equivalent. Suppose that E_1 is less than E_2. The two roots of Eq. (20.4) are still explicitly given by Eq. (20.5). To inspect the roots in a qualitative way we write Eq. (20.4) as

$$(E_1 - E)(E_2 - E) = (H_{12} - ES)^2 > 0 \qquad (20.11)$$

so that either $E < E_1$ or $E > E_2$.

Symmetry

If ψ_1 and ψ_2 differ in symmetry then $H_{12} = 0$ and $S = 0$ with the results

(i) $\phi = \psi_1$, $E = E_1$

and (20.12)

(ii) $\phi = \psi_2$, $E = E_2$.

If ψ_1 and ψ_2 are of the same symmetry then $(H_{12} - ES)$ is generally not equal to zero which gives rise to the results (i) for $E_+ < E_1$, $|c_1| > |c_2|$ and $\phi \approx \psi_1$; and (ii) for $E_- > E_2$, $|c_2| > |c_1|$ and $\phi \approx \psi_2$. We find from Eq. (20.11),

$$E_+ \approx E_1 - (E_2 - E_1)^{-1}(H_{12} - E_1S)^2$$

and (20.13)

$$E_- \approx E_2 + (E_2 - E_1)^{-1}(H_{12} - E_1S)^2.$$

These two are remarkable observations, for they show that the energy shifts $(E_1 - E_+)$ and $(E_- - E_2)$ as well as the level splitting $(E_- - E_+)$ increase as the energy difference $(E_2 - E_1)$ decreases and vice versa. This is the basis of the non-crossing rule.

Correlation diagram and crossing of states

Let ψ_1, ψ_2, etc. be the eigenstates in a particular limit of molecular geometry. If the geometry of the system changes, these functions change accordingly, and they also mix, (that is, linearly combine) with each other to form new state functions which are mutually orthogonal and normalized. Suppose that the change is continued until the geometry of the molecular system reaches another limit. The plot of E as function of the molecular geometry between two limits is called a *correlation diagram*. The non-crossing rule refers to the non-crossing of some of the energy curves in a correlation diagram. This rule is stated as follows: *In the correlation diagram, the energy curves for states of the same symmetry do not cross each other, whereas states differing in symmetry may cross each other*. That crossing is avoided is a consequence of Eq. (20.13): as the states ψ_1 and ψ_2 come closer in energy, they would mix more strongly if they have same symmetry and the resulting (matrix) eigenvalues show a large splitting. This rule is illustrated in Figure 4.2. States of different symmetries do not mix with each other as shown by the results in Eq. (20.12). Hence their energy curves cross each other without any restriction.

Two types of correlation diagrams are very popular — orbital correlation diagrams and state correlation diagrams. The former are usually drawn to illustrate the change of the diatomic molecular orbitals from *united atom limit* to *separated atom limit*, and for discussing the mechanism and outcome of pericyclic reactions. The state correlation diagrams too are used to discuss the course (such as the thermally allowed or the photochemically allowed path) of pericyclic reactions. Tanabe-Sugano diagrams in crystal field theory (CFT) are state energy curves drawn as functions of the crystal field strength.

The state correlation diagram is in fact more general than the specific diagrams mentioned above. Even an ordinary potential energy surface is basically a state correlation diagram with V_{nn} added to the energy of the electronic motion. Energy surfaces used in the transition state theory of chemical reactions and those used for catalysis are in reality state correlation diagrams in which the ground state potential energy surface for the combined system of the reacting species (and the catalyst) is plotted as a function of the reaction coordinate.

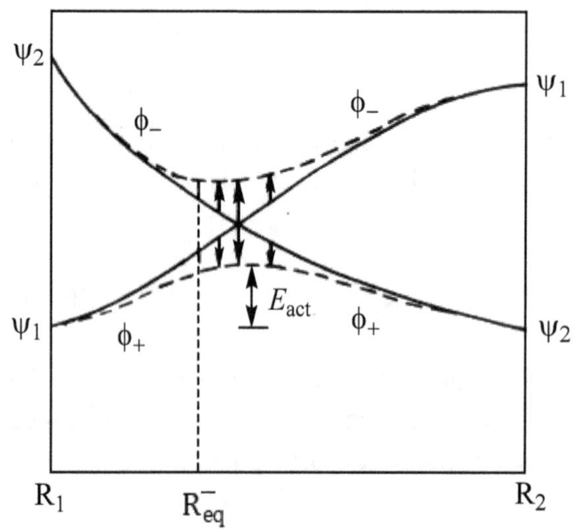

Figure 4.2 The non-crossing phenomenon: Two states ψ_1 and ψ_2 of the same symmetry mix together to form two new state vectors ϕ_+ and ϕ_- in the region between the two limits R_1 and R_2 of the molecular geometry. The corresponding energy levels are shown by solid lines (ψ_1 and ψ_2) and dashed lines (ϕ_+ and ϕ_-). The ϕ_+ state is associated with an energy barrier while the ϕ_- state has an energy minimum. So the reaction $\psi_1(R=R_1) \rightarrow \psi_2(R=R_2)$ has to overcome the activation energy E_{act} whereas starting from $\psi_2(R_1)$ or from $\psi_1(R_2)$ the system would be stabilized at the equlibrium configuration $R=R_{eq}^-$.

Texts Suggested for Further Reading

S. T. Epstein, "Variation Method in Quantum Chemistry", Chaps. 1-2, (Academic Press, New York, 1974).

R. McWeeny, "Coulson's Valence", 3rd Ed., Chap. 3, (Oxford University Press, Oxford, 1979).

H. L. Strauss, "Quantum Mechanics - An Introduction", Chap. 6, (Prentice-Hall India, New Delhi, 1972).

PROBLEMS

1. Use the LCAO-MO scheme to find the qualitative molecular orbitals of the H_2^+ molecular ion. (Hint: Choose the 1s orbital of each hydrogen atom as a basis function. Apply Ritz variation. See Example 4 for the results expected.)
2. Show that the H_2 molecule has the same qualitative molecular orbitals though the orbital energies are different from those for H_2^+.
3. Can you explain why both these species have the same set of qualitative molecular orbitals?

CHAPTER 5

APPLICATION OF THE VARIATION METHOD: ELECTRONIC STRUCTURE OF ATOMS

Lecture 21: Characteristics of a Many-Electron Wave Function

While discussing the application of the variation technique to the helium atom in a previous lecture, we wrote the two-electron wave function as a product of two one-electron wave functions. This was done because the probability density of simultaneously finding one electron at point $\mathbf{r}_1$ while it is in the single particle state i and another at point $\mathbf{r}_2$ while it stays in the single particle state j must equal the product of the probability densities for the individual events of finding each electron separately,

$$P(\mathbf{r}_1, \mathbf{r}_2) = p_i(\mathbf{r}_1)p_j(\mathbf{r}_2). \qquad (21.1)$$

In other words,

$$|\Psi(\mathbf{r}_1, \mathbf{r}_2)|^2 = |\psi_i(\mathbf{r}_1)|^2\, |\psi_j(\mathbf{r}_2)|^2$$

such that

$$\Psi(\mathbf{r}_1, \mathbf{r}_2) = e^{i\eta}\, \psi_i(\mathbf{r}_1)\psi_j(\mathbf{r}_2). \qquad (21.2)$$

The phase η is completely arbitrary. Physical quantities are expressed as expectation values. Expressions for these values involve the square modulus of the wave function. Thus the expectation values are independent of the phase of the wave function. Therefore, without any loss of generality, one may take $\eta = 0$ to write

$$\Psi(\mathbf{r}_1, \mathbf{r}_2) = \psi_i(\mathbf{r}_1)\psi_j(\mathbf{r}_2). \qquad (21.3)$$

This argument can be easily generalized to the case of an arbitrary number of electrons or fundamental particles of any other type. We state that the many-particle wave function should have the primary characteristic of being a product of single particle functions.

Antisymmetry

Another basic feature of the many-particle wave function arises from the fundamental particles being indistinguishable from each other. We cannot distinguish one electron from another. All we can say is that there is a certain number of these particles, like ten electrons in a neon atom, each having the same rest mass and same charge. Let us first confine ourselves to the discussion of two indistinguishable particles in a system, one at position $\mathbf{r}_1$ and the other at $\mathbf{r}_2$. Since it is not possible to make a distinction between the two particles, the probability density of finding the first particle at point $\mathbf{r}_1$ and the second one at $\mathbf{r}_2$ must be the same as the probability density for finding the second one at $\mathbf{r}_1$ and the first one at $\mathbf{r}_2$. That is,

$$P(\mathbf{r}_2, \mathbf{r}_1) = P(\mathbf{r}_1, \mathbf{r}_2) \quad (21.4)$$

that implies

$$\Psi(\mathbf{r}_2, \mathbf{r}_1) = \mathbf{P}_{12}\Psi(\mathbf{r}_1, \mathbf{r}_2) = c\Psi(\mathbf{r}_1, \mathbf{r}_2) \quad (21.5)$$

with $|c| = 1$. Here the operator $\mathbf{P}_{12}$ interchanges the coordinates of the two particles. Let us effect a second exchange,

$$\Psi(\mathbf{r}_1, \mathbf{r}_2) = \mathbf{P}_{12}\Psi(\mathbf{r}_2, \mathbf{r}_1) = c\Psi(\mathbf{r}_2, \mathbf{r}_1). \quad (21.6)$$

If we combine Eq.(21.5) with Eq.(21.6) we see that

$$\Psi(\mathbf{r}_1, \mathbf{r}_2) = c^2\Psi(\mathbf{r}_1, \mathbf{r}_2) \quad (21.7)$$

so that $c = \pm 1$. Therefore,

$$\Psi(\mathbf{r}_2, \mathbf{r}_1) = \mathbf{P}_{12}\Psi(\mathbf{r}_1, \mathbf{r}_2) = \pm\, \Psi(\mathbf{r}_1, \mathbf{r}_2). \quad (21.8)$$

As Eq. (21.8) arises only from indistinguishability, all types of fundamental particles share this feature.

Therefore, the fundamental particles are classified into two clearly defined categories. For some of them, the many-particle wave function remains symmetric [with the plus sign in Eq. (21.8)] under the interchange of the coordinates of any two particles in the system. For the others, the many-particle wave function is antisymmetric under the exchange of coordinates [with minus sign in Eq. (21.8)]. Particles of the former type (with symmetric wave functions) are called Bose particles or bosons while those of the latter kind (with antisymmetric wave functions) are called Fermi particles or fermions. The thermal behavior of a collection of bosons is described by Bose-Einstein statistics. S. N. Bose, an Indian physicist, developed this statistics for a collection of photons in 1924. Einstein (1925) extended it to the treatment of gases of matter. Photons (light particles - quantized excitations of the electromagnetic field), a quantum of vibrational excitation in a molecule, phonons (quantized excitations of lattice vibrations in a crystal), excitons (quantized excitations of the electronic states in a crystal) and so forth (in general the so-called elementary excitations) are the examples of bosons. Other examples include pions, He^4 nuclei, etc. Bose particles tend to condense at a point or in the same region of space. This phenomenon is known as Bose-Einstein condensation. Superfluid helium and electron-hole drops in crystals are examples of Bose-Einstein condensate. The thermal behavior of an ensemble of fermions is described by Fermi-Dirac statistics. E. Fermi (1926) was the first to develop this statistics. His development was based on Pauli's exclusion principle. Dirac (1928) discovered the same statistics by using antisymmetric wave functions. The most important property of the fermions is that they obey the Pauli exclusion principle (Pauli, 1925). Fermi particles tend to exclude each other from the same region of space. Electrons, protons, neutrons and in general the half-spin particles are fermions.

Slater Determinant (Slater, 1929)

It is clear from the discussion given in the above that the simplest form for a many-electron wave function is an anti-symmetrized product which is a determinant. This can be visualized with the aid of the product in Eq. (21.3). The definitions

$$\mathbf{P}_{12}[\psi_a(\mathbf{r}_1)\psi_b(\mathbf{r}_2)] = \psi_a(\mathbf{r}_2)\psi_b(\mathbf{r}_1) \quad (21.9)$$

$$\mathbf{P}_{12}[\psi_a(\mathbf{r}_2)\psi_b(\mathbf{r}_1)] = \psi_a(\mathbf{r}_1)\psi_b(\mathbf{r}_2)$$

yield

$$\mathbf{P}_{12}[\psi_a(\mathbf{r}_1)\psi_b(\mathbf{r}_2) \pm \psi_a(\mathbf{r}_2)\psi_b(\mathbf{r}_1)] = \pm [\psi_a(\mathbf{r}_1)\psi_b(\mathbf{r}_2) \pm \psi_a(\mathbf{r}_2)\psi_b(\mathbf{r}_1)]. \quad (21.10)$$

Thus a two-boson wave functon is written as

$$\Psi_+(\mathbf{r}_1, \mathbf{r}_2) = N_+[\psi_a(\mathbf{r}_1)\psi_b(\mathbf{r}_2) + \psi_a(\mathbf{r}_2)\psi_b(\mathbf{r}_1)] \quad (21.11)$$

where N_+ is the normalization constant, whereas the simplest wave function for two fermions (such as a two-electron wave function) has the form

$$\Psi_-(\mathbf{r}_1, \mathbf{r}_2) = \|\psi_a(\mathbf{r}_1)\psi_b(\mathbf{r}_2)\| \equiv N_- \begin{vmatrix} \psi_a(\mathbf{r}_1) & \psi_a(\mathbf{r}_2) \\ \psi_b(\mathbf{r}_1) & \psi_b(\mathbf{r}_2) \end{vmatrix} \quad (21.12)$$

with the normalization constant N_-.

The determinantal functions are called Slater determinants (Slater, 1929). The simplest N-electron wave function is a Slater determinant of order N,

$$\Psi(\mathbf{r}_1, \mathbf{r}_2, \dots, \mathbf{r}_N) = \|\psi_a(\mathbf{r}_1)\psi_b(\mathbf{r}_2) \dots \psi_v(\mathbf{r}_N)\|$$

$$\equiv N \begin{vmatrix} \psi_a(\mathbf{r}_1) & \psi_a(\mathbf{r}_2) & \cdots & \psi_a(\mathbf{r}_N) \\ \psi_b(\mathbf{r}_1) & \psi_b(\mathbf{r}_2) & \cdots & \psi_b(\mathbf{r}_N) \\ \vdots & \vdots & \vdots & \vdots \\ \psi_v(\mathbf{r}_1) & \psi_v(\mathbf{r}_2) & \cdots & \psi_v(\mathbf{r}_N) \end{vmatrix} \quad (21.13)$$

where N is the normalization constant. When expanded, the N-th order determinant is seen to be a combination of $N!$ terms, each being a product of N single particle functions. In case the one-electron functions $\psi_a, \psi_b, \dots, \psi_v$ are normalized and orthogonal to each other, each of the $N!$ terms in the determinant becomes normalized, and the terms become orthogonal to each other so that the normalization constant is found as

$$N = (N!)^{-1/2}. \quad (21.14)$$

This form will be used in some of the later lectures.

Lecture 22: General Angular Momentum; The Electron Spin

The general quantum mechanical definition of an angular momentum is given in terms of the commutation relationships among its cartesian components,

$$[M_x, M_y] = i\hbar M_z, \qquad (x, y, z \text{ cyclic}). \quad (22.1)$$

Although the components do not commute with each other it is easy to establish that each of them commutes with the square of the angular momentum operator, $\boldsymbol{M}^2$ $(= M_x^2 + M_y^2 + M_z^2)$. Thus one may choose a set of common eigenfunctions of any of these components and the squared operator. By convention one seeks the common eigenfunctions of $\boldsymbol{M}^2$ and M_z. In one of the latter lectures (in Part II of this book) we will discuss how to find the eigenvalues of the operators $\boldsymbol{M}^2$ and M_z, and how to construct the common eigenfunctions of these two operators. Operators for the angular momentum and its components are of course hermitean. Here we take it for granted that

1. the common eigenfunctions are characterized by the total angular momentum quantum number (Λ) and the quantum number (m) for the z component of the angular momentum;
2. the eigenvalues equal $\hbar^2\Lambda(\Lambda+1)$ and $\hbar m$ respectively:

   $\boldsymbol{M}^2 f_{\Lambda m} = \hbar^2\Lambda(\Lambda+1)\, f_{\Lambda m}$,

   and (22.2)

   $\boldsymbol{M}_z\, f_{\Lambda m} = \hbar m\, f_{\Lambda m}$;
3. the quantum number Λ has positive integral or half-integral values, and
4. the values assigned to m are

   $m = -\Lambda, -(\Lambda-1), \ldots, (\Lambda-1), \Lambda.$ (22.3)

This completes the description of the angular momentum in quantum mechanics. One may easily compare the expressions (22.1), (22.2) and (22.3) with the corresponding expressions for the orbital angular momentum in Chapter 3.

There are two reasons for defining the angular momentum in an abstract form by Eq. (22.1). First, all the properties of an angular momentum except the explicit functional form of the eigenfunctions are determined by the algebra of commutation. Second, in many cases one is unable to give a classical or a picturesque interpretation of the angular momentum as it is done for the orbital angular momentum. One such instance where a simple physical interpretation is lacking is the case of the electron spin.

Spin is an angular momentum. In 1921 O. Stern and W. Gerlach carried out a unique experiment. When a stream of silver atoms having only one s electron in the valence shell and thus devoid of any orbital angular momentum is shot through an externally imposed homogeneous magnetic field, it divides into two streams moving in curved paths as illustrated in Figure 5.1. G. E. Uhlenbeck and S. Goudsmit (1925) interpreted this observation in terms of an additional angular momentum of the electron with two possible states. The radius of curvature indicated that this angular momentum is half-integral, that is, the quantum numbers involved in the z-component are $\pm\frac{1}{2}$.

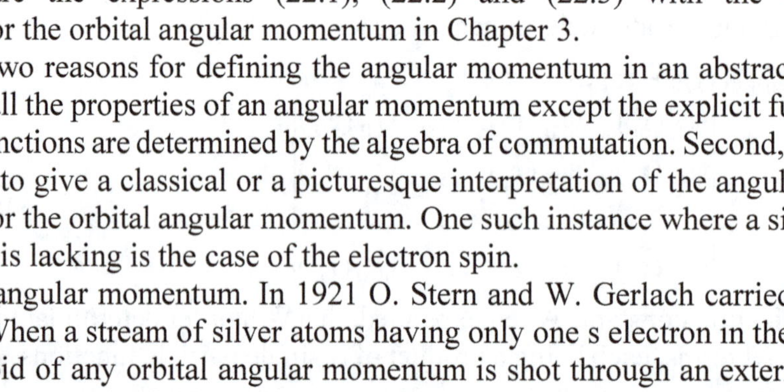

Figure 5.1 Schematic representation of the experiment of O. Stern and W. Gerlach. S is a stream of gaseous Ag atoms. After passing through a magnetic field of strength B and applied along a direction perpendicular to the direction of the stream, the original beam splits into two new streams S_1 and S_2 which strike the screen SC.

The new angular momentum was called the electron spin (**S**) in analogy with a spinning top or the spinning earth. Nevertheless, as the electron is considered as a point charge with a specific rest mass but no structure, its spin angular momentum cannot be classically interpreted.

The mathematical prescription for the interpretation put forward by Uhlenbeck and Goudsmit is as follows. Let $\boldsymbol{S}$ be the spin angular momentum operator for one electron. The components of $\boldsymbol{S}$ satisfy

$$[S_x, S_y] = i\hbar S_z, \qquad (x, y, z \text{ cyclic}), \tag{22.4}$$

$$S^2 = S_x^2 + S_y^2 + S_z^2. \qquad [\text{with eigenvalue } \hbar^2 s(s+1)].$$

The Stern-Gerlach experiment suggests that for one electron the quantum numbers s and m_s (eigenvalue of S_Z in unit $\hbar$) have the following values: $s = \frac{1}{2}$ and $m_s = \frac{1}{2}$ or $-\frac{1}{2}$. There are two eigenstates, α and β, which satisfy the following relations

$$S^2\alpha = \frac{3}{4}\hbar^2\alpha, \qquad S^2\beta = \frac{3}{4}\hbar^2\beta, \tag{22.5}$$

$$S_Z\alpha = \frac{1}{2}\hbar\alpha, \qquad S_Z\beta = -\frac{1}{2}\hbar\beta.$$

Explicit forms for the functions α and β are not known but a matrix representation can be easily constructed. This will be discussed in Chapter 11.

Further support of the spinning electron was obtained from the splitting of the optical spectra of alkali atoms. The electron spin is now considered to be a basic property of the electron like its rest mass and charge.

The existence of spin gives rise to two specific aspects of the electronic structure of atoms and molecules. The first of these is that there exists the total angular momentum

$$J = L + S. \tag{22.6}$$

The operator $\boldsymbol{J}$ has Cartesian components satisfying the commutation relations

$$[J_x, J_y] = i\hbar J_z. \qquad (x, y, z \text{ cyclic}). \tag{22.7}$$

We then seek the eigenfunctions of J^2 and J_Z. These operators are decomposable as

$$J^2 = J \cdot J = L^2 + S^2 + 2L \cdot S$$

and (22.8)

$$J_Z = L_Z + S_Z.$$

If we try to describe the eigenfunctions of J^2 and J_Z in terms of the spherical harmonics (which are the common eigenfunctions of L^2 and L_Z) and the spin functions α and β (the common eigenfunctions of S^2 and S_Z), we need to do a careful analysis of the interaction $L \cdot S$. The effect of this term vanishes for the s electron as $l = 0$ for it, and one can write the eigenfunctions as the simple products $Y_{00}\alpha$ and $Y_{00}\beta$. For other types of electrons, one has to consider linear combinations of products of types $Y_{lm}\alpha$ and $Y_{lm}\beta$. This aspect will be quantitatively discussed in Chapter 11. In any case, the products of the spatial orbitals and the spin functions such as $\phi_{nlm}(\mathbf{r})\alpha(s)$ and $\phi_{nlm}(\mathbf{r})\beta(s)$ form a basis for the single particle functions, the so-called spin orbitals.

The second aspect is as follows. In terms of the spin orbitals $\psi(\mathbf{x})$'s where $\mathbf{x}$ stands for the space-spin coordinates $(\mathbf{r}, s)$, the Slater determinant is written as

$$\Psi(\mathbf{x}_1, \mathbf{x}_2, \ldots, \mathbf{x}_N) = \|\psi_a(\mathbf{x}_1)\psi_b(\mathbf{x}_2) \ldots \psi_\nu(\mathbf{x}_N)\|$$

$$\equiv N!^{-1/2}\begin{vmatrix} \psi_a(\mathbf{x}_1) & \psi_a(\mathbf{x}_2) & \ldots & \psi_a(\mathbf{x}_N) \\ \psi_b(\mathbf{x}_1) & \psi_b(\mathbf{x}_2) & \ldots & \psi_b(\mathbf{x}_N) \\ \vdots & \vdots & \vdots & \vdots \\ \psi_\nu(\mathbf{x}_1) & \psi_\nu(\mathbf{x}_2) & \ldots & \psi_\nu(\mathbf{x}_N) \end{vmatrix}. \tag{22.9}$$

In principle the atomic or molecular Hamiltonian contains the spin-orbit interaction which is approximately of the form $\lambda(\mathbf{r})\boldsymbol{L}\bullet\boldsymbol{S}$ for each electron in the species. This interaction is not normally included in the Hamiltonian as it serves as a very small perturbation for ordinary atoms, and its effect becomes pronounced only for atoms with $Z > 30$. It is possible to effectively separate the spatial part of the orbitals from the spin-dependent part when the spin-orbit coupling is negligibly small. The separable spin orbitals $\phi\alpha$ and $\phi\beta$ are degenerate in the absence of an external magnetic field as the corresponding one-electron Hamiltonian is independent of the electron spin. We can write the N-electron Slater determinant in terms of the separable spin orbitals as

$$\Psi(\mathbf{x}_1, \mathbf{x}_2, \ldots, \mathbf{x}_N) = \|\phi_a(\mathbf{r}_1)\alpha(s_1)\ \phi_a(\mathbf{r}_2)\beta(s_2)\ \ldots\ \phi_{\nu/2}(\mathbf{r}_N)\beta(s_N)\| \tag{22.10}$$

$$\equiv N!^{-1/2}\begin{vmatrix} \phi_a(\mathbf{r}_1)\alpha(s_1) & \phi_a(\mathbf{r}_2)\alpha(s_2) & \ldots & \phi_a(\mathbf{r}_N)\alpha(s_N) \\ \phi_a(\mathbf{r}_1)\beta(s_1) & \phi_a(\mathbf{r}_2)\beta(s_2) & \ldots & \phi_a(\mathbf{r}_N)\beta(s_N) \\ \vdots & \vdots & \vdots & \vdots \\ \phi_{\nu/2}(\mathbf{r}_1)\beta(s_1) & \phi_{\nu/2}(\mathbf{r}_2)\beta(s_2) & \ldots & \phi_{\nu/2}(\mathbf{r}_N)\beta(s_N) \end{vmatrix}$$

where s_i represents the spin coordinate of the *ith* electron and $\mathbf{r}_i$ as usual stands for the spatial coordinates of the same.

The Slater determinant is the mathematical representation of Pauli's exclusion principle (Pauli, 1925) which states that no two electrons in the same species may remain in the same spin orbital. If two electrons remain in the same single particle state, two rows of the determinant become identical and the determinant becomes equal to zero. Then the determinantal wave function vanishes at every point in space, yielding zero probability of finding the electrons. Thus one can never find two electrons in the same spin orbital of a system, which is the exclusion principle. Conversely, one can say that the exclusion principle originates from the requirement of antisymmetry.

Lectures 23 and 24: Hartree and Hartree-Fock Models; Self-Consistent-Field Treatment

Lecture 23: The Hartree Models

The simplest quantitative model for atoms was constructed by Hartree and it was subsequently improved by Fock. These are now recognized as Hartree and Hartree-Fock models. Although these were initially proposed for the investigation of the atomic structure, it was very soon realized that the same methodologies apply equally well to the investigation of the electronic structure of molecules and solids. Though the models are strictly mathematical in nature, a lucid physical interpretation can be given for each of them. The rigorous mathematical derivations of these models will be given in Chapter 12. In this lecture we introduce the Hartree and the Hartree-Fock treatments only in a conceptual framework that makes the average physics of atomic or molecular electrons explicitly clear.

The electronic Hamiltonian for a general N-electron atom (or for a general N-electron molecule under the Born-Oppenheimer approximation) is written as

$$\mathbf{H}_{\text{el}} = \sum_{i=1}^{N} \mathbf{h}(i) + \sum_{1 \le i < j \le N} \mathbf{v}(i, j) \tag{23.1}$$

where $\mathbf{h}(i)$ is the one-electron Hamiltonian for the *ith* electron,

$$\mathbf{h}(i) = -\frac{1}{2}\nabla_i^2 - \sum_{\substack{m=1 \\ (nuclei)}}^{M} Z_m |\mathbf{r}_i - \mathbf{R}_m|^{-1} \tag{23.2}$$

and the interaction $\mathbf{v}(i,j)$ between the *ith* and the *jth* electrons is electrostatic in nature,

$$\mathbf{v}(i,j) = |\mathbf{r}_i - \mathbf{r}_j|^{-1}. \tag{23.3}$$

Atomic units have been used in writing the expressions above. There is only one nucleus in an atom, that is, the number of nucleus M is 1 for an atom. Because of the presence of the interaction term $\mathbf{v}$, the Hamiltonian $\mathbf{H}_{\text{el}}$ is not separable and it is not possible to derive an exact analytical solution of the wave equation

$$\mathbf{H}_{\text{el}}\Psi_{\text{el}}(\mathbf{x}_1, \mathbf{x}_2, \dots, \mathbf{x}_N) = E_{\text{el}}\Psi_{\text{el}}(\mathbf{x}_1, \mathbf{x}_2, \dots, \mathbf{x}_N). \tag{23.4}$$

Therefore, what one needs to do is to take up a trial wave function for the N electrons, calculate the expectation value of $\mathbf{H}_{\text{el}}$ by using this function, and find the condition for which this average value would be minimum (rather, stationary) for variation of the trial function. This task was precisely carried out by Hartree (1928) and the treatment was subsequently extended by Fock (1930) to take account of the indistinguishability of electrons. The trial functions employed in Hartree and Hartree-Fock models are of course somewhat different from each other.

Hartree model

In the Hartree model the trial wave function is written as a single product

$$\Phi(\mathbf{x}_1, \mathbf{x}_2, \dots, \mathbf{x}_N) = [\phi_1(\mathbf{r}_1)\alpha(s_1)][\phi_1(\mathbf{r}_2)\beta(s_2)] \dots [\phi_{N/2}(\mathbf{r}_N)\beta(s_N)] \tag{23.5}$$

for a closed-shell atom or a molecule. For an open-shell species, it is to be written as a linear combination of a few (nearly) degenerate single products so as to obtain the appropriate angular momentum states. The spatial orbitals ϕ_is are assumed to form an orthonormal set. The corresponding closed-shell total energy turns out to be

$$E = 2\sum_{i=1}^{N/2} h_{ii} + \sum_{i=1}^{N/2} J_{ii} + 2\sum_{i=1}^{N/2}\sum_{\substack{j=1 \\ (i \ne j)}}^{N/2} J_{ij} \tag{23.6}$$

where h_{ii} and the *direct integral* J_{ij} are the expectation values

$$h_{ii} = \int dv\, \phi_i^*(\mathbf{r})\mathbf{h}(\mathbf{r})\phi_i(\mathbf{r}),$$

and (23.7)

$$J_{ij} = e^2 \int dv_1 \int dv_2\, \phi_i^*(\mathbf{r}_1)\phi_j^*(\mathbf{r}_2)|\mathbf{r}_1 - \mathbf{r}_2|^{-1}\phi_i(\mathbf{r}_1)\phi_j(\mathbf{r}_2).$$

The indices i and j are now orbital indices. Hartree showed that if one makes E stationary for variation of ϕ_1, ϕ_2, etc. subject to the constraint that the single particle wave functions remain normalized and orthogonal to each other, one *eventually* obtains the *effective one-electron Schrödinger equation*

$$\mathbf{h}_{\text{Hartree}}(\mathbf{r})\,\phi_k(\mathbf{r}) = \varepsilon_k\,\phi_k(\mathbf{r}). \tag{23.8}$$

This wave equation is called the *Hartree equation.* The index k is an orbital index.

The *Hartree operator* is given for a closed shell species by

$$\mathbf{h}_{\text{Hartree}}(\mathbf{r}) = \mathbf{h}(\mathbf{r}) + \mathbf{J}_k(\mathbf{r}) + 2\sum_{\substack{j=1\\(j\neq k)}}^{N/2} \mathbf{J}_j(\mathbf{r}) \tag{23.9a}$$

for operation on an occupied orbital ($k \leq N/2$), and

$$\mathbf{h}_{\text{Hartree}}(\mathbf{r}) = \mathbf{h}(\mathbf{r}) + 2\sum_{j=1}^{N/2} \mathbf{J}_j(\mathbf{r}) \tag{23.9b}$$

for operation on an unoccupied orbital ($k > N/2$)

where $\mathbf{J}_j(\mathbf{r})$ is the average potential energy operator for the coulombic repulsion of the electron at point $\mathbf{r}$ by the average charge distribution due to another electron in orbital ϕ_j:

$$\mathbf{J}_j(\mathbf{r}) = e^2\int dv_2\, \rho_j(\mathbf{r}_2)|\mathbf{r} - \mathbf{r}_2|^{-1}, \qquad (e = -1 \text{ in a.u.}). \tag{23.10}$$

The distribution of the electron in space is characterized by the probability density $\rho_j(\mathbf{r}) = |\phi_j(\mathbf{r})|^2$ that yields the charge density $e\rho_j(\mathbf{r})$. The average repulsion energy can be visualized from Figure 5.2. The operator $\mathbf{J}_j$ is called the direct operator (for coulombic repulsion). Thus the effective one-electron Hamiltonian is obtained by adding the potential energy operator for the repulsion of the electron by the charge distribution due to all other electrons to the one-electron Hamiltonian that consists of the kinetic energy operator of the electron and the potential energy of interaction of the electron with the external charges (the nuclei). Among the other electrons, one is in the same spatial orbital though with a different spin, and two are in every other spatial orbital occupied in the Hartree-Fock ground state. This helps in counting the direct interaction terms (the $\mathbf{J}$'s) in equations (23.9). One can realize that the integral J_{ij} is merely the expectation value of the direct operator,

$$J_{ij} = \int dv_1\, \phi_i^*(\mathbf{r}_1)\mathbf{J}_j(\mathbf{r}_1)\phi_i(\mathbf{r}_1) = \int dv_2\, \phi_j^*(\mathbf{r}_2)\mathbf{J}_i(\mathbf{r}_2)\phi_j(\mathbf{r}_2). \tag{23.11}$$

The *orbital energy* is an eigenvalue of the operator $\mathbf{h}_{\text{Hartree}}$ so that it can be calculated as an expectation value over the eigenfunction. Thus one finds

$$\varepsilon_k = h_{kk} + J_{kk} + 2\sum_{\substack{j=1\\(j\neq k)}}^{N/2} J_{jk}\,. \tag{23.12}$$

In terms of the orbital energies, total energy may be written as

$$E = \sum_{i=1}^{N/2} (h_{ii} + \varepsilon_i). \tag{23.13}$$

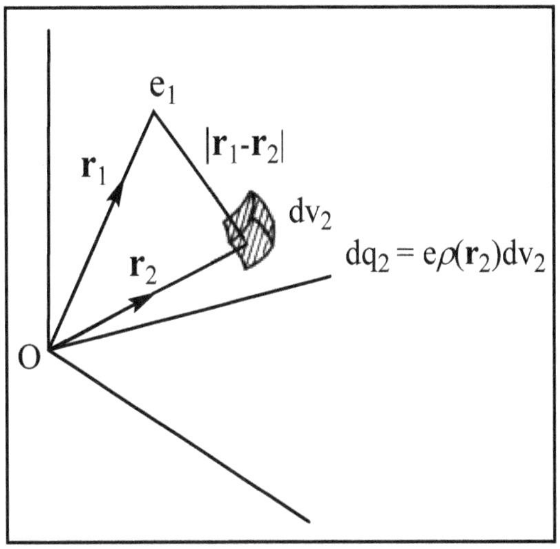

Figure 5.2 Illustration of the direct coulomb repulsion operator $\mathbf{J}_k(\mathbf{r}_1)$: The position vector of the second electron is denoted by $\mathbf{r}_2$. The probability density at point $\mathbf{r}_2$ is $\rho(\mathbf{r}_2) = |\phi_k(\mathbf{r}_2)|^2$ while the second electron stays in the *kth* orbital. The charge inside the infinitesimally small volume element dv_2 is given by $dq_2 = e\rho(\mathbf{r}_2)dv_2$. The first electron at point $\mathbf{r}_1$ faces an instantaneous coulombic repulsion by the charge dq_2. This gives the infinitesimally small potential energy $edq_2\,|\mathbf{r}_1 - \mathbf{r}_2|^{-1} = e^2 dv_2 \rho(\mathbf{r}_2)|\mathbf{r}_1 - \mathbf{r}_2|^{-1}$. The potential energy of repulsion by the second electron is obtained by integrating over all space which yields the direct operator in Eq. (23.10). This interaction term is a space average.

These results have very simple but picturesque interpretations. The orbital energy equals the sum of the kinetic energy of the electron, the nuclear attraction energies, and the average potential energy of coulombic repulsion by $(N-1)$ other electrons distributed over the occupied orbitals. The total energy E is equal to the sum of all one-electron energies and all appropriately averaged interaction energies. It differs from the sum of the occupied-orbital energies $2\sum_{i=1}^{N/2} \varepsilon_i$ in which each average interaction energy term occurs twice. Note that the sum of occupied orbital energies is an eigenvalue of the many-electron Hamiltonian operator ($\mathbf{H}_H$) that is the sum of the Hartree operators ($\mathbf{h}_{Hartree}$) for all the N electrons,

$$\mathbf{H}_H = \sum_{i=1}^{N} \mathbf{h}_{Hartree}(i). \tag{23.14}$$

The operator $\mathbf{H}_H$ is called the *Hartree Hamiltonian.* The optimized function Φ_{opt} is an eigenfunction of this operator. Other eigenfunctions are obtained by forming the products of other sets of orbitals. In every case the eigenvalue of the Hartree Hamiltonian is given by the sum of the energies of the orbitals which are involved in writing the eigenfunction.

Lecture 24: The Hartree-Fock Model

The Hartree model neglects the correlation between the instantaneous positions of the electrons. It also ignores the indistinguishability of the electrons from each other, but this important point was taken into account by Fock. In the Hartree-Fock model, the trial many-electron function is written as a single Slater determinant like that shown in Eq.(22.10) for a closed-shell species for which N must be even,

$$\Phi(\mathbf{x}_1,\mathbf{x}_2,\ldots,\mathbf{x}_N) = \frac{1}{\sqrt{N!}}\begin{vmatrix} \phi_1(\mathbf{r}_1)\alpha(s_1) & \phi_1(\mathbf{r}_2)\alpha(s_2) & \ldots & \phi_1(\mathbf{r}_N)\alpha(s_N) \\ \phi_1(\mathbf{r}_1)\beta(s_1) & \phi_1(\mathbf{r}_2)\beta(s_2) & \ldots & \phi_1(\mathbf{r}_N)\beta(s_N) \\ \vdots & \vdots & \vdots & \vdots \\ \phi_{N/2}(\mathbf{r}_1)\beta(s_1) & \phi_{N/2}(\mathbf{r}_2)\beta(s_2) & \ldots & \phi_{N/2}(\mathbf{r}_N)\beta(s_N) \end{vmatrix}, \tag{24.1}$$

and otherwise an appropriate linear combination of a few Slater determinants differing from each other only in the open shell part. In the latter case, the linear combination of the open shell substates is chosen in such a way that the resulting trial function is an eigenfunction of some of the total angular momentum operators. Again, it is assumed that the spatial functions ϕ's form an orthonormal set.

The closed-shell total energy is

$$E_\Phi = 2\sum_{i=1}^{N/2} h_{ii} + \sum_{i=1}^{N/2} J_{ii} + \sum_{i=1}^{N/2}\sum_{\substack{j=1 \\ (i\neq j)}}^{N/2} (2J_{ij}-K_{ij}) \tag{24.2}$$

where the *exchange integral* K_{ij} is given by

$$K_{ij} = e^2\int dv_1\int dv_2\, \phi_i^*(\mathbf{r}_1)\phi_j^*(\mathbf{r}_2)|\mathbf{r}_1-\mathbf{r}_2|^{-1}\phi_i(\mathbf{r}_2)\,\phi_j(\mathbf{r}_1). \tag{24.3}$$

If one considers E_Φ to be stationary for arbitrary variations of the ϕ's subject to the constraint of orthonormality of these single particle functions, one can *eventually* obtain the *Hartree-Fock* equation

$$\mathbf{F}\phi_k(\mathbf{r}) = \varepsilon_k\phi_k(\mathbf{r}). \tag{24.4}$$

The operator $\mathbf{F}$ is called the Fock operator. The Fock operator for the closed-shell system is given by

$$\mathbf{F}(\mathbf{r}) = \mathbf{h}(\mathbf{r}) + \mathbf{J}_k(\mathbf{r}) + \sum_{\substack{j=1 \\ (j\neq k)}}^{N/2} [2\mathbf{J}_j(\mathbf{r}) - \mathbf{K}_j(\mathbf{r})] \tag{24.5a}$$

for operation on an occupied orbital ϕ_k corresponding to $k \le N/2$. For operation on an unoccupied orbital ϕ_k with $k > N/2$ the Fock operator is written as

$$\mathbf{F}(\mathbf{r}) = \mathbf{h}(\mathbf{r}) + \sum_{j=1}^{N/2} [2\mathbf{J}_j(\mathbf{r}) - \mathbf{K}_j(\mathbf{r})]. \tag{24.5b}$$

The Fock operator in equations (24.5) differs from the Hartree operator in equations (23.9) by possessing the additional term $-\sum_{j=1}^{N/2} \mathbf{K}_j(\mathbf{r})$. The *exchange operator* $\mathbf{K}_j(\mathbf{r})$ is the potential energy operator for the average coulombic repulsion of the electron at point $\mathbf{r}$ by an electron in the *jth* orbital with the effect that the two electrons are exchanged between the concerned orbitals as

$$\mathbf{K}_j(\mathbf{r})\phi_k(\mathbf{r}) = e^2\int dv_2\, \phi_j^*(\mathbf{r}_2)|\mathbf{r}-\mathbf{r}_2|^{-1}\phi_k(\mathbf{r}_2)\phi_j(\mathbf{r}). \tag{24.6}$$

The *exchange integral* K_{ij} can be written as the expectation value of the exchange operator,

$$K_{ij} = \int dv_1\, \phi_i^*(\mathbf{r}_1)\mathbf{K}_j(\mathbf{r}_1)\phi_i(\mathbf{r}_1) = \int dv_2\, \phi_j^*(\mathbf{r}_2)\mathbf{K}_i(\mathbf{r}_2)\phi_j(\mathbf{r}_2). \qquad (24.7)$$

Note that the direct operators in equations (24.5) are the same as those in Eq. (23.10) only in appearance, as the eigenfunctions (and also the eigenvalues) of $\mathbf{h}_{\text{Hartree}}$ are clearly different from those of **F**.

With the definition of the exchange operator given by Eq. (24.6) one finds $\mathbf{J}_k\phi_k = \mathbf{K}_k\phi_k$. Hence Eq. (24.5b) is the general form of the Fock operator. It is valid even for operation on the occupied orbitals.

For the closed-shell system one obtains the Hartree-Fock orbital energy as

$$\varepsilon_i = h_{ii} + \sum_{j=1}^{N/2} (2J_{ij} - K_{ij}) \qquad (24.8a)$$

which for an occupied orbital $(i \leq N/2)$ reduces to

$$\varepsilon_i = h_{ii} + J_{ii} + \sum_{\substack{j=1 \\ (j \neq i)}}^{N/2} (2J_{ij} - K_{ij}). \qquad (24.8b)$$

The direct interaction inherent in the expression for ε_i can be counted term by term as it was done for the Hartree model. We now discuss how to count the exchange interactions.

The exchange operator and the exchange integral arise from the anti-symmetry of the many-electron wave function. Since a single interchange of the coordinates of a pair of electrons alters the sign of the wave function, the exchange integrals in equations (24.2) and (24.8) and the exchange operators in equations (24.5) appear with a negative sign. Actually the functions which are at first involved in writing both direct and exchange integrals are the spin orbitals. These should replace the spatial functions in equations (23.7) and (24.3); furthermore, integration over the spin coordinates is to be included in each equation. For two electrons in the same spin state (both of α spin or both β spin) the spin integration in direct as well as exchange integrals lead to the result 1 for each spin integral. So what is left is precisely Eq. (23.7) for the direct integral and Eq. (24.3) for the exchange integral. For electrons in different spin states (one with α function and the other with β function) the spin integral present in the exchange integral becomes zero due to the orthogonality of the two spin functions while the direct integrals again reduce to the form of Eq.(23.7). Thus there are twice as many direct interactions as the exchange interactions are. This leads to the (2J–K) terms in the Fock operator in equations (24.5), the $(2J_{ij}-K_{ij})$ terms in total energy given by Eq.(24.2) and similar terms in the expression for orbital energy given by equations (24.8a) and (24.8b).

The closed-shell total energy can be written as

$$E = \sum_{i=1}^{N/2} (h_{ii} + \varepsilon_i). \qquad (24.9)$$

Each interaction occurs once and only once in Hartree-Fock total energy in Eq. (24.9), just as we have seen it to happen in the Hartree case. The exchange interaction appears only in the Hartree-Fock treatment.

Self-consistent field

We now discuss a very important aspect of the Hartree and the Hartree-Fock models. The Hartree equation, Eq. (23.8), and the Hartree-Fock equation, Eq. (24.4), are not truly eigenvalue equations. These are *pseudo-eigenvalue equations* as the operators involved, $\mathbf{h}_{\text{Hartree}}$ and $\mathbf{F}$, depend upon their own eigenfunctions. These equations can be solved by an iterative procedure that involves the four steps as detailed below:

1. Make an intelligent guess of the one-electron function.
2. Form the operators $\mathbf{J}_j$'s and $\mathbf{K}_j$'s, and hence form the Hartree operator $\mathbf{h}_{\text{Hartree}}$ or the Fock operator $\mathbf{F}$.
3. Solve the Hartree or the Hartree-Fock equation. The solutions obtained form a new set of the one-electron functions.
4. Use the new set in step 2 and iterate to self-consistency, that is, until the one-electron functions which has been used to form the effective one-electron Hamiltonian becomes identical with the eigenfunctions of the same Hamiltonian.

These steps are fairly simple but steps 1 and 4 merit special comments. For the initial one-electron functions required in step 1 one may choose approximate solutions such as the atomic orbitals (for a molecular Hartree or Hartree-Fock calculation) or one may select the eigenfunctions of the one-electron Hamiltonian $\mathbf{h}$ (or $\mathbf{h}$ with some sort of average interaction). In step 4, it would be exceedingly difficult to compare the newly found solutions with the previous ones as one needs to compare each pair of the new and the previous ϕ's at every point of space. Hence the total energy is normally taken as the criterion for convergence of the solutions. When the absolute magnitude of the difference between the total energies calculated in two successive steps becomes less than a sufficiently small pre-assigned positive value, one considers the results to have converged and become self-consistent.

The iterative procedure leads to a situation where the electric field created by the average distribution of the charge of all the electrons is consistent with itself, that is, the electric field gives rise to an effective Hamiltonian that has eigenfunctions from which one may calculate the same average charge distribution that in turn creates the same field. For this reason the iterative procedure is called the Hartree or Hartree-Fock self-consistent-field (HF SCF) procedure.

Lecture 25: Atomic Orbitals and Molecular Orbitals

Definition

In Lecture 16, we have defined an atomic orbital for a one-electron atom as the electronic state represented by an eigenfunction of the electronic Hamiltonian. Orbitals are single particle states. For a many-electron species, one has to choose an effective one-electron Hamiltonian. Orbitals are represented by the solutions of the effective Schrödinger equation for one electron in an atom or a molecule. In the Hartree model the atomic or molecular orbitals are represented by the eigenfunctions of the Hartree operator. These are called *Hartree orbitals*. In the Hartree-Fock model, the atomic or molecular orbitals are represented by the eigenfunctions of the Fock operator. These are called *Hartree-Fock orbitals* or more simply as atomic or molecular orbitals. Thus, if nothing else is stated, orbitals of a many-electron species normally mean Hartree-Fock orbitals.

Physical interpretation of orbital energies

Any rigorous definition of the orbital must accompany an exemplary physical interpretation without which the definition may remain simply as a mathematical construct. The orbital energy is interpreted by

Theorem 11. Koopmans' Theorem, that states

$$\varepsilon_i = -I_i \qquad (\text{for } i \le \frac{N}{2}) \tag{25.1}$$

when orbital relaxation effects are neglected, I_i being the orbital ionization energy.

The orbital ionization energy is the *minimum* energy necessary to remove an electron from the *ith* orbital leaving behind a positively charged ion. The energy requirement for the ionization process is illustrated by Figure 5.3. Any excess input of energy turns into the kinetic energy of the freed electron. The validity of Eq. (25.1) rests upon the assumption that the orbitals do not change by the ionization process. But the orbitals do change after ionization, since the Fock operator for the ionized species is different from that for the neutral atom or molecule. In a time-dependent description the orbitals would steadily relax into the orbitals of the positive ion, creating a further difference in energy. Hence the relationship (25.1) is only approximately valid. These points will be discussed in detail in Chapter 12.

Virtual orbitals

An electron in an occupied orbital faces repulsion from (N–1) other electrons [see Eq. (24.5a)] whereas an electron residing in an orbital unoccupied in the Hartree-Fock ground state would be repelled by all the N electrons [according to Eq. (24.5b)]. The energy of repulsion is always positive. This is why the energies of the unoccupied orbitals are always found to be considerably high. These energies are often found near zero and some of them even turn out to be positive. This feature is an artifact of Hartree and Hartree-Fock treatments. Hence these orbitals are referred to as *virtual orbitals*: they represent the bound states of the electron (for a fixed nuclear geometry in the case of a molecule), yet their energies are often extremely close to or even fall inside the range of energies ($\varepsilon > 0$) for the unbound single particle states (the continuum states).

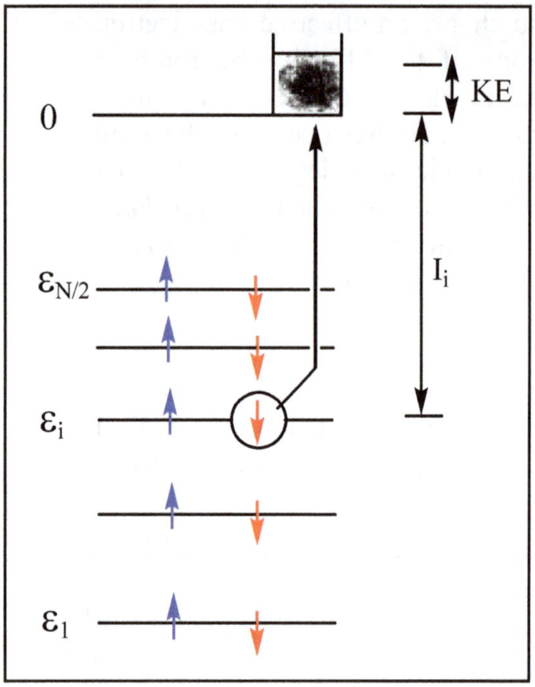

Figure 5.3 Illustrating Koopmans' theorem: The ionization energy is the minimum energy required to separate one electron by taking it out of a specific orbital of an atom or a molecule to an infinite distance from the positively charged ion. The excess energy supplied turns into the kinetic energy of the freed electron (and the negligibly small kinetic energy of the positive ion). The translational levels are on top of the $\varepsilon = 0$ line. These levels form a continuum.

It is difficult to give a direct physical interpretation of the virtual orbital energy. An interpretation analogous to that of the Eq.(25.1) cannot be given. A somewhat similar relationship – but an entirely untrustworthy one – may be attempted by writing

$$\varepsilon_i = -EA_i \qquad \text{(for } i > N/2\text{)} \qquad (25.2)$$

where EA_i is the orbital electron affinity, that is, the minimum energy evolved while a negative ion is formed from the neutral species by the addition of one electron to the *ith* orbital that was unoccupied in the Hartree-Fock ground state. Again this would be valid under the assumption that the orbitals do not change while the negative ion is formed. This assumption does not work well when one electron is added to a neutral atom or molecule. In fact, ground state electron affinities are usually very small quantities. Equation (25.2) always gives a large deviation in electron affinity for $\varepsilon_i < 0$, and for $\varepsilon_i > 0$, it leads to a negative electron affinity indicating that energy is actually absorbed in the process of forming the negative ion. This is clearly not a happy situation. At the present moment we would disregard Eq. (25.2) except mentioning that this equation has found partial merit in the preparation of matrix elements in semi-empirical treatments.

Notwithstanding the difficulties discussed above, one can always give an indirect interpretation to the virtual orbital energies in the form that the difference $[\varepsilon_i(i{>}N/2) - \varepsilon_j(j{\le}N/2)]$ approximates the spectroscopic transition energy for the excitation of an electron from an occupied orbital ψ_j to an unoccupied one ψ_i:

$$\varepsilon_i(i{>}N/2) - \varepsilon_j(j{\le}N/2) \approx \Delta E_{j\to i} = h\nu_{j\to i}. \qquad (25.3a)$$

This relationship is found to be very approximate for transitions to the lowest-lying excited states, and is used in qualitative discussions, for instance, in the discussions on $\pi \leftrightarrow \pi^*$ transitions. It is certainly not correct. See Figure 5.4 for an illustration of the approximate relationship (25.3a). Actual energies for transitions to the low-lying excited states are known to be about half the difference of the respective orbital energies, or even smaller. A qualitatively correct result for a single-electron transition from a singlet state to another singlet state is of the form

$$\Delta E_{j\to i} = \varepsilon_i(i > N/2) - \varepsilon_j(j \leq N/2) - J_{ij} + 2K_{ij}. \qquad (25.3b)$$

But different excited singlet configurations interact with each other and they also interact with doubly- and triply-excited configurations. These interactions affect the energy of the excited state. This aspect will be discussed in a later lecture, in Part II of this text.

Figure 5.4 Interpreting the energy of a virtual orbital in terms of spectroscopic transition energy. The translational continuum is also shown to indicate the correspondence with Koopman's theorem.

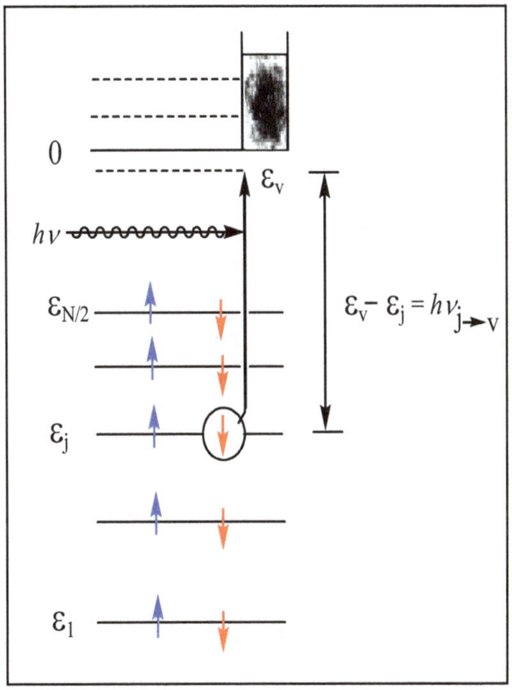

Shapes of Atomic Orbitals

The operator $\mathbf{h}(\mathbf{r})$ commutes with L^2 and L_Z. The operator for the average interaction energy in the Hartree-Fock ground state, $\sum_{j=1}^{N/2} (2\mathbf{J}_j - \mathbf{K}_j)$, also commutes with these operators when the total charge distribution is spherically symmetric — a situation that is generally found in atoms in their ground states. The spherically symmetric atom is a consequence of the isotropy of space. Gaseous atoms are known to have heat capacities C_V typical of only the translational degrees of freedom, indicating that rotational and vibrational characteristics are absent in isolated atoms. The Fock operator consists of the one-electron hamiltonian $\mathbf{h}(\mathbf{r})$ and the operator for the average interaction energy. Hence it too commutes with the operators L^2 and L_Z. This implies that it is possible to have common eigenfunctions of $\mathbf{F}$, L^2 and L_Z, and the eigenfunctions of $\mathbf{F}$ can be separated as $R_{nl}(r)Y_{lm}(\theta,\varphi)$. Thus the full orbital angular momentum symmetries can be retained in the orbitals of a many-electron atom, and these orbitals can be classified as 1s, 2s, $2p_{m=0,\pm1}$, 3s, $3p_{m=0,\pm1}$, $3d_{m=0,\pm1,\pm2}$, etc. Because of the degeneracy for the same (n,l) but different m, the atomic orbitals may also be chosen as 1s, 2s, $2p_x$, $2p_y$, $2p_z$, 3s, $3p_x$, $3p_y$, $3p_z$, $3d_{z^2}$, $3d_{x^2-y^2}$, $3d_{xy}$, $3d_{xz}$, $3d_{yz}$, and so on. The angular probability distributions remain same as those discussed in the case of the hydrogen atom.

As discussed in the case of the one-electron atom, the s orbitals have nonzero values and all the other orbitals have zero values at $r = 0$. Thus there is a finite though small probability of finding even the outermost s electron in the core region of space, near the nucleus. As the quantum number l increases from zero, the probability of finding the electron near the nucleus vanishes. Because of the repulsion by the core electrons (by electrons in the inner shells) and other inner electrons, and the constraint of orthogonality to the orbitals of lower (n,l) values, orbitals of the higher l values are progressively spread in the outward direction. The orthogonality constraint manifests itself in the form of an increasing centrifugal force that acts on an electron belonging to a high-l orbital. The s orbital is a penetrating orbital. The p electrons also penetrate but to a lesser extent. The d and f orbitals are considerably spread outside and are called diffused orbitals. The influence of the nuclear spin on the electronic distribution is evidenced in nuclear magnetic resonance (n.m.r.) and electron paramagnetic resonance (e.p.r.) spectroscopies. These influences are basically derived from the *contact interaction*, that is, the interaction that comes into play while the s electron remains in contact with the nucleus. The contact interaction of the valence s electron changes with the chemical environment, leading to chemical shifts and a splitting of the n.m.r. lines. It also causes hyperfine splitting of the e.p.r. spectrum.

Lecture 26: The Building Up of Elements

The periodic behavior of elements was known to chemists long before the birth of quantum theory. Mendeleev (1869, 1871) and Lothar Meyer (1869) organised elements along columns and rows (periods) according to similar chemical and physical properties and by following the smooth variation in these properties. The reason for the empirical arrangement of elements in the periodic table remained unclear until the methodology for the investigation of the electronic structure of atoms was established through the work of Schrödinger, Heisenberg, Dirac, Slater, Pauli, Hartree and Fock. The introduction of the concept of atomic orbitals by the quantum mechanical treatment of atomic electrons made it possible to understand the periodic behavior of elements.

Three important features of atomic orbitals came out of Hartree and Hartree-Fock treatments. *First*, atomic orbitals of same principal quantum number n and same orbital angular momentum quantum number l but different magnetic quantum number m are energetically degenerate. *Second*, the orbital energies for a many electron atom depend not only on the principal quantum number n but also on the azimuthal quantum number l. These energies follow the trend

$$\varepsilon_{1s} < \varepsilon_{2s} < \varepsilon_{2p} < \varepsilon_{3s} < \varepsilon_{3p} < \varepsilon_{3d} \approx \varepsilon_{4s} < \varepsilon_{4p} < \varepsilon_{4d} \approx \varepsilon_{5s} < \varepsilon_{5p} < \varepsilon_{4f} \approx \varepsilon_{5d} \approx \varepsilon_{6s}$$
$$< \varepsilon_{6p} < \varepsilon_{5f} \approx \varepsilon_{6d} \approx \varepsilon_{7s} < \varepsilon_{7p} \ldots \tag{26.1}$$

Third, the relative energies of the 3d and 4s orbitals follow a systematic pattern of change. Before the beginning of the first transition series, (for $Z < 20$), $\varepsilon_{3d} \approx \varepsilon_{4s}$. As Z increases, ε_{4s} decreases but ε_{3d} decreases at a faster rate. For $21 \leq Z \leq 30$ corresponding to the elements of the first transition series, ε_{3d} is clearly less than ε_{4s}. We will see later Hartree-Fock orbital energies correct through the third decimal point in atomic units for Z=1 – Z=36 in Table 5.4. After the first transition elements (for $Z > 30$), the energy gap ($\varepsilon_{4s} - \varepsilon_{3d}$) becomes very large and it increases at a fast rate. Similar trends are found with the 4d, 5s energies; the 4f, 5d, 6s energies; and the 5f, 6d, 7p energies.

The knowledge of the ordering of the orbital energies is not enough. One must also have a prescription for the way the electrons are to be assigned to these orbitals. The atomic orbitals may be filled with electrons by following Pauli's *exclusion principle* (Pauli, 1925) which states that no two electrons in the same atom may have the same quantum numbers n, l, m_l and m_s. Stated differently, the exclusion principle decrees that no two electrons may be put in the same spin orbital of an atom (or a molecule). When the spin orbitals are separable into spatial and spin functions, there cannot be more than two electrons sharing the same spatial orbital in an atom.

Using the energy ordering of atomic orbitals and the exclusion principle, we can systematically present the electronic configurations of elements in their ground states. The electronic configuration by default implies the electronic configuration (that is, orbital occupancy) in the ground state. The ground state electronic configurations for various elements are listed in the Table 5.1. It is anticipated that the reader is already familiar with the *building up* of atoms. Hence we restrict our discussion here only to the finer aspects of the *aufbau* principle.

Atomic orbitals of a general many-electron atom are approximately of the form of *Slater type orbitals* (STOs),

$$\psi_{nlm}(\mathbf{r}) = N_{nl}\, r^{n-1} \exp(-\zeta_{nl}\, r)\, Y_{lm}(\theta,\varphi). \tag{26.2}$$

The exponent ζ_{nl} is written as

$$\zeta_{nl} = Z_{nl}/n, \tag{26.3}$$

and the normalization constant N_{nl} is given by

$$N_{nl} = [(2n)!]^{-1/2}\,(2\zeta_{nl})^{n+1/2}. \tag{26.4}$$

The quantity Z_{nl} is the *effective nuclear charge* faced by the electron in the atomic orbital characterized by quantum numbers n and l (and m). It can be written as

$$Z_{nl} = Z - \sigma_{nl} \tag{26.5}$$

where σ_{nl} is the corresponding screening constant.

Actually the STOs may be the chosen single particle basis functions for the Hartree-Fock calculation. The resulting atomic orbitals are found as linear combinations of the STOs of the appropriate l and m quantum numbers. Nevertheless, in the outer region of space, the linear combination for a particular orbital is found to be well-approximated by the STO ψ_{nlm}. The inner components of the linear combination become more prominent in the inner region of space. They do not altogether vanish. The requirement that the higher energy orbitals must be orthogonal to all the lower energy ones forces the linear combination coefficients to have nonzero values.

Table 5.1 Electronic configurations of elements in their ground states.

Atomic Number	Atomic Symbol	Electronic Configuration	Name
1	H	1s	Hydrogen
2	He	$1s^2$	Helium
3	Li	[He]2s	Lithium
4	Be	$[He]2s^2$	Beryllium
5	B	$[He]2s^22p$	Boron
6	C	$[He]2s^22p^2$	Carbon
7	N	$[He]2s^22p^3$	Nitrogen
8	O	$[He]2s^22p^4$	Oxygen
9	F	$[He]2s^22p^5$	Fluorine
10	Ne	$[He]2s^22p^6$	Neon
11	Na	[Ne]3s	Sodium
12	Mg	$[Ne]3s^2$	Magnesium
13	Al	$[Ne]3s^23p$	Aluminum
14	Si	$[Ne]3s^23p^2$	Silicon
15	P	$[Ne]3s^23p^3$	Phosphorus
16	S	$[Ne]3s^23p^4$	Sulphur
17	Cl	$[Ne]3s^23p^5$	Chlorine
18	Ar	$[Ne]3s^23p^6$	Argon
19	K	[Ar]4s	Potassium
20	Ca	$[Ar]4s^2$	Calcium
First Transition Series begins			
21	Sc	$[Ar]3d4s^2$	Scandium
22	Ti	$[Ar]3d^24s^2$	Titanium
23	V	$[Ar]3d^34s^2$	Vanadium
24	Cr	$[Ar]3d^54s$	Chromium
25	Mn	$[Ar]3d^54s^2$	Manganese
26	Fe	$[Ar]3d^64s^2$	Iron
27	Co	$[Ar]3d^74s^2$	Cobalt
28	Ni	$[Ar]3d^84s^2$	Nickel
29	Cu	$[Ar]3d^{10}4s$	Copper
30	Zn	$[Ar]3d^{10}4s^2$	Zinc
First Transition Series ends			
31	Ga	$[Ar]3d^{10}4s^24p$	Gallium
32	Ge	$[Ar]3d^{10}4s^24p^2$	Germanium
33	As	$[Ar]3d^{10}4s^24p^3$	Arsenic
34	Se	$[Ar]3d^{10}4s^24p^4$	Selenium
35	Br	$[Ar]3d^{10}4s^24p^5$	Bromine
36	Kr	$[Ar]3d^{10}4s^24p^6$	Krypton

Table 5.1 Continued

Atomic Number	Atomic Symbol	Electronic Configuration	Name
37	Rb	[Kr]5s	Rubidium
38	Sr	$[Kr]5s^2$	Strontium
Second Transition Series starts			
39	Y	$[Kr]4d5s^2$	Yttrium
40	Zr	$[Kr]4d^25s^2$	Zirconium
41	Nb	$[Kr]4d^35s^2$	Niobium
42	Mo	$[Kr]4d^55s$	Molybdenum
43	Tc	$[Kr]4d^55s^2$	Technetium
44	Ru	$[Kr]4d^65s^2$	Ruthenium
45	Rh	$[Kr]4d^75s^2$	Rhodium
46	Pd	$[Kr]4d^85s^2$	Palladium
47	Ag	$[Kr]4d^{10}5s$	Silver
48	Cd	$[Kr]4d^{10}5s^2$	Cadmium
Second Transition Series terminates			
49	In	$[Kr]4d^{10}5s^25p$	Indium
50	Sn	$[Kr]4d^{10}5s^25p^2$	Tin
51	Sb	$[Kr]4d^{10}5s^25p^3$	Antimony
52	Te	$[Kr]4d^{10}5s^25p^4$	Tellurium
53	I	$[Kr]4d^{10}5s^25p^5$	Iodine
54	Xe	$[Kr]4d^{10}5s^25p^6$	Xenon
55	Cs	[Xe]6s	Cesium
56	Ba	$[Xe]6s^2$	Barium
Third Transition Series begins			
57	La	$[Xe]5d6s^2$	Lanthanum
First Inner Transition or Lanthanide Series starts			
58	Ce	$[Xe]4f^26s^2$	Cerium
59	Pr	$[Xe]4f^36s^2$	Praseodymium
60	Nd	$[Xe]4f^46s^2$	Neodymium
61	Pm	$[Xe]4f^56s^2$	Promethium
62	Sm	$[Xe]4f^66s^2$	Samarium
63	Eu	$[Xe]4f^76s^2$	Europium
64	Gd	$[Xe]4f^75d6s^2$	Gadolinium
65	Tb	$[Xe]4f^96s^2$	Terbium
66	Dy	$[Xe]4f^{10}6s^2$	Dysprosium
67	Ho	$[Xe]4f^{11}6s^2$	Holmium
68	Er	$[Xe]4f^{12}6s^2$	Erbium
69	Tm	$[Xe]4f^{13}6s^2$	Thulium
70	Yb	$[Xe]4f^{14}6s^2$	Ytterbium
71	Lu	$[Xe]4f^{14}5d6s^2$	Lutetium
Lanthanide Series terminates			
72	Hf	$[Xe]4f^{14}5d^26s^2$	Hafnium

Table 5.1 Continued

Atomic Number	Atomic Symbol	Electronic Configuration	Name
73	Ta	$[Xe]4f^{14}5d^{3}6s^{2}$	Tantalum
74	W	$[Xe]4f^{14}5d^{4}6s^{2}$	Tungsten
75	Re	$[Xe]4f^{14}5d^{5}6s^{2}$	Rhenium
76	Os	$[Xe]4f^{14}5d^{6}6s^{2}$	Osmium
77	Ir	$[Xe]4f^{14}5d^{7}6s^{2}$	Iridium
78	Pt	$[Xe]4f^{14}5d^{8}6s^{2}$	Platinum
79	Au	$[Xe]4f^{14}5d^{10}6s$	Gold
80	Hg	$[Xe]4f^{14}5d^{10}6s^{2}$	Mercury
Third Transition Series ends			
81	Tl	$[Xe]4f^{14}5d^{10}6s^{2}6p$	Thallium
82	Pb	$[Xe]4f^{14}5d^{10}6s^{2}6p^{2}$	Lead
83	Bi	$[Xe]4f^{14}5d^{10}6s^{2}6p^{3}$	Bismuth
84	Po	$[Xe]4f^{14}5d^{10}6s^{2}6p^{4}$	Polonium
85	At	$[Xe]4f^{14}5d^{10}6s^{2}6p^{5}$	Astatine
86	Rn	$[Xe]4f^{14}5d^{10}6s^{2}6p^{6}$	Radon
87	Fr	$[Rn]7s$	Francium
88	Ra	$[Rn]7s^{2}$	Radium
Fourth Transition Series begins			
89	Ac	$[Rn]6d7s^{2}$	Actinium
Second Inner Transition or Actinide Series starts			
90	Th	$[Rn]6d^{2}7s^{2}$	Thorium
91	Pa	$[Rn]5f^{2}6d7s^{2}$	Protactinium
92	U	$[Rn]5f^{3}6d7s^{2}$	Uranium
93	Np	$[Rn]5f^{4}6d7s^{2}$	Neptunium
94	Pu	$[Rn]5f^{6}7s^{2}$	Plutonium
95	Am	$[Rn]5f^{7}7s^{2}$	Americium
96	Cm	$[Rn]5f^{7}6d7s^{2}$	Curium
97	Bk	$[Rn]5f^{9}7s^{2}$	Berkelium
98	Cf	$[Rn]5f^{10}7s^{2}$	Californium
99	Es	$[Rn]5f^{11}7s^{2}$	Einsteinium
100	Fm	$[Rn]5f^{12}7s^{2}$	Fermium
101	Md	$[Rn]5f^{13}7s^{2}$	Mendelevium
102	No	$[Rn]5f^{14}7s^{2}$	Nobelium
103	Lr	$[Rn]5f^{14}6d7s^{2}$	Lawrencium
Actinide Series terminates			
104	Unq	$[Rn]5f^{14}6d^{2}7s^{2}$	Unnilquadium
105	Unp	$[Rn]5f^{14}6d^{3}7s^{2}$	Unnilpentium
106	Unh	$[Rn]5f^{14}6d^{4}7s^{2}$	Unnilhexium
107	Uns	$[Rn]5f^{14}6d^{5}7s^{2}$	Unnilseptium
Fourth Transition Series continues			

Slater's Rules

A list of the effective nuclear charge for a few light atoms is given in Table 5.2. These are obtained by using one STO for each occupied spatial orbital and by minimizing the computed Hartree-Fock total energy for the system. Slater (1930) proposed a few simple rules to estimate the screening constants σ_{nl} and the effective nuclear charge Z_{nl}. These rules yield results which are extremely good from a qualitative and conceptual standpoint, and they can be used as a mnemonic device for writing down the approximate ground state wave function for a light atom. These rules are as follows.

1) *Atomic orbitals are grouped into the subshells* 1s, (2s,2p), (3s,3p), 3d, (4s,4p),4d, 4f, *etc. Orbitals belonging to the same subshell have the same screening constant.* As ns and np belong to the same subshell (ns, np), they have the same screening constant. This is nearly correct. One may check from Table 5.2 that σ_{2s} and σ_{2p} for the second-row elements in the periodic table (first-row elements in the language of quantum chemistry) differ only in the third significant digit.
2) *Electrons in the same or the inner subshells screen the nucleus effectively whereas the screening by the charge distribution due to the electrons in the outer subshells is negligible.* This happens as the electrons in the outer sub- shells remain almost always outside. Hence the electronic population only in the concerned subshell and the inner subshells need be taken into account. Thus only one 1s electron shields the nuclear charge from another 1s electron. Screening of a 2s or a 2p electron obtains contributions from the 1s and the other (2s, 2p) electrons.
3) Contributions from various sources are evaluated by the following rules.
   - *For an electron belonging to the* (ns, np) *subshell, the screening constant receives a contribution of 0.35 for every other electron belonging to the same subshell, 0.85 for each electron in the next inner subshell of type* (s, p) *but 1.00 for each electron if the next inner subshell is of any other type, and 1.00 for each electron belonging to all the other inner subshells.*
   - *An exception is made in the case of the 1s electrons. Each 1s electron contributes 0.30 to the shielding of the nuclear charge such that the other 1s electron sees a net positive charge that is* $0.30|e|$ *less than the charge of the naked nucleus.*
   - *For electrons belonging to the* d, f, *etc. orbitals, the contribution is 0.35 for every other electron in the same subshell and 1.00 for each electron belonging to all the inner subshells.*

Example 22: Fluorine atom with the electronic configuration $1s^2 2s^2 2p^5$. Rule 1 prescribes the subshells (1s) and (2s, 2p). Rule 2 indicates that σ_{1s} receives only one contribution, that from the presence of the other 1s electron, whereas σ_{2s} and σ_{2p}, equal to each other, receive contributions from the two 1s electrons and six other electrons belonging to the (2s, 2p) subshell. By Rule 3 we get

$\sigma_{1s} = 0.30$ such that $Z_{10} = 9 - 0.30 = 8.70$;

$\sigma_{2s} = \sigma_{2p} = 2\times 0.85 + 6\times 0.35 = 3.80$ such that $Z_{20} = Z_{21} = 9 - 3.80 = 5.20$.

These estimates are in strikingly good agreement with the single-zeta Hartree-Fock results given in Table 5.2.

Example 23: Chromium that has the electronic configuration $1s^22s^22p^63s^23p^63d^54s$. The effective nuclear charges for the orbitals occupied in the ground state are worked out as follows:

$Z_{10} = 24 - 0.30 = 23.70;$

$Z_{20} = Z_{21} = 24 - (2\times 0.85 + 7\times 0.35) = 19.85;$

$Z_{30} = Z_{31} = 24 - (2\times 1 + 8\times 0.85 + 7\times 0.35) = 12.75;$

$Z_{32} = 24 - (2\times 1 + 8\times 1 + 8\times 1 + 4\times 0.35) = 4.6;$

$Z_{40} = 24 - (2\times 1 + 8\times 1 + 8\times 1 + 5\times 1) = 1.0.$

Note that the incomplete contribution of 0.85 to screening by the next inner electron when both the concerned electron and the inner one are s or p electrons represents the penetrating power of s and p orbitals. As Z increases, the discrepancy between the actual effective nuclear charge and the value estimated by Slater's rules increases.

An alternative definition of the screening parameter has been given in the following way. In the hydrogenic case the expectation values of the radial distance is found as

$$<r>_{nl}^{H} = n^2 a_0 Z^{-1}\left[1 + \frac{1}{2}\left(1 - \frac{l(l+1)}{n^2}\right)\right]. \tag{16.7}$$

One may adopt the same expression for many-electron atoms but with Z replaced by Z_{eff}. In that case one obtains

$$Z^{-1}(Z - \sigma_{nl}) = <r>_{nl} / <r>_{nl}^{H} \tag{26.6}$$

that gives

$$\sigma_{nl} = Z(1 - <r>_{nl} / <r>_{nl}^{H}). \tag{26.7}$$

This screening parameter has been determined by Froese-Fischer (1972, 1973, 1977) who solved the Hartree-Fock equations very accurately by a numerical method. The Hartree-Fock equations are in reality integro-differential equations which may be solved by numerical procedures.

Table 5.2 Effective nuclear charge from single-zeta Hartree-Fock calculations and from Slater's rules for the elements belonging to the first and the second rows of the periodic table.

Atom	Z	Computed Z_{eff}			Z_{eff} from Slater's rules	
		1s	2s	2p	1s	2s, 2p
He	2	1.6875			1.70	
Li	3	2.6906	1.2792		2.70	1.30
Be	4	3.6848	1.9120		3.70	1.95
B	5	4.6795	2.5762	2.4214	4.70	2.60
C	6	5.6727	3.2166	3.1358	5.70	3.25
N	7	6.6651	3.8474	3.8340	6.70	3.90
O	8	7.6579	4.4916	4.4532	7.70	4.55
F	9	8.6501	5.1276	5.1000	8.70	5.20
Ne	10	9.6421	5.7584	5.7584	9.70	5.85

The screening parameters thus obtained are compared with the screening constants from Slater's rules in Table 5.3. It is found that Eq.(26.7) shows a considerable deviation from the energy-based definition of the screening constants. For instance, the 2p orbitals are considerably diffused compared to the 2s orbitals in spite of what one would tend to believe from Slater's rules or from a merely single-zeta STO calculation.

Lecture 27: Trends of the Effective Nuclear Charge

From the last lecture we can note a few trends of the effective nuclear charge. We notice that the screening is never fully effective. For elements with the same valence subshell, the effective nuclear charge Z_{eff} increases as the nuclear charge Z increases. This happens along a general period in the periodic table. Along a column, Z_{eff} increases but at a considerably slow rate; in fact the increase is slower than the increase in the principal quantum number n. For example, Slater's rules predict the effective nuclear charges 1.0 for the 1s orbital of hydrogen, 1.3 for the 2s orbital of lithium, 2.2 for the 3s orbital of sodium, 2.2 for the 4s orbital of potassium and so forth. For the first transition series, Z_{eff} remains more or less unchanged for 4s and 4p electrons but it increases steadily for the 3d electrons as the nuclear charge Z increases. A similar pattern of change is observed with elements belonging to the second and third transition series and also with the inner transition elements. These three trends can account for the periodic behavior of elements in a handsome way.

We are now ready to discuss a few typical aspects. These discussions would be based on the estimation of the valence orbital energies by $\varepsilon \approx -Z_{eff,nl}{}^2/2n^2$ in atomic unit. Another energy parameter is of crucial importance. This is the energy of the coulombic repulsion among electrons. We know that the Hartree-Fock total energy is

Table 5.3 Comparison of the screening constants from Slater's rules (SR) with those from Hartree-Fock calculations employing single-zeta basis functions (STO) and the screening parameters from numerical Hartree-Fock calculations (HF).

Atomic Number	Element	SR	σ_{1s} STO	HF	SR	σ_{2s} STO	HF	SR	σ_{2p} STO	HF
2	He	0.30	0.313	0.382						
3	Li	0.30	0.309	0.383	1.70	1.721	1.451	1.70		
4	Be	0.30	0.315	0.385	2.05	2.088	1.735	2.05		
5	B	0.30	0.321	0.397	2.40	2.424	1.965	2.40	2.579	2.732
6	C	0.30	0.327	0.412	2.75	2.783	2.225	2.75	2.864	3.084
7	N	0.30	0.335	0.430	3.10	3.153	2.496	3.10	3.166	3.453
8	O	0.30	0.342	0.447	3.45	3.508	2.746	3.45	3.547	3.942
9	F	0.30	0.350	0.465	3.80	3.872	3.007	3.80	3.900	4.391
10	Ne	0.30	0.358	0.484	4.15	4.242	3.274	4.15	4.242	4.820

$$E = \sum_{i=1}^{N_C} (h_{ii} + \varepsilon_i) + f_O \sum_{i=N_C+1}^{N_C+N_O} (h_{ii} + \varepsilon_i) \tag{27.1}$$

where N_C is the number of the closed-shell orbitals, N_O is the number of orbitals in the partially occupied open shell, and f_O is the fractional occupancy of the open shell. The equation given above can be written as

$$E = 2[\sum_{i=1}^{N_C} \varepsilon_i + f_O \sum_{i=N_C+1}^{N_C+N_O} \varepsilon_i] - [\sum_{i=1}^{N_C} (\varepsilon_i - h_{ii}) + f_O \sum_{i=N_C+1}^{N_C+N_O} (\varepsilon_i - h_{ii})]. \tag{27.2}$$

Remember that $(\varepsilon_i - h_{ii})$ stands for the average repulsive potential energy faced by the electron in the *ith* orbital. This energy is comparatively large for the penetrating orbitals (s, p, etc.) and small for the diffused ones (d, f, etc.). Terms within the second set of parentheses in Eq.(27.2) are to be subtracted as the net electron-electron interaction energy has been counted twice while summing the orbital energies for the occupied orbitals.

The Aufbau principle

The building up pattern can be understood following the arguments in support of the least total energy. Of course the least energy value is obtained by populating the lowest-lying orbitals in accordance with the exclusion principle. This task is accomplished in the following steps.

1) First the s orbitals are to be filled. Electrons would enter into the np orbitals in the ground state of the atom only when the ns orbitals are fully occupied.
2) Orbitals in a degenerate set are at first filled by a single electron in each of them. The assignment of two electrons to different orbitals minimizes the coulomb repulsion energy; two electrons in the same spatial orbital would have the same spatial distribution and as a consequence, they would strongly repel each other. The orbitals are doubly filled only when there are additional electrons to enter them.
3) The filling in is accomplished in a way that the unpaired electrons have parallel spins. This creates the state of maximum spin multiplicity which is accompanied by the greatest magnitude of exchange energy that generally appears as a negative contribution to total energy. Thus this state corresponds to the minimum energy. Hund (1925, 1931, 1932) proposed the rule that the ground state must be of maximum spin multiplicity.
4) Consider the pattern of orbital occupation in the transition elements. The popular perception is that 3d orbitals lie higher in energy than 4s orbital but much lower than 4p orbitals, and so 3d orbitals are filled right after 4s orbital is fully occupied – a process that leads to the transitional properties until the 3d level is completely filled. Contrary to the popular belief, even for scandium that heralds the beginning of the first transition series the 3d Hartree-Fock orbital energy (–0.344 a.u.) is actually lower than the 4s

orbital energy (– 0.210 a.u.). Table 5.4 makes it abundantly clear that the same situation exists for all transition elements. The one-electron energy h_{ii} is of course considerably lower for 4s. The Coulomb energy for the repulsion of the valence electron by the core electrons (electrons in 1s, 2s, 2p, 3s, and 3p orbitals) is much larger for 4s orbital as the s orbitals are much more penetrating than the diffused d orbitals. Whichever way we may look, at Eq. (27.1) or at Eq.(27.2), the total energy is seen to be lower when the valence electrons first occupy 4s orbital and then 3d orbitals. This is certainly not the same as saying that the 4s orbital has less energy than the 3d orbitals and therefore the electron enters 4s at first! The 4s orbital has less one-electron energy but greater orbital energy than the 3d orbitals even at scandium (Z = 21). The electron-electron interaction is clearly at play, and the explanation is not as simple and straight as the popular myth would lead us to believe. The myth propagates because it has been inadvertently put into several great texts of inorganic and physical chemistry, and it becomes apparent only when one consults the results of rigorous calculations on the electronic structure of atoms.

5) As Z increases the effective nuclear charge seen by both the 3d and the 4s electrons go on increasing, but Z_{eff} for the 3d electrons increases at a faster rate. As a result both the 3d and the 4s orbital energies decrease with an increasing Z, but the 3d orbital energy slides down faster (Table 5.4). At copper (Z = 29) the energy gap becomes so large that the electrons would enter the 3d orbitals first and then the 4s orbital, and the 3d set clearly becomes a part of the inner shell.
6) For the positively charged ions of the transition metals, the 3d orbitals are much lower in energy and they are filled first. Electronic configurations of Fe^{2+} and Fe^{3+} in the ground state are $3d^6$ and $3d^5$ respectively.
7) The electronic configuration of chromium is $3d^54s$ in lieu of $3d^44s^2$ and that of copper is $3d^{10}4s$ instead of $3d^94s^2$. For half-filled shells (and also for closed shells) the total orbital angular momentum becomes zero. Half-filled shells are exchange-stabilized with the maximum spin multiplicity. They are thus especially stable configurations, such as $3d^54s^1$ for Cr. Closed shells are exceedingly stable since the removal of one electron from the closed shell would require the electron to be put back into an orbital of a larger principal quantum number and a greater orbital energy.
8) Trends similar to those discussed so far are also observed in the case of other transition series and the inner transition elements.

The orbital energies given in Table 5.4, though truncated through three or four significant digits, correspond to a very accurate solution of the Hartree-Fock equations for the ground state atoms. Such accurate solutions were first obtained by Froese-Fischer (1972) from a numerical Hartree-Fock calculation, and later by H. Partridge (1989) who relied on analytical basis sets (using Gaussian functions).

The Aufbau approach manifests the shell structure of atoms. Two electrons in 1s orbital completes the K shell; the configuration $2s^22p^6$ completes the L shell; after the first transition series ends and before the second transition series starts, the M shell becomes completely filled at krypton with a total of 18 electrons in it corresponding to the electronic configuration $3s^2 3p^6 3d^{10}$.

The shell structure becomes apparent when one plots the total radial probability density in the ground state,

$$P_{tot}(r) = \sum_{n=1}^{\infty} \sum_{l=0}^{n-1} n_{nl} \left(r^2 R_{nl}^2(r)\right), \tag{27.3}$$

Table 5.4 Orbital energies from Hartree-Fock calculations on isolated atoms (up to Z=36). Valence orbital energies are in bold.

Z	Element	Orbital energy in atomic unit							
		1s	2s	2p	3s	3p	3d	4s	4p
1	H	**−0.500**							
2	He	**−0.918**							
3	Li	−2.478	**−0.196**						
4	Be	−4.733	**−0.309**						
5	B	−7.695	**−0.495**	**−0.310**					
6	C	−11.33	**−0.706**	**−0.433**					
7	N	−15.63	**−0.945**	**−0.568**					
8	O	−20.67	**−1.244**	**−0.632**					
9	F	−26.38	**−1.573**	**−0.730**					
10	Ne	−32.77	**−1.930**	**−0.850**					
11	Na	−40.48	−2.797	−1.518	**−0.182**				
12	Mg	−49.03	−3.768	−2.282	**−0.253**				
13	Al	−58.50	−4.911	−3.218	**−0.393**	**−0.210**			
14	Si	−68.81	−6.157	−4.256	**−0.540**	**−0.297**			
15	P	−79.97	−7.511	−5.401	**−0.696**	**−0.392**			
16	S	−92.00	−9.004	−6.683	**−0.880**	**−0.437**			
17	Cl	−104.9	−10.61	−8.072	**−1.073**	**−0.506**			
18	Ar	−118.6	−12.32	−9.571	**−1.277**	**−0.591**			
19	K	−133.5	−14.49	−11.52	−1.749	−0.954		**−0.147**	
20	Ca	−149.4	−16.82	−13.63	−2.245	−1.341		**−0.196**	
21	Sc	−165.9	−19.08	−15.67	−2.567	−1.575	**−0.344**	**−0.210**	
22	Ti	−183.3	−21.42	−17.79	−2.873	−1.795	**−0.441**	**−0.221**	
23	V	−201.5	−23.87	−20.02	−3.183	−2.019	**−0.510**	**−0.231**	
24	Cr	−220.6	−26.44	−22.37	−3.500	−2.249	**−0.569**	**−0.240**	
25	Mn	−240.5	−29.11	−24.81	−3.817	−2.480	**−0.639**	**−0.248**	
26	Fe	−261.4	−31.94	−27.41	−4.169	−2.742	**−0.647**	**−0.258**	
27	Co	−283.7	−34.87	−30.12	−4.524	−3.006	**−0.675**	**−0.267**	
28	Ni	−305.6	−37.92	−32.94	−4.888	−3.278	**−0.707**	**−0.276**	
29	Cu	−329.0	−41.08	−35.88	−5.261	−3.557	**−0.740**	**−0.285**	
30	Zn	−353.3	−44.36	−38.92	−5.638	−3.839	**−0.783**	**−0.293**	
31	Ga	−378.8	−48.17	−42.49	−6.395	−4.482	−1.193	**−0.425**	**−0.208**
32	Ge	−405.2	−52.15	−46.24	−7.191	−5.162	−1.635	**−0.553**	**−0.287**
33	As	−432.6	−56.31	−50.15	−8.030	−5.881	−2.113	**−0.686**	**−0.369**
34	Se	−460.9	−60.67	−54.27	−8.932	−6.662	−2.650	**−0.837**	**−0.403**
35	Br	−490.1	−65.20	−58.55	−9.872	−7.478	−3.220	**−0.993**	**−0.457**
36	Kr	−520.2	−69.90	−63.01	−10.85	−8.332	−3.825	**−1.153**	**−0.524**

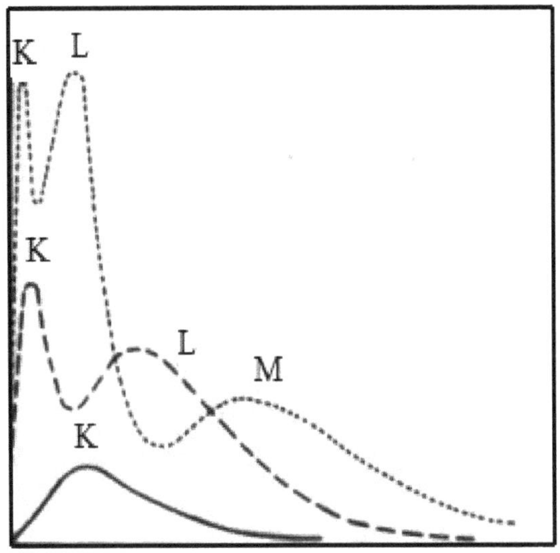

Figure 5.5 Shell structures from atomic Hartree-Fock calculations. The total radial probability density is indicated by the bold line for Li^+ that has electrons only in the K shell, by the dashed line for Na^+ having electrons in both K and L shells, and by the small-dashed line for K^+ with fully occupied K and L shells and a partly filled M shell. The radial distance is plotted along the x-axis, and $P_{tot}(r)$ along the y-axis, both in arbitrary scale.

where n_{nl} is the number of electrons in the subshell (n, l). Both Hartree and Hartree-Fock calculations yield results which may be used to illustrate the shell structure.

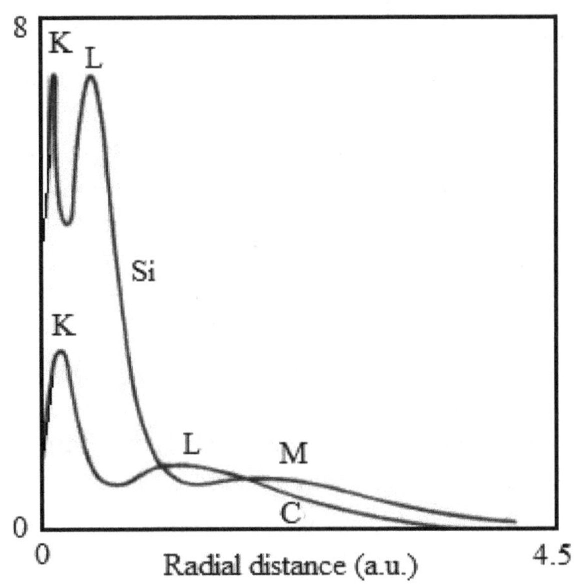

Figure 5.6 Shell structures from atomic Hartree-Fock calculations on carbon and silicon.

The Hartree-Fock shell structures of the mono-positive ions of alkali metals are shown in Figure 5.5. Figure 5.6 shows the shell structures for the elements carbon and silicon. Both elements are known to form strong covalent bonds with other atoms. On the experimental side, atomic spectroscopy (uv-visible) and X-ray spectroscopy have lent credence to the existence of shell structures.

Lecture 28: Ionization Energy, Electron Affinity, Atomic Radius and the Diagonal Rule

The first ionization potentials are smallest for the Group I elements (the alkali metals) and largest for the Group VIII elements (the noble gases). In general it increases along a period – a general consequence of the increase of the effective nuclear charge Z_{eff}, (see Table 5.2). Nevertheless, bumps are observed when the l quantum number of the valence orbital increases with the accompaniment of a general rise in the valence orbital energy. This happens at boron. Beryllium has the electronic configuration $2s^2$. For boron, the valence orbitals are 2p orbitals which are partially filled. Although the effective nuclear charge increases from beryllium to boron, the larger l quantum number makes the 2p orbital energy of boron almost the same as the 2s orbital energy of beryllium (see Table 5.4). By Koopmans' theorem beryllium and boron would have the same ionization potential. But Koopmans' theorem is only approximately valid as orbital relaxation effects cannot be completely neglected. In fact boron has a slightly smaller first ionization potential. The first few ionization potentials of elements up to $Z = 89$ are given in Table 5.5. Kinks are also found when the configuration changes from a considerably stable half-filled shell. The valence shell 2p is half-filled in nitrogen. The 2p orbital energy for oxygen, –0.632 a.u., is less than the 2p orbital energy for nitrogen, –0.568 a.u. Yet the exceptional stability of nitrogen relative to the positive ion N^+ and the exceptional stability of O^+ relative to oxygen atom makes the ionization potential of oxygen smaller than that of nitrogen. The variation of ionization energy with Z is shown in Figure 5.7. Closed shell atoms of the noble gases have very low valence orbital energies and very high ionization potentials. Alkali metal atoms have one additional electron in an outer orbital with a larger principal quantum number and a significantly high orbital energy. These are characterized by very small ionization potentials. In fact these elements have the smallest ionization energies. They can easily ionize and the energy required to do so is greatly surpassed by the energy liberated while an atom of another element turns into a negative ion by acquiring the liberated electron. This leads to a great decrease in the total energy of the reactants and the alkali metals derive their typical properties. Remember that *chemistry is always a subject of competitive processes*: when one element turns into a positive ion by absorbing energy, another turns into a negative ion and liberates energy, and there is a tug of war between the two elements. The winner of an electron is decided by whether a net amount of energy is released or not. Halogen atoms have very high ionization potentials, and their negative ions are exceptionally stable owing to their closed shell configurations. Halogens are among the most electronegative elements.

Along a column in the periodic table the valence orbital n value increases but Z_{eff} increases only slightly, and the overall result is an increase in the valence orbital energy. The ionization potential decreases downward in a group. Cesium is known to be the most electropositive among the naturally occurring elements. Fluorine is the least electro-positive halogen and the most electronegative element.

There can be a second, a third, etc. ionization energies corresponding to the liberation of one electron from the mono-positive ion, the dipositive ion, etc. See Table 5.5. It is also possible to define the orbital ionization energy corresponding to the loss of one electron from a specific orbital of a neutral atom. The orbital ionization energies are usually obtained from the interpretation of the photoelectron spectra of atoms and molecules. The photoelectron spectra really give information on the *state ionization* energies which, sometimes, can be approximately interpreted as the orbital ionization energies.

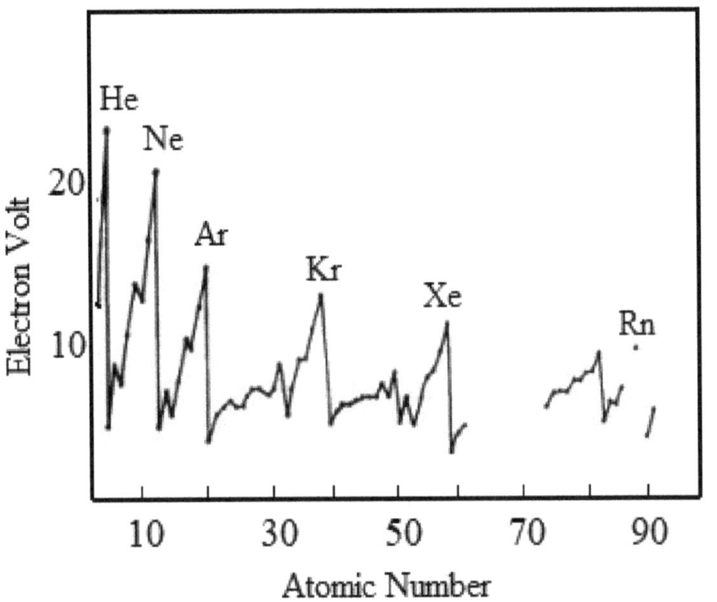

Figure 5.7 Variation of ionization potential with Z.

The successive ionization energies I_1, I_2, I_3 and so on can be used to calculate the total electronic energy of an isolated atom. These ionization energies are determined for gaseous atoms at extremely low pressures so that these atoms may be considered effectively isolated from each other and also isolated from other species such as the atoms or molecules of the container or even the impurity atoms or molecules in the gas.

Using Table 5.5, one can calculate total electronic energy for small atoms and total valence electronic energy for larger atoms.

Example 24: Calculation of total electronic energy
$E(\text{H}) = -I_1 = -13.605$ eV $= -0.500$ a.u.
$E(\text{He}) = -(I_1 + I_2) = -79.00$ eV $= -2.903$ a.u.
$E(\text{Li}) = -(I_1 + I_2 + I_3) = -203.5$ eV $= -7.480$ a.u.
$E(\text{Be}) = -(I_1 + I_2 + I_3 + I_4) = -399.3$ eV $= -14.673$ a.u.

Example 25: Calculation of *total valence electronic energy*
$E_V(\text{K}) = E(\text{K}) - E(\text{K}^+) = -I_1 = -4.343$ eV $= -0.160$ a.u.
$E_V(\text{Ca}) = E(\text{Ca}) - E(\text{Ca}^{2+}) = -(I_1 + I_2) = -17.99$ eV $= -0.661$ a.u.
$E_V(\text{Sc}) = E(\text{Sc}) - E(\text{Sc}^{3+}) = -(I_1 + I_2 + I_3) = -44.23$ eV $= -1.625$ a.u.
$E_V(\text{Ti}) = E(\text{Ti}) - E(\text{Ti}^{4+}) = -(I_1 + I_2 + I_3 + I_4) = -91.91$ eV $= -3.378$ a.u.

It will be instructive to compare these numbers with those estimated from Slater's rules.

Table 5.5 Approximate ionization energies of gaseous atoms in electron volts.

Atomic Number	Element	I_1	I_2	I_3	I_4
1	H	13.6009			
2	He	24.5944	54.4035		
3	Li	5.3926	75.6616	122.4729	
4	Be	9.3232	18.2169	153.9262	217.7874
5	B	8.2993	25.1627	37.9393	259.4360
6	C	11.2625	24.3904	47.8959	64.5119
7	N	14.5380	29.6226	47.4620	77.4837
8	O	13.6226	35.1670	54.9675	77.4403
9	F	17.4317	34.9978	62.6898	87.2885
10	Ne	21.5705	41.0933	65.0759	97.2234
11	Na	5.141	47.3319	71.6703	98.9154
12	Mg	7.6486	15.0369	80.1735	109.3709
13	Al	5.9870	18.8330	28.4555	120.0434
14	Si	8.1518	16.3471	33.4794	45.1627
15	P	11.0195	19.6616	30.1735	51.3666
16	S	10.3644	23.4273	35.0108	47.3319
17	Cl	13.0512	23.8134	39.9219	53.3623
18	Ar	15.7657	27.6356	40.9241	59.8265
19	K	4.3427	31.8265	47.7223	60.9554
20	Ca	6.1128	11.8785	51.2364	67.2451
21	Sc	6.5640	12.8980	24.7636	73.7527
22	Ti	6.8547	13.6356	28.1562	43.2625
23	V	6.7245	14.2299	29.7180	47.7223
24	Cr	6.7679	16.4989	30.9675	49.4577
25	Mn	7.4360	15.6486	33.7093	
26	Fe	7.8959	16.1909	30.6594	
27	Co	7.8525	17.0586	33.5098	
28	Ni	7.6356	18.1605	35.1800	
29	Cu	7.7267	20.2993	36.8503	
30	Zn	9.3970	17.9696	39.7223	
31	Ga	5.9870	20.5206	30.7158	64.2082
32	Ge	7.8959	15.9393	34.2299	45.5531
33	As	9.8048	20.2169	28.3297	50.3254
34	Se	9.7614	21.5184	32.0174	42.9067
35	Br	11.8438	21.6052	35.9219	
36	Kr	14.0043	24.5727	36.9197	

1 eV = 23.06 kcal mol^{-1}.

Table 5.5 Continued ...

Atomic Number	Element	I_1	I_2	I_3	I_4
37	Rb	4.1783	27.5054	39.9132	
38	Sr	5.6963	11.0325	56.3991	
39	Y	6.3774	12.2386	20.5206	
40	Zr	6.8547	13.1367	22.9935	34.3514
41	Nb	6.8850	14.3297	25.1540	38.3080
42	Mo	7.1150	16.1605	27.1453	46.4208
43	Tc	7.2885	15.2668		
44	Ru	7.3666	16.7679	28.4772	
45	Rh	7.4620	18.0781	31.0672	
46	Pd	8.3297	19.4317	32.9371	
47	Ag	7.5792	21.4924	34.8416	
48	Cd	8.9978	16.9154	37.4924	
49	In	5.7874	18.8720	28.0477	54.2299
50	Sn	7.3449	14.6377	30.5076	40.7419
51	Sb	8.6421	16.49	25.29	44.25
52	Te	9.0239	18.6117	31.2364	38.1779
53	I	10.4599	19.1019		
54	Xe	12.1345	21.2147	32.1041	
55	Cs	3.8950	25.1193		
56	Ba	5.2147	10.0087		
57	La	5.5965	11.4360	19.1800	
72	Hf	6.9419	14.9241		
73	Ta	7.8959	16.2256		
74	W	7.9826	17.7006		
75	Re	7.8959	16.6160		
76	Os	8.6768	16.9197		
77	Ir	8.6768			
78	Pt	9.1106	185.6833		
79	Au	9.2408	20.5206		
80	Hg	10.4338	18.7636	34.2299	
81	Tl	6.1085	20.4295	29.8048	50.7592
82	Pb	7.4186	15.0369	31.9479	42.3384
83	Bi	7.2928	16.6898	25.5748	45.1193
84	Po	8.4165			
85	At				
86	Rn	10.7505			
87	Fr				
88	Ra	5.2798	10.1518		
89	Ac	6.9414	12.1041		

Example 26: Estimates of ionization energies of lithium

$I_1 = \frac{1}{8} \times 1.3^2 = 0.211$ a.u. = 5.748 eV; (Experimental: 5.393 eV)

$I_2 + I_3 = 2 \times \frac{1}{2} \times 2.7^2 = 7.290$ a.u. = 198.36 eV.

But $I_3 = \frac{1}{2} \times 3^2 = 4.5$ a.u. = 122.4 eV; (122.473 eV)

Therefore, $I_2 = 2.790$ a.u. = 75.92 eV; (75.662 eV)

Besides, $E = -(I_1 + I_2 + I_3) = -7.501$ a.u. = –204.1 eV; (–203.5 eV).

Thus the total electronic energy estimated by a careful application of Slater's rules is in very good agreement with that measured by experiments. The difference of energy for lithium atom is found within 0.3%. However, the agreement found in this case is largely fortuitous. In general Slater's rules are found to yield results which are only qualitatively correct.

Electron affinity

The electron affinity is defined as the minimum amount of energy required to remove an electron from an isolated (gaseous, at low pressure) negative ion such that the neutral species is left in its ground state, as in

$$F^-(g) = F(g) + e^-(g). \qquad (28.1)$$

By this definition the electron affinity would be a positive quantity. In the older literature it used to be defined as the maximum energy absorbed for an isolated neutral atom or molecule to combine with an electron thus forming an isolated negative ion, both the neutral and charged species being in their respective ground states. As the mono-negative ions are more stable than the neutral species, energy is actually evolved while the ion is formed and the electron affinity was treated as a negative quantity. The maximum energy absorbed is equal to the negative of the minimum energy evolved and it corresponds to zero kinetic energy for the free electron.

Halogens, the elements of Group VII A, are enormously stabilized by turning into negative ions with noble gas electronic configurations. These elements have very high electron affinities. They are also known to be extremely electronegative elements. Electronegativity decreases downward in a group. Fluorine is the most electronegative element known. Alkali metals have very low electron affinities. Rare gases have nearly zero electron affinities.

Just as one may assign ionization energies to orbitals and states, one can assign orbital electron affinities and state electron affinities. The determination of electron affinities of elements is much more difficult than the measurement of ionization potentials. This is because of the intricate complexity of the process of preparing gaseous negative ions. Often ions with different charges, molecular fragments and clusters are produced. It is also difficult to separate ions of various types. It is extremely difficult to hold ions of the same type together in a considerably large concentration. Electron affinities of a few elements are given in Table 5.6.

Table 5.6 Electron affinities of gaseous atoms in electron Volt.

Atomic Number	Element	Electron affinity	Atomic Number	Element	Electron affinity
Experimentally determined					
1	H	0.755	35	Br	3.35
6	C	1.26	47	Ag	1.31
8	O	1.48	53	I	3.06
9	F	3.45	78	Pt	2.13
11	Na	0.542	79	Au	2.30
13	Al	0.521	Calculated from quantum mechanics		
16	S	2.08	3	Li	0.61
17	Cl	3.62	5	B	0.30
19	K	0.499	7	N	0.039
29	Cu	1.23	14	Si	1.4
34	Se	2.02	15	P	0.78

Atomic Radius

As the effective nuclear charge increases, the orbitals and so the electrons become squeezed towards the nucleus. Following the trend of Z_{eff}, atomic radius generally decreases from left to right along a period. It increases downward along a group in the periodic table as the elements in this direction have successively outer orbitals as valence orbitals, and though both n and Z_{eff} increase, the ratio n^2/Z_{eff} continues to increase. See Eq. (16.7). The ionic radius exhibits a slightly different trend. Like the atomic radius it increases downward along a column. One may see this trend by comparing the radii of the alkali metal ions in Table 5.7. But it decreases sharply along a period: compare the ionic radii of Li^+, Be^{2+} and B^{3+}. The sharp decrease is the result of the increase in Z_{eff} by almost 1 from one group to the next, as in ordinary cases the stable ions have invariably the noble gas configuration.

Note that the transition elements and their ions have slowly varying ionization potential and ionic radius. They exhibit a smooth transition of physical and chemical properties. The 3d orbital energy is very much less than the 4s energy for ions of the first transition series. So these ions have valence electrons decidedly in the d orbitals. This is why the transition metals generally exhibit variable valency.

Elements of the inner transition series exhibit a remarkable trend. For these elements the orbitals being filled are 4f orbitals (lanthanides) or 5f orbitals (actinides). As Z increases the f orbitals are progressively filled and Z_{eff} increases. After the completion of the series the net increase in Z_{eff} is so large that the elements following an inner transition series do not show a marked increase in atomic radius over the previous elements of the respective groups. Zinc has an atomic radius of $1.25\mathring{A}$ and the radius of cadmium, the next element in group IIB, is larger by $0.16\mathring{A}$. But mercury that belongs to the same group and follows the rare earth series has about the same atomic radius as cadmium.

The progressive decrease of the atomic and the ionic size along the lanthanide series is known as the *lanthanide contraction.* This influences the chemical properties of elements belonging to the third transition series in certain ways. In Table 5.7, figures in bold before and after the lanthanide series can be compared to directly visualize the lanthanide contraction.

The diagonal rule

Successive elements along the direction of the upper left-lower right diagonal in the periodic table have somewhat similar chemical properties. The sets of elements (i) Be and Al, (ii) B, Si, As, Te and At, (iii) W, S and Br, (iv) O and Cl, etc., are found as examples. That a similar properties persist along the diagonal is referred to as the diagonal rule. All elements along a particular diagonal axis have comparable electropositivity and electronegativity. This is a consequence of the simultaneous increase in n and Z, and a further increase in Z_{eff} due to the lateral displacement along the lower period. The changes in n and Z_{eff} also make the diagonally placed elements of approximately same size. The diagonal rule is merely a qualitative rule – it holds only approximately. For instance, it indicates that the N...H hydrogen bond is about as strong as the S...H hydrogen bond although nitrogen is known to be predominantly in the oxidation state -III whereas sulphur is better known for its oxidation state -II.

Table 5.7 Atomic and ionic radii in Angstroms for tetrahedral coordination. Figures in parentheses are the average radii in metallic crystals (atomic radii) or in NaCl type alkali halide crystals (ionic radii).†

Atom	Radius	Ion	Radius	Atom	Radius	Ion	Radius
Li	(1.55)	Li^+	0.68 (0.94)	Ti	(1.32)	Ti^{2+}	0.90
Be	1.06 (0.89)	Be^{2+}	0.30			Ti^{4+}	0.60
B	0.88 (0.80)	B^{3+}	0.16	V	(1.22)	V^{2+}	0.88
C	0.77			Cr	(1.17)	Cr^{2+}	0.84
N	0.70			Mn	(1.17)	Mn^{2+}	0.80
O	0.66	O^{2-}	1.46	Fe	(1.16)	Fe^{2+}	0.76
F	0.64	F^-	1.33 (1.16)	Co	(1.16)	Co^{2+}	0.74
Ne	1.58 (1.90)			Ni	(1.15)	Ni^{2+}	0.72
		Na^+	0.98 (1.24)	Cu	1.35 (1.17)	Cu^+	0.96
Mg	1.40 (1.36)	Mg^{2+}	0.65	Zn	1.31 (1.25)	Zn^{2+}	0.83
Al	1.26 (1.25)	Al^{3+}	0.45	Ga	1.26 (1.25)	Ga^{3+}	0.62
Si	1.17	Si^{4+}	0.38	Ge	1.22	Ge^{4+}	0.44
P	1.10			As	1.18	As^{3+}	0.69
S	1.04	S^{2-}	1.90	Se	1.14		
Cl	0.99	Cl^-	1.81	Br	1.11	Br^-	1.95 (1.76)
Ar	1.88			Kr	2.00		
K	(2.53)	K^+	1.33 (1.54)	Rb	(2.48)	Rb^+	1.48 (1.68)
Ca	(1.74)	Ca^{2+}	0.94	Sr	(1.91)	Sr^{2+}	1.10
Sc	(1.44)	Sc^{3+}	0.68				

(Table 5.7. continued)†

Atom	Radius	Ion	Radius		
		Y^{3+}	**0.88**		
Zr	(**1.45**)	Zr^{4+}	0.77		
Nb	(**1.34**)	Nb^{5+}	0.67		
Mo	(**1.29**)				
Ru	(**1.24**)				
Rh	(**1.25**)				
Pd	(**1.28**)	Pd^{2+}	0.86		
Ag	1.52 (**1.34**)	Ag^{+}	1.13		
Cd	1.48 (**1.41**)	Cd^{2+}	1.03		
In	1.44 (1.50)	In^{3+}	0.92		
Sn	1.40	Sn^{4+}	0.74		
Sb	1.36	Sb^{3+}	0.90		
Te	1.32				
I	1.28	I^{-}	2.16 (1.97)		
Xe	2.17				
Cs	(2.67)	Cs^{+}	1.67 (1.83)		
Ba	(1.98)	Ba^{2+}	1.29		
La	(1.69)	La^{3+}	1.06		
		Ce^{3+}	1.03	**L**	
		Ce^{4+}	0.92	**a**	
		Pr^{3+}	1.01	**n**	
		Nd^{3+}	1.00	**t**	
		Pm^{3+}	0.98	**h**	**c**
		Sm^{3+}	0.96	**a**	**o**
		Eu^{3+}	0.95	**n**	**n**
		Gd^{3+}	0.94	**i**	**t**
		Tb^{3+}	0.92	**d**	**r**
		Dy^{3+}	0.91	**e**	**a**
		Ho^{3+}	0.89		**c**
		Er^{3+}	0.88		**t**
		Tm^{3+}	0.87		**i**
		Yb^{3+}	0.86		**o**
		Lu^{3+}	**0.85**		n
Hf	(**1.44**)				
Ta	(**1.34**)				
W	(**1.30**)				
Re	(**1.28**)				
Os	(**1.26**)				
Ir	(**1.26**)				
Pt	(**1.29**)				
Au	(**1.34**)	Au^{+}	1.37		
Hg	**1.48** (**1.44**)	Hg^{2+}	1.12		
Tl	(1.55)	Tl^{3+}	1.05		

(Table 5.7. continued)†

Pb^{4+}	0.84
Fr^{+}	1.75
Ra^{2+}	1.37
Ac^{3+}	1.11
Th^{4+}	0.99
Pa^{5+}	0.90
U^{4+}	1.05
U^{6+}	0.83

† M. P. Tosi and F. G. Fumi, J. Phys. Chem. Solids, **25**, 31, 45 (1964).

Lecture 29: The Density Functional Approach

We have already encountered the quantitative aspects of the electronic structure of atoms up to the level of the mean field approximations. We now discuss an interesting feature of atomic structure. Our objectives here are to make a crude estimate of the electronic energy of an atom in order to gain a qualitative understanding of the electronic structure, and finally to discuss a few quantitative aspects.

Example 27: The Crude Atom

The total electronic energy of a closed shell atom is seen to be approximately proportional to $Z^{7/3}$. It will be nice to understand why this is so.

Consider an atom with N closed shells. In average each electron in the *nth* closed shell has energy $-Z^2/2n^2$ in atomic units. There are $2n^2$ electrons in this shell, and all of them together contribute $-Z^2$ a.u. Hence for N closed shells,

$$E = -\mathrm{N}Z^2. \tag{29.1}$$

Because of the electroneutrality of the atom,

$$Z = \sum_{n=1}^{\mathrm{N}} 2n^2 = \frac{1}{3}\mathrm{N}(\mathrm{N}+1)(2\mathrm{N}+1) \tag{29.2}$$

and for $\mathrm{N} >> 1$, the atomic number Z must be much larger than 1 and one obtains from Eq.(29.2) the approximate relationship

$$\mathrm{N} \approx \left(\frac{3}{2}\right)^{\frac{1}{3}} Z^{1/3}. \tag{29.3}$$

If we substitute this result in Eq. (29.1) we get

$$E \approx -1.144 Z^{7/3}. \tag{29.4}$$

This establishes the $Z^{7/3}$ dependence of total energy. However, the proportionality constant is known to be about –0.6, significantly different from –1.14 in Eq.(29.4). The reason for this difference is that during the estimation of energy we have neglected the mutual repulsion of electrons. The electron-electron interaction contributes a positive term to total energy, and thus the absolute magnitude of the actual proportionality constant should be considerably less than 1.14.

Example 28: The Thomas-Fermi model

A better estimation of the total electronic energy for both closed-shell and open-shell atoms is made by a simple statistical model due to Thomas (1927) and Fermi (1927, 1928). This is a qualitative model of atom. In this model the Coulomb potential $-Ze^2/r$ is replaced by the screened potential

$$V(\mathbf{r}) = - Z_{SC}(r)\frac{e^2}{r} \tag{29.5}$$

where $Z_{SC}(r) \to Z$ as $r \to 0$, and $Z_{SC}(r) \to 0$ as $r \to \infty$. The screened potential varies slowly in an electron wavelength, and many electrons can be localized within a volume over which $V(r)$ changes only by a little amount. The electrons are treated by Fermi-Dirac statistics. At an ordinary temperature the thermal energy k_BT is very small compared to $V(r)$ except at the edge of the atom. The single particle states are then filled in order of increasing energy up to the Fermi level that is characterized by Fermi momentum $p_F = \hbar k_F$ and Fermi energy $\varepsilon_F = p_F^2/2m$.

We have seen earlier from the box normalization of the free particle wave functions that there is one k value for each $2\pi/L$ in one dimension. There is one $\boldsymbol{k}$ value for each $(2\pi/L)^3$ in the three-dimensional space. The number of states below and up to the momentum p_F is

$$N_F = 2(L^3/8\pi^3)\int_0^{p_F/\hbar} dk\, k^2 \int_0^{\pi} d\theta \sin\theta \int_0^{\pi} d\varphi = \frac{p_F^3 L^3}{3\pi^2\hbar^3} \tag{29.6}$$

where we have considered a factor of 2 to account for the two possible spin states for each $\boldsymbol{k}$ mode. This result yields the number density

$$n = \frac{N_F}{L^3} = \frac{p_F^3}{3\pi^2\hbar^3}. \tag{29.7}$$

The maximum kinetic energy at any radial distance r is $-V(r)$. If we put

$$p_F^2/2m = -V(r) \tag{29.8}$$

in Eq. (29.7), we get the number density at radial distance r as a function of the screened potential $V(\mathbf{r})$,

$$n(\mathbf{r}) = \frac{[-2mV(\mathbf{r})]^{3/2}}{3\pi^2\hbar^3}. \tag{29.9}$$

Since the potential $V(\mathbf{r})$ is electrostatic in origin, it is also determined from the number density by Poisson's equation

$$\nabla^2 V = - 4\pi e^2 n(\mathbf{r}). \tag{29.10}$$

Equations (29.9) and (29.10) are to be simultaneously solved by using the boundary conditions for V stated after Eq. (29.5). To solve these, one usually chooses the dimensionless variables χ and x defined by

$$V(\mathbf{r}) = -\frac{Ze^2}{r}\chi, \tag{29.11}$$

$$r = bx$$

where

$$b = \frac{1}{2}\left(\frac{3\pi}{4}\right)^{\frac{2}{3}} \frac{\hbar^2}{me^2 Z^{1/3}} = -\frac{0.885a_0}{Z^{1/3}}, \tag{29.12}$$

and $a_0 = \frac{\hbar^2}{me^2}$ is Bohr radius (the radius of the first Bohr orbit). If one substitutes Eq. (29.11) into equations (29.9) and (29.10) and combine the results together, one obtains

$$\frac{d^2\chi}{dx_2} = \frac{\chi^{3/2}}{x^{1/2}}. \tag{29.13}$$

Subject to the boundary conditions

$$\chi = 1 \text{ at } x = 0, \text{ and } \chi = 0 \text{ at } x = \infty. \tag{29.14}$$

The solutions of the nonlinear equation have been tabulated by Bush and Caldwell (1931). The total energy can be expressed as

$$E = \frac{3}{5}\int d^3r \, \frac{p_F^2(r)}{2m} \mathrm{n}(\mathbf{r}) + \int d^3r \, \mathrm{V}(\mathbf{r})\mathrm{n}(\mathbf{r}) \tag{29.15}$$

which by equations (29.8) and (29.9) reduces to

$$E = \frac{3\hbar^2(3\pi^2)^{2/3}}{10m}\int d^3r \, \mathrm{n}^{5/3}(\mathbf{r}) + \int d^3r \, \mathrm{V}(\mathbf{r})\mathrm{n}(\mathbf{r}) = -0.5125 Z^{7/3}\frac{e^2}{a_0}. \tag{29.16}$$

But in this calculation the electron-electron interaction has been counted twice. If one corrects for the double counting one obtains

$$E = -0.7687 Z^{7/3}\frac{e^2}{a_0} \tag{29.17}$$

which is somewhat larger in absolute magnitude than what one observes by studying the ionization process.

Single particle density

The density operator is defined by

$$\rho(\mathbf{x}) = \sum_{i=1}^{N} \delta(\mathbf{x}_i - \mathbf{x}). \tag{29.18}$$

The expectation value of $\rho(\mathbf{x})$ in an N-electron state that is represented by the (antisymmetric) function $\Psi(\mathbf{x}_1, \mathbf{x}_2, \ldots, \mathbf{x}_N)$ gives the *single particle density*

$$\rho^{(1)}(\mathbf{x}_1) = N\int dx_2 \int dx_3 \ldots \int dx_N \, \Psi(\mathbf{x}_1, \mathbf{x}_2, \ldots, \mathbf{x}_N)\, \Psi^*(\mathbf{x}_1, \mathbf{x}_2, \ldots, \mathbf{x}_N). \tag{29.19a}$$

In the above $\mathbf{x}$ stands for both space and spin coordinates. The single particle density at a spatial point is given by

$$\rho^{(1)}(\mathbf{r}_1) = \int ds_1 \, \rho^{(1)}(\mathbf{x}_1). \tag{29.19b}$$

Note that the integral of the single particle density equals the total number of electrons:

$$\int dx_1 \, \rho^{(1)}(\mathbf{x}_1) = N = \int d^3r_1 \, \rho^{(1)}(\mathbf{r}_1). \tag{29.20}$$

The single particle density can be generalized to define the *first-order reduced density matrix* or *the one-particle reduced density matrix*:

$$\rho^{(1)}(\mathbf{x}_1, \mathbf{x}_1') = N\int dx_2 \int dx_3 \ldots \int dx_N \, \Psi(\mathbf{x}_1, \mathbf{x}_2, \ldots, \mathbf{x}_N)\, \Psi^*(\mathbf{x}_1', \mathbf{x}_2, \ldots, \mathbf{x}_N), \tag{29.21}$$

$$\rho^{(1)}(\mathbf{r}_1, \mathbf{r}_1') = \int ds_1 \, [\rho^{(1)}(\mathbf{x}_1, \mathbf{x}_1')]_{s_1 = s_1'}.$$

If Ψ is a single product like the Hartree function, $\rho^{(1)}$ is called the Dirac density matrix and it is given by

$$\rho_D^{(1)}(\mathbf{x}, \mathbf{x}') = \sum_{i=1}^{occ.} \psi_i(\mathbf{x})\psi_i^*(\mathbf{x}'). \tag{29.22a}$$

When Ψ is a single configuration like the Hartree-Fock wave function, $\rho^{(1)}$ is called the Fock-Dirac density matrix and it is given by

$$\rho_{FD}^{(1)}(\mathbf{x}, \mathbf{x}') = \sum_{\alpha} f_\alpha \psi_\alpha(\mathbf{x}) \, \psi_\alpha^*(\mathbf{x}') \tag{29.22b}$$

where f_α is the occupancy of the spin-orbital α. The Fock-Dirac density is

$$\rho_{FD}^{(1)}(\mathbf{x}) = \sum_{\alpha} f_\alpha |\psi_\alpha(\mathbf{x})|^2. \tag{29.22c}$$

The single particle density is basically the sum of the probability densities due to all the electrons present in the system. It plays a very important role as discussed in the following.

The density functional approach

The Thomas-Fermi model is undoubtedly the simplest density functional approach. However, the density functional theory is usually associated with the name of Hohenberg and Kohn (1964) who established a two-step theorem. The latter is essentially a derivative of the variation theorem.

Theorem 12. The Hohenberg-Kohn theorem: For a nondegenerate ground state Ψ_g of the quantum system, the external potential energy function $\mathbf{V}$ is to within a constant a *unique* functional of the ground state single particle density $\rho_g^{(1)}(\mathbf{r})$. For a given external potential $\mathbf{V}$, the correct single particle density $\rho_g^{(1)}(\mathbf{r})$ minimizes energy to produce the ground state energy E_g which is a unique functional of $\rho_g^{(1)}$.

The proof of this theorem is by *reductio ad absurdum*. Consider two different external potentials $\mathbf{V}$ and $\mathbf{V}'$ (for all the electrons) corresponding to the Hamiltonians $\mathbf{H}$ and $\mathbf{H}'$, ground state wave functions Ψ_g and Ψ_g', and ground state energies E_g and E_g'. Let Ψ_g and Ψ_g' correspond to the same single particle density $\rho^{(1)}(\mathbf{r})$. Since Ψ_g and Ψ_g' are eigenfunctions of different operators ($\mathbf{H}$ and $\mathbf{H}'$), they are in general different from each other. Note that

$$\mathbf{H} = \mathbf{T} + \mathbf{U} + \mathbf{V}$$

and (29.23)

$$\mathbf{H}' = \mathbf{T} + \mathbf{U} + \mathbf{V}'$$

where $\mathbf{T}$ and $\mathbf{U}$ are the kinetic energy operator and the two-electron interaction energy operator respectively.

The two Hamiltonian operators are related to each other by the very simple relationships

$$\mathbf{H}' = \mathbf{H} + (\mathbf{V}' - \mathbf{V})$$
and (29.24)
$$\mathbf{H} = \mathbf{H}' + (\mathbf{V} - \mathbf{V}').$$
We know from the variation theorem that
$$Eg' = \int d\tau\, \Psi g'^{*}\mathbf{H}'\Psi g' < \int d\tau\, \Psi g^{*}\mathbf{H}'\Psi g. \tag{29.25}$$
But by Eq. (29.24),
$$\int d\tau\, \Psi g^{*}\mathbf{H}'\Psi g = Eg + \int d^3r\, \rho g^{(1)}(\mathbf{r})\, [\mathbf{v}'(\mathbf{r}) - \mathbf{v}(\mathbf{r})] \tag{29.26}$$
where $\mathbf{v}(\mathbf{r})$ and $\mathbf{v}'(\mathbf{r})$ are the corresponding external potentials for one electron and the integration represented by $\int d\tau$ stands for the integration over the space and spin coordinates of all the electrons involved. Therefore,
$$Eg' < Eg + \int d^3r\, \rho g^{(1)}(\mathbf{r})\, [\mathbf{v}'(\mathbf{r}) - \mathbf{v}(\mathbf{r})]\,. \tag{29.27}$$
In an exactly similar way but beginning with Eg one can write
$$Eg < Eg' + \int d^3r\, \rho g'^{(1)}(\mathbf{r})\, [\mathbf{v}(\mathbf{r}) - \mathbf{v}'(\mathbf{r})]. \tag{29.28}$$
Combining Eq. (29.27) with Eq. (29.28) one finds
$$Eg + Eg' < Eg + Eg' \tag{29.29}$$
which is an impossibility. Hence *either* **V** and **V'** are same in which case Ψg and $\Psi g'$ may be degenerate with the *less than* sign (<) replaced by the *equal to* sign (=) in equations (29.27), (29.28) and (29.29), *or* **V** and **V'** are different and Ψg and $\Psi g'$ must not correspond to the same single particle density. Thus Ψg, and to within an additive constant Eg, $\mathbf{v}(\mathbf{r})$ and **V** must be *unique* functionals of $\rho g^{(1)}(\mathbf{r})$.

The second part of the theorem has practical significance. Since Ψg is a unique functional of $\rho g^{(1)}(\mathbf{r})$, all ground state expectation values are functionals of $\rho g^{(1)}(\mathbf{r})$. In particular, T, the expectation value of kinetic energy, and U, the expectation value of two-electron interaction energy, are *universal* functionals of $\rho g^{(1)}$:
$$\mathrm{T}[\rho g^{(1)}(\mathbf{r})] = \int d\tau\, \Psi g^{*}\mathbf{T}\Psi g,$$
$$\mathrm{U}[\rho g^{(1)}(\mathbf{r})] = \int d\tau\, \Psi g^{*}\mathbf{U}\Psi g. \tag{29.30}$$
These two are added together to form another universal functional F:
$$\mathrm{F}[\rho g^{(1)}(\mathbf{r})] = \mathrm{T}[\rho g^{(1)}(\mathbf{r})] + \mathrm{U}[\rho g^{(1)}(\mathbf{r})]. \tag{29.31}$$
Let the energy expectation value of **H** be written as E. The ground state energy is given by
$$E[\rho g^{(1)}(\mathbf{r})] = \mathrm{F}[\rho g^{(1)}(\mathbf{r})] + \int d^3r\, \rho g^{(1)}(\mathbf{r})\mathbf{v}(\mathbf{r}). \tag{29.32}$$
The variation theorem tells us that for another wave function Ψ,
$$E[\Psi] = \mathrm{F}[\rho^{(1)}(\mathbf{r})] + \int \mathrm{d}^3\mathrm{r}\, \rho^{(1)}(\mathbf{r})\mathbf{v}(\mathbf{r}) > E[\rho g^{(1)}(\mathbf{r})]. \tag{29.33}$$
Thus the minimum property of $E[\rho g^{(1)}(\mathbf{r})]$ is established. **QED**

Applications of the density functional theorem are known as the density functional treatment (DFT). Quantitative applications have been quite successful for general properties in the ground state – especially thermochemical properties. However, if one cares for accuracy in calculations, then there are problems. The main problem is that of *N-representability*, that is, one cannot guarantee that the single particle density indeed corresponds to a properly anti-symmetrized wave function for N electrons. This gives rise to the second difficulty in the form that one does not have an exact analytical expression for the universal functional U. The latter consists of the Coulomb repulsion energy and the so-called exchange-correlation functional.

The Coulomb repulsion energy is exactly known as the sum of the J integrals, and accounts for the average direct repulsion between electron densities due to different electrons. The exchange-correlation functional is a combination of the exchange energy that is also an average term but non-local in nature, and difficult to evluate. There is an additional energy arising from the instantaneous – rather than average – nature the Coulomb interaction that leads to an explicitly correlated movement of the electrons.

Many simple expressions for U have been suggested. These expressions approximately correspond to anti-symmetrized wave functions of particular types. The most notable among them is the one suggested by Slater long before the density functional theorem was stated and proved by Hohenberg and Kohn. This U contains the direct Coulomb integral and the so-called Slater exchange term as a functional of the single particle density. The Hartree-Fock-Slater calculation, a wave function-based approach, was quite popular although it does not represent a very accurate treatment.

DFT may be carried out with Slater exchange (and even Slater exchange-correlation) term(s). This DFT is essentially comparable to a Hartree-Fock-Slater treatment (with possible correlation corrections introduced in an average way). As it is a density-based approach it is easier to work with although due to the N-representability problem one would not be certain about the accuracy of the resulting ground state single particle density. Also, it should be strictly applied to atomic systems as the Slater terms owe their origin to the Thomas-Fermi model. The ease with which the calculation can be carried out makes this approach almost irresistible.

Recent years have seen the density functional treatment to be applied to the investigation of metal complexes with considerably large organic ligands. This is in a way stretching the theory a little bit too far into empiricism, but then the researchers in this field are satisfied with establishing several qualitative and quantitative trends of the molecular and physicochemical properties of complexes of moderate size. Various sophisticated exchange-correlation functionals have been designed. The most successful functionals include the Becke-3-functional exchange and Lee-Yang-and-Parr correlation (B3LYP) functional, and the functionals designed by the University of Minnesota group under the leadership of Truhlar (such as M06-2X). These are known to contain a large fraction of the Hartree-Fock exchange, and reproduce very good thermochemistry. They also give a good account of magnetic properties. As a general trend, the number of terms involved in the exchange-correlation functional increases at a rapid pace as it becomes increasingly more effective.

Because it has been designed for the ground state, DFT has been less successful for reproducing excited state properties, and therefore, general spectroscopic characteristics. In most spectroscopic applications it has been used for the ground states of different space and spin symmetries. The Hohenberg-Kohn theorem applies to the ground state of each type of state. Still, there are problems. The exchange-correlation functional is normally designed for the global ground state. Hence the spectroscopic calculations are not reliable enough.

DFT has found spectacular success in qualitatively explaining a large number of fundamental aspects, such as electronegativity. The subject is, therefore, rich in concept.

Texts Suggested for Further Reading

C. F. Fischer, "The Hartree-Fock Method for Atoms – A Numerical Approach", (Wiley, New York, 1977).
D. R. Hartree, "The Calculation of Atomic Structures", (Wiley, New York, 1957).
R. M. Hochstrasser, "Behavior of Electrons in Aoms", (Benjamin, New York, 1964).
M. H. Shamos, "Great Experiments in Physics", (Holt, New York, 1959).

PROBLEMS

1. Prove that the total angular momentum operator $\boldsymbol{J}$ commutes with $\boldsymbol{L}\bullet\boldsymbol{S}$.
2. Show that $\boldsymbol{L}\bullet\boldsymbol{S} = \frac{1}{2}(\boldsymbol{J}^2 - \boldsymbol{L}^2 - \boldsymbol{S}^2)$. Calculate the expectation value $<\boldsymbol{L}\bullet\boldsymbol{S}>$.
3. The fine structure constant is given by $\alpha = \frac{e^2}{\hbar c}$ where c is the speed of light in vacuum. It is a universal constant, a dimensionless quantity. Its inverse, α^{-1}, equals 137.037 in every system of units. What is the speed of light c in atomic unit? Given $c = 2.9978 \times 10^8$ m s^{-1}, determine the atomic unit of time.
4. Use Slater's rules to estimate the effective nuclear charge, the average radius of each electron and the successive ionization potentials of the sodium atom.
5. Prepare a plot of the radial density of electrons versus the radial distance for this atom.
6. The 1s electron of a many-electron atom is equally likely to be found at a distance from the nucleus greater than R and less than R. Find R in terms of the atomic number Z.
7. Express the Thomas-Fermi electron density n(**r**) in terms of the dimensionless functions x and χ.

CHAPTER 6

APPLICATION OF THE LINEAR VARIATION METHOD: MOLECULAR STRUCTURE

An understanding of the molecular electronic structure does not necessarily require the application of Ritz variation principle. The Schrödinger equation is basically a second-order differential equation and there are well-known numerical procedures for solving a differential equation. The Hartree or Hartree-Fock equations are integro-differential equations. In recent years solutions of the electronic wave equation for small molecules of light elements have been computed at different levels of theory by the numerical technique. But the numerical method of solution generally requires a vast computing effort, and even now it is not practicable for most of the molecules. Besides, an analytical method of solution has the advantage of bringing forth physical pictures with which we can gain a conceptual understanding of the bonds and molecular geometry.

The self-consistent-field methodology is undoubtedly the backbone of a systematic investigation of the electronic structure of a molecule. The methodology itself is an outcome of the application of the variation principle. The SCF equations are most of the time solved by employing the linear variation technique. Before one encounters this application, one must learn how the electronic Schrödinger equation is extracted from the Schrödinger equation that involves the movement of not only the electrons but also the nuclei belonging to the molecule.

Lecture 30: Born-Oppenheimer Approximation

We have seen in Lecture 8 that the Hamiltonian for a general molecule with N electrons and M nuclei is written as

$$\mathbf{H} = (\mathbf{T}_e + \mathbf{T}_n) + (\mathbf{V}_{en} + \mathbf{V}_{ee} + \mathbf{V}_{nn}). \tag{8.5}$$

The kinetic energy of the nuclei is of the order of $\frac{m_e}{m_n}$, about 10^{-4} times the kinetic energy of the electrons. This allows $\mathbf{T}_n$ to be treated as a small perturbation with which one must solve the Schrödinger equation

$$\mathbf{H}\Psi(\{\mathbf{x}\};\{\mathbf{R}\}) = \mathrm{E}\Psi(\{\mathbf{x}\};\{\mathbf{R}\}). \tag{8.8}$$

In writing Eq. (8.8) we have ignored nuclear spins.

If one neglects $\mathbf{T}_n$, the nuclear coordinates $\{\mathbf{R}\}$ are no longer dynamical variables in the Hamiltonian, but they are only parameters. The Schrödinger equation (8.8) reduces to

$$[\mathbf{H}_{el}(\{\mathbf{r}\};\{\mathbf{R}\})+\mathbf{V}_{nn}(\{\mathbf{R}\})]\Phi_i(\{\mathbf{x}\};\{\mathbf{R}\})=[E_{el,i}(\{\mathbf{R}\})+\mathbf{V}_{nn}(\{\mathbf{R}\})]\Phi_i(\{\mathbf{x}\};\{\mathbf{R}\}) \tag{30.1}$$

where $\mathbf{H}_{el}$ is the electronic Hamiltonian (that is usually taken to be free from any spin-dependent term in the absence of a magnetic field),

$$\mathbf{H}_{el}(\{\mathbf{r}\};\{\mathbf{R}\}) = \mathbf{T}_e + \mathbf{V}_{en}(\{\mathbf{r}\});\{\mathbf{R}\}) + \mathbf{V}_{ee}(\{\mathbf{r}\}). \tag{30.2}$$

The energy eigenvalue $E_{el,i}$ depends upon the parameters $\{\mathbf{R}\}$. For a specific set of parameters $\{\mathbf{R}\}$ representing the internal (vibrational) coordinates of motion, the term $\mathbf{V}_{nn}$ is a constant. If we drop the constant term from both sides of Eq. (30.1) we obtain the electronic Schrödinger equation

$$\mathbf{H}_{el}\Phi_i(\{\mathbf{x}\};\{\mathbf{R}\}) = E_{el,i}(\{\mathbf{R}\})\Phi_i(\{\mathbf{x}\};\{\mathbf{R}\}) \tag{30.3}$$

which offers an orthonormal complete set of solutions.

The exact solution of Eq. (8.8) can be expanded in the basis of the complete set of eigenfunctions of $\mathbf{H}_{el}$ as

$$\Psi(\{\mathbf{x}\};\{\mathbf{R}\}) = \sum_i \Phi_i(\{\mathbf{x}\};\{\mathbf{R}\})\, \chi_i(\{\mathbf{R}\}). \tag{30.4}$$

The expansion coefficients χ_i's depend on the parameters $\{\mathbf{R}\}$. If we substitute Eq. (30.4) for the wave function in Eq. (8.8), we get

$$\sum_i [E_{el,i}(\{\mathbf{R}\}) + \mathrm{V}_{nn}(\{\mathbf{R}\}) + \mathbf{T}_n]\, \Phi_i(\{\mathbf{x}\};\{\mathbf{R}\})\, \chi_i(\{\mathbf{R}\})$$

$$=E\sum_i \Phi_i(\{\mathbf{x}\};\{\mathbf{R}\})\chi_i(\{\mathbf{R}\}). \tag{30.5}$$

Let us multiply both sides of Eq. (30.5) by $\Phi_j^*(\{\mathbf{x}\};\{\mathbf{R}\})$ and integrate over the electronic coordinates. Then, by using the orthonormality of the electronic wave functions Φ's, we get

$$\sum_i \int dx_1 \int dx_2 ... \int dx_N\, \Phi_j^*(\{\mathbf{x}\};\{\mathbf{R}\})\, \mathbf{T}_n\, \Phi_i(\{\mathbf{x}\};\{\mathbf{R}\})\, \chi_i(\{\mathbf{R}\})$$

$$+ [E_{el,j}(\{\mathbf{R}\}) + \mathrm{V}_{nn}(\{\mathbf{R}\})]\, \chi_j(\{\mathbf{R}\}) = E\chi_j(\{\mathbf{R}\}). \tag{30.6}$$

As the operator $\mathbf{T}_n$ involves second derivatives with respect to nuclear coordinates, we write

$$[\mathbf{T}_n + E_{el,j}(\{\mathbf{R}\}) + \mathrm{V}_{nn}(\{\mathbf{R}\})]\, \chi_j(\{\mathbf{R}\}) = E\chi_j(\{\mathbf{R}\}) - \sum_i C_{ji}\chi_i(\{\mathbf{R}\}) \tag{30.7}$$

where the coupling constants are given by

$$C_{ji} = -\hbar^2 \sum_p \frac{1}{2m_p} \int dx_1 \int dx_2 ... \int dx_N\, \Phi_j^*(\{\mathbf{x}\};\{\mathbf{R}\})$$

$$\times [2\nabla_p \chi_i(\{\mathbf{R}\})\cdot\nabla_p \Phi_i(\{\mathbf{x}\};\{\mathbf{R}\}) + \chi_i(\{\mathbf{R}\})\nabla_p^2 \Phi_i(\{\mathbf{x}\};\{\mathbf{R}\})]. \tag{30.8}$$

To the lowest order in C we have

$$[\mathbf{T}_n + E_{el,i}(\{\mathbf{R}\}) + V_{nn}(\{\mathbf{R}\})]\ \chi_i(\{\mathbf{R}\}) = E\chi_i(\{\mathbf{R}\}) \tag{30.9}$$

that corresponds the wave function

$$\Psi(\{\mathbf{x}\};\{\mathbf{R}\}) = \chi_i(\{\mathbf{R}\})\ \Phi_i(\{\mathbf{x}\};\{\mathbf{R}\}). \tag{30.10}$$

This wave function is referred to as the wave function within the framework of the *electronic-vibrational coupling*. Under Born-Oppenheimer approximation, the energy $(E_{el,i}+V_{nn})$ acts as the potential energy for movement of nuclei as shown by Eq. (30.9). Hence $(E_{el,i}+V_{nn})$ is called as the potential energy surface. For bound states of the molecule, $(E_{el,i}+V_{nn})$ will have a set of minima in the relative coordinates which describe the vibrations of the molecular frame. To obtain the vibrational energy one expands $(E_{el,i}+V_{nn})$ about the global minimum that represents the equilibrium geometry of the nuclei. This procedure has been discussed already in Lecture 3. The treatment of the deviation of the potential curve $(E_{el,i}+V_{nn})$ from the harmonic oscillator form will be taken up later in Chapter 7.

The coupling terms C_{ji} are about 10^4 times less than the electronic energy. They represent the coupling between the electronic states and the molecular vibrational motion. If these terms are not neglected then the solution of Eq. (30.7) will yield eigenfunctions of the form shown in Eq. (30.4). The resulting states are called *vibronic states*. The C_{ji} terms represent the *vibronic interaction*. The latter is responsible for the broadening of electronic spectra and for effects like fluorescence (internal conversion) and phosphorescence (intersystem crossing); so it cannot be truly ignored. Actually the terms C_{ji} can be neglected when the variation of the electronic wave function with respect to the relative coordinates of nuclei is extremely slow.

Nonrigid molecules have a few very small frequencies of vibration. These vibrations are associated with large-amplitude oscillations which result in the fluxional behavior. For such molecules the electronic-vibrational coupling is not at all a good description. Besides, the ro-vibronic coupling (coupling between the rotational and vibrational or vibronic states) is quite strong in these species. Because of these effects, the microwave and infrared spectra of these molecules are often modified and develop interesting features.

Born and Oppenheimer (1927) originally adopted a perturbative procedure. If the operator $\mathbf{T}_n$ is treated as a small perturbation, Eq. (30.3) will stand for the zero-order problem with Eq. (30.9) appearing as an outcome of the first order correction. These two equations together describe the molecular electronic treatment under Born-Oppenheimer approximation.

In the following lectures we proceed as follows. First we discuss an exceedingly simple and empirical molecular orbital treatment (*Hückel* molecular orbital method for π electrons and the *extended Hückel* method for π as well as σ electrons). Next we learn the qualitatively correct molecular orbital theory (Qualitative molecular orbitals or *QMO*s) including the molecular orbital treatments for H_2^+, H_2 and other diatomic molecules of light elements. We then consider the important concept of *hybridization* and the formation of molecular orbitals from the hybrid orbitals, and explore how the orbital energies and the energetics of the molecular system may change with molecular symmetry.

Lecture 31 and 32: The Hückel Molecular Orbital Treatment
Lecture 31: Basic Principles, Parametrization and Simple Polyenes

The Hückel treatment is essentially a one-electron treatment where the mutual interaction of the electrons is not explicitly considered. The Hamiltonian $\mathbf{H}_{el}$ for the π electrons in a conjugated polyene is considered as the sum of the one-electron Hamiltonian operators, $\sum_{i}^{electron} \mathbf{H}(i)$ which obviously makes the electronic wave equation separable into a set of one-electron Schrödinger equations each of the form

$$\mathbf{H}(i)\pi_k(i) = E_k\pi_k(i). \tag{31.1}$$

It is assumed that the σ orbitals are so much low-lying in energy that the influence of the σ electrons on the dynamics of the π electrons may be safely approximated to cause an average potential. The sp^2 hybrid orbitals form the σ framework, and in the Hückel treatment only the valence p_z orbitals of the carbon atoms are used to build up the π orbitals of the conjugated molecule. Thus one writes

$$\pi = \sum_{i}^{atom} c_i p_{zi} \tag{31.2}$$

where p_{zi} is the $2p_z$ orbital of the *ith* carbon atom and the summation extends over all the conjugated carbon atoms. The expansion coefficients c_i's are the parameters of variation, and Eq. (31.1) is solved by the Ritz variation technique that eventually leads to the algebraic equation

$$\| H_{ij} - ES_{ij}\| = 0. \tag{31.3}$$

In the above the H_{ij} is the *Hückel energy integral*

$$H_{ij} = \int dv\, p_{zi}\mathbf{H}p_{zj} \tag{31.4}$$

and S_{ij} is the *overlap integral*

$$S_{ij} = \int dv\, p_{zi}p_{zj}. \tag{31.5}$$

One proceeds to solve Eq. (31.3), then obtains for each root the ratios of the expansion coefficients and finally determines the coefficients uniquely by normalizing each orbital. The normalization condition is

$$\sum_i \sum_j c_i c_j S_{ij} = 1. \tag{31.6}$$

This is the basic principle of the Hückel method. To simplify the treatment further, Hückel (1930, 1931) introduced drastic approximations for the energy integral and the overlap integral in the form of parameters.

Hückel Parametrization: (31.7)

(i) $H_{ii} = \alpha$ for all i.
The energy α is called the *coulomb integral*. It represents the energy of the $2p_z$ electron in each carbon atom in the state of sp^2 hybridization.

(ii) $H_{ij} = \beta$ if C_i and C_j are nearest neighbors.

This energy, β, is called the resonance integral since it leads to the phenomenon of resonance. It is in general a negative quantity. Furthermore,

$H_{ij} = 0$ if $i \neq j$, and C_i and C_j are not nearest-neighbor atoms.

(iii) $S_{ij} = \delta_{ij}$. That $S_{ii} = 1$ implies that each p_z orbital is normalized. Because they are directed along the z-axis, the overlap between any two of them is small; hence $S_{ij}=0$ for $i \neq j$ is a good approximation.

Using these parameters, the linear homogeneous equations and the secular determinant resulting from the linear variation technique can be written in greatly simplified forms. For instance, because of (iii), the normalization condition (31.6) simplifies to

$$\sum_i c_i^2 = 1. \tag{31.8}$$

Once the π orbitals and the corresponding orbital energies are obtained, one can draw the energy level diagram and populate each orbital in accordance with Pauli's exclusion principle and determine the ground-state configuration.

The *π-electron ground-state energy*, E_π, is then calculated as the sum of the energies of the electrons in the occupied orbitals,

$$E_\pi = \sum_k n_k E_k. \tag{31.9}$$

We have used n_k to denote the population of the *kth* π orbital.

For a neutral species, the number of the π electrons $\sum_i^{electron} n_i$ equals N, the number of the conjugated carbon atoms. Total energy of the atomic p_z orbitals is $N\alpha$. So the *π-electron binding energy* is

$$\Delta E_\pi = E_\pi - N\alpha. \tag{31.10}$$

This is generally negative (as β is a negative quantity), indicating that the formation of π-orbitals leads to stability. In other words the p_z electrons acquire stability by delocalization, that is, by spreading over more than one atom.

In organic chemistry, resonance is normally referred to as the excess delocalization compared to the delocalization for a carbon-carbon double bond. The *resonance energy* for a conjugated polyene with N conjugated carbon atoms is calculated as the difference of ΔE_π and the sum of the *π-electron binding energies* for $N/2$ ethylenic bonds,

$$\text{Resonance Energy} = \Delta E_\pi - \frac{N}{2}(\Delta E_\pi)_{\text{ethylene}}. \tag{31.11}$$

The whole treatment can be best understood in terms of a few simple examples which are discussed here.

Example 29: Ethylene (N=2). The linear homogeneous equations are

$$c_1(H_{11} - ES_{11}) + c_2(H_{12} - ES_{12}) = 0,$$
$$c_1(H_{12} - ES_{12}) + c_2(H_{22} - ES_{22}) = 0.$$

Introduction of the Hückel parameters convert these into

$$c_1(\alpha - E) + c_2\beta = 0,$$
$$c_1\beta + c_2(\alpha - E) = 0. \qquad (31.12)$$

For the existence of nontrivial solutions the secular determinant must be zero, that is,

$$\begin{vmatrix} \alpha - E & \beta \\ \beta & \alpha - E \end{vmatrix} = 0. \qquad (31.13)$$

This equation has the two roots

$$E = \alpha + \beta \quad \text{and} \quad E = \alpha - \beta. \qquad (31.14)$$

For $E = \alpha + \beta \equiv E_1$, *both* linear homogeneous equations give $c_1=c_2$. The normalization condition is $c_1^2+c_2^2 = 1$ which yields $c_1=c_2= 2^{-1/2}$ and so

$$\pi = 2^{-1/2}(p_{z1} + p_{z2}) \equiv \pi_1. \qquad (31.15)$$

For the second root $E = \alpha - \beta \equiv E_2$, the linear homogenous equations yield $c_1=-c_2$ so that from the normalization condition one obtains $c_1=-c_2=2^{-1/2}$ and

$$\pi = 2^{-1/2}(p_{z1} - p_{z2}) \equiv \pi_2. \qquad (31.16)$$

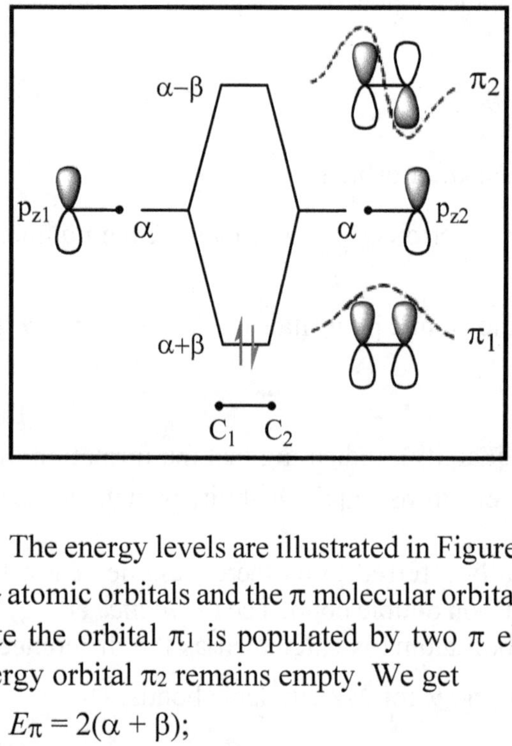

Figure 6.1 Hückel π orbital energy level diagram for ethylene – shaded and unshaded areas are of different signs.

The energy levels are illustrated in Figure 6.1 that includes nodal representations of the $2p_z$ atomic orbitals and the π molecular orbitals. As β is negative, $E_1 < \alpha < E_2$. In the ground state the orbital π_1 is populated by two π electrons with opposite spins and the higher-energy orbital π_2 remains empty. We get

$$E_\pi = 2(\alpha + \beta);$$

and (31.17)

$$\Delta E_\pi = E_\pi - 2\alpha = 2\beta < 0.$$

Thus π_1^2 is a bound state. The orbital π_1 is a bonding molecular orbital and the orbital π_2 is an antibonding one. The π-electron binding energy is 2β for the ethylene molecule in its ground state. This energy will be used for calculating the resonance energy in other conjugated polyenes, linear as well as cyclic.

Example 30: The Allyl System (N=3). The linear homogeneous equations are

$$\begin{aligned} c_1(\alpha - E) + \quad c_2\beta \qquad\qquad\qquad &= 0, \\ c_1\beta \quad + c_2(\alpha - E) + \quad c_3\beta \quad &= 0, \\ c_2\beta \quad + c_3(\alpha - E) &= 0. \end{aligned} \tag{31.18}$$

The normalization condition is $c_1^2+c_2^2+c_3^2 = 1$. We get, for nontrivial solutions to exist, the condition

$$\begin{vmatrix} \alpha - E & \beta & 0 \\ \beta & \alpha - E & \beta \\ 0 & \beta & \alpha - E \end{vmatrix} = 0 \tag{31.19}$$

that yields the roots

$$\begin{aligned} E_1 &= \alpha+\sqrt{2}\,\beta, \\ E_2 &= \alpha, \\ E_3 &= \alpha-\sqrt{2}\,\beta. \end{aligned} \tag{31.20}$$

The corresponding π molecular orbitals are

$$\begin{aligned} \pi_1 &= 2^{-1}[p_{z1}+\sqrt{2}\,p_{z2}+p_{z3}], \\ \pi_2 &= 2^{-1/2}[p_{z1}-p_{z3}], \\ \pi_3 &= 2^{-1}[p_{z1}-\sqrt{2}\,p_{z2}+p_{z3}]. \end{aligned} \tag{31.21}$$

The energy level diagram is shown in Figure 6.2 . The orbital π_1 is a bonding orbital, π_2 is a nonbonding one and π_3 is an antibonding orbital.

The allyl radical has three π electrons. In Hückel theory, allyl cation, allyl radical and allyl anion all have the same π electron binding energies, ΔE_π=2.828β. This is not strictly correct: the binding energies should be only approximately equal, and not exactly equal as the Hückel model predicts.

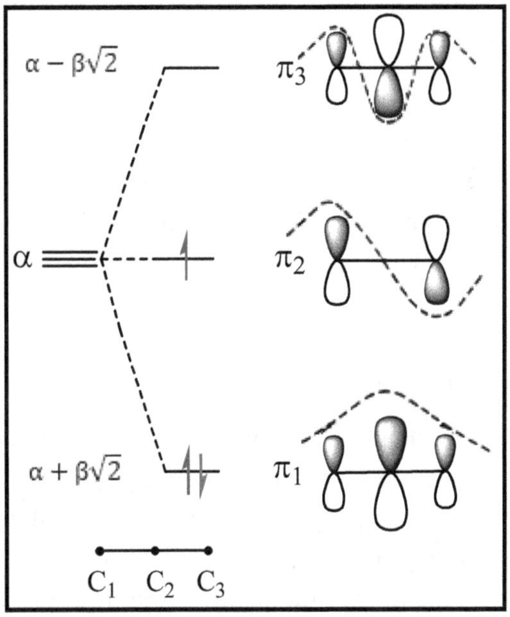

Figure 6.2 Hückel π orbital energy level diagram for the allyl system.

This slightly wrong conclusion arises from the fact that the Hückel model does not take into account the effects of the mutual repulsion of electrons in an explicit way. Nevertheless, the calculated values serve to remind us that one π electron in the radical and two π electrons in the anion occupy the nonbonding orbital with comparatively small contributions to the binding energy. The cation with only two π electrons has resonance energy 0.828β whereas the anion having four π electrons is greatly destabilized compared to *two* ethylenic π bonds.

Example 31: The Cyclopropenyl System (N=3). Every carbon atom is the nearest neighbor of the other two carbon atoms in this molecule. Therefore, the linear homogeneous equations are

$$\begin{aligned} c_1(\alpha - E) + \quad c_2\beta \quad + \quad c_3\beta \quad &= 0, \\ c_1\beta \quad + c_2(\alpha - E) + \quad c_3\beta \quad &= 0, \\ c_1\beta \quad + \quad c_2\beta \quad + c_3(\alpha - E) &= 0. \end{aligned} \tag{31.22}$$

The normalization condition is as in the previous example, that is, $c_1^2+c_2^2+c_3^2=1$. The existence of nontrivial solutions requires

$$\begin{vmatrix} \alpha - E & \beta & \beta \\ \beta & \alpha - E & \beta \\ \beta & \beta & \alpha - E \end{vmatrix} = 0 \tag{31.23}$$

which yields the roots

$$E_1 = \alpha + 2\beta \quad \Leftrightarrow \quad \pi_1 = 3^{-1/2}(p_{z1} + p_{z2} + p_{z3}) \tag{31.24}$$

and

$$E_2 = E_3 = \alpha - \beta. \tag{31.25}$$

If Eq. (31.25) is substituted in equations (31.22) one obtains the same result $c_1+c_2+c_3=0$ three times. This condition along with the normalization is not sufficient to yield unique values for the three coefficients. Actually we need to find two orbitals and hence two sets of coefficients. If one of the coefficients belonging to the first set is assumed, the other two can be uniquely determined. Furthermore, the coefficients belonging to the second set can be determined by using the additional condition of orthogonality of the two degenerate orbitals. Let us choose $c_1=0$ in the expansion of π_2. Then we get, $c_2=-c_3$ and because of the normalization, $c_2=2^{-1/2}$, so that

$$\pi_2 = 2^{-1/2}(p_{z2} - p_{z3}). \tag{31.26}$$

The orbital π_3 is written as $\pi_3= c_1'p_{z1}+ c_2'p_{z2}+ c_3'p_{z3}$. Then,

$$\begin{aligned} &c_1'+c_2'+c_3' = 0 \qquad \text{(because of the energy root)} \\ &c_1'^2+c_2'^2+c_3'^2 = 1 \qquad \text{(normalization)} \end{aligned} \tag{31.27}$$

and

$$c_2'-c_3' = 0.$$

The last result is a consequence of the orthogonality, $\int d\tau\, \pi_2\pi_3 = 0$. Solving the three equations in (31.27) for the coefficients one finds

$$\pi_3 = 6^{-1/2}(2p_{z1} - p_{z2} - p_{z3}). \tag{31.28}$$

Thus the assumption $c_1= 0$ for π_2 has yielded two unique solutions, Eq. (31.26) for π_2 and Eq. (31.28) for π_3.

One may argue that the assumption $c_1= 0$ has been arbitrary. But we have already seen from Theorem 1 that the functions belonging to a degenerate set can linearly combine with each other with the result that the combinations are also eigenfunctions with the same eigenvalue and with little difficulty they can be made orthogonal to each other. The expansion coefficients c_1, c_2 or c_3 can be chosen in an infinite number of ways. Had we made a different assumption, we would have obtained a different set of π orbitals, but these new orbitals can be recognized as two normalized and mutually orthogonal linear combinations of π_2 and π_3.

For example, one may start with $c_2 = 0$ and find $\pi_2' = 2^{-1/2}(p_{z3} - p_{z1})$, $\pi_3' = 6^{-1/2}(2p_{z2} - p_{z3} - p_{z1})$. It is easy to verify $\pi_2' = 2^{-3/2}(2^{1/2}\pi_2 + 6^{1/2}\pi_3)$, $\pi_3' = 2^{-3/2}(6^{1/2}\pi_2 - 2^{1/2}\pi_3)$. Thus the set (π_2', π_3') can be derived from (π_2, π_3) by using a linear transformation. The transformation is unitary. Hence the solutions (π_2', π_3') are not truly different from (π_2, π_3). The situation is just like the case of atomic orbitals with the same n and the same l $(l \neq 0)$: one has to choose either the orbitals $2p_{m=1}$, $2p_{m=0}$, $2p_{m=-1}$, or the orbitals $2p_x$, $2p_y$, $2p_z$. Both sets should lead to the same physical result. Here too one may consider the orbitals (π_1, π_2, π_3), or orbitals of the set (π_1, π_2', π_3') or of any other set. Whatever combination may be chosen, physical results like stability, charge distribution, transition energies, transition probabilities, etc. remain unaltered. This has to be the case; otherwise, the theory would not have been acceptable.

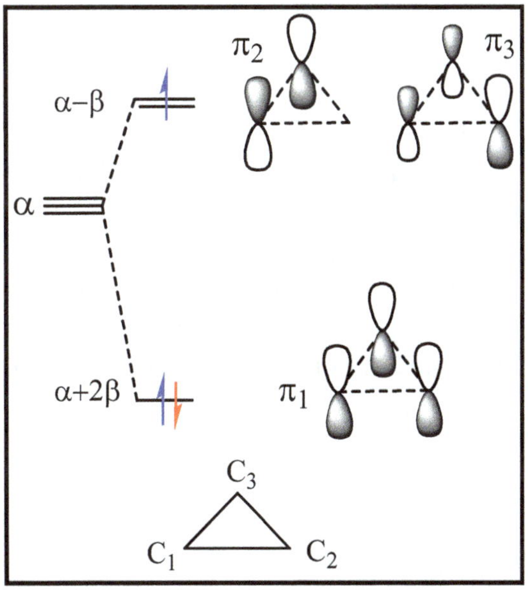

Figure 6.3 Hückel π orbital energy level diagram for the cyclopropenyl system.

Lecture 32: More Conjugated Polyenes

Example 32: For *linear butadine* we write

$$\begin{aligned}
c_1(\alpha - E) + c_2\beta &= 0 \\
c_1\beta + c_2(\alpha - E) + c_3\beta &= 0 \\
c_2\beta + c_3(\alpha - E) + c_4\beta &= 0 \\
c_3\beta + c_4(\alpha - E) &= 0
\end{aligned} \qquad (32.1)$$

which give the orbital energies and the molecular orbitals

$$\begin{aligned}
E_1 &= \alpha + 1.618\beta & \pi_1 &= 0.372p_{z1} + 0.602p_{z2} + 0.602p_{z3} + 0.372p_{z4}, \\
E_2 &= \alpha + 0.618\beta & \pi_2 &= 0.602p_{z1} + 0.372p_{z2} - 0.372p_{z3} - 0.602p_{z4},
\end{aligned} \qquad (32.2)$$

$E_3 = \alpha - 0.618\beta \quad \pi_3 = 0.602p_{z1} - 0.372p_{z2} - 0.372p_{z3} + 0.602p_{z4},$
$E_4 = \alpha - 1.618\beta \quad \pi_4 = 0.372p_{z1} - 0.602p_{z2} + 0.602p_{z3} - 0.372p_{z4}.$

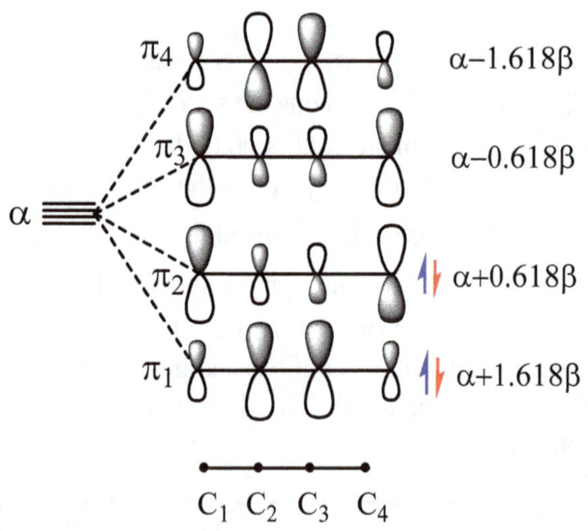

Figure 6.4 Hückel π-orbital energy level diagram for the linear butadiene.

In the ground state,

$$E_\pi = 4\alpha + 4.472\beta,\ \Delta E_\pi = 4.472\beta,$$

and (32.3)

$$\text{Resonance energy} = 0.472\beta\ .$$

The cis- and the trans- isomers are not distinguished by the simple Hückel treatment.

Example 33: For *cyclobutadiene* we obtain

$$\begin{aligned}
c_1(\alpha - E) + c_2\beta \quad\quad\quad + c_4\beta &= 0\\
c_1\beta + c_2(\alpha - E) + c_3\beta \quad\quad\quad &= 0\\
c_2\beta + c_3(\alpha - E) + c_4\beta &= 0\\
c_1\beta \quad\quad\quad + c_3\beta + c_4(\alpha - E) &= 0
\end{aligned} \tag{32.4}$$

which yield

$$\begin{aligned}
E_1 &= \alpha + 2\beta & \pi_1 &= 2^{-1}(p_{z1} + p_{z2} + p_{z3} + p_{z4})\\
E_2 &= \alpha & \pi_2 &= 2^{-1/2}(p_{z1} - p_{z3})\\
E_3 &= \alpha & \pi_3 &= 2^{-1/2}(p_{z2} - p_{z4})\\
E_4 &= \alpha - 2\beta & \pi_4 &= 2^{-1}(p_{z1} - p_{z2} + p_{z3} - p_{z4})
\end{aligned} \tag{32.5}$$

The second and the third π-orbitals are now nonbonding in nature. For the ground state,

$$E_\pi = 4\alpha + 4\beta$$

$$\Delta E_\pi = 4\beta \tag{32.6}$$

and hence compared to two ethylenic bonds, the resonance energy is zero.

Figure 6.5 Hückel π-orbital energy level diagram for cyclobutadiene.

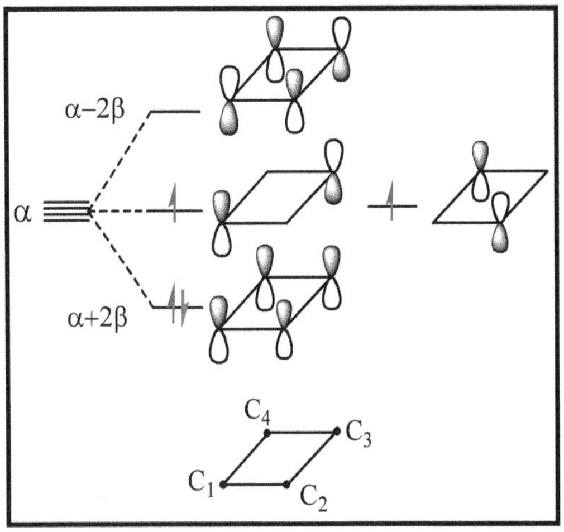

Example 34: For *benzene* one writes

$$\begin{aligned}
c_1(\alpha - E) + c_2\beta + c_6\beta &= 0\\
c_1\beta + c_2(\alpha - E) + c_3\beta &= 0\\
c_2\beta + c_3(\alpha - E) + c_4\beta &= 0\\
c_3\beta + c_4(\alpha - E) + c_5\beta &= 0\\
c_4\beta + c_5(\alpha - E) + c_6\beta &= 0\\
c_1\beta + c_5\beta + c_6(\alpha - E) &= 0
\end{aligned} \qquad (32.7)$$

The solutions are

$$\begin{aligned}
E_1 &= \alpha + 2\beta & \pi_1 &= 6^{-1/2}(p_{z1} + p_{z2} + p_{z3} + p_{z4} + p_{z5} + p_{z6})\\
E_2 &= \alpha + \beta & \pi_2 &= 2^{-1}3^{-1/2}(p_{z1} + 2p_{z2} + p_{z3} - p_{z4} - 2p_{z5} - p_{z6})\\
E_3 &= \alpha + \beta & \pi_3 &= 2^{-1}(p_{z1} - p_{z3} - p_{z4} + p_{z6})\\
E_4 &= \alpha - \beta & \pi_4 &= 2^{-1}3^{-1/2}(p_{z1} - 2p_{z2} + p_{z3} + p_{z4} - 2p_{z5} + p_{z6})\\
E_5 &= \alpha - \beta & \pi_5 &= 2^{-1}(p_{z1} - p_{z3} + p_{z4} - p_{z6})\\
E_6 &= \alpha - 2\beta & \pi_6 &= 6^{-1/2}(p_{z1} - p_{z2} + p_{z3} - p_{z4} + p_{z5} - p_{z6})
\end{aligned} \qquad (32.8)$$

There are no nonbonding pi molecular orbitals. Three are bonding orbitals (π), and the other three are antibonding (π*). The ground state characteristics are as follows:

$$\begin{aligned}
E_\pi &= 6\alpha + 8\beta;\\
\Delta E_\pi &= 8\beta;\\
\text{Resonance Energy} &= 2\beta.
\end{aligned} \qquad (32.9)$$

The resonance energy of benzene is 155 kJ mol^{-1}. It is normally used to set the value of the resonance integral β. Table 6.1 shows resonance energies for selected polyenes. There is a qualitative agreement between the calculated values and the experimental values. Figures 6.4, 6.5 and 6.6 can be used to illustrate the energy levels and the π-orbitals in different species.

Table 6.1 Resonance energies of a few conjugated polyenes. The Hückel values have been calculated with $2\beta = -155$ kJ mol^{-1}.

Molecule	Hückel result	Empirically calculated value (kJ mol^{-1})	Experimental value (kJ mol^{-1})
Butadiene	0.472β	36	15
Benzene	2.00β	155	155
Diphenyl	4.38β	339	330
Naphthalene	3.68β	285	314
Anthracene	5.32β	412	439
Phenanthrene	5.45β	422	460

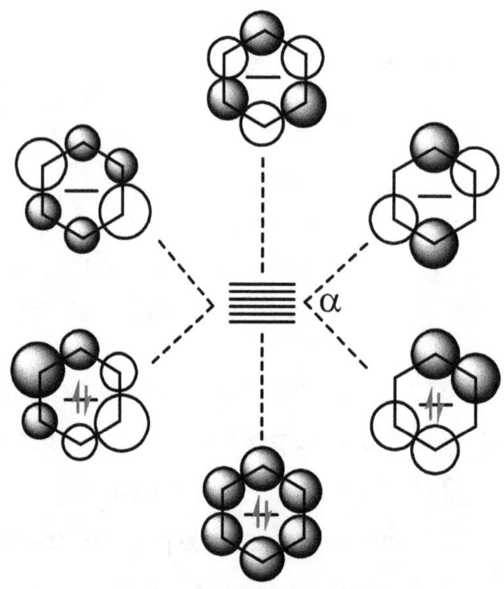

Figure 6.6 Hückel π-orbital energy level diagram for benzene. Shaded circles represent positive lobes of atomic functions above the plane of paper. Unshaded circles stand for negative lobes above and positive lobes below the plane of paper.

A few general observations are made now.

Example 35: One can show that the orbital energies for a *linear polyene* of *N* conjugated carbon atoms are given by

$$E_k = \alpha + 2\beta\cos\left(\frac{k\pi}{N+1}\right) \qquad (k = 1, 2, \dots, \mathrm{N}). \tag{32.10}$$

The corresponding pi molecular orbitals are

$$\pi_k = \left(\frac{2}{\mathrm{N}+1}\right)^{1/2} \sum_{m=1}^{N} \sin\left(\frac{mk\pi}{N+1}\right) \mathrm{p}_{zm}. \tag{32.11}$$

There is a nonbonding orbital whenever N is an odd number. As N increases, separation between successive energy levels decreases and a band of width $4|\beta|$ tends to be formed. This has been illustrated in Figure 6.7.

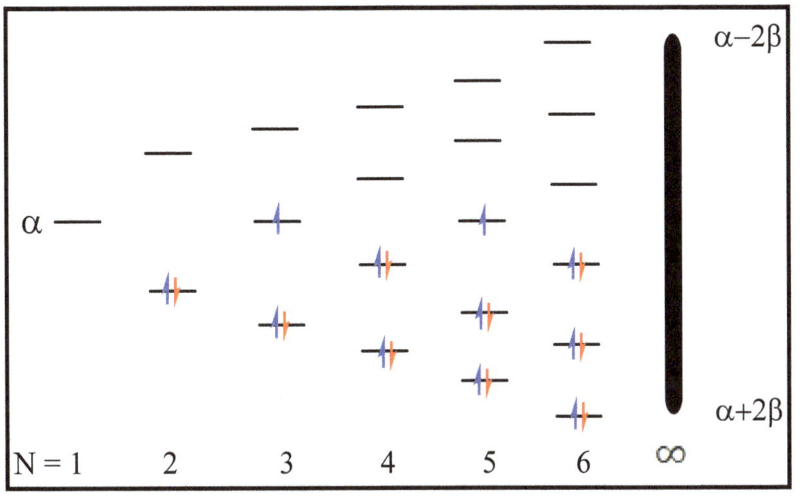

Figure 6.7 Formation of the energy band in C_N linear conjugated polyene as $N \rightarrow \infty$

Example 36: For a *cyclic polyene* of N conjugated carbon atoms one obtains slightly different expressions for the orbital energies,

$$E_k = \alpha + 2\beta \cos\left(\frac{2k\pi}{N}\right) \qquad (k = 0, 1, 2, \ldots, N-1), \qquad (32.12)$$

and different expressions for the coefficient of the *mth* atomic orbital (of orbital p_{zm}) in the *kth* molecular orbital (orbital π_k):

$$c_{m,0} = N^{-1/2};\ c_{m,(N/2)} = (-1)^{m+1} N^{-1/2}; \qquad (32.13)$$

and for $k \neq 0$ as well as $k \neq N/2$, two sets of coefficients corresponding to double degeneracy,

$$c_{mk} = \left(\frac{2}{N}\right)^{1/2} \cos\left(\frac{2\pi mk}{N}\right) \qquad \text{(cosine pattern)}, \qquad (32.14)$$

$$c_{mk} = \left(\frac{2}{N}\right)^{1/2} \sin\left(\frac{2\pi mk}{N}\right) \qquad \text{(sine wave)}.$$

There is a pair of nonbonding orbitals whenever N is a multiple of 4. Again a band is formed with width 4β as N increases. Figure 6.8 illustrates the band formation in the case of cyclic systems. From this Figure we see that the following rules are valid for a cyclic polyene:

Hückel's rules:
i) The molecule is highly stable when $N = 4n + 2$, n being an integer.
ii) The molecule is a free radical when N is odd, ($N = 4n+1$ or $N = 4n+3$).
iii) The molecule has a highly unstable triplet ground state when $N = 4n$.
This possibility follows from Hünd's rule of maximum spin multiplicity.

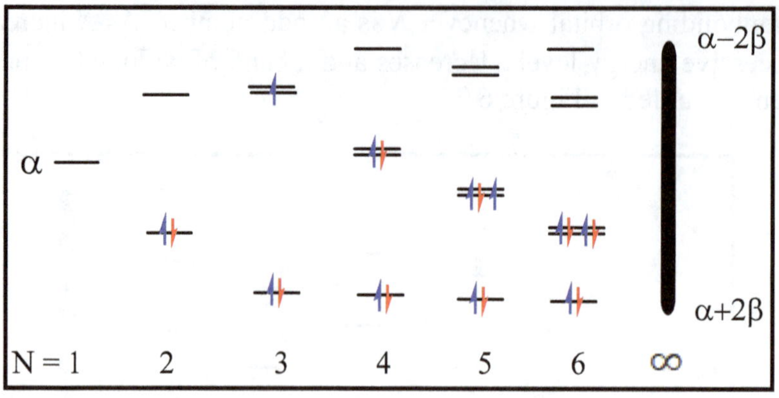

Figure 6.8 Formation of the energy band in C_N cyclic conjugated polyene as $N \to \infty$. Notice that for N=4 the singlet electronic population has been shown. The triplet has parallel spins in the degenerate HOMOs, and as an isolated molecule it is more stable than the singlet. It is also more reactive, hence not stable in the proximity of other molecules.

The above rules are known as *Hückel's rules*, especially the first one that is known as the Hückel's *4n+2 rule* for aromaticity.

In the two examples above, the $N=\infty$ cases are correct as far as the Hückel theory that is based on standing waves is concerned, and a band gap may exist for the cyclic polyene. These $N=\infty$ solutions are not strictly correct as the existence of translational symmetry in an infinite lattice forces the wave function to the form of Bloch function that is a product of a plane wave times a function with the periodicity of the lattice. A band gap between π and π^* bands becomes possible for both linear and cyclic polyenes.

Lecture 33: Pi-Electron Properties

The *Mulliken population matrix* is calculated as follows. The probability that one electron in orbital π_k is shared by the *mth* and *nth* atomic orbitals together is given by the product $c_{mk}c_{nk}$. If n_k is the population of π_k, then $n_k c_{mk} c_{nk}$ is the probability that the n_k electrons in orbital π_k are shared by the p_z orbitals on carbon atoms C_m and C_n. The total probability of sharing, called the *Mulliken population*, is given by

$$P_{mn} = \sum_{k}^{\pi\ mo} n_k c_{mk} c_{nk}. \tag{33.1}$$

The *π-electron charge* of the *mth* carbon atom is defined by

$$q_m = P_{mm} = \sum_{k}^{\pi\ mo} n_k c_{mk}^2, \tag{33.2}$$

and the *overlap population* by

$$q_{mn} = 2S_{mn}P_{mn} = 2S_{mn} \sum_{k}^{\pi\ mo} n_k c_{mk} c_{nk} \qquad (m \neq n). \tag{33.3}$$

Each carbon atom has a core charge of +1 in the π-electron model. This leads to a *local atomic charge*

$$\delta_m = 1 - q_m. \tag{33.4}$$

The *bond order* for the binding of C_m with C_n through a π bond is given by

$$p_{mn} = P_{mn}. \tag{33.5}$$

The free valence index is another important concept. The free valence index for the *mth* carbon atom is defined by

$$F_V(m) = C_{nh}(m) - \sum_{n(\neq m)}^{C\ atoms} p_{mn} \tag{33.6}$$

where $C_{nh}(m)$ is a constant that depends on the state of hybridization of the *mth* atom. The constant $C_{nh}(m)$ is taken equal to 1.732 when the atom C_m is the state of sp^2 hybridization. The higher the free valence index is, the greater the π-electron reactivity of the carbon atom is. Free valence index greater than 0.732 indicates a center for electrophilic attack. Nucleophilic substitutions take place at centers with F_V less than 0.732. The free valence index can be regarded as a quantitative expression for Thiele's *partial valence* or Werner's *residual affinity*.

Let us check these molecular properties for some of the molecules considered as examples in the previous lectures.

Example 37: Ethylene (Figure 6.1)

$P_{11} = 2\times2^{-1/2}\times2^{-1/2} = 1$ $\quad P_{22} = 1$ $\quad P_{12} = 1$
$q_1 = q_2 = 1$ $\quad q_{12} = 0$ $\quad \delta_1 = \delta_2 = 0$ $\quad$ (See atomic charges in Figure 6.1)
$p_{12} = 1$ (bond order)
$F_V(1) = F_V(2) = 0.732$

Example 38: The allyl system (Figure 6.2)

(a) Radical $\pi_1^2\ \pi_2$
$q_1 = q_2 = q_3 = 1$ $\quad \delta_1 = \delta_2 = \delta_3 = 0$
$p_{12} = p_{23} = 0.707$ $\quad p_{13} = 0$
$F_V(1) = F_V(3) = 1.025$ $\quad F_V(2) = 0.318$
C_2 is a center for nucleophilic attack whereas C_1 and C_3 are centers for electrophilic reactions.

(b) Cation π_1^2
$q_1 = q_3 = 0.5$ $\quad q_2 = 1.0$
$\delta_1 = \delta_3 = 0.5$ $\quad \delta_2 = 0$
$p_{12} = p_{23} = 0.707$ $\quad p_{13} = 0.5$
$F_V(1) = F_V(3) = 0.525$ $\quad F_V(2) = 0.318$
All three carbon atoms are susceptible to nucleophilic attack.

(c) Anion $\pi_1^2\ \pi_2^2$
$q_1 = q_3 = 1.5$ $\quad q_2 = 1.0$
$\delta_1 = \delta_3 = -0.5$ $\quad \delta_2 = 0$
$p_{12} = p_{23} = 0.707$ $\quad p_{13} = -0.5$

$F_V(1) = F_V(3) = 1.525$ $F_V(2) = 0.318$
C_1 and C_3 are prominent centers for electrophillic attack, but C_2 remains as before a center for nucleophilic attack.

For both charged species we find that the overall charge is distributed over the atoms at the two ends of the chain. This is a general feature of electrostatics: the charge is always distributed over the outer surface. When the charge resides at the extremities, the system becomes stable as the charge-charge repulsion is minimized. This is also the general trend found for the results from quantum chemical calculations, and is best exemplified by the cyclic systems.

Example 39: The cyclopropenyl system (Figure 6.3)

(a) Radical $\frac{1}{\sqrt{2}}(\pi_1^2\,\pi_2 + \pi_1^2\,\pi_3)$

$q_1 = q_2 = q_3 = 1.0$ $\delta_1 = \delta_2 = \delta_3 = 0$
$p_{12} = p_{23} = p_{31} = 0.5$
$F_V(1) = F_V(2) = F_V(3) = 0.732$

(b) Cation π_1^2

$q_1 = q_2 = q_3 = 0.667$ $\delta_1 = \delta_2 = \delta_3 = 0.333$
$p_{12} = p_{23} = p_{31} = 0.667$
$F_V(1) = F_V(2) = F_V(3) = 0.399$
Nucleophilic attack can take place with equal ease at any carbon atom.

(c) Anion $\pi_1^2\,\pi_2\pi_3$

$q_1 = q_2 = q_3 = 1.333$ $\delta_1 = \delta_2 = \delta_3 = -0.333$
$p_{12} = p_{23} = p_{31} = 0.333$
$F_V(1) = F_V(2) = F_V(3) = 1.066$
Electrophilic attack can take place with equal ease at any carbon atom.

All the three carbon atoms have the same charge — zero in the neutral radical, +1/3 in the cation and −1/3 in the anion.

Alternant Hydrocarbons

Conjugated polyenes have a few common properties:
(i) For each bonding molecular orbital of energy $(\alpha + x\beta)$ there is an antibonding orbital of energy $(\alpha - x\beta)$ for a linear polyene.
(ii) For a linear polyene with an odd number of conjugated carbon atoms, there is one nonbonding orbital with energy α.
(iii) Each nonbonding orbital of a cyclic polyene is singly occupied to minimize the electron-electron repulsion energy in the ground state. Whether there are nonbonding orbitals or not, the π charge on each atom is 1 in a cyclic polyene in its ground state.

Consider an alternant radical. This must have an odd number of carbon atoms – see Figure 6.9 – and at least one nonbonding molecular orbital. One finds, for each nonbonding molecular orbital, a linear equation of the form

$$c_{i-1} + c_{i+1} = 0 \qquad (i = 1, 2, \ldots, N) \qquad (33.7)$$

that serves as a mnemonic device for guessing the form of the orbital. Let us star alternate carbon atoms in the alternant molecule such that the unstarred carbon atoms are fewer in number. There are $(N-1)/2$ node(s) in the nonbonding orbital, and these are located on the unstarred carbon atoms.

Example 40: Nonbonding orbitals of alternant hydrocarbons

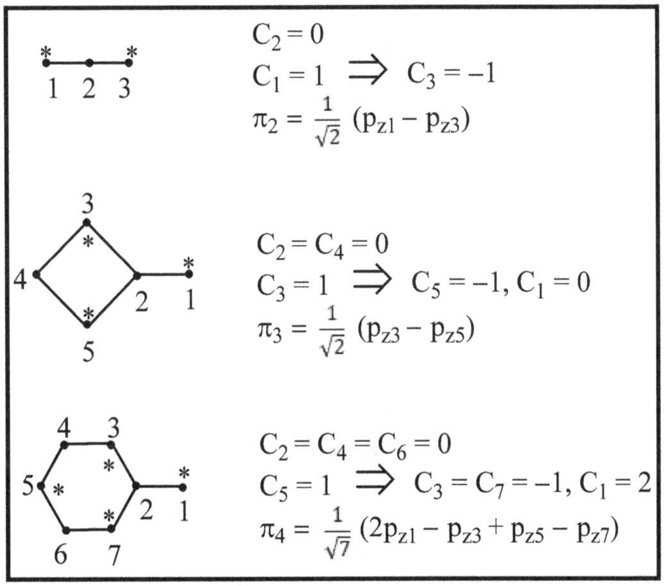

Figure 6.9 Finding non-bonding orbitals of alternant hydrocarbons.

The linear expansion coefficient of each atomic orbital corresponding to the unstarred set is zero, and the sum of the coefficients of the neighboring atomic orbitals is also zero. This observation helps us to write the nonbonding orbital as illustrated in the above.

The calculations shown here are in agreement with the observed radical reactivities and the hyperfine structure of the electron spin resonance (E.S.R.) signal. The hyperfine structure is generated from the coupling of the electron spin with the spins of the protons attached to the carbon atoms. The hyperfine structure shows that in the benzyl radical the odd electron is likely to be found in the terminal, ortho and para positions in the ratio 4:1:1, and never in the meta positions. These probabilities are consistent with the linear expansion for π_4 of benzyl radical as shown above.

The knowledge of the nonbonding orbital allows us to write the most important canonical structures involved in resonance by following the location of the unpaired electron. This is illustrated in Figure 6.10. Alternatively, from these most common canonical structures, one can easily write down the expression for the nonbonding orbital in the alternant hydrocarbon.

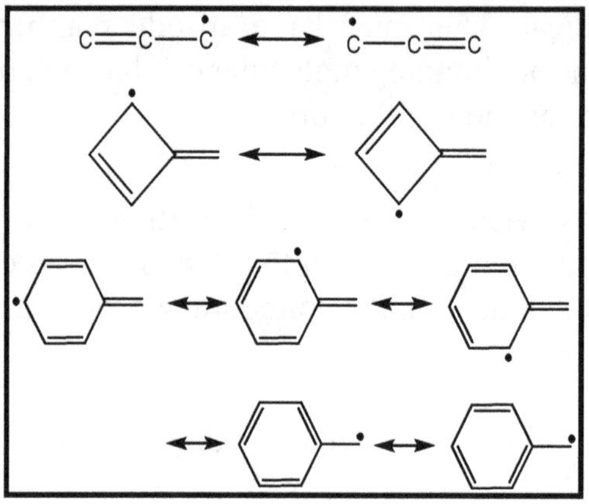

Figure 6.10 The most important canonical covalent structures involved in resonance.

Lecture 34: The Extended Hückel Method

Heteroatom substitutions

The Hückel method is modified by using additional parameters α_X, β_{CX} and β_{XX} where X is a heteroatom in the conjugated system. This modification leads to somewhat different values for the orbital energies. The linear expansion for the π orbital now contains the $2p_z$ orbital of the heteroatom X in addition to the $2p_z$ orbitals of the carbon atoms. The most important difference is created by the number of $2p_z$ electrons in the heteroatom. Thus a nitrogen atom in sp^2 hybridized state has two $2p_z$ electrons and three σ orbitals, while an oxygen atom in a similar state has two $2p_z$ electrons, two σ orbitals and a pair of electrons in a nonbonding orbital oriented along the third sp^2 direction.

The difference $(\beta_{CN} - \beta)$ is small and it introduces a very small second-order effect on the charge distribution. Hence one may employ the same parameter β for conjugated molecules with one or more carbon atoms substituted by nitrogen or nitrogen-containing groups. Representative values for α_X have been listed by Streitwieser (1961). The parameter α_X is approximated by

$$\alpha_X = \alpha + k_X\beta. \tag{34.1}$$

For nitrogen, $k_N = \frac{1}{2}$. The π-electron charges estimated for pyrrole, pyridine and aniline are shown in Figure 6.11.

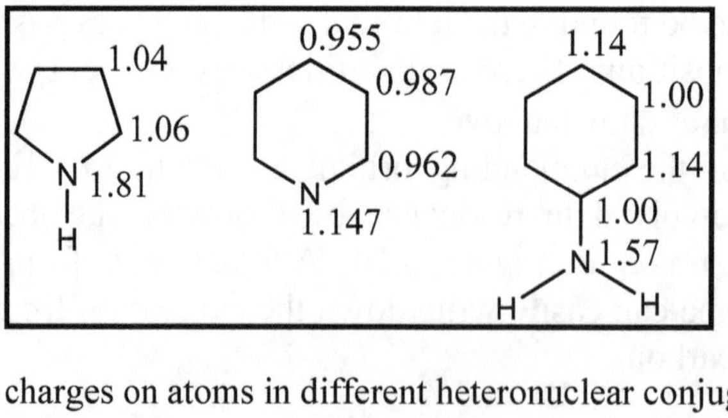

Figure 6.11 Pi electron charges on atoms in different heteronuclear conjugated organic molecule

General improvement.

The Hückel calculation can be generally improved by bringing in a nonzero overlap integral for two p_z orbitals on neighboring atoms:

$$S_{i,i\pm 1} = S \neq 0. \qquad (34.2)$$

If we express each $2p_z$ orbital by a single STO we get

$$S = \frac{\zeta_{2p}^5}{\pi} \int dv\, z^2 \exp\left[-\zeta_{2p}\,|\mathbf{r} - \mathbf{R}/2| - \zeta_{2p}|\mathbf{r}+\mathbf{R}/2|\right]$$

$$= \frac{1}{\pi} \int dv\, z^2 \exp\left[-|\mathbf{r} - \zeta_{2p}\mathbf{R}/2| - |\mathbf{r} + \zeta_{2p}\mathbf{R}/2|\right]. \qquad (34.3)$$

The exponent of the STO is taken as $\zeta_{2p} = 3.25$. The absolute magnitude of the argument of the exponential function in the integral in Eq. (34.3) is at least equal to $\zeta_{2p}R$ which for $R \sim 2.5$ a.u. approximates 8.1. The argument increases in magnitude as z increases. Thus S is a rather small quantity for the π electron treatment, and the results calculated with Eq. (34.2) do not greatly differ from those calculated with Hückel's original approximation $S_{i,i\pm 1} = 0$.

There are two major difficulties with the Hückel treatment. *First*, the Hückel theory is an empirical theory with α and β parameters adjusted by comparing the results of calculations with specific experimental results for specific molecules. For instance, the parameter β is set equal to half the resonance energy of benzene. The parameter α may be calculated from the first ionization potential of the same molecule, or of the methylene radical, or it may even be equated to the negative of the first ionization potential of carbon. Normally one sets the value of α at $-I_{2p}$ for the carbon atom. There is no reason why the α and β parameters, once chosen from the data for specific atoms or molecules, will lead to accurate results for all other molecules. *Second*, the Hückel theory does not explicitly consider the electron repulsion effects. If these effects are taken to be inherent in the orbital energies, then the total energy suffers from a double-counting of these repulsions. If these effects are inherent in the total energy, then the orbital energies suffer. If α is chosen to satisfy the orbital energies and β is taken to satisfy the total energy of a particular molecule, the parametrization scheme gives rise to elements of inaccuracy in the orbital energies as well as the total energy. So the Hückel energies are at best qualitatively correct. A *third* type of inaccuracy enters through the choice of only the valence p_z orbital of each conjugated atom, that is, through the neglect of the contribution of p_z atomic orbitals of higher principal quantum numbers. But this is often found to be a minor source of error for the π orbitals.

The best point of the Hückel theory is that the Hamiltonian matrix has the symmetry of the molecule and so the π-orbitals calculated are quite good, (although contributions from the $3p_z$, $4p_z$ etc. orbitals are missing). Hence the π-molecular properties discussed in Lecture 33 are quite nicely evaluated.

The integral S is about 0.1 for σ orbitals. The Hückel scheme along with the provision of nonzero nearest-neighbor overlap integrals as in Eq. (34.2) may be extended to the additional treatment of the σ electrons. The resulting method is known as the *extended Hückel treatment*.

Extended Hückel Treatment (EHT)

The ordinary Hückel treatment can be extended in two ways – (i) by including the σ electrons in the treatment, and (ii) by including the valence (σ and π) electrons of all other atoms (other than the carbon atoms) in the treatment. The most popular parametrization scheme for the extended Hückel treatment was suggested by Wolfsberg and Helmholz (1952). It involves three rules:

Wolfsberg-Helmholz Rules:

(i) $H_{rr} = -I_r$ (34.4)

where I_r is the valence state ionization energy for a specific orbital of the metal atom or the ligand atom orbital. A list of H_{rr} matrix elements is given in Table 6.2.

(ii) $H_{rs} = \frac{1}{2} kS_{rs} (H_{rr} + H_{ss})$ (34.5)

where orbitals r and s are on two different atoms and S_{rs} is the group overlap integral. The exponents used for the calculation of the latter integrals are given in Table 6.3. The constant k has the value 0.75 for σ overlap and 1.0 for π overlap. If r and s are orbitals of the same atom, then $H_{rs} = 0$.

(iii) $S_{rr} = 1$, (34.6a)

and $S_{rs} = 0$ (34.6b)

if r and s are on the same atom. Otherwise S_{rs} is calculated from the group (σ or π) overlap integrals and the relative geometry of the two atoms containing orbitals r and s.

Standard computer programs for the extended Hückel calculation are available from various sources. The extended Hückel calculations have been very successful in explaining in qualitative terms the structural features, spectroscopic characteristics and chemical reactivities of transition metal complexes, particularly those complexes which contain organic ligands, and organometallic compounds. With the advent of Gaussian codes for quantum chemical calculations and the increasing speed of numerical computing, the extended Hückel calculations have become old fashioned. It still remains good for understanding the basics of inorganic chemistry.

Table 6.2 The extended Hückel H_{rr} for different elements in electron Volt. The (p, p) components may be used in a simple Hückel π-electron treatment.

Element	1s	2s	2p
H	−13.60585		
Li		−5.34227	−3.49861
Be		−8.41572	−5.63154
B		−13.46176	−8.43172
C		−19.37633	−11.07236
N		−26.22301	−13.84126
O		−34.02433	−16.76756
F		−42.79131	−19.86494

Table 6.3 Exponents used for calculating the group overlap integrals.

Element	1s	2s	2p
H	1.00000		
Li		0.64538	0.52377
Be		0.94360	0.87543
B		1.26450	1.13391
C		1.57648	1.43470
N		1.88490	1.72816
O		2.19159	2.01763
F		2.49727	2.30460

Orbital energies of a few homonuclear diatomic molecules of light elements which have been calculated by the extended Hückel method are given in Table 6.4. This Table should be used to check only the qualitative trends. The values obtained from an empirical calculation can be considerably inaccurate and these may be used only to develop the concept of the structure of specific molecules or to identify such features as the mechanism of a specific reaction.

Table 6.4 Extended Hückel orbital energies in atomic unit for homonuclear diatomics of elements of the first and second rows in the periodic table.

	$1\sigma_g$	$1\sigma_u$	$2\sigma_g$	$2\sigma_u$	$1\pi_u$	$3\sigma_g$	$1\pi_g$	$3\sigma_u$
H_2	−0.661	0.639						
Li_2			−0.253	−0.176	−0.163	−0.086	−0.013	2.477
Be_2			−0.389	−0.269	−0.249	−0.194	−0.116	1.381
B_2			−0.621	−0.422	−0.375	−0.282	−0.163	2.289
C_2			−0.894	−0.579	−0.493	−0.367	−0.208	3.653
N_2			−1.198	−0.760	−0.609	−0.472	−0.297	3.433
O_2			−1.471	−1.021	−0.695	−0.635	−0.496	0.659
F_2			−1.707	−1.425	−0.769	−0.791	−0.685	−0.433

Thus we find that a lot of structural characteristics and molecular properties can be explained even by an empirical theory that is based on Ritz variation principle. We are now ready for a fully quantitative approach that is termed *ab-initio* as it "starts from the first principles". Although *ab-initio* in nature, the method discussed hereafter is better known as the *qualitative molecular orbital* treatment. The reason for this nomenclature is that the basis functions used for the linear expansion of the molecular orbitals are grossly insufficient in number and so the Ritz variation technique yields molecular orbitals and orbital energies which are only qualitatively correct.

Lecture 35: The H_2^+ Molecule-Ion

The electronic Hamiltonian is [see Figure 2.3(iii)]

$$\mathbf{H}_{el} = -\frac{1}{2}\nabla^2 - \frac{Z}{|\mathbf{r}-\mathbf{R}_\mathrm{A}|} - \frac{Z}{|\mathbf{r}-\mathbf{R}_\mathrm{B}|} \tag{35.1}$$

where $Z = 1$ for H_2^+. For the sake of convenience we take $\mathbf{R}_A = -\mathbf{k}\frac{R}{2}$ and $\mathbf{R}_B = \mathbf{k}\frac{R}{2}$ such that the z axis is the molecular axis and $R_{AB} = |\mathbf{R}_A - \mathbf{R}_B| = R$. If ψ is the solution of the Schrödinger equation with $\mathbf{H}_{el}$ as Hamiltonian, and ϕ_A and ϕ_B are similar orbitals of atoms A and B, we must have for each specific value of R,

$$\psi \to \phi_A \text{ as } \mathbf{r} \to \mathbf{R}_A$$

and

$$\psi \to \pm\,\phi_B \text{ as } \mathbf{r} \to \mathbf{R}_B\,.$$

That ϕ_A and ϕ_B should be similar functions is a consequence of the symmetry of the molecule. Therefore, we write the molecular orbital ψ as

$$\psi = c_1\phi_A + c_2\phi_B. \tag{35.2}$$

The lowest-lying molecular orbitals correspond to the choice $\phi_A = 1s_A$ and $\phi_B = 1s_B$ where $1s_A$ and $1s_B$ are 1s-like orbitals on atoms A and B and they should reduce to the 1s orbitals of isolated atoms A and B as R becomes infinitely large.

The linear variation technique using the expansion (35.2) yields

$$\psi_+ = [2(1+S)]^{-1/2}(\phi_A + \phi_B); \qquad E_+^{el} = \frac{H_{11}+H_{12}}{1+S}$$

and

$$\psi_- = [2(1-S)]^{-1/2}(\phi_A - \phi_B); \qquad E_-^{el} = \frac{H_{11}-H_{12}}{1-S} \tag{35.3}$$

where

$$H_{11} = H_{22} = \int dv\; \phi_A(\mathbf{r})\mathbf{H}_{el}\phi_A(\mathbf{r}),$$

$$H_{12} = H_{21} = \int dv\; \phi_A(\mathbf{r})\mathbf{H}_{el}\phi_B(\mathbf{r}), \tag{35.4}$$

and

$$S = \int dv\; \phi_A(\mathbf{r})\phi_B(\mathbf{r}).$$

Note that the combinations $c_1 = \pm\, c_2$ can be obtained directly from the symmetry of the molecule ion. Under 180° rotation around the molecular center the orbital ψ is transformed as

$$\psi \to \psi' = c_1\phi_B + c_2\phi_A. \tag{35.5}$$

As the molecule attains an equivalent configuration,

$$\psi^2 = \psi'^2 \tag{35.6}$$

which immediately leads to the solutions $c_1 = c_2$ for ψ_+ and $c_1 = -c_2$ for ψ_-.

If we fix ϕ_A and ϕ_B to be the hydrogenic 1s functions corresponding to the hydrogenic energy $E_H^Z = -\frac{1}{2}Z^2$ then

$$\mathbf{H}_{el}\phi_A = E_H^Z\,\phi_A - \frac{Z}{|\mathbf{r}-\mathbf{R}_B|}\phi_A \tag{35.7}$$

with which we can simplify H_{11} and H_{12} in equations (35.4) as

$$H_{11} = H_{22} = E_H^Z + \boldsymbol{J}$$

$$H_{12} = H_{21} = E_H^Z S + \boldsymbol{K} \tag{35.8}$$

where the integrals $\boldsymbol{J}$ and $\boldsymbol{K}$ are given by

$$J = -Z\int dv\ \phi_A(\mathbf{r})\,|\,\mathbf{r}-\mathbf{R}_B|^{-1}\phi_A(\mathbf{r}),$$

$$K = -Z\int dv\ \phi_B(\mathbf{r})\,|\,\mathbf{r}-\mathbf{R}_B|^{-1}\phi_A(\mathbf{r}). \qquad (35.9)$$

The orbital energies are now written as

$$E_{+}^{el} = E_{H}^{Z} + \frac{J+K}{1+S}$$

$$E_{-}^{el} = E_{H}^{Z} + \frac{J-K}{1-S}. \qquad (35.10)$$

The level splitting is shown in Figure 6.12.

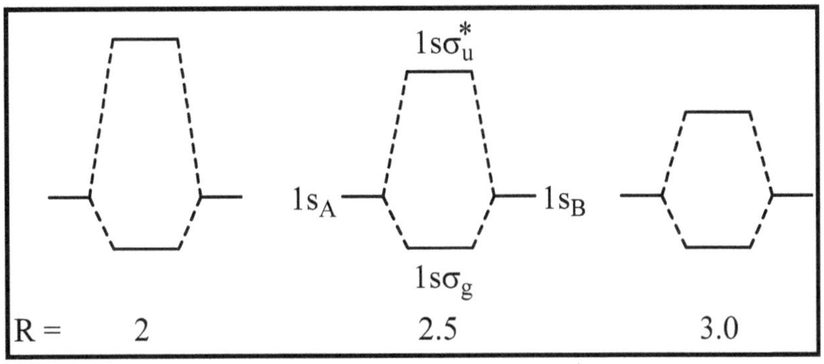

Figure 6.12 Level splitting in the QMO treatment of the H_2^+ molecule-ion with internuclear separation 2.0, 2.5 and 3.0 a. u. The internuclear repulsion energy (R^{-1} in a. u.) has been added to the orbital energy.

The orbitals ψ_+ and ψ_- are symmetric around the molecular axis, that is, they are cylindrically symmetric. The corresponding z-component of angular momentum operator Λ_z has the eigenvalue $\lambda\hbar$, (λ= 0 here). The molecular orbitals are usually named according to the quantum number λ. Thus

$$\begin{aligned} &\lambda = 0 \leftrightarrow \sigma \quad \text{(greek for s, s} \leftrightarrow l=0),\\ &\lambda = 1 \leftrightarrow \pi \quad \text{(greek for p, p} \leftrightarrow l=1),\\ &\lambda = 2 \leftrightarrow \delta \quad \text{(greek for d, d} \leftrightarrow l=2),\\ &\lambda = 3 \leftrightarrow \phi \quad \text{(greek for f, f} \leftrightarrow l=3), \end{aligned} \qquad (35.11)$$

etc. The orbitals ψ_+ and ψ_- are both σ orbitals. Since they are formed from 1s atomic orbitals, they are both called 1sσ. Since $E_{+}^{el} < E_H$ and $E_{-}^{el} > E_H$, ψ_+ and ψ_- are termed 1sσ (for bonding) and 1sσ^* (for antibonding). The molecule H_2^+ has a center of inversion as one of its symmetry elements. Under reflection through this center, ψ_+ remains thoroughly unchanged and ψ_- remains unchanged in the absolute magnitude but it changes its sign:

$$\psi_+ = [2(1+S)]^{-1/2}(\phi_A + \phi_B) \xrightarrow{\hat{i}} [2(1+S)]^{1/2}(\phi_B + \phi_A) = \psi_+,$$

and (35.12)

$$\psi_- = [2(1-S)]^{-1/2}(\phi_A - \phi_B) \xrightarrow{\hat{i}} [2(1-S)]^{-1/2}(\phi_B - \phi_A) = -\psi_-.$$

Hence ψ_+ is gerade (german for symmetric) and ψ_- is ungerade (german for antisymmetric) under the inversion operation, and these states are finally termed as $1s\sigma_g$ (g for gerade) and $1s\sigma_u^*$ (u for ungerade) orbitals.

Both Figure 6.13 and Figure 6.14 can explain why ψ_+ is a bonding state and ψ_- is an antibonding one. Both the diagrams show that in the ψ_+ state the probability of finding the electron in the region between the two nuclei increases from the atomic value. Since the electron pulls the two nuclei towards itself and the two nuclei repel each other, an equilibrium internuclear distance is reached where the opposing forces exactly balance each other. The increased charge distribution in the internuclear region "cements" the two nuclei. The antibonding state is associated with a cancellation of the wave function between the nuclei. This cancellation results in a decrease of the electron cloud in the internuclear region and a build-up of the charge cloud outside the two nuclei. The effect is now opposite of the phenomenon of "cementing" as discussed in the previous case. The pull by the increased charge distribution in the outward direction and the mutual repulsion of the two nuclei now assist each other. Hence the nuclei move apart. This situation is opposite to chemical binding.

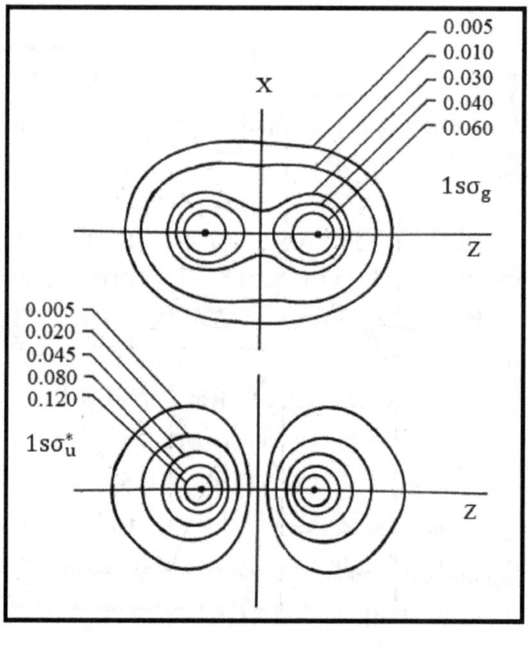

Figure 6.13 Surfaces of constant electron density in H_2^+ molecule-ion.

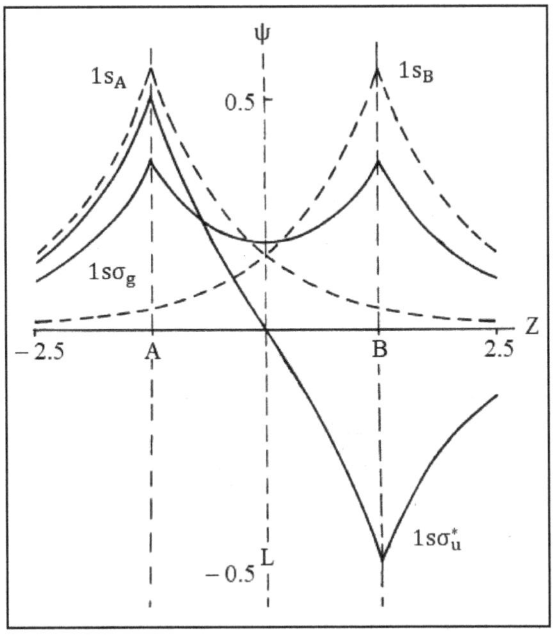

Figure 6.14 Another way of understanding the bonding and antibonding effects in terms of the wave function along the z-axis (the molecular axis). The $1s\sigma_g$ has a large value in the internuclear region whereas the $1s\sigma_u{}^*$ is more prominent in the outward directions.

A pictorial representation of the formation of the molecular orbitals $1s\sigma_g$ and $1s\sigma_u{}^*$ is given in Figure 6.15. This diagram may be considered similar to Figure 6.13 except that the constant probability density surface is now drawn for the atomic orbital at the most probable distance and that the sign of ψ has been indicated in every region of space.

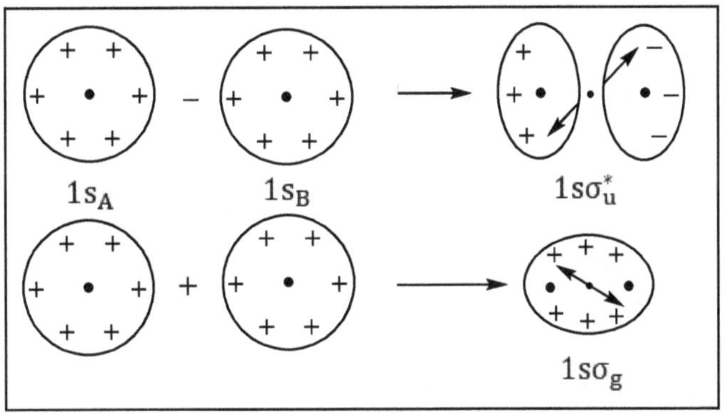

Figure 6.15 Formation of the QMOs $1s\sigma_g$ and $1s\sigma_u{}^*$ from the 1s orbitals of two participating atoms – a pictorial representation.

The bonding and antibonding nature of ψ_+ and ψ_- can also be explained by the potential energy curves

$$E_+ = E_+^{el} + \frac{Z^2}{R}, \quad E_- = E_-^{el} + \frac{Z^2}{R}. \tag{35.13}$$

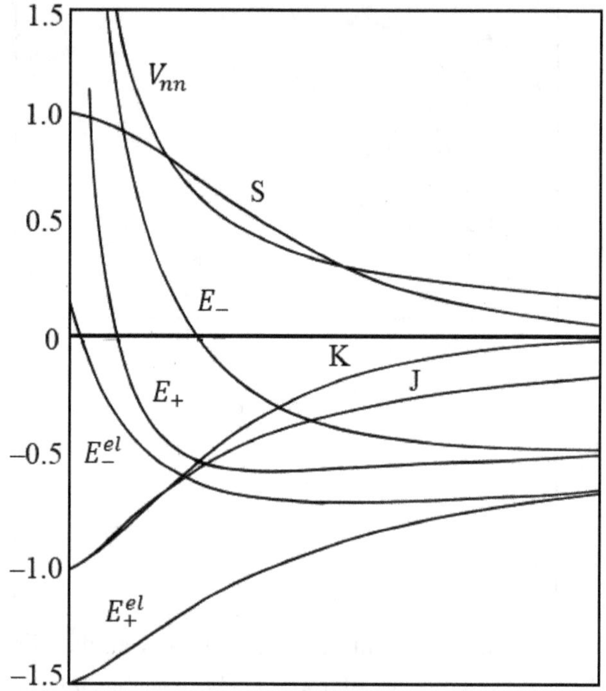

Figure 6.16 Energy components calculated for the H_2^+ molecule-ion.

It is shown in Figure 6.16 that E_+^{el} monotonically increases with R whereas both E_-^{el} and $V_{nn} = \dfrac{Z^2}{R}$ monotonically decreases as R increases. The outcome is an energy minimum for the bonding surface $E_+(R)$ and an accentuated monotonic fall-off for the antibonding curve $E_-(R)$.

The bonding and antibonding nature of ψ_+ and ψ_-respectively can now be ascertained with the help of the Hellmann-Feynman theorem that was introduced in Lecture 3 (Chapter 1).

Lecture 36: Improvement of Calculation; The H_2 Molecule

Improvement

The qualitative molecular orbital treatment explains the bonding of H_2^+. Notwithstanding its success, this treatment fails to generate quantitative accuracy for molecular properties such as equilibrium bond length and dissociation energy. The calculated bond length is 1.32Å and the calculated dissociation energy is 1.77eV. The experimental values are 1.06 Å and 2.78eV respectively. The latter values must correspond to the accurate solution of the Schrödinger equation. Hence it is necessary to have a short discussion on the ways to improve the molecular orbital calculation.

A quick improvement is obtained from the consideration of two more limiting behaviors of $\psi_\pm$. In united atom limit, that is, as $R \to 0$,

$S \to 1$ and $H_{12} \to H_{11}$ (36.1)

so that

$\psi_+ \to \phi_A \equiv \phi_B$ and $E_+ \to H_{11} = 3\,E_H^Z$,

and (36.2)

$\psi_- \to 0.$

The combined atomic number is $2Z$. The energy value $3\,E_H^Z$ is the expectation value calculated with the wave function $\phi(Z)$. It is different from the expected eigenvalue $4\,E_H^Z$. So the calculated wave function is not truly representative of the united system. In fact the exponent of the wave function has not been varied, but it has been maintained at Z. The true ground state wave function of the united system has the exponent $2Z$. Thus the correct united atom limit may be obtained only after the exponent is allowed to change. Calculation of the separated atom limit is another story. As $R \to \infty$,

$S \to 0$ and $H_{12} \to 0$ (36.3)

so that in separated atom limit

$E_+ \to H_{11} = E_H^Z,\ E_- \to H_{11} = E_H^Z;$

$2^{-1/2}(\psi_+ + \psi_-) \to \phi_A,\ 2^{-1/2}(\psi_+ - \psi_-) \to \phi_B.$ (36.4)

All of the separated atom results are correct. This happens as the basis functions employed have correct exponents for the separated atom limit. This indicates that the orbitals ϕ_A and ϕ_B must be characterized by an effective nuclear charge that varies from $2Z$ to Z as R changes from 0 to ∞. We should also stick to using basis functions of the form of 1s. Hence one writes

$$\phi_A = \frac{\zeta^{3/2}}{\pi^{1/2}} e^{-\zeta r_a}$$

and (36.5)

$$\phi_B = \frac{\zeta^{3/2}}{\pi^{1/2}} e^{-\zeta r_b},$$

calculates the energy $E(\zeta)$ which is a function of the exponent ζ, and finally minimizes $E(\zeta)$ by varying ζ at each value of R. This treatment leads to the proper behavior of $\psi_\pm$ in united and separated atom limits. The potential energy curve is now vastly improved with equilibrium bond length 1.06Å and dissociation energy 2.25eV. See the dashed curve in Figure 6.17. But the calculated dissociation energy is still a lot different from the experimental energy, and the calculation of a better number within the QMO framework takes an enormous effort in the form of very complicated basis functions.

For a steady improvement of the calculated results one must reach beyond the frame work of the qualitative molecular orbitals. The systematic approach is to expand the molecular orbital ψ in a larger basis set that includes 1s, 2s, $2p_z$, 3s, $3p_z$, etc. atomic orbitals for each atom. The $2p_x$, $2p_y$, etc. orbitals do not contribute to the σ molecular orbitals in the present case; they form the π orbitals as the molecular axis is parallel to the z-axis. By accommodating more and more atomic orbitals as basis functions in the calculation one makes the basis set less and less incomplete, and the calculated energy curve approaches the experimental one.

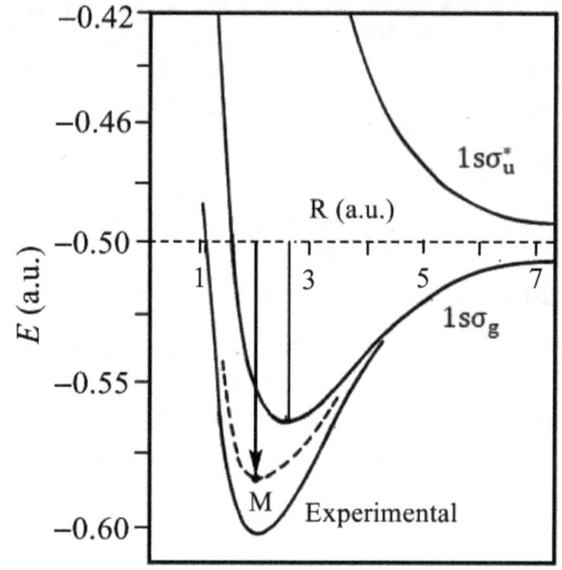

Figure 6.17 Energy curves computed for the H_2^+ molecule-ion.

The H2 molecule

The electronic Hamiltonian for this molecule is written as [see Figure 2.3(iv)]

$$\mathbf{H}_{el} = \left(-\frac{1}{2}\nabla_1^2 - \frac{Z}{r_{1A}} - \frac{Z}{r_{1B}}\right) + \left(-\frac{1}{2}\nabla_2^2 - \frac{Z}{r_{2A}} - \frac{Z}{r_{2B}}\right) + \frac{1}{r_{12}} \tag{36.6}$$

where $Z = 1$ for H_2. The molecular geometry of H_2^+ is retained here. There is more than one electron and the molecular orbitals should be found by solving the Hartree-Fock equation. The qualitative molecular orbitals are, however, determined from the linear combination of just the 1s orbitals of the two atoms. Since the symmetry of H_2 is the same as that of H_2^+, the qualitative molecular orbitals of H_2 are same as those of H_2^+ :

$$\psi_+ = [2(1 + S_{AB})]^{-1/2}\,(\phi_A + \phi_B) \equiv 1s\sigma g;$$

$$\psi_- = [2(1 - S_{AB})]^{-1/2}\,(\phi_A - \phi_B) \equiv 1s\sigma u^*. \tag{36.7}$$

The qualitative orbital energies are calculated as expectation values of the Fock operator. Because the ground state electronic configuration is $1s\sigma g^2$, the QMO wave function is the corresponding Slater determinant [shown later in (36.12)], and the Fock operator is given by

$$\mathbf{F} = -\frac{1}{2}\nabla^2 - \frac{Z}{|\mathbf{r} - \mathbf{R}_A|} - \frac{Z}{|\mathbf{r} - \mathbf{R}_B|} + (2\mathbf{J}_\sigma - \mathbf{K}_\sigma) \tag{36.8}$$

where the direct and the exchange coulomb repulsion operators are written as

$$\mathbf{J}_\sigma(\mathbf{r}_1) = \int d^3r_2\, \psi_+(\mathbf{r}_2)\frac{1}{r_{12}}\psi_+(\mathbf{r}_2)$$

and (36.9)

$$\mathbf{K}_\sigma(\mathbf{r}_1)\psi(\mathbf{r}_1) = \int d^3r_2\, \psi_+(\mathbf{r}_2)\,\frac{1}{r_{12}}\,\psi(\mathbf{r}_2)\,\psi_+(\mathbf{r}_1).$$

This gives us the orbital energies

$$\varepsilon_+ = E_+^{el} + J_{\sigma\sigma}, \qquad (\text{as } K_{\sigma\sigma} = J_{\sigma\sigma}),$$

and (36.10)

$$\varepsilon_- = E_-^{el} + (2J_{\sigma\sigma^*} - K_{\sigma\sigma^*})$$

where $E_\pm^{el}$ are the energy values defined in Eq.(35.3). The level splitting is shown in Figure 6.18. The total electronic energy as an expectation value of $\mathbf{H}_{el}$ is

$$E_{el} = 2E_+^{el} + J_{\sigma\sigma} \tag{36.11}$$

in the ground state. The total energy in the ground state (also called *the ground state total energy* or even as *the electronic energy* in spectroscopy) is obtained by adding the internuclear repulsion energy to the total electronic energy: $E_{tot} = E_{el} + Z^2R^{-1}$. These quantities, E_{el} and E_{tot}, are functions of the inter-nuclear distance R. In particular, the plot of E_{tot} against R is known as the potential energy curve. The minimum of E_{tot} is also referred to as the electronic energy in the ground state.

The choice of $\phi_A = 1s_A$ and $\phi_B = 1s_B$ yields $D_e = 2.65\text{eV}$, $R_{eq} = 0.83\,\text{Å}$ and $\bar{\nu} = 4160\text{cm}^{-1}$. The experimental values are $D_e = 4.747\text{eV}$, $R_{eq} = 0.74\text{Å}$ and $\bar{\nu} = 4401\text{cm}^{-1}$. Again, the qualitative molecular orbitals can be improved by using atomic orbitals of the form (36.5) and minimizing the calculated total energy for variation of ζ. Thus one obtains $D_e = 3.489\text{eV}$, $R_{eq} = 0.73\text{Å}$ and $\bar{\nu} = 4580\text{cm}^{-1}$. The corresponding exponent, the optimized ζ, is 1.197 at the equilibrium separation. Kolos and Roothaan (1960) have shown that if the basis size is increased by including more and more atomic functions in it and simultaneously ζ is varied, one gets $D_{el} = 3.635\text{eV}$ that corresponds to the optimized value of 0.95 for ζ at the equilibrium bond length.

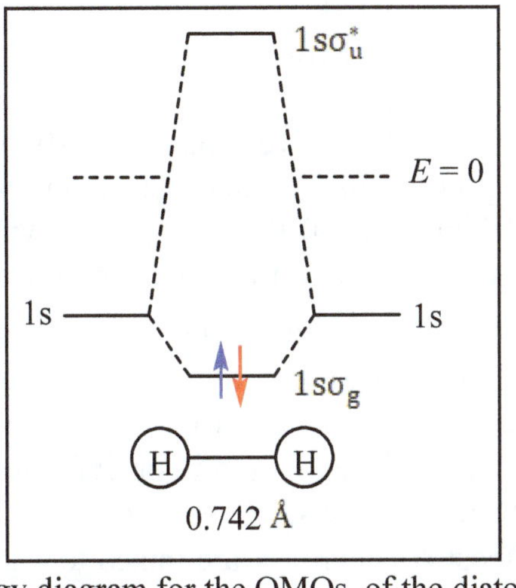

Figure 6.18 Orbital energy diagram for the QMOs of the diatomic hydrogen molecule.

The Slater determinant corresponding to the configuration $1s\sigma_g^2$ is

$$\Phi = 2^{-1/2}\begin{vmatrix} 1s\sigma_g(\mathbf{r}_1)\alpha(s_1) & 1s\sigma_g(\mathbf{r}_2)\alpha(s_2) \\ 1s\sigma_g(\mathbf{r}_1)\beta(s_1) & 1s\sigma_g(\mathbf{r}_2)\beta(s_2) \end{vmatrix}. \tag{36.12}$$

This factorizes into a coordinate-dependent part

$$\Psi(\mathbf{r}_1, \mathbf{r}_2) = 1s\sigma_g(\mathbf{r}_1)1s\sigma_g(\mathbf{r}_2) \tag{36.13}$$

that is symmetric under the interchange of spatial coordinates, and a two-electron spin function

$$X(s_1, s_2) = 2^{-1/2}\,[\alpha(s_1)\beta(s_2) - \alpha(s_2)\beta(s_1)] \tag{36.14}$$

that is antisymmetric under the interchange of spin coordinates.

The antisymmetric spin function is associated with the eigenvalue zero for $\boldsymbol{S}^2$ where $\boldsymbol{S} = \boldsymbol{S}_1 + \boldsymbol{S}_2$ is the sum of the spin angular momenta for the two electrons. Thus $S = 0$. The S_z eigenvalue is also zero. This spin state is nondegenerate: the spin multiplicity $2S+1$ is equal to 1. Therefore, it is referred to as a *Singlet* state.

Expression (36.13) makes obvious one drawback of the molecular orbital treatment. If we use Eq. (36.7) for the expansion of the molecular orbitals we find

$$\Psi(\mathbf{r}_1, \mathbf{r}_2) = [2(1 + S_{AB})]^{-1} \times [\{\phi_A(\mathbf{r}_1)\phi_B(\mathbf{r}_2) + \phi_A(\mathbf{r}_2)\phi_B(\mathbf{r}_1)\} + \{\phi_A(\mathbf{r}_1)\phi_A(\mathbf{r}_2) + \phi_B(\mathbf{r}_2)\phi_B(\mathbf{r}_1)\}]. \tag{36.15}$$

The term $\phi_A(\mathbf{r}_1)\phi_B(\mathbf{r}_2)$ represents a structure with electron 1 in atom A and electron 2 in atom B. The term $\phi_A(\mathbf{r}_2)\phi_B(\mathbf{r}_1)$ similarly indicates electron 2 with atom A and electron 1 with atom B. These two terms together represent a sharing of the electrons 1 and 2 by the atoms A and B. So the first part of the two-electron spatial function, which in the normalized form is

$$\Psi_{\text{cov}} = [2(1 + S_{AB}{}^2)]^{-1}\{\phi_A(\mathbf{r}_1)\phi_B(\mathbf{r}_2) + \phi_A(\mathbf{r}_2)\phi_B(\mathbf{r}_1)\}, \tag{36.16}$$

represents a covalent structure. The two terms belonging to the second part of Ψ represent the ionic structures A^-B^+ and A^+B^-. The net ionic wave function is

$$\Psi_{\text{ion}} = [2(1 + S_{AB}{}^2)]^{-1}\{\phi_A(\mathbf{r}_1)\phi_A(\mathbf{r}_2) + \phi_B(\mathbf{r}_2)\phi_B(\mathbf{r}_1)\}. \tag{36.17}$$

Hence the molecular orbital wave function $\Psi(\mathbf{r}_1, \mathbf{r}_2)$ is seen to interpret the chemical bond H—H as an equal combination of covalent and ionic structures. The ionicity of the bond is thus predicted to be 50%, although experiment tends to suggest a bond polarity of about 2%. The molecular orbital description always overemphasizes ionicity in primarily covalent molecules. This is a general feature of the simple molecular orbital method.

Lecture 37: Homonuclear Diatomic Molecules

We now turn our attention to the qualitative molecular orbitals of homonuclear diatomic molecules of elements belonging to the first and the second periods of Mendelev's table. The qualitative molecular orbital (QMO) theory does not yield accurate energies, but it establishes all the major trends of the properties of homonuclear diatomic molecules. The formalism is as follows. We have already discussed the formation of $1s\sigma_g$ and $1s\sigma_u{}^*$ orbitals from the 1s atomic orbitals of two atoms of the same element. We have also encountered two remarkable properties while discussing the noncrossing rule:

- Atomic orbitals combine only with atomic orbitals of the same symmetry to form (the qualitative) molecular orbitals.
- Atomic orbitals of almost equal energy are mixed to a great extent. The extent of the mixing decreases as the difference in energy of the two participating orbitals increases. Two orbitals differing from each other by a large amount of energy will hardly combine.

These two observations lead to a set of principles for forming the qualitative molecular orbitals of a homonuclear diatomic molecule.

Principles for the formation of QMOs of homonuclear diatomics

(i) *Linearly combine atomic 1s orbitals to form the molecular orbitals* $1s\sigma_g$ *and* $1s\sigma_u^*$ as

$$1s\sigma_g = [2(1 + S_{1s_A,1s_B})]^{-1/2}(1s_A + 1s_B)$$

$$1s\sigma_u^* = [2(1 - S_{1s_A,1s_B})]^{-1/2}(1s_A - 1s_B) \qquad (37.1)$$

For the second-row elements, 1s orbital is a low-lying core orbital. It turns out that $1s_A$ and $1s_B$ have negligible overlap and interaction so that the bonding and the antibonding QMOs have almost the same energy,

$$\varepsilon_+ \approx \varepsilon_{1s} \approx \varepsilon_- . \qquad (37.2a)$$

The $1s\sigma$ and $1s\sigma^*$ orbitals are completely filled in the ground state of the second-row diatomics. The charge density distribution due to the core electrons in the molecule (two electrons in $1s\sigma$ and two more in $1s\sigma^*$) is almost equal to that due to the core electrons in the atoms (two electrons in $1s_A$ and two more in $1s_B$) as $S_{1s_A,1s_B}$ is negligibly small:

$$2[(1s\sigma)^2 + (1s\sigma^*)^2] \approx 2[(1s_A)^2 + (1s_B)^2]. \qquad (37.2b)$$

Thus the core orbitals can be considered to be "hardly changed" due to the formation of a molecule. This is the so-called *frozen-core* approximation and it leads to the following rule.

(ii) *In the qualitative molecular orbital theory, the molecular core electrons can be represented as the atomic core electrons. Thus for the second-row homonuclear diatomics, the core electronic configuration is written as* KK *instead of* $1s\sigma_g^2 1s\sigma_u^{*2}$.

(iii) *Since the molecular axis is chosen parallel to the z-axis, the other* σ *orbitals are formed as follows:*

$$2s\sigma_g = [2(1 + S_{2s_A,2s_B})]^{-1/2}(2s_A + 2s_B),$$

$$2s\sigma_u^* = [2(1 - S_{2s_A,2s_B})]^{-1/2}(2s_A - 2s_B); \qquad (37.3a)$$

and

$$2p\sigma_g = [2(1 - S_{2p_{zA},2p_{zB}})]^{-1/2}(2p_{zA} - 2p_{zB}),$$

$$2p\sigma_u^* = [2(1 + S_{2p_{zA},2p_{zB}})]^{-1/2}(2p_{zA} + 2p_{zB}). \qquad (37.3b)$$

Note the formation of $2s\sigma$ and $2p\sigma$ orbitals in Figure 6.19. Check the inversion symmetries.

(iv) *The* $2p_x$ *atomic orbitals form the bonding* π_x *molecular orbital and the antibonding* π_x^* *molecular orbitals*:

$$2p\pi_{x,u} = [2(1 + S_{2p_{xA},2p_{xB}})]^{-1/2}(2p_{xA} + 2p_{xB}),$$

$$2p\pi_{x,g}^* = [2(1 - S_{2p_{xA},2p_{xB}})]^{-1/2}(2p_{xA} - 2p_{xB}). \qquad (37.4)$$

Similarly, $2p\pi_{y,u}$ *and* $2p\pi_{y,g}^*$ *orbitals are built up from* $2p_y$ *atomic orbitals.*

The orbitals $2p\pi_{x,u}$ and $2p\pi_{y,u}$ are mutually degenerate, as the antibonding $2p\pi_{x,g}{}^*$ and $2p\pi_{y,g}{}^*$ orbitals are. Notice here that the π orbitals are ungerade whereas the π^* orbitals are symmetric under inversion. See Figure 6.19.

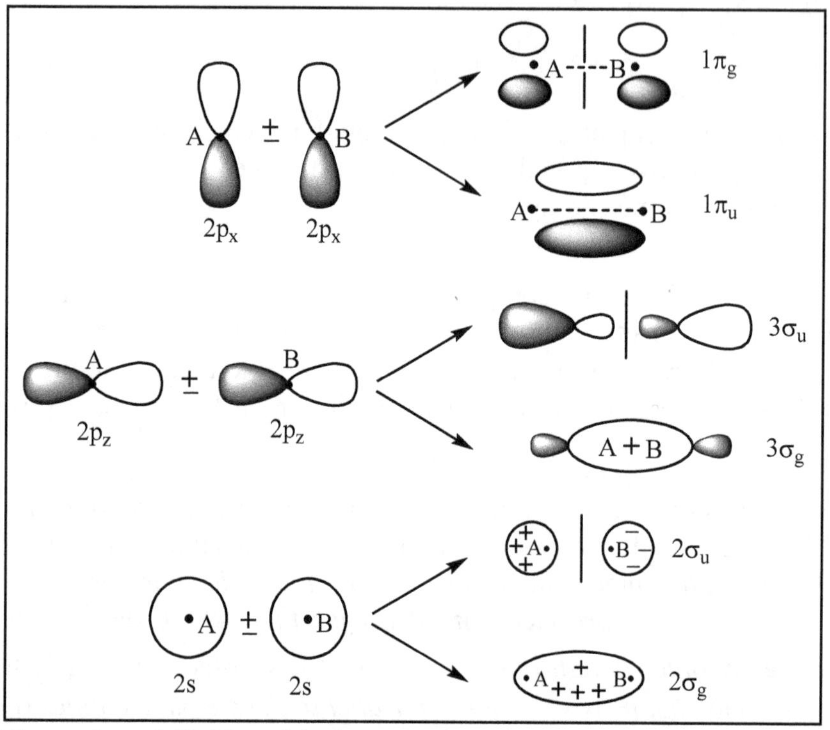

Figure 6.19 Formation of QMOs of the homonuclear diatomics from atomic valence 2s and 2p orbitals.

*(v) The treatment can be extended to the homonuclear diatomics of higher row elements, but the possible occurrence of inner-shell atomic orbitals (*d *and* f *orbitals) progressively complicates the situation. The qualitative molecular orbital theory has been most successful in giving a qualitative description of the bonding in homonuclear diatomics of the first and second row elements.*

The qualitative scheme yields the following order of energy levels:

$1s\sigma_g < 1s\sigma_u{}^* << 2s\sigma_g < 2s\sigma_u{}^* << 2p\pi_u < 2p\sigma_g < 2p\pi_g{}^* < 2p\sigma_u{}^*$ (for diatomics up to N_2); and (37.5)

$1s\sigma_g < 1s\sigma_u{}^* << 2s\sigma_g < 2s\sigma_u{}^* << 2p\sigma_g < 2p\pi_u < 2p\pi_g{}^* < 2p\sigma_u{}^*$ (for diatomics O_2 onwards).

The orbitals are filled in accordance with Pauli's exclusion principle. In the open shell case we follow Hund's rule of maximum spin multiplicity, that is, electrons are first fed into as many degenerate orbitals as possible in order to minimize the electron-electron interaction energy, and then *all the spin-unpaired electrons are assigned parallel spins* so as to reduce the electron-electron interaction energy further by exchange stabilization. Figures 6.20 and 6.21 show the orbital energy diagrams.

Figure 6.20 QMO energy diagram for the homonuclear diatomics Li_2 to N_2. Orbital population has been shown for the nitrogen molecule. For these species the π orbital is lower in energy than the 2pσ orbital.

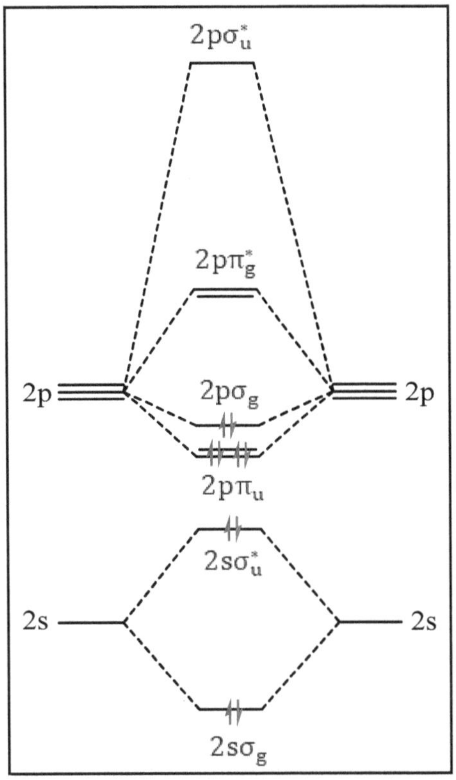

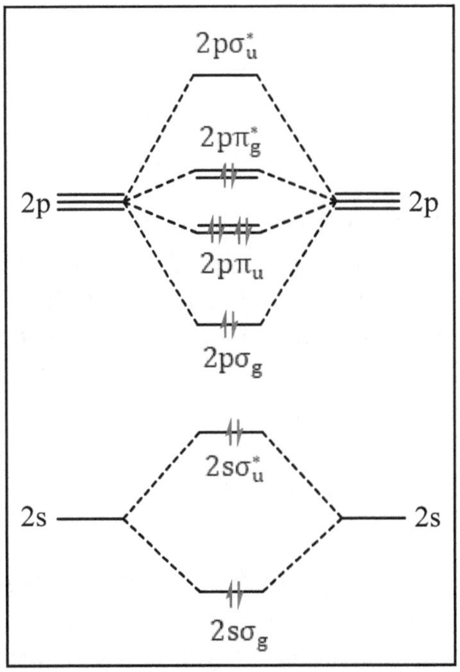

Figure 6.21 QMO energy diagram for the homonuclear diatomics from O_2 to Ne_2. Orbital population has been shown for O_2.

The ground state electronic configuration for the neutral homonuclear diatomics of first and second rows elements are listed in Table 6.2. The appropriately filled QMO energy level diagrams give a lot of information on the chemical structure and ground state molecular properties. To start with, the bond order (n) is calculated as

$$n = \frac{1}{2}(\text{Number of electrons in bonding molecular orbitals} - \text{Number of electrons in antibonding orbitals}). \quad (37.6)$$

This formula results in a bond order zero for He_2, Be_2 and Ne_2. Indeed, stable homonuclear diatomic molecules of helium, beryllium and neon are not known. There have been conflicting reports of the existence of a highly unstable Be_2 molecule in the gaseous phase. The bond orders of H_2, Li_2, B_2 and F_2 are all equal to 1. These bond orders are valid for isolated diatomic species. Molecules in gas phase at low pressure can be viewed as approximately isolated. From a study of general chemical behavior of the elements we already know that there is a single covalent or electron-sharing bond (a σ bond) in H_2 as well as F_2.

Spectroscopic studies tell us that the gaseous Li_2 is covalent and has a single bond. The B_2 molecule also has a single bond, but it is paramagnetic in the ground state. The paramagnetism corresponds to a spin triplet, that is, two unpaired electrons. So B_2 has really two half bonds, and the QMO treatment makes clear why this should be so. Molecules C_2 and O_2 have bond order 2 each, and N_2 has bond order 3 corresponding to a triple bond (one σ bond and two π bonds).

The molecule C_2 is interesting as it has a carbon-carbon double bond, but both the bonds are π bonds and not one σ-bond and one π-bond as is conventionally found for unsaturated organic molecules. This indicates that there is a low-lying excited state for this molecule. This is indeed observed by spectroscopic methods.

The qualitative molecular orbital treatment is very successful in pointing out the relative stabilities. For instance, O_2^+ has bond order 2.5 and O_2^- has bond order 1.5. So the bond energy follows the pattern $O_2^+ > O_2 > O_2^-$. With only a little effort, the reader should be able to explain why the dissociation energy of C_2^- is greater than that for C_2. Although stable Be_2 is not observed, Be_2^+ is known from mass spectrometry.

Another success of the qualitative theory is in explaining the correct ground state spin. Except B_2 and O_2, all the diatomics in Table 6.2 have singlet ground states. The molecules B_2 and O_2 are spin triplets. A net nonzero spin manifests itself as spin paramagnetism. Spin-zero diatomics like H_2, Li_2, C_2, N_2 and F_2 are diamagnetic species. Molecules B_2 and O_2 are paramagnetic and spin triplets as their ground state electronic configurations are

$$[KK]\ 2s\sigma_g^2 2s\sigma_u^{*2} 2p\pi_{x,u}^1 2p\pi_{y,u}^1$$

and

$$[KK]\ 2s\sigma_g^2 2s\sigma_u^{*2} 2p\sigma_g^2 2p\pi_u^4 2p\pi_{x,g}^{*1} 2p\pi_{y,g}^{*1}$$

respectively. That is two electrons in two different orbitals so as to minimize the electron-electron repulsion energy, thereby acquiring stability as explained in the earlier lectures on atomic structure and the Aufbau principle. Furthermore the unpaired electrons have parallel spins to have exchange energy that gives a small negative contribution and increases stability (Hund's rule).

Table 6.5 Ground state electronic configurations of neutral homonuclear diatomic molecules of elements belonging to the first and second rows of the periodic table.

Molecule	Electronic configuration	Bond order	Spin Number
H_2	$1s\sigma_g{}^2$	1	0
He_2	$1s\sigma_g{}^2 1s\sigma_u{}^{*2}$	0	0
Li_2	[KK] $2s\sigma_g{}^2$	1	0
Be_2	[KK] $2s\sigma_g{}^2 2s\sigma_u{}^{*2}$	0	0
B_2	[KK] $2s\sigma_g{}^2 2s\sigma_u{}^{*2}\ 2p\pi_u{}^2$	1	1
C_2	[KK] $2s\sigma_g{}^2 2s\sigma_u{}^{*2}\ 2p\pi_u{}^4$	2	0
N_2	[KK] $2s\sigma_g{}^2 2s\sigma_u{}^{*2}\ 2p\pi_u{}^4 2p\sigma_g{}^2$	3	0
O_2	[KK] $2s\sigma_g{}^2 2s\sigma_u{}^{*2}\ 2p\sigma_g{}^2 2p\pi_u{}^4 2p\pi_g{}^{*2}$	2	1
F_2	[KK] $2s\sigma_g{}^2 2s\sigma_u{}^{*2}\ 2p\sigma_g{}^2 2p\pi_u{}^4 2p\pi_g{}^{*4}$	1	0
Ne_2	[KK] $2s\sigma_g{}^2 2s\sigma_u{}^{*2}\ 2p\sigma_g{}^2 2p\pi_u{}^4 2p\pi_g{}^{*4} 2p\sigma_u{}^{*2}$	0	0

Lecture 38: Heteronuclear Diatomic Molecules

The most important difference between heteronuclear and homonuclear diatomics is that in the heteronuclear case the two sets of atomic orbitals to be used in the formation of the qualitative molecular orbitals do not match in energy. Linear combinations are made by using those atomic orbitals which have the same symmetry and which are energetically closest to each other. This procedure leads to two effects: (1) The linear expansion coefficients no longer equal each other in absolute magnitude, and the molecular orbitals appear as

$$\psi_+ = N_+(\phi_A + \lambda'\phi_B),$$
$$\psi_- = N_-(\lambda''\phi_A - \phi_B) \qquad (38.1)$$

with the normalization constants

$$N_+ = (1 + 2\lambda' S_{AB} + \lambda'^2)^{-1/2},$$
$$N_- = (1 - 2\lambda'' S_{AB} + \lambda''^2)^{-1/2}. \qquad (38.2)$$

If we assume $\varepsilon_A < \varepsilon_B$ and $\varepsilon_+ < \varepsilon_-$ then $|\lambda'| < 1$ and $|\lambda''| < 1$, that is, the lower-energy molecular orbital is relatively close to the lower-energy atomic orbital and the molecular orbital of higher energy is relatively close in appearance to the atomic orbital of higher energy. There is every possibility that another orbital of atom A, say ϕ'_A, is close to ϕ_B in energy in which case ψ_- will have a substantial contribution from ϕ'_A. This effect complicates the form of the linear expansion.

The consequence of these effects is that the molecular orbitals can no longer be identified by their generic names. Since the angular momentum symmetry around the molecular axis is preserved, the orbitals are still called as σ, π, etc. orbitals. Molecular orbitals are counted from the lowest energy level upward, and so they are identified as 1σ,

2σ, 3σ, etc.; 1π, 2π, 3π, etc.; and so on. The g and u specifications are no longer applicable as the heteronuclear diatomic does not have an inversion center. A schematic diagram illustrating the formation of the molecular orbital energy levels is given in Figure 6.22. Only the valence orbitals are considered, as the core orbitals have been assumed frozen. As Figure 6.22 shows, 2s and $2p_z$ orbitals of carbon combine with 2s and $2p_z$ orbitals of oxygen to form four σ orbitals. The 1σ is almost exclusively formed by the 2s orbital of oxygen. Also, the 4σ mostly appears as the $2p_z$ orbital of carbon. The $2p_x$ and $2p_y$ orbitals of carbon and similar orbitals of oxygen form the doubly-degenerate sets 1π and 2π. The atomic energy level ordering $\varepsilon(2s_O) < \varepsilon(2s_C) < \varepsilon(2p_O) < \varepsilon(2p_C)$ leads to the ordering of molecular energy levels

$$\varepsilon(1\sigma) < \varepsilon(2\sigma) < \varepsilon(1\pi) < \varepsilon(3\sigma) < \varepsilon(4\sigma) < \varepsilon(2\pi)$$

The electronic configuration of the carbon monoxide molecule is [KK] $1\sigma^2 2\sigma^2 1\pi^4 3\sigma^2$.

The qualitative treatment has the difficulty that for heteronuclear diatomics it is not always easy to guess the bond order. The electronic configuration of CO does not directly manifest that the molecule has bond order 3. To know the bond order, one must know the exact expressions for the qualitative molecular orbitals. Notwithstanding this problem, sometimes (as in the case of carbon monoxide) one can guess the bond order from a very useful but qualitative process of comparison. The corresponding guiding principle is known as the isoelectronic rule.

The isoelectronic rule

Diatomic molecules having the same number of valence electrons have similar orbital energy diagrams. They resemble each other in many of their properties.

This rule is best illustrated by the energy level diagrams of N_2 and CO. Atomic orbitals of carbon are somewhat higher in energy than the corresponding orbitals of the nitrogen atom. The oxygen atomic orbitals are in turn somewhat lower lying orbitals. The relative energy ordering is a direct consequence of the steady increase of Z_{eff} from carbon to oxygen. A consequence of this order is that the qualitative molecular orbitals of CO and those of N_2 have about the same energy. This can be seen from Figure 6.22.

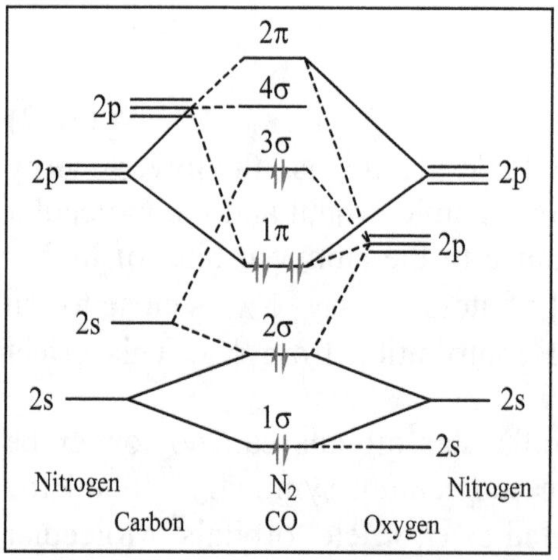

Figure 6.22 Illustration of the isoelectronic rule: the CO energy levels are approximately same as the orbital energy levels of N_2. The 4σ-2π energy ordering turns out to be reverse in the case of N_2 (not illustrated in this diagram).

The molecules CN^- and NO^+ are isoelectronic with CO and N_2. Each of these species has bond order 3. All of them except N_2 can easily bind itself to a transition metal ion.

A parallel is often found between the molecular orbitals for diatomics and the occupied molecular orbitals for bicentric hydrides of the forms HAAH, H_2AAH_2 and H_3AAH_3. A simple illustration is achieved from the successive isoelectronic substitutions of CH_3 for NH_2, NH_2 for OH, and OH for F. The common description is often found to be a useful starting point for the analysis of bonding and structure. As another example of the isoelectronic series one may quote the molecules O_2, HNO, HNNH, H_2CO, H_2CNH, and H_2CCH_2 which have similar molecular orbitals and an approximately similar description of the electronic structure in terms of one double bond and four bond pairs or lone pairs. We close this discussion with the caveat that one must not be tempted to extend this qualitative rule too far, e.g., to the comparison of properties of O_2 with those of C_2H_4, or to the distinction between cis- and trans- forms of N_2H_2 vis-a-vis the structure of HNO or H_2CO.

Polarity

As the molecular orbitals of a heteronuclear diatomic molecule consist of unequal contributions from the orbitals of the two different atoms, the bond develops a polarity. The valence electronic charge is not symmetrically distributed on two sides of the *xy* plane that bisects the bond. In effect one atom is associated with a larger concentration of the valence electrons than what it would have if it was not bonded. Consider a molecular orbital

$$\psi = c_1\phi_A + c_2\phi_B \tag{38.3}$$

with $|c_1| \neq |c_2|$. If there is one electron in this orbital, the probability that the electron will be found with atom A (or B) is c_1^2 (or c_2^2). If there are n_i electrons in the *ith* molecular orbital expressed in terms of coefficients c_{1i} and c_{2i}, and so on, the valence electron population on the two atoms and the overlap population are

$$\begin{aligned} P_{AA} &= -q_A = \sum_i n_i c_{1i}^2, \\ P_{BB} &= -q_B = \sum_i n_i c_{2i}^2, \\ P_{AB} &= 2\sum_i n_i c_{1i} c_{2i} S_{ABi}. \end{aligned} \tag{38.4}$$

Let N_A and N_B be the number of valence electrons in neutral atoms A and B. Then the local atomic charges in the molecule are

$$\begin{aligned} \delta_A &= N_A - (P_{AA} + \frac{1}{2}P_{AB}), \\ \delta_B &= N_B - (P_{BB} + \frac{1}{2}P_{AB}). \end{aligned} \tag{38.5}$$

The normalization of each ψ_i stipulates $c_{1i}^2 + c_{2i}^2 + 2c_{1i}c_{2i}S_{ABi} = 1$ so that $P_{AA} + P_{AB} + P_{BB} = N_A + N_B$ when the molecule is electrically neutral. From equations (38.5), we get $\delta_A = \delta$ and $\delta_B = -\delta$. Thus for a neutral species $(P_{BB} - N_B) - (P_{AA} - N_A) = 2\delta$ where δ is the net charge on atom A.

The *ith* bond becomes completely ionic if $c_{1i} = 0$ and $c_{2i} = 1$. A convenient parameter for the polarity of a bond (orbital) was introduced by Klessinger and McWeeny (1965). These authors defined the *fractional ionic character* of the *ith* bond as the difference of the probabilities of finding two electrons directly associated with the two atoms, divided by a factor of 2 as two electrons are required to form a single (covalent) bond:

$$(FIC)_i = |c_{2i}^2 - c_{1i}^2|. \tag{38.6}$$

The FIC varies from 0 for a covalent bond to 1 for a completely ionic bond. The electric dipole moment associated with *ith* bond (orbital) can be easily calculated. We have

$$\begin{aligned}\mu_i &= -\, n_i \int dv\,(\mathbf{r} - \overline{\mathbf{R}})\psi_i^2(\mathbf{r}) \\ &= -\, n_i\,[\boldsymbol{k}\{c_{1i}^2 z_A + c_{2i}^2 z_B + 2c_{1i}c_{2i}z_{AB}\} - \overline{\mathbf{R}}\,]\end{aligned} \tag{38.7}$$

where $\overline{\mathbf{R}}$ is the center of core charges, $\overline{\mathbf{R}} = (N_A + N_B)^{-1}(N_A\mathbf{R}_A + N_B\mathbf{R}_B)$, and

$$z_A = \int dv\, z\phi_A^2(\mathbf{r}), \tag{38.8}$$

etc. The reason that only these integrals occur in Eq. (38.7) is that the centers of the two atoms lie along the *z*-axis (the molecular axis). If the atoms A and B are centered at points $(0, 0, -R/2)$ and $(0, 0, R/2)$, respectively, we get $z_A = -R/2$, $z_B = R/2$, and $z_{AB} \approx 0$. By substituting these values in Eq. (38.7) one obtains

$$\mu_i = -\frac{1}{2} n_i R (FIC)_i + n_i \overline{R} \tag{38.9}$$

where

$$\overline{R} = \frac{(N_A - N_B)R}{2(N_A + N_B)}. \tag{38.10}$$

This expression for the orbital dipole moment can be used to calculate the net dipole moment of the molecule. We obtain

$$\mu_{AB} = \sum_i \mu_i = -\frac{1}{2} R \sum_i n_i (FIC)_i + \frac{1}{2} R(N_B - N_A). \tag{38.11}$$

The first sum on the right side is equal to $(P_{BB} - P_{AA})$ so that

$$\mu_{AB} = -\frac{1}{2} R\{(P_{BB} - N_B) - (P_{AA} - N_A)\}. \tag{38.12}$$

But we have seen, following Eq. (38.5), that the term within the second bracket on the right equals 2δ when the molecule is neutral. Therefore, the neutral molecule has the dipole moment

$$\mu_{AB} = -R\delta. \tag{38.13}$$

This is, of course, an approximation as its derivation relies on a qualitative theory, and also requires $z_{AB} = 0$. Still, this relationship is extremely popular and experimentalists use it to interpret observed dipole moments by assigning effective partial charges $+\delta$ and $-\delta$ to the two atoms.

Lecture 39: Hybridization

Molecules which are highly symmetric possess spatially equivalent bonds. These bonds are also energetically equivalent. The spatial and energetic equivalences can be established by isotopic substitution or by molecular spectroscopy. Because of their spatial and energetic equivalence, these bonds are also associated with definite features of orientation relative to each other, and led to Sidgwick's concept of directed valency. A straight-forward molecular orbital calculation by the Hartree-Fock method with geometry optimization is capable of yielding specific relative orientations such as an angle of about 120° between the C–C bond and the C–H bonds in the methylene terminal (C)–CH_2. The spatial and energetic equivalences are also obtainable from an empirical (such as Hückel or extended Hückel) or a qualitative molecular orbital calculation providing that these calculations are equipped with enough flexibility to distinguish total energy values for two different geometries of the same molecule. For instance, if one adds the V_{nn} term to the Hückel π-energy one may distinguish between the cis- and trans- forms of linear butadiene, and in case one also makes α and β parameters dependent on the distance and relative orientation of two carbon atoms one can optimize the geometry of a conjugated molecule simply by Hückel calculations. Normally, however, one performs empirical or qualitative molecular orbital calculations with the known (experimental) geometries.

It is clear that in principle the molecular orbital treatment with geometry optimization can produce the correct molecular geometry that is associated with directed bonds. Some visualize how the valence atomic orbitals p_x, p_y, p_z, etc. which are oriented along specific Cartesian axes may combine together to form the directed bonds which are oriented relative to each other and give rise to the equilibrium molecular structure. A conceptual breakthrough is achieved when one realizes that the directed bonds must be formed from correspondingly directed atomic orbitals. The construction of atomic orbitals directed along the chemical bonds is called hybridization.

Hybridization is defined as the process of forming spatially equivalent and hence energetically degenerate atomic orbitals by linearly combining the conventional atomic orbitals. The resulting orbitals are called *hybrid orbitals*. They form an orthonormal set. The hybrid orbitals combine with the ligand orbitals to form the molecular orbitals. Hence we have two possible schemes for obtaining the molecular orbitals:

(Scheme of brute force numerical calculation)
Central atom orbitals + Ligand orbitals → (straight-forward Linear Combination) → Molecular orbitals (39.1)
and
(Scheme based on conceptual understanding)
Central atom orbitals → (L.C.) → Hybrid orbitals
Hybrid orbitals + Ligand orbitals → (L.C.) → Bond orbitals
Bond orbitals → (Further L.C.) → Molecular orbitals (39.2)

Hybridization involves the mixing of at least two orbitals of different energy but of the same atom. The energy of hybrid orbitals is the average of the energies of all participating atomic orbitals.

The construction of hybrid orbitals is essentially very simple. These orbitals should be as far away from each other as possible. These orbitals are also spatially equivalent, so these should be equally apart from each other. This means that when the hybrid consists of only two orbitals the latter would be directed along the same line (usually chosen as the z-axis) but in opposite directions, and when the hybrid has three orbitals the latter would be directed along three axes in a plane with the axes inclined at an angle of 120° with respect to each other. Four hybrid orbitals are either directed along the tetrahedral coordination, or towards the corners of a square. Six hybrid orbitals are directed at the corners of an octahedron. This type of geometric consideration allows one to evaluate the contributions of the participating orbitals to a specific hybrid orbital merely as the projections of the atomic functions along the specific axis.

Table 6.6 Hybridization and the geometry of hybrids.

Geometry	Coordination Number	Hybridization
Linear	2	sp, dp
Bent	2	sd
Trigonal plane	3	sp^2, dp^2, d^2s
Trigonal pyramid	3	d^2p
Tetrahedral	4	sp^3, d^3s
Tetragonal plane	4	dsp^2
Trigonal bipyramid	5	dsp^3, d^3sp
Tetragonal pyramid	5	d^4s, sd^4
Octahedral	6	d^2sp^3, sp^3d^2
Trigonal prism	6	d^4sp

The spherically symmetric orbitals (s orbitals) contribute equally to each hybrid orbital. The hybrid orbitals are to be normalized to unity and they should be mutually orthogonal. These criteria are sufficient to determine the hybrid orbitals uniquely. Some examples are worked out in the following.

Example 41: The sp hybrid orbitals are written as

$$h_1 = c_1\,s + c_2\,p_z, \qquad (39.3)$$

$$h_2 = c_1\,s - c_2\,p_z.$$

The total contribution of the s orbital, $2c_1^2$, must equal 1. Therefore, $c_1 = 2^{-1/2}$. Then h_1 and h_2 would be normalized if $c_2 = 2^{-1/2}$. Thus we get

$$h_1 = 2^{-1/2}(s + p_z), \qquad (39.4)$$

$$h_2 = 2^{-1/2}(s - p_z).$$

Contour diagrams for these orbitals are shown in Figure 6.23. Notice that h_1 and h_2 are already mutually orthogonal. This has happened because of the geometrical construction and the orthonormality of s and p_z orbitals. The hybrid orbitals h_1 and h_2 combine with ligand orbitals to form two bond orbitals of the bonding type and two more of the antibonding type. Normally, only those bond or molecular orbitals which are of the bonding type are populated in the ground state of the neutral molecule. The resulting complex has a linear geometry like the linear environment around every carbon atom in acetylene.

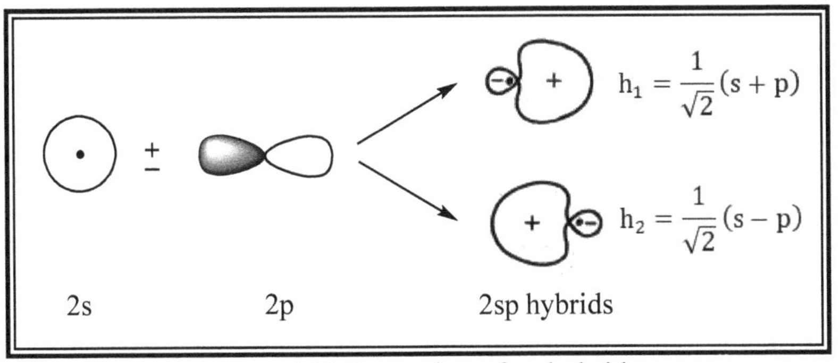

Figure 6.23 Formation of sp hybrids.

Example 42: The sp^2 hybrid may be formed in the *xy* plane as

$$h_1 = c_1\,s + c_2\,p_x,$$

$$h_2 = c_1\,s + c_2\left(-\frac{1}{2}p_x + \frac{\sqrt{3}}{2}p_y\right), \qquad (39.5)$$

$$h_3 = c_1\,s + c_2\left(-\frac{1}{2}p_x - \frac{\sqrt{3}}{2}p_y\right).$$

Considering the *total weightage* of the s orbital we get $3c_1^2 = 1$ or $c_1 = 3^{-1/2}$. Normalization requires $c_2 = \left(\frac{2}{3}\right)^{1/2}$ so that

$$h_1 = 3^{-1/2}s + (2/3)^{1/2}\,p_x,$$

$$h_2 = 3^{-1/2}s - 6^{-1/2}\,p_x + 2^{-1/2}\,p_y, \qquad (39.6)$$

$$h_3 = 3^{-1/2}s - 6^{-1/2}\,p_x - 2^{-1/2}\,p_y.$$

See the lobes in Figure 6.24. The orbitals in equations (39.6) are mutually orthogonal.

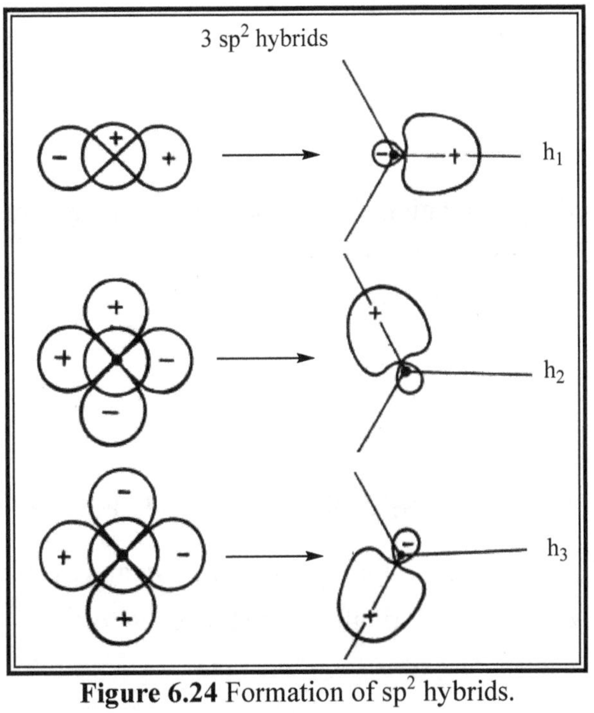

Figure 6.24 Formation of sp^2 hybrids.

Example 43: The sp^3 hybrid combinations are calculated by drawing a tetrahedron in a convenient way. The latter task is accomplished by joining the diagonally opposite corners of a cube by straight lines. As shown in Figure 6.25, the geometry allows one to write

$$\begin{aligned} h_1 &= c_1\,s + c_2\,3^{-1/2}(p_x + p_y + p_z),\\ h_2 &= c_1\,s + c_2\,3^{-1/2}(p_x - p_y - p_z),\\ h_3 &= c_1\,s + c_2\,3^{-1/2}(-p_x + p_y - p_z),\\ h_4 &= c_1\,s + c_2\,3^{-1/2}(-p_x - p_y + p_z). \end{aligned} \tag{39.7}$$

As $4c_1^2 = 1$, we get $c_1 = ½$. Normalization of any of the hybrid orbitals indicates that $c_2 = 3^{1/2}/2$. Therefore,

$$\begin{aligned} h_1 &= \tfrac{1}{2}(s + p_x + p_y + p_z),\\ h_2 &= \tfrac{1}{2}(s + p_x - p_y - p_z),\\ h_3 &= \tfrac{1}{2}(s - p_x + p_y - p_z),\\ h_4 &= \tfrac{1}{2}(s - p_x - p_y + p_z). \end{aligned} \tag{39.8}$$

These orbitals are orthogonal to each other.

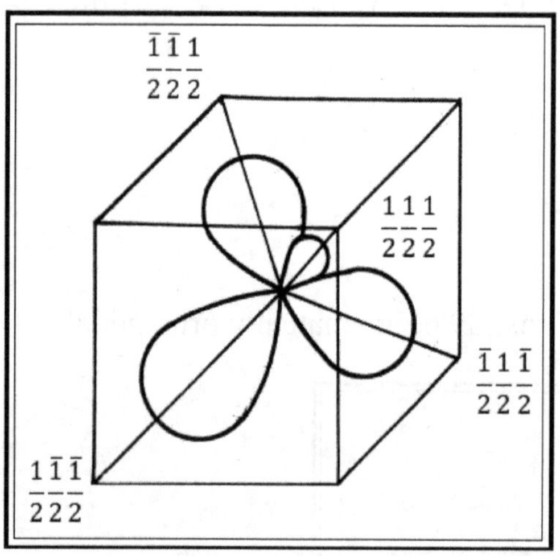

Figure 6.25 The sp^3 hybrid orbitals from the geometry of a cube.

Example 44: The dsp^2 hybridization for the square planar geometry is figured out as follows. Consider Figure 6.26 and write

$$\begin{aligned} h_1 &= c_1\,s + c_2\,2^{-1/2}(p_x + p_y) + c_3 d_{xy},\\ h_2 &= c_1\,s + c_2\,2^{-1/2}(-p_x + p_y) + c_3' d_{xy},\\ h_3 &= c_1\,s + c_2\,2^{-1/2}(-p_x - p_y) + c_3'' d_{xy},\\ h_4 &= c_1\,s + c_2\,2^{-1/2}(p_x - p_y) + c_3''' d_{xy}. \end{aligned} \tag{39.9}$$

Now, $4c_1^2 = 1$ so that $c_1 = ½$. Orthogonality of h_1 and h_2 requires $c_3c_3' = -\frac{1}{4}$. But c_3^2 must equal $c_3'^2$ so that $c_3 = ½$, $c_3' = -½$. Similarly from the h_2-h_3 and h_3-h_4 othogonality relations one finds $c_3'' = ½$ and $c_3''' = -½$. Normalization shows that $c_2 = 2^{-1/2}$. Therefore, one gets the combinations

$$h_1 = \tfrac{1}{2}(s + p_x + p_y + d_{xy}),$$
$$h_2 = \tfrac{1}{2}(s - p_x + p_y - d_{xy}),$$
$$h_3 = \tfrac{1}{2}(s - p_x - p_y + d_{xy}), \qquad (39.10)$$
$$h_4 = \tfrac{1}{2}(s + p_x - p_y - d_{xy}).$$

With these combinations each hybrid orbital is orthogonal to all other hybrid orbitals.

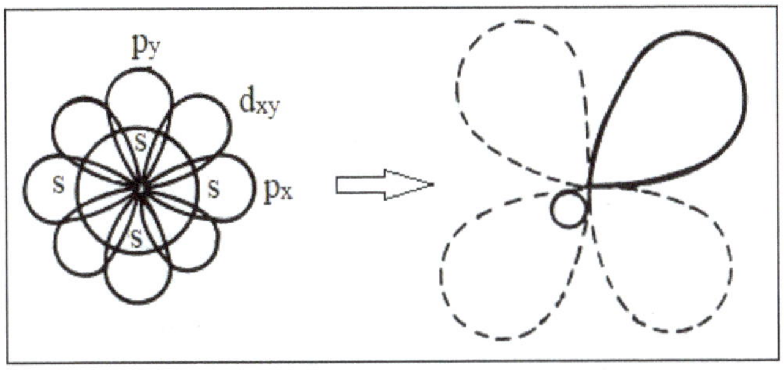

Figure 6.26 Formation of the dsp^2 hybrid orbitals from s, p_x, p_y and d_{xy} orbitals of the central atom.

An energy level description for the formation of molecular orbitals via the combination of hybrid orbitals with ligand orbitals followed by the combination of bond orbitals with themselves is given in Figure 6.27. The final molecular orbitals are not as degenerate as one would expect from the equivalence of bonds.

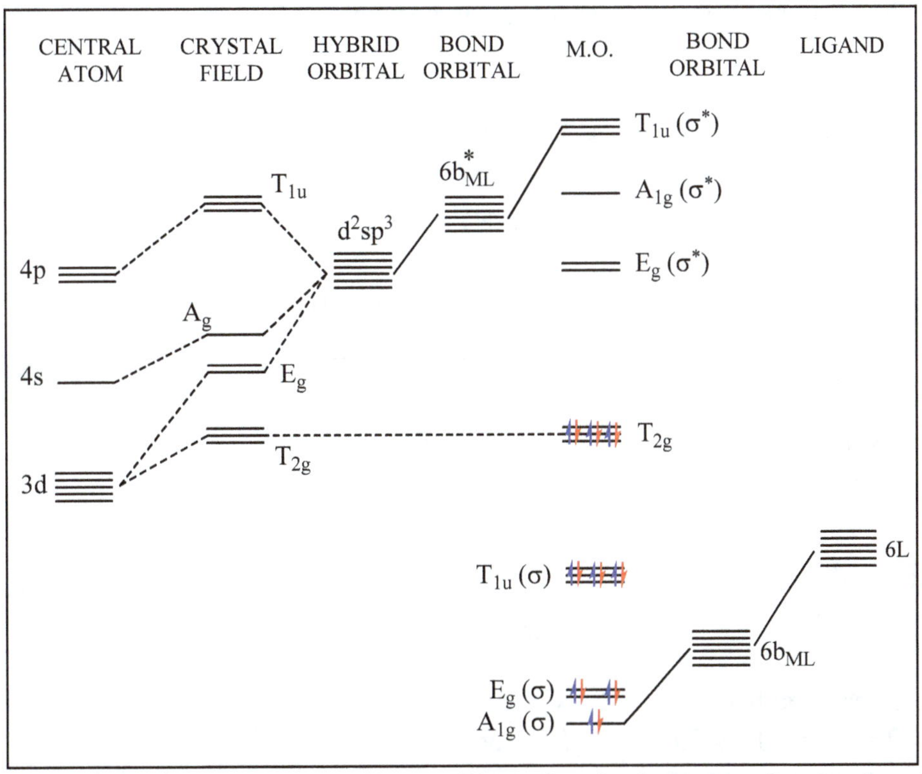

Figure 6.27 Schematic energy level diagram illustrating the formation of molecular orbitals through the hybrid orbital route in an octahedral complex.

Molecular orbital calculations usually yield a very symmetric structure at the equilibrium geometry. However, if one wants to interpret the chemical bonds directly in terms of the molecular orbitals or the molecular orbital wave function, one can do so only in the form of one-electron probabilities designated as bond orders. This is in sharp contrast with the so-called *valence bond theory* where the wave function directly represents all chemical bonds. The valence bond theory will be discussed in Chapter 8.

Lecture 40: Walsh M.O. Diagrams

Following Mulliken's work on the AB_2 systems (1942), Walsh (1953) presented a very picturesque analysis of the change of qualitative molecular orbital energy with the change in molecular geometry. This analysis results in a simple explanation of the dependence of the equilibrium structure of molecules on the number of valence electrons. The diagram of molecular orbital energy levels as functions of the changing molecular geometry is known as the Walsh molecular orbital diagram.

The Walsh rule is easily demonstrated for a system that deviates from linearity. Consider a linear species of type H—A—A—H. The atom A has only four valence orbitals, namely, s, p_x, p_y and p_z. There are ten atomic valence orbitals which combine linearly to generate ten molecular orbitals for A_2H_2. If the species is linear then each A atom is in the state of sp hybridization. Let us identify the atoms as H′—A′—A″—H″. In the following we write down the bond orbitals for the molecule A_2H_2 and also show the qualitative molecular orbitals which can be prepared from these bond orbitals.

$$\begin{array}{lcl}
\underline{\text{Bond orbitals}} & & \underline{\text{Molecular orbitals}} \\
\sigma(A'\text{–}H') & \longrightarrow & 1\sigma(A\text{–}H) \equiv \sigma_g(A\text{–}H) \\
\sigma(A''\text{–}H'') & & 2\sigma(A\text{–}H) \equiv \sigma_u(A\text{–}H) \\
\\
\sigma^*(A'\text{–}H') & \longrightarrow & 5\sigma(A\text{–}H) \equiv \sigma_g^*(A\text{–}H) \\
\sigma^*(A''\text{–}H'') & & 6\sigma(A\text{–}H) \equiv \sigma_u^*(A\text{–}H) \\
\\
\sigma(A'\text{–}A'') & \longrightarrow & 3\sigma(A\text{–}A) \equiv \sigma_g(A\text{–}A) \\
\sigma^*(A'\text{–}A'') & & 4\sigma(A\text{–}A) \equiv \sigma_u^*(A\text{–}A) \\
\\
\pi_{x,y}(A'\text{–}A'') & \longrightarrow & 1\pi_{x,y}(A\text{–}A) \equiv \pi_u(A\text{–}A) \\
\pi_{x,y}^*(A'\text{–}A'') & & 2\pi_{x,y}(A\text{–}A) \equiv \pi_g^*(A\text{–}A)
\end{array} \tag{40.1}$$

As may be observed from Figure 6.28, the energy level ordering is

$$1\sigma < 2\sigma < 3\sigma < 1\pi < 2\pi < 4\sigma < 5\sigma < 6\sigma, \tag{40.2}$$

and for ten valence electrons, the electronic configuration is $1\sigma^2 2\sigma^2 3\sigma^2 1\pi^4$.

Next consider the bending of the molecule in the *xz* plane. The hybridization changes from sp to somewhat like sp^2, requiring the additional participation of p_x orbitals of the two A atoms. This raises the energy of A–H bond orbitals and hence the 1σ and 2σ orbital energies increase. Similarly the 3σ energy increases. The most important effect is the increase in the energy of $1\pi_x$ and a greater decrease of the energy of $2\pi_x$. If the molecule is bent out of the plane, both $1\pi_x$ and $1\pi_y$ increase in energy while both $2\pi_x$ and $2\pi_y$ become more stable. This phenomenon is illustrated in Figure 6.28. Assuming that the lowering of 2π energy is the dominating effect, the ground state geometries for 12 and 14 valence electrons are determined to be the planar bent structure and the nonplanar bent structure respectively. This conclusion is supported by the simple examples of C_2H_2, N_2H_2 and O_2H_2. Although the species N_2H_2 is not known, a similar molecule N_2F_2 is found to possess a planar *trans* geometry in its ground state.

Example 45: Number of valence electrons and predicted geometry

Species	Valence electrons	Electronic configuration	Predicted geometry
C_2H_2	10	$\sigma_g(A{-}H)^2\sigma_u(A{-}H)^2\sigma_g(A{-}A)^2\pi_{ux}^2\pi_{uy}^2$	Linear
N_2H_2	12	$\sigma_g(A{-}H)^2\sigma_u(A{-}H)^2\sigma_g(A{-}A)^2\pi_{ux}^2\pi_{uy}^2\pi_{gy}^2$	Planar bent
O_2H_2	14	$\sigma_g(A{-}H)^2\sigma_u(A{-}H)^2\sigma_g(A{-}A)^2\pi_{ux}^2\pi_{uy}^2\pi_{gy}^2\pi_{ux}^2$	Nonplanar bent

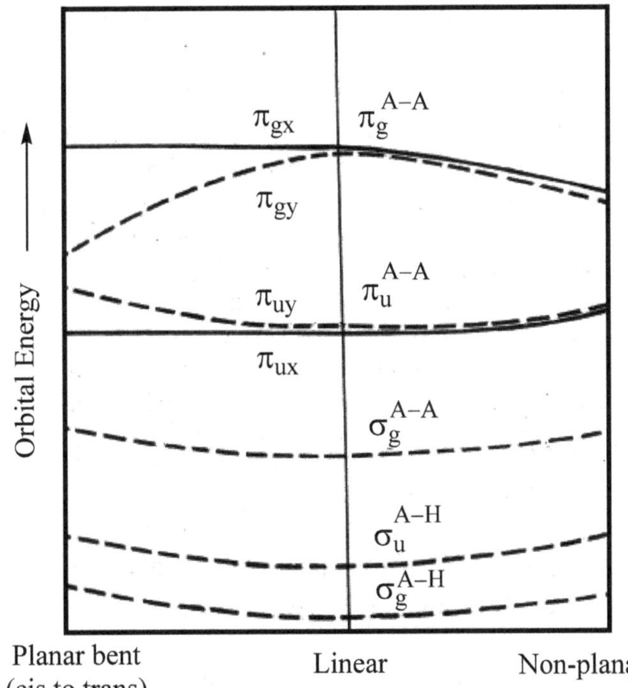

Figure 6.28 Walsh molecular orbital energy level diagram showing the effect of distortion on the orbital energies of HAAH.

The Walsh m. o. treatment can be extended to a few low-lying excited states. The acetylene molecule has a planar bent structure, actually the *trans-* configuration, in its lowest excited state. This geometry can be easily predicted from the Walsh m. o. diagram. The explanation is not so easy in the case of other excited states. The main difficulty in the treatment is that without an explicit calculation it is not possible to decide which of the orbital energies will play the dominant role.

The Walsh m. o. treatment gives rise to a simple and effective electron-counting rule. For the AH_2 systems with four valence electrons one obtains linear geometries while bent geometries are favored with six or eight electrons. The treatment leaves the details of the molecular geometry such as bond lengths, bond angles, dissociation energy, vibrational frequencies, etc. unexplained. First row hydrides such as H_2O and $-NH_2$ have bond angles in the range 100°-109°. These are somewhat less than the tetrahedral angle. The second- and third-row hydrides like H_2S, $-AsH_2$, etc. have bond angles nearly equal to 90°. The difference emerges from a less effective hybridization involving the s orbital of any element of the second and subsequent rows (third and later periods in the periodic table). The p orbitals of the central atom are quite diffused and they directly participate in the formation of bonds in hydrides of elements belonging to the second and latter rows. The sp^3 hybridization is in principle energetically more expensive than the p^3 hybridization whenever bonding with ligands leaves out a lone electron or a lone pair on the central atom. Yet it is found to occur in molecules of the first-row elements. This happens because of the proximity of ligands to the central atom, which is required to lead to the bonding situation and consequent stability. It also gives rise to the possibility of ligand-ligand repulsion forcing a tetrahedral disposition of lone pairs around a first-row atom. If the central atom belongs to other rows, the ligands are less proximate to each other and the p^3 hybridization leads to the more stable equilibrium configuration for the molecule.

The Walsh treatment has also been successful in determining the geometries of certain types of polyatomic molecules. Gimarc (1979) has presented a discussion on molecules of the type AB_2 and AB_3. Albright, Burdette and Whangbo (1985) have used Walsh-type m. o. diagrams to investigate transition metal complexes and organometallic systems like ML_n.

Lecture 41: The Hellmann-Feynman Theorem

The bonding or antibonding nature of an electronic state is conveniently interpreted by using the Hellmann-Feynman theorem. The theorem is in fact quite general, and it can be rather easily proved.

Consider the Hamiltonian as a function of a parameter μ. Assume that the eigenfunction of $\mathbf{H}(\mu)$ is known. This wave function will also be a function of μ as it satisfies

$$\mathbf{H}(\mu)\psi(\mu) = E(\mu)\psi(\mu). \tag{41.1}$$

The eigenvalue $E(\mu)$ can be written as the expectation value

$$E(\mu) = \int dv\, \psi^*(\mu)\mathbf{H}(\mu)\psi(\mu) \tag{41.2}$$

when $\psi(\mu)$ is normalized. The rate of change of E with respect to the parameter μ is given by

$$\frac{dE}{d\mu} = \int dv\, \psi^*(\mu)\frac{d\mathbf{H}}{d\mu}\psi(\mu) + \int dv \left[\frac{d\psi^*}{d\mu}\mathbf{H}\psi(\mu) + \{\mathbf{H}\psi(\mu)\}^* \frac{d\psi}{d\mu}\right] \tag{41.3}$$

where we have used the hermitian property of **H**. If one uses Eq.(41.1) and its complex conjugate, and the fact that E is real as **H** is hermitean, one obtains

$$\frac{dE}{d\mu} = \int dv\, \psi^*(\mu)\frac{d\mathbf{H}}{d\mu}\psi(\mu) + E\frac{d}{d\mu}\int dv\, \psi^*(\mu)\psi(\mu). \tag{41.4}$$

Since the function $\psi(\mu)$ is normalized, the second part on the right hand side becomes zero and one obtains

Theorem 13. The Hellmann-Feynman theorem:

$$\frac{dE}{d\mu} = \int dv\, \psi^*(\mu)\frac{d\mathbf{H}}{d\mu}\psi(\mu) = \left\langle\frac{d\mathbf{H}}{d\mu}\right\rangle. \tag{41.5}$$

Its proof here depends on $\psi(\mu)$ being an eigenfunction of $\mathbf{H}(\mu)$.

The interpretation of chemical binding is reached by considering the electronic Hamiltonian as a function of the internulear separation. We then have

$$\frac{dE_{el}(R_{ab})}{dR_{ab}} = \left\langle\frac{\partial H_{el}(R_{ab})}{\partial R_{ab}}\right\rangle = \left\langle\frac{\partial(\mathrm{V}_{en}+\mathrm{V}_{ee})}{\partial R_{ab}}\right\rangle \tag{41.6}$$

as $\partial \mathrm{T}/\partial R_{ab} = 0$. This leads to the expectation value of the force

$$\langle F\rangle = -\left\langle\frac{\partial(\mathrm{V}_{en}+\mathrm{V}_{ee}+\mathrm{V}_{nn})}{\partial R_{ab}}\right\rangle = -\frac{\partial(E_{el}+\mathrm{V}_{nn})}{\partial R_{ab}}. \tag{41.7}$$

The result holds not only for the eigenfunctions of $\mathbf{H}_{el}$ as proved here, but also when the energy surface has been optimized for variation of the scale factor (cf. variation of exponents). The interpretation discussed in Chapter 1 and elsewhere follows suit.

Texts Suggested for Further Reading

E. Heilbronner and H. Bock, "The HMO Model and Its Application", (J. Wiley, New York, 1976) Vol. 1-3.

I. Levine, "Quantum Chemistry", (Allyn and Bacon, Boston, 1983).

L. Pauling, "The Nature of the Chemical Bond and the Structure of Molecules and Crystals. An Introduction to the Modern Structural Theory of Chemistry", 3rd edition, (Cornell Univ. Press, Ithaca, 1960).

L. Salem, "The Molecular Orbital Theory of Conjugated Systems", (W. A. Benjamin, New York, 1966).

R. B. Woodward and R. Hoffmann, "The Conservation of Orbital Symmetry", (Verlag Chemie - Academic Press, Weinheim, 1970).

PROBLEMS

1. Solve the linear homogeneous equations given in Eqs. (32.1), (32.4) and (32.7), and find the solutions discussed in the text.
2. Set up the secular determinant for π electrons in pyrrole, pyridine and aniline, and verify the charge distributions shown in Figure 6.11.
3. The occupied molecular orbitals of LiH are approximately given by

$$1\sigma = \phi_{1s}^{Li}, \quad 2\sigma = 0.32\phi_{2s}^{Li} + 0.23\phi_{2p_z}^{Li} + 0.69\phi_{1s}^{H}$$

What is the hybridization in lithium? Can you guess the forms for 3σ and 4σ orbitals?
4. Calculate the π-electron bond orders in pyrrole.
5. Use the following data to calculate the fractional ionic character and find the approximate form of the involved molecular orbital, assuming overlap integral ~ 0.3.

Molecule	HF	HCl	HBr	HI
μ in a.u.	0.716	0.405	0.326	0.176
R in a.u.	1.74	2.40	2.66	3.04

6. Write the molecular orbital wave function for the ground state of the diatomic oxygen molecule.
7. Obtain expressions for the dsp hybrid orbitals from a geometrical consideration.
8. Determine the orbitals of the AH_2 molecule and verify the electron-counting rule given in the text.
9. Look at the diagram below (Figure 6.29). It shows an approximate reproduction of the experimental photoelectron spectrum of the N_2 molecule. The QMO energy levels have been shown in such a way that the transitions can be immediately assigned to the peaks. Interpret the diagram. (Hint: Instead of holding the book straight in the horizontal plane, hold it in the vertical direction such that the ionization energy increases downward).

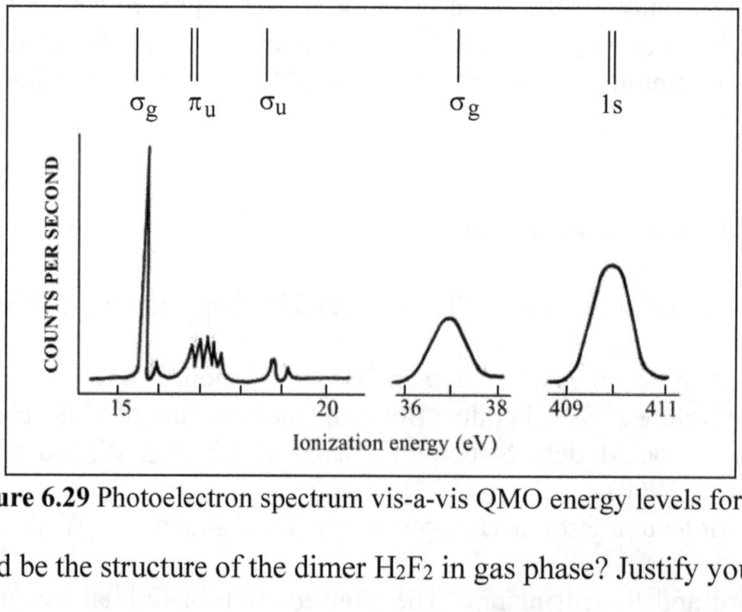

Figure 6.29 Photoelectron spectrum vis-a-vis QMO energy levels for N_2.

10. What would be the structure of the dimer H_2F_2 in gas phase? Justify your answer.

CHAPTER 7

STATIONARY STATE PERTURBATION THEORY

We have already discussed in Chapter 4 that there are only a few physical systems for which the wave equation can be exactly solved. In almost all applications of theory one must rely on approximate methods for solving the wave equation. The variation method is one very successful approximation method. Perturbation formalisms constitute another viable approach. Perturbation theory has an additional advantage. By the virtue of its design, it makes explicit the effects of a progressively large number of physical interactions, which leads to a clear understanding of the physics of the quantum problem. The variation method basically culminates in a brute-force numerical calculation that yields the net result, but the details of physics normally remain hidden in the treatment.

The stationary state perturbation theory is concerned with the evaluation of the discrete eigenvalues and corresponding eigenfunctions of a Hamiltonian operator that is explicitly independent of time. We assume that it is possible to divide the Hamiltonian $\mathbf{H}$ into two parts, $\mathbf{H}_0$ and $\mathbf{H}'$, such that the structure of $\mathbf{H}_0$ is sufficiently simple to permit one to exactly solve the corresponding Schrödinger equation. The second part $\mathbf{H}'$ is small enough so that it can be viewed as a *perturbation.* The eigenfunctions of $\mathbf{H}_0$ form the "zeroth-order" stationary states of the system, and the effect of $\mathbf{H}'$ is to perturb these states. Thus

$$\mathbf{H} = \mathbf{H}_0 + \mathbf{H}'. \tag{42.1}$$

The Schrödinger equation involving $\mathbf{H}_0$ is

$$\mathbf{H}_0 \psi_i^{(0)} = E_i^{(0)} \psi_i^{(0)} \tag{42.2}$$

and we assume that the $E_i^{(0)}$'s and the $\psi_i^{(0)}$'s are known. Our goal is to find the solutions (E_i and ψ_i) of the problem

$$\mathbf{H}\psi_i = E_i\psi_i \tag{42.3}$$

in terms of the zeroth-order solutions $E_j^{(0)}$'s and $\psi_j^{(0)}$'s.

The most popular form of the stationary state perturbation theory is known as Rayleigh-Schrödinger perturbation theory (RSPT). A different form of perturbation theory, known as Brillouin-Wigner perturbation theory (BWPT), reveals the structure of the perturbed states. There are also several variation-perturbation methods. Here we limit our discussion to the simplest perturbation methodology, namely, the Rayleigh-Schrödinger perturbation theory.

Lecture 42: Rayleigh-Schrödinger Perturbation Theory (The Nondegenerate Case); Degenerate Perturbation Theory

The crucial point in this theory is that as the perturbation is small, it is possible to express both ψ_i and E_i as power series in $\mathbf{H}'$. To obtain the series expansion we make use of a perturbation parameter λ and replace $\mathbf{H}'$ by $\lambda\mathbf{H}'$. The parameter λ is allowed to vary continuously from $\lambda = 0$ to $\lambda = 1$. This gives us the parametric operator

$$\mathbf{H}(\lambda) = \mathbf{H}_0 + \lambda\mathbf{H}', \qquad (0 \le \lambda \le 1). \tag{42.4}$$

Note that for $\lambda = 0$ we have $\mathbf{H}(\lambda) = \mathbf{H}_0$ and the zeroth-order problem, Eq. (42.2), for which exact solutions are assumed to be known. For $\lambda = 1$ we have $\mathbf{H}(\lambda) = \mathbf{H}_0 + \mathbf{H}' = \mathbf{H}$ and the actual problem, Eq. (42.3), which we want to solve. Hence the immediate objective is to seek a power series solution for the parametric wave equation

$$\mathbf{H}(\lambda)\psi_i(\lambda) = E_i(\lambda)\psi_i(\lambda) \tag{42.5}$$

valid for all λ in the range $0 \le \lambda \le 1$. Solutions of Eq. (42.3) are then obtained by inserting $\lambda = 1$.

To accomplish this task, the perturbed wave function and the corresponding energy eigenvalue are expanded as

$$\psi_i(\lambda) = \psi_i^{(0)} + \lambda\psi_i^{(1)} + \lambda^2\psi_i^{(2)} + \dots, \tag{42.6}$$

and

$$E_i(\lambda) = E_i^{(0)} + \lambda E_i^{(1)} + \lambda^2 E_i^{(2)} + \dots . \tag{42.7}$$

Negative powers of λ are not admitted as $\psi_i(\lambda)$ and $E_i(\lambda)$ *must remain finite* and they *must tend to* $\psi_i^{(0)}$ and $E_i^{(0)}$ respectively as λ approaches zero. When Equations (42.4), (42.6) and (42.7) are substituted in Eq. (42.5), one obtains

$$(\mathbf{H}_0+\lambda\mathbf{H}')\,[\psi_i^{(0)} + \lambda\psi_i^{(1)} + \lambda^2\psi_i^{(2)} + \dots]$$
$$= [E_i^{(0)} + \lambda E_i^{(1)} + \lambda^2 E_i^{(2)} + \dots]\,[\psi_i^{(0)} + \lambda\psi_i^{(1)} + \lambda^2\psi_i^{(2)} + \dots]. \tag{42.8}$$

This equation is to remain valid for the continuous range of λ $(0 \le \lambda \le 1)$. Therefore, the coefficients of equal powers of λ on both sides must be equal. Equating the coefficients of λ^0, λ^1, λ^2, etc., and in general of λ^n (for $n \ge 0$) on both sides of Eq. (42.8), we get the zeroth-order, first-order, second-order, etc., and in general the nth-order equations as shown below:

$$\mathbf{H}_0\psi_i^{(0)} = E_i^{(0)}\psi_i^{(0)} \qquad \text{(order = 0)}, \tag{42.9}$$

$$\mathbf{H}_0\psi_i^{(1)} + \mathbf{H}'\psi_i^{(0)} = E_i^{(0)}\psi_i^{(1)} + E_i^{(1)}\psi_i^{(0)} \qquad \text{(order = 1)}, \tag{42.10}$$

$$\mathbf{H}_0\psi_i^{(2)} + \mathbf{H}'\psi_i^{(1)} = E_i^{(0)}\psi_i^{(2)} + E_i^{(1)}\psi_i^{(1)} + E_i^{(2)}\psi_i^{(0)} \qquad \text{(order = 2)}, \tag{42.11}$$

and in general

$$\mathbf{H}_0\psi_i^{(n)} + \mathbf{H}'\psi_i^{(n-1)} = \sum_{j=0}^{n} E_i^{(j)}\psi_i^{(n-j)} \qquad \text{(for order} = n \ge 1). \tag{42.12}$$

The zeroth-order problem, Eq. (42.9), is assumed to be already solved. The Rayleigh-Schrödinger technique involves determining the corrections to wave function and energy in successive orders. This is accomplished by considering a few general points. First, the zeroth-order functions are normalized and these are mutually orthogonal as $\mathbf{H}_0$ is hermitian. Second, from Equations (42.9)-(42.12) we find that one may add an arbitrary multiple of $\psi_i^{(0)}$ to $\psi_i^{(n)}$ without affecting the *nth* order equation. The chosen multiple is such that $\psi_i^{(n)}$ becomes independent of $\psi_i^{(0)}$ (for $n \geq 1$), that is,

$$\int d\tau\, \psi_i^{(0)*}\psi_i^{(n)} = 0. \tag{42.13}$$

Third, after multiplying both sides of the *nth* order equation (42.12) by $\psi_i^{(0)\,*}$, integrating over all space and using the choice in Eq. (42.13) one finds

$$E_i^{(n)} = \int d\tau\, \psi_i^{(0)*}\mathbf{H}'\psi_i^{(n-1)} \qquad \text{(for } n > 0\text{)}. \tag{42.14}$$

This implies that the $(n-1)st$ order correction to wave function is sufficient to determine the *nth* order correction to energy.

However, Löwdin has shown (1964, 1965) by using the condition (42.13) that in reality the *sth* order correction to wave function $\psi_i^{(s)}$ is necessary and sufficient to determine the (2s+1)*st* order correction to energy, $E_i^{(2s+1)}$.

First-order perturbation effects

If we multiply both sides of Eq. (42.10) by $\psi_j^{(0)*}$, integrate over all space and make the identification

$$\int d\tau\, \psi_j^{(0)*}\mathbf{H}_0\psi_i^{(1)} = \{\int d\tau\, \psi_i^{(1)*}\mathbf{H}_0\psi_j^{(0)}\}^* \qquad \text{(since } \mathbf{H}_0 \text{ is hermitian)}$$

$$= E_j^{(0)} \int d\tau\, \psi_j^{(0)*}\psi_i^{(1)}, \tag{42.15}$$

we get

$$(E_j^{(0)} - E_i^{(0)})\int d\tau\, \psi_j^{(0)*}\psi_i^{(1)} = E_i^{(1)}\,\delta_{i,j} - \int d\tau\, \psi_j^{(0)*}\mathbf{H}'\psi_i^{(0)}. \tag{42.16}$$

For $j = i$, Eq. (42.16) gives

$$E_i^{(1)} = \int d\tau\, \psi_i^{(0)*}\mathbf{H}'\psi_i^{(0)} \tag{42.14a}$$

which can be directly obtained from Eq. (42.14) for n=1. For $j \neq i$, one obtains

$$\int d\tau\, \psi_j^{(0)*}\psi_i^{(1)} = \frac{H'_{ji}}{E_i^{(0)} - E_j^{(0)}} \tag{42.17}$$

where

$$H'_{ji} = \int d\tau\, \psi_j^{(0)*}\mathbf{H}'\psi_i^{(0)}.$$

Because of Eq. (42.13) we write $\psi_i^{(1)}$ as

$$\psi_i^{(1)} = \sum_{j(\neq i)} c_j^{(1)}\psi_j^{(0)}. \tag{42.18}$$

The coefficient $c_j^{(1)}$ is found as

$$c_j^{(1)} = \int d\tau\, \psi_j^{(0)*} \psi_i^{(1)} \tag{42.19}$$

so that by using equations (42.17) and (42.18) we get

$$\psi_i^{(1)} = \sum_{j(\neq i)} \frac{H'_{ji}}{E_i^{(0)} - E_j^{(0)}} \psi_j^{(0)}. \tag{42.20}$$

This treatment breaks down in the degenerate case as some of the $E_j^{(0)}$'s become equal to $E_i^{(0)}$.

Second-order perturbation effects

If one places Eq. (42.20) into Eq. (42.14) for $n = 2$ one obtains

$$E_i^{(2)} = \sum_{j(\neq i)} \frac{|H'_{ji}|^2}{E_i^{(0)} - E_j^{(0)}}. \tag{42.21}$$

To find the second-order correction to wave function we write

$$\psi_i^{(2)} = \sum_{j(\neq i)} c_j^{(2)} \psi_j^{(0)} \tag{42.22}$$

and substitute this expression in the second-order equation, (42.11). This procedure leads to the equation

$$\sum_{j(\neq i)} c_j^{(2)} (E_j^{(0)} - E_i^{(0)}) \psi_j^{(0)} = \sum_{j(\neq i)} c_j^{(1)} (E_i^{(1)} - \mathbf{H}') \psi_j^{(0)} + E_i^{(2)}\, \psi_i^{(0)}.$$

We multiply both sides by $\psi_k^{(0)*}$ and integrate over all space to obtain

$$c_k^{(2)} = \sum_{j(\neq i)} \frac{H'_{kj} H'_{ji}}{(E_i^{(0)} - E_k^{(0)})(E_i^{(0)} - E_j^{(0)})} - \frac{H'_{ki} H'_{ii}}{(E_i^{(0)} - E_k^{(0)})^2}. \tag{42.23}$$

The energy and the wave function correct through second order in perturbation are

$$E_i = E_i^{(0)} + H'_{ii} + \sum_{j(\neq i)} \frac{|H'_{ji}|^2}{E_i^{(0)} - E_j^{(0)}}, \tag{42.24}$$

and

$$\psi_i = \psi_i^{(0)} + \sum_{j(\neq i)} \Big[\frac{H'_{ji}}{E_i^{(0)} - E_j^{(0)}} \Big(1 - \frac{H'_{ii}}{(E_i^{(0)} - E_j^{(0)})}\Big) + \sum_{k(\neq i)} \frac{H'_{jk} H'_{ki}}{(E_i^{(0)} - E_j^{(0)})(E_i^{(0)} - E_k^{(0)})} \Big] \psi_j^{(0)}. \tag{42.25}$$

The wave function in the above is not normalized to unity. One finds

$$\int d\tau\, \psi_i^* \psi_i = 1 + \sum_{j(\neq i)} \frac{|H'_{ji}|^2}{(E_i^{(0)} - E_j^{(0)})^2} \tag{42.26}$$

through second order in perturbation. This calls for a *renormalization* of the wave function. The requirement for renormalization is a common occurrence in perturbation theory.

The degenerate case

For a set of degenerate orthogonal functions we attempt to find the linear combinations which diagonalize the Hamiltonian. This procedure yields the first-order correction to energy and wave function. Thus we write

$$\psi = \sum_j c_j \psi_j^{(0)} \tag{42.27}$$

where the sum is over the degenerate functions, and obtain linear homogeneous equations of the form

$$\sum_j (H_{ij} - ES_{ij})c_j = 0 \tag{42.28}$$

for each i belonging to the degenerate set. The matrix elements are given by

$$H_{ij} = E_i^{(0)}\, \delta_{i,j} + H'_{ij},$$

$$S_{ij} = \delta_{i,j},$$

and (42.29)

$$E = E_i^{(0)} + E^{(1)}$$

so that

$$H_{ij} - ES_{ij} = H'_{ij} - E^{(1)}\, \delta_{i,j}. \tag{42.30}$$

Thus the linear equations are

$$\sum_j (H'_{ij} - E^{(1)}\, \delta_{i,j})c_j = 0 \tag{42.31}$$

for every i. The corresponding secular determinant is formed and equated to zero whence the first order correction $E^{(1)}$ and the linear expansion coefficients are found out. The treatment will be made explicit while we discuss Stark effect in the next lecture.

Lecture 43: Zeeman and Stark Effects

Zeeman effect

When an atom stays in an externally applied uniform magnetic field, the atomic magnetic moment produced by the orbital motion of the charged electrons and their spins interact with the field. Consequently, the energy levels undergo a change. The degenerate states become nondegenerate with an approximately constant energy difference between them while the average energy of the group of degenerate states remains unaltered. This phenomenon is known as the *Zeeman effect.*

The magnetic field is given as the curl of the vector potential **A**,

$$\boldsymbol{H} = \nabla \times \mathbf{A}. \tag{43.1}$$

When the vector potential is not zero the effective linear momentum of an electron, called the mechanical momentum, is given by the operator

$$\boldsymbol{\pi} = \mathbf{p} - \frac{e}{c}\mathbf{A} \tag{43.2}$$

where c is the speed of light. This yields the kinetic energy operator

$$T = \frac{1}{2m_e} p^2 - \frac{e}{2m_e c}(\mathbf{p} \cdot \mathbf{A} + \mathbf{A} \cdot \mathbf{p}) + \frac{e^2}{2m_e c^2} A^2 . \tag{43.3}$$

where m_e is as usual the rest mass of the electron. For a uniform magnetic field

$$\mathbf{A} = \frac{1}{2} \boldsymbol{H} \times \mathbf{r} \tag{43.4}$$

in symmetric gauge, so that

$$T = \frac{1}{2m_e} p^2 - \frac{e}{2m_e c} \boldsymbol{L} \cdot \boldsymbol{H} + \frac{e^2}{8m_e c^2} |\boldsymbol{H} \times \mathbf{r}|^2 . \tag{43.5}$$

The second and the third terms on the right side of this expression becomes a perturbation on the electronic Hamiltonian.

We will evaluate perturbation correction only through first order in the magnetic field strength (H), as the latter is considered to be very small and associated with the ratio v/c. Therefore we choose the perturbation operator

$$\mathbf{H}' = -\frac{e}{2m_e c} \boldsymbol{L} \cdot \boldsymbol{H} . \tag{43.6}$$

This perturbation represents the interaction of the orbital magnetic moment with operator $\boldsymbol{\mu}_L = \frac{e}{2m_e c} \boldsymbol{L}$ and the magnetic field $\boldsymbol{H}$: $\mathbf{H}' = -\boldsymbol{\mu}_L \cdot \boldsymbol{H}$. The orbital magnetic moment arises from the orbital angular motion of the electron. The magnetic field is considered parallel to the z-axis. So,

$$\mathbf{H}' = -\frac{e}{2m_e c} H L_z . \tag{43.7}$$

The first order correction to energy is

$$E^{(1)\prime} = -m_l \beta_e H \tag{43.8}$$

where m_l is the magnetic quantum number of the orbital and β_e is the Bohr magneton,

$$\beta_e = \frac{e\hbar}{2m_e c} . \tag{43.9}$$

As the electron is negatively charged, the magneton β_e is negative.

Similarly, the spin angular momentum gives rise to the spin magnetic moment

$$\boldsymbol{\mu}_S = \frac{1}{\hbar} g_e \beta_e \boldsymbol{S} \tag{43.10}$$

where g_e is the so-called ESR (electron spin resonance) *g factor*. Experiment shows that $g_e \approx 2$; this is validated by Dirac's relativistic electron theory. Interaction of the spin magnetic moment with the external magnetic field causes the perturbation

$$\mathbf{H}'' = -\boldsymbol{\mu}_S \cdot \boldsymbol{H} = -g_e \frac{e}{2m_e c} \boldsymbol{S} \cdot \boldsymbol{H} \tag{43.11}$$

This perturbation yields the first-order correction

$$E^{(1)\prime\prime} = -m_S g_e \beta_e H \tag{43.12}$$

where m_S is the z-component of the spin angular momentum quantum number. Since β_e is negative, $E^{(1)\prime\prime}$ increases as m_S increases.

Considering both orbital and spin angular momenta, the total magnetic moment is

$$\boldsymbol{\mu}_L + \boldsymbol{\mu}_S = \hbar^{-1}\beta_e(\boldsymbol{L}+2\boldsymbol{S}).$$

This would give rise to the interaction

$$\mathbf{H}' + \mathbf{H}'' = -\hbar^{-1}\beta_e H(\boldsymbol{L}_z + 2\boldsymbol{S}_z)$$

However, there is some complication. The total angular momentum **J**= **L**+**S** has j values ranging from $|l-s|$ to $(l+s)$, and two different j values may correspond to the same m_l+2m_s. In general **L**+2**S** = g_L**J** where g_L is the *Landé g factor*. One has to look for the correct total magnetic moment operator, which is

$$\boldsymbol{\mu}_J = \hbar^{-1} g_L \beta_e \boldsymbol{J} \tag{43.13}$$

The net perturbation is

$$\mathbf{H}_{\text{pert}} = -\boldsymbol{\mu}_J \bullet \boldsymbol{H} = -\hbar^{-1} g_L \beta_e H \boldsymbol{J}_z. \tag{43.14}$$

The corresponding change in the energy level is

$$E^{(1)} = -g_L \beta_e H m_j. \tag{43.15}$$

The factor g_L may be calculated. One finds

$$g_L = \frac{3}{2} + \frac{s(s+1) - l(l+1)}{2j(j+1)}. \tag{43.16}$$

Notice that for $s = 0, j = l$ and g_L equals the orbital-only value of unity while for $l = 0$, it equals the spin-only value of 2.

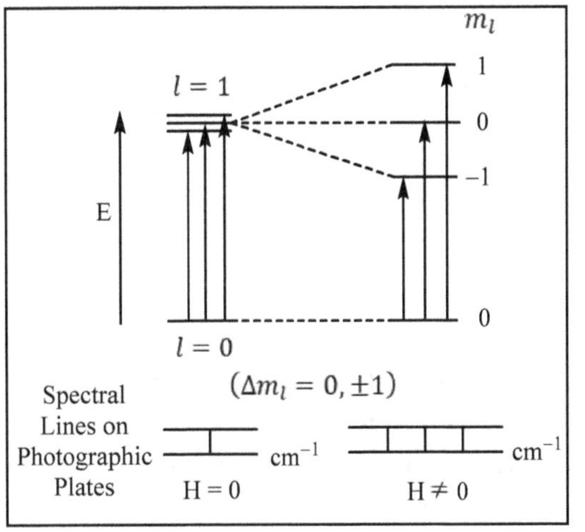

Figure 7.1 The "orbital-only" Zeeman effect. Observe the corresponding (fictitious) spectral pattern (Δm_l= 0, ±1).

These changes in energy levels are detected by the magnetic resonance spectroscopies. Figure 7.1 shows the "orbital-only" Zeeman effect. The spin-only Zeeman effect and the total Zeeman effect are illustrated by Figure 7.2. The experimental technique based on the spin-only effect is known as *electron spin resonance* (e.s.r.) spectroscopy. A non-zero magnetic moment μ_J makes the atom paramagnetic. The nuclear contribution to atomic paramagnetism is negligibly small, but it gives rise to nuclear magnetic resonance (NMR) which is an important tool in chemistry. The spectroscopy involved with the "total" electronic effect is called as *electron paramagnetic resonance* (e.p.r.) spectroscopy. The e.p.r. transition may or may not involve a change in the l quantum number whereas in e.s.r. the l quantum number remains strictly unchanged. Hence the e.p.r. transitions generally occur at higher frequencies. Note that e.s.r. is a special case of e.p.r. Also note the ordering of the energy levels: energy increases as m_j increases. This is an outcome of the negative charge of the electron such that the Bohr magneton is a negative quantity.

The nuclear charge is positive, and so is the nuclear magneton. The NMR levels decrease with increasing M_I values. Besides, the nuclear mass is several thousands of times larger than the electronic mass. Hence the NMR splitting is much smaller, and appears in the radio frequency range for normally used strength of the "uniform" magnetic field.

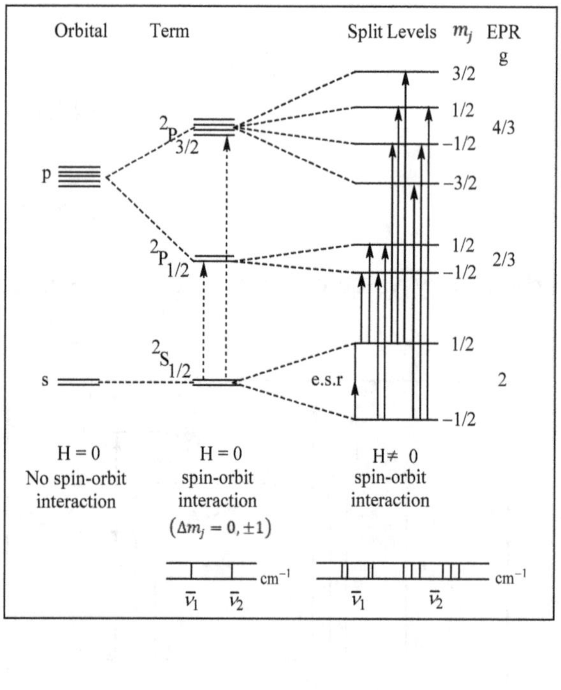

Figure 7.2 Zeeman splitting for both esr and epr spectroscopies. Note the varying e.p.r. g values. For e.s.r., g = 2. The pattern of spectral lines on photographic plates is shown under the energy level diagrams ($\Delta m_j = 0, \pm 1$).

The orbital angular motion gives rise to an effective magnetic field within the atom. This is an internal field. The spin magnetic moment interacts with this field. The corresponding (spin-orbit) interaction operator for an atom is of the form.

$$\mathbf{H}_{SO} = f(r)\boldsymbol{L}\cdot\boldsymbol{S} \qquad (43.17)$$

where f(r) is a function of radial distance, varying as r^{-3} to the lowest order in $1/r$. An alternative view of the same phenomenon is that the orbital magnetic moment interacts with the magnetic field that is internally generated by the spinning electron. Dirac's (1928) relativistic electron treatment for a one-electron atom shows that

$$f(r) = \frac{Ze^2}{2m_e^2c^2}\langle\frac{1}{r^3}\rangle$$

The spin-orbit interaction term H_{SO} is to be treated as an additional perturbation and its effect is to complicate the Zeeman splitting further. The s orbitals do not undergo the spin-orbit splitting as for these orbitals $l = 0$. The integral $<r^{-3}>$ is sufficiently small for the valence orbitals or any outer orbital. Consequently the spin-orbit splitting is found to be weak compared to the normal Zeeman splitting caused by the application of an external magnetic field for the study of e.s.r. or e.p.r. spectra.

Stark effect

The change of atomic energy levels while the atom is placed in a uniform electric field $\boldsymbol{E}$ is called Stark effect. The perturbation is now

$$\mathbf{H}' = -\boldsymbol{\mu}\cdot\boldsymbol{E} \tag{43.18}$$

where $\boldsymbol{\mu}$ is the electric dipole moment vector. Again the direction of the applied (electric) field is chosen as the z-axis so that the perturbation $\mathbf{H}'$ can be written as

$$\mathbf{H}' = -\mu_z E. \tag{43.19}$$

The origin in the relative coordinate system is also the center of the positive charge of the nucleus. Therefore, the electric dipole moment operator for each electron is given by

$$\boldsymbol{\mu} = e\mathbf{r} \tag{43.20}$$

from which we get

$$\mathbf{H}' = -eEz \tag{43.21}$$

for each electron in the atom.

Here we are interested in evaluating the first-order correction to the energy of the n = 2 orbitals of hydrogen atom. The first-order correction to the 1s energy vanishes as the expectation value of z over any ns orbital is equal to zero. There are only two nonzero integrals involving the operator z and the orbitals for $n = 2$. These are

$$\int d^3r\, \psi_{2s}{}^*(\mathbf{r})\, z\, \psi_{2p_z}(\mathbf{r}) = 3a_0 \tag{43.22}$$

and its complex conjugate that is identical with the original integral since the integrand is real and it does not contain any derivative operator. The use of the trial function

$$\psi = c_1\psi_{2s} + c_2\psi_{2p_z} + c_3\psi_{2p_x} + c_4\psi_{2p_y} \tag{43.23}$$

leads to the algebraic equation

$$\begin{vmatrix} -E^{(1)} & -3eEa_0 & 0 & 0 \\ -3eEa_0 & -E^{(1)} & 0 & 0 \\ 0 & 0 & -E^{(1)} & 0 \\ 0 & 0 & 0 & -E^{(1)} \end{vmatrix} = 0. \tag{43.24}$$

This has the solutions

$$E^{(1)} = 0 \Leftrightarrow \psi = \psi_{2p_x}, \psi_{2p_y};$$

$$E^{(1)} = \mp 3eEa_0 \Leftrightarrow \psi = 2^{-1/2}\,(\psi_{2s} \pm \psi_{2p_z}). \tag{43.25}$$

These solutions indicate that there are three possible orientations for the electric dipole moment of the hydrogen atom in its lowest excited state in the presence of an externally applied electric field. In one state the dipole is parallel to the applied field. In another state it is antiparallel. In two other states the net dipole moment is zero. The level splitting is shown in Figure 7.3.

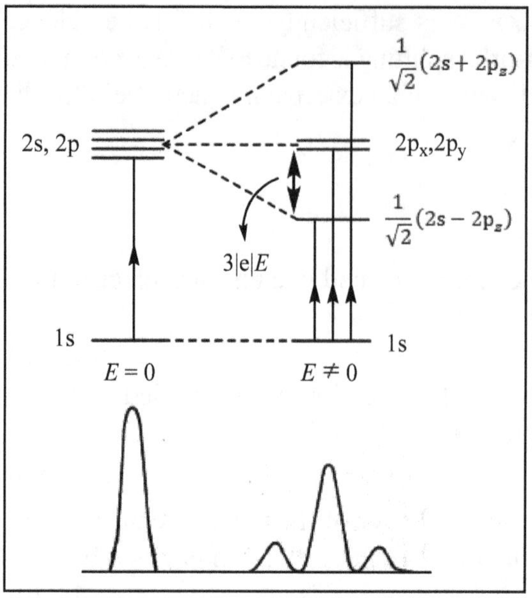

Figure 7.3 Stark effect: the splitting of the $n = 2$ levels of hydrogen atom in a uniform external electric field. The transition 1s↔2p gives rise to the spectra shown below the energy level diagram.

These two effects, Zeeman splitting and Stark effect, are examples of the first-order perturbation treatments in nondegenerate and degenerate cases respectively. We now turn our attention to a simple yet very elegant perturbative formalism in one of the major areas of chemical binding, viz., the crystal field treatment. Crystal field theory (CFT) is routinely used to discuss the bonding in transition metal complexes. Aided by the concept of crystal and ligand fields, the e.p.r. spectroscopy has turned into a powerful tool for the exploration of the realm of metal ions in molecules of biological interest. Another magnetic resonance spectroscopy – nuclear magnetic resonance (n.m.r.) – has been spectacularly successful in the investigation of the structure of molecules, especially those molecules which contain both carbon and hydrogen atoms.

Lecture 44: Crystal Field Splitting

When a point charge Qe is placed at each of the m corners of a polyhedron that surrounds an atomic ion, the ionic electron experiences an additional electric field due to the external potential

$$V_{CF}(\mathbf{r}) = Qe^2 \sum_{i=1}^{m} |\mathbf{r} - \mathbf{R}_i|^{-1} \tag{44.1}$$

where $\mathbf{R}_i$ is the position vector of the *ith* vertex. If each ligand like X^- is considered as a point charge located in one of the vertices of the polyhedron, the potential V_{CF} is called as the *crystal field potential* in normal chemical parlance. This is because a metal ion in a crystal of the metal complex would have a similar geometrical arrangement.[1] The crystal potential would be quite different, for we have considered only the ligands of the metal ion while writing V_{CF}. The term crystal field potential strictly applies to the complex ML_m in vacuum. In solids, or in solution, the state of the complex ML_m will be perturbed by interaction with the medium.

For a tetrahedral coordination as shown in Figure 7.4 one finds

$$V_{T_d}(\mathbf{r}) = \frac{4Qe^2}{R} - \frac{35Qe^2}{9R^5}(x^4 + y^4 + z^4 - \frac{3}{5}r^4) + ... \quad \text{(for } r < R\text{)}, \qquad (44.2)$$

$$\frac{4Qe^2}{r} \quad \text{(for } r \geq R\text{)}.$$

Similarly, for the octahedral coordination shown in the same Figure one gets

$$V_{O_h}(\mathbf{r}) = \frac{6Qe^2}{R} + \frac{35Qe^2}{4R^5}(x^4 + y^4 + z^4 - \frac{3}{5}r^4) + ... \quad \text{(for } r < R\text{)}, \qquad (44.3)$$

$$\frac{6Qe^2}{r} \quad \text{(for } r \geq R\text{)}.$$

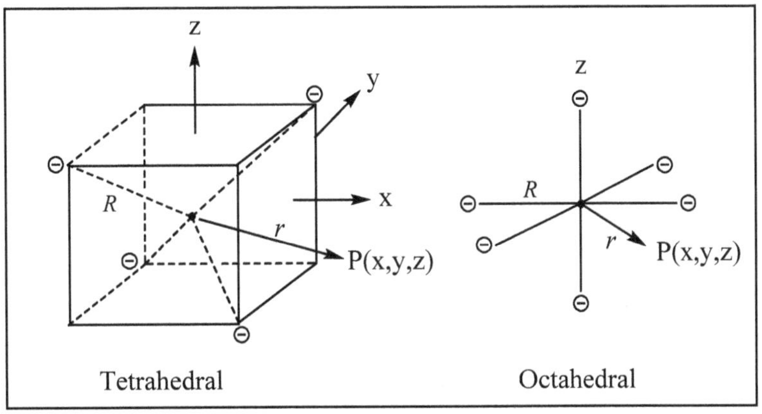

Figure 7.4 Tetrahedral and octahedral coordination.

In a general discussion of crystal field theory, the crystal field potential is treated as a perturbation and the form for $r < R$ is used to estimate the first order correction to energy. In quantum chemical treatments, V_{CF} is generally not approximated but it (or its ligand field version) is fully absorbed in the Hamiltonian, and the optimized geometry is found from the Hartree-Fock self-consistent-field methodology.

The first order level shift can be estimated by calculating the expectation value of $(x^4+y^4+z^4)$ for different orbitals. The following results can be easily established from the angular dependence of different atomic (ionic) orbitals:

$$\langle x^4+y^4+z^4\rangle_s = \frac{3}{5}\langle r^4\rangle_s$$

$$\langle x^4+y^4+z^4\rangle_{p_x} = \langle x^4+y^4+z^4\rangle_{p_y} = \langle x^4+y^4+z^4\rangle_{p_z} = \frac{3}{5}\langle r^4\rangle_p$$

$$\langle x^4+y^4+z^4\rangle_{d_{xy}} = \langle x^4+y^4+z^4\rangle_{d_{yz}} = \langle x^4+y^4+z^4\rangle_{d_{xz}} = \frac{11}{21}\langle r^4\rangle_d \qquad (44.4)$$

$$\langle x^4+y^4+z^4\rangle_{d_{z^2}} = \langle x^4+y^4+z^4\rangle_{d_{x^2-y^2}} = \frac{5}{7}\langle r^4\rangle_d.$$

Therefore, one obtains for the tetrahedral coordination

$$\varepsilon_s^{(1)} = \frac{4Qe^2}{R},$$

$$\varepsilon^{(1)}_{p_x} = \varepsilon^{(1)}_{p_y} = \varepsilon^{(1)}_{p_z} = \frac{4Qe^2}{R},$$

$$\varepsilon^{(1)}_{d_{xy}} = \varepsilon^{(1)}_{d_{yz}} = \varepsilon^{(1)}_{d_{xz}} = \frac{4Qe^2}{R} + \frac{2}{5}\Delta_t,$$

(44.5)

$$\varepsilon^{(1)}_{d_{z^2}} = \varepsilon^{(1)}_{d_{x^2-y^2}} = \frac{4Qe^2}{R} - \frac{3}{5}\Delta_t$$

where

$$\Delta_t = \frac{40}{9}Dq, \quad D = \frac{35Qe^2}{4R^5}, \quad q = \frac{2}{105}\langle r^4\rangle_d. \tag{44.6}$$

Similarly, one finds for the octahedral coordination

$$\varepsilon^{(1)}_{s} = \varepsilon^{(1)}_{p_x} = \varepsilon^{(1)}_{p_y} = \varepsilon^{(1)}_{p_z} = \frac{6Qe^2}{R},$$

$$\varepsilon^{(1)}_{d_{xy}} = \varepsilon^{(1)}_{d_{yz}} = \varepsilon^{(1)}_{d_{xz}} = \frac{6Qe^2}{R} - \frac{2}{5}\Delta_o, \tag{44.7}$$

$$\varepsilon^{(1)}_{d_{z^2}} = \varepsilon^{(1)}_{d_{x^2-y^2}} = \frac{6Qe^2}{R} + \frac{3}{5}\Delta_o$$

where

$$\Delta_o = 10Dq. \tag{44.8}$$

For the same R and the same $\langle r^4\rangle_d$, the level splitting in the two cases are related by

$$\Delta_t = \frac{4}{9}\Delta_o. \tag{44.9}$$

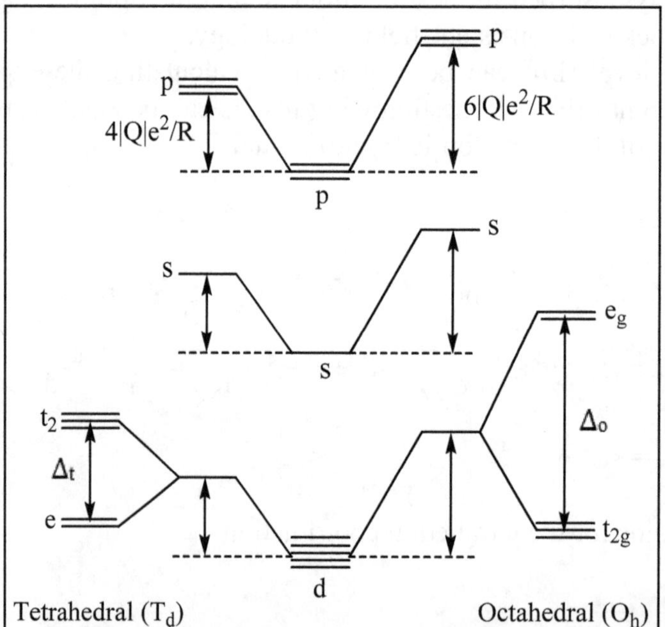

Figure 7.5 Level shift and splitting in tetrahedral and octahedral crystal fields.

The center of the previously degenerate d states is also shifted by an amount of $\frac{4Qe^2}{R}$ for tetrahedral coordination and $\frac{6Qe^2}{R}$ in the case of octahedral geometry. These level shifts and splitting patterns are illustrated by Figure 7.5. The splitting Δ is geometry-dependent. The relative magnitudes of Δ in different well-known crystal-field symmetries are compared with each other in Figure 7.6, but the level shift is not indicated in this Figure.

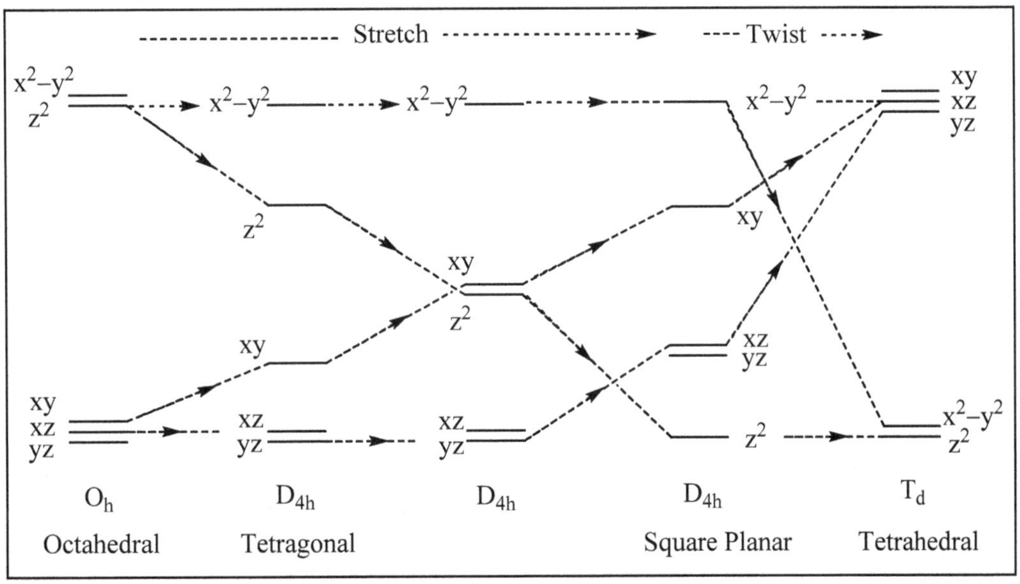

Figure 7.6 Comparison of the splitting pattern in crystal fields of different symmetries for the same metal ion and the same metal-ligand distance (but different ligand charges Qe). Level shifts have not been illustrated here.

The usual interpretation given for the t_{2g} states (d_{xy}, d_{yz} and d_{xz} orbitals) being lower in energy than the e_g states ($d_{x^2-y^2}$ and d_{z^2} orbitals) in an octahedral environment and the reverse energy ordering $\varepsilon(e) < \varepsilon(t_2)$ in a tetrahedral coordination is as follows.[1] When negatively charged ligands ($Q > 0$) approach along Cartesian axes, electrons in orbitals oriented along the same axes face greater repulsion and the electrons which occupy orbitals oriented between the axes face less repulsion. The electrostatic repulsion causes a positive level shift and compels the e_g states to be of higher energy than the t_{2g} set of orbitals. In a tetrahedral coordination the ligands approach along directions which are equally inclined to all three Cartesian axes, and this results again in a positive level shift but a reverse energy ordering for the e and the t_2 sets.

[1] See Appendix-I for the nomenclature e, t_2, e_g, t_{2g} etc. These are group-theoretical symbols. They indicate the irreducible representations for the transformation of the set of orbitals under the symmetry operations of point groups T_d, O_h, etc.

Ligands are really not point charges. Some ligands may be ionic, but still the electronic charge is distributed over the space surrounding the positively charged nuclei at the atomic centers. Some ligands may even be neutral species like H_2O, NH_3, etc. which have permanent electric dipole moment. This is why ligands vary in their capability of causing the crystal field splitting. The first order splitting is of course proportional to the strength of the crystal field. If we arrange the ligands in an increasing order of crystal field strength we get what is known as the *spectrochemical series*:

$$I^- < Br^- < S^{-2} < Cl^- < F^- < OH^- < C_2H_5OH < H_2O < NH_3 < en < NO_2^- < CN^- < CO \quad (44.10)$$

The weak-field ligands like halide ions form high-spin complexes in which the electrons occupy as many d orbitals as possible with maximum number of unpaired electrons having parallel spins. Conversely, the strong-field ligands favor the formation of low-spin complexes. These aspects are well known in inorganic chemistry.

The spin of a particle moving in a curved trajectory precesses around the axis of the internal magnetic field that is generated by the orbital angular momentum. The spin-orbit interaction term is really derived from the interaction between an internally present electric field $\boldsymbol{E}$ and the spin magnetic moment μ_S,

$$\mathbf{H}_{SO} = -\frac{e\hbar}{2m_e^2c^2}\boldsymbol{S}\bullet(\boldsymbol{E}\times \mathbf{p}). \quad (44.11)$$

The electric field experienced by an atomic electron is in the radial direction which leads to the $\boldsymbol{L}\bullet\boldsymbol{S}$ form in (43.17) for $\mathbf{H}_{SO}$. Repeated occurrence of groups of atoms, functional groups or molecules aligned in a particular direction leads to the existence of a considerably strong electric field along one of the axes in some of the crystals. This field leads to a *strong* spin-orbit interaction and a large splitting is observed. The level splitting is found even in the absence of an externally applied magnetic field. Hence the phenomenon is called the *zero-field splitting.*

Lecture 45: Ligand Field Theory

Since the ligands are not exactly point charges, it is better to make an explicit consideration of the electronic distribution within them. Therefore, one should resort to a molecular orbital treatment as the first step in a more systematic investigation of the electronic structure of metal complexes. A valence bond treatment is also possible, and some of the aspects of this methodology will be discussed in the next chapter. The metal-ligand bonds in transition metal complexes retain significant ionicity. As the simple molecular orbital treatment results in an equal weightage of covalent and ionic structures, the same procedure yields initial results that are quite satisfactory for these complexes.

The theory of formation of molecular orbitals from the valence orbitals of central metal ion and the valence orbitals of the atoms or molecules serving as ligands is called *adjusted crystal field theory* (ACFT) or *ligand field theory* (LFT). The latter name is given since in the formalism the crystal field is effectively determined from the distribution of charge within the ligands.

Figure 7.7 Direct formation of molecular orbitals from the symmetry-adapted linear combinations for an octahedral complex.

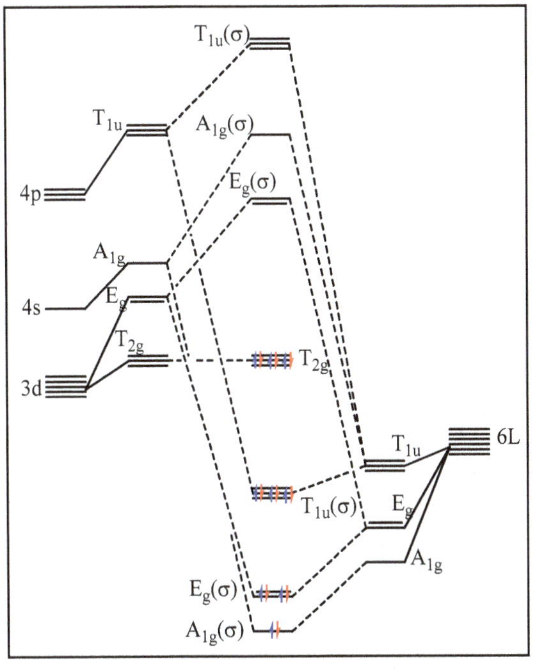

Normally one follows either of the two possible routes for the ligand field treatment. The first way is based on a direct, brute-force, numerical calculation of the molecular orbital coefficients and orbital energies. This task may be carried out at various levels of approximation by adopting an empirical (EHMO), a semiempirical (CNDO, INDO, etc.) or even an *ab-initio* approach.

The second route can be used for qualitative as well as quantitative calculations, and it clarifies the systematics of the formation of molecular orbitals. So it provides the necessary interpretation for the results obtained from either approach.

The second procedure involves the following sequence of steps:

1. Group the orbitals of the central atom which have been split by the crystal field according to their symmetry. For example, group the transition metal ion orbitals into those belonging to the t_{2g} set ($3d_{xy}$, $3d_{xz}$, $3d_{yz}$), those of the e_g set ($3d_{z^2}$, $3d_{x^2-y^2}$), those in the a_{1g} (4s) and t_{1u} ($4p_x$, $4p_y$, $4p_z$) sets, etc., for an octahedral coordination. For a qualitative treatment one may even estimate the first-order correction to the energy levels.
2. Similarly group the ligand orbitals according to the point group symmetry of the complex.
3. Linearly combine the metal ion orbitals with the ligand orbitals of the same symmetry. For example, combine the a_{1g} orbitals of the central metal ion and ligands together to form molecular orbitals which transform under the symmetry operations according to the a_{1g} irreducible representation of the point group O_h.

An alternative sequence is also in vogue:

1. Form the appropriate hybrid orbitals of the central atom, such as the d^2sp^3 hybrid orbitals for an octahedral complex with a transition metal ion at its center

or the sp^3d^2 hybrid orbitals for the primarily covalent octahedral complexes like SF_6.

2. Combine each hybrid orbital with one or more than one orbital of the nearest ligands to form bond orbitals of both bonding and antibonding types.
3. Combine the bond orbitals to form symmetry orbitals (like a_{1g}, e_g, etc.) typical of the complex. Symmetry orbitals of each type may be combined together to form the molecular orbitals.

Both the sequences should yield the same result (same m.o.s and orbital energies). This has been illustrated for an octahedral complex in Figure 7.7 (for hybrid orbital-symmetry orbital-m.o. route) and 6.27 (for hybrid orbital-bond orbital-symmetry/molecular orbital approach).

What does a bond orbital look like? The answer is already known to most of the readers. A bond orbital looks like a σ orbital, a π orbital, etc. as shown in Figure 7.8. A few more examples are shown in Figures 7.9 (a-c). Besides, there are quite a few π-electron ligands [Figure 7.10]. Acetylene, ethylene and C_5H_5 are typical π-electron ligands. Sometimes these ligands take part in the dσ-pπ and dπ-pπ back-bonding. The latter situation has been illustrated in Figure 7.11. The bond orbitals formed in the back-bonding are of type σ and π respectively.

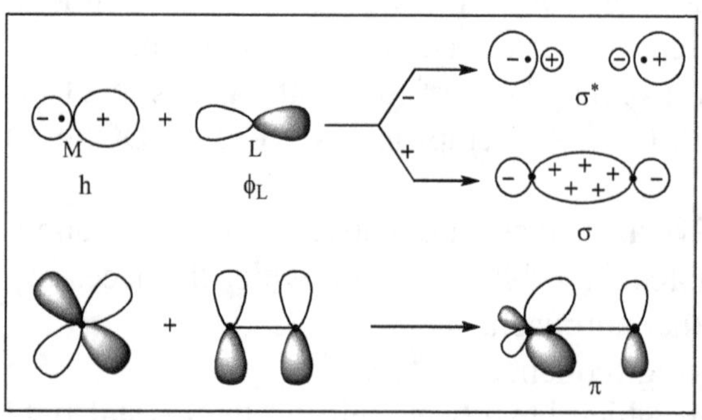

Figure 7.8 Bond orbitals of σ and π type.

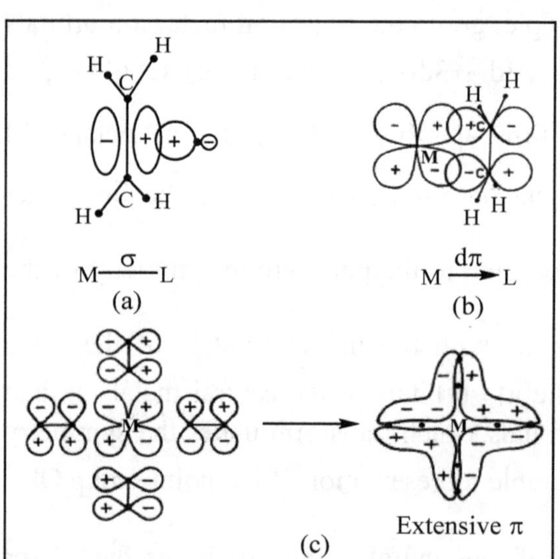

Figure 7.9 A few interesting cases of the formation of bond orbitals: (a) metal-ligand σ bond (a covalent affair); (b) metal-ligand π bond with metal as the electron donor (an instance of coordinate bonding); (c) extensive π bond formation.

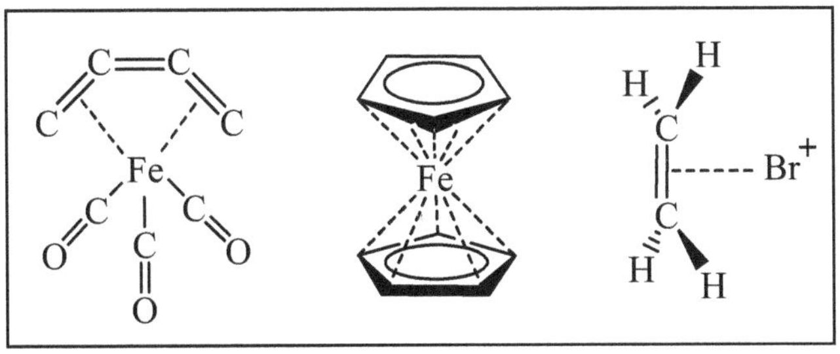

Figure 7.10 Some of the π-electron ligands involved in bonding.

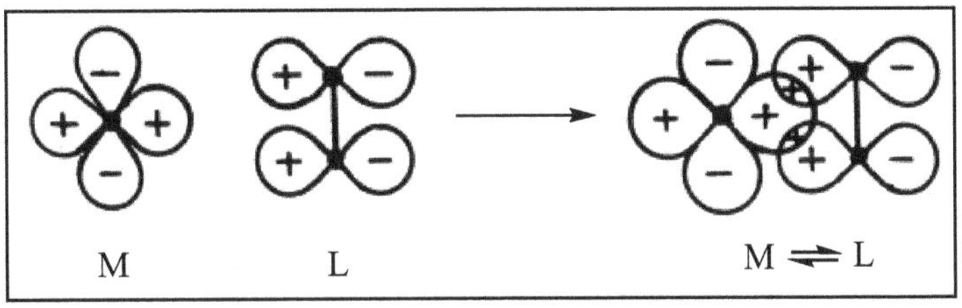

Figure 7.11 Pi electron ligands taking part in back-bonding.

To gain an understanding of exactly how LFT works one may consult the examples of an octahedral complex and mild distortions from the octahedral geometry. In the octahedral complex ML_6 (point group O_h) the valence orbitals of the central atom M span the irreducible species a_{1g}, t_{1u}, and $e_g \oplus t_{2g}$. Suppose the central atom is a transition metal ion with valence orbitals (n −1)d's, ns and np's. In the absence of a crystal field the degenerate (n −1)d manifold lies below the ns and np levels. Each of the six ligands has a sigma hybrid orbital ready for participation in the metal-ligand bonding. These six hybrid orbitals are *accidentally degenerate*. Together they span the irreducible species a_{1g}, e_g and t_{1u}. They would linearly combine and form ligand m.o.s of symmetries a_{1g}, e_g and t_{1u}. Orbitals compatible in symmetry mix with each other. Such mixing between the ligand m.o.s and metal ion orbitals gives rise to a low-lying bonding a_{1g} and a very high-lying antibonding a_{1g} levels. The ligand e_g and t_{1u} levels are lower in energy than the corresponding metal levels. The ligand levels are stabilized at the expense of the metal e_g and t_{1u} levels. The metal e_g levels are elevated and split from the t_{2g} levels, as in Figures 6.27 and 7.7. Thus the mixing generates six net bonding levels corresponding to six metal-ligand σ bonds. Excess electrons go into the split metal levels, first entering into the lower-lying nonbonding d orbitals belonging to the t_{2g} symmetry. Thus up to six electrons may be accommodated in the metal d orbitals of t_{2g} symmetry without destabilizing the system, and in fact actually stabilizing the complex by a relatively small extent. This is the basis of the so-called "18 electron rule" that states that transition metal complexes with 18 valence electrons are particularly stable.

Complexes MX_4Y_2 can be viewed as primarily octahedral complexes with slight departures from the perfect symmetry. These can have two geometrical isomers, *cis* and *trans*. The *cis* form has the symmetry of the point group $\mathbf{C}_{2v}$ and the point group for the *trans* form is $\mathbf{D}_{4h}$. The change in the crystal-field potential from the octahedral form has the symmetires t_{1u} and t_{2g} in the two cases respectively. If we treat this change as an additional perturbation, we may find the changes in the energy levels starting from the levels for the octahedral symmetry. The levels corresponding to the octahedral geometry split under distortion as shown in Figure 7.6. As long as the departure from the octahedral geometry is not great, the additional splitting remains small enough to guarantee the 18 electron rule. Deviations from this rule are noticed when the distortions are much larger.

Lecture 46: The Anharmonic Oscillator

The total energy curve under Born-Oppenheimer approximation generally deviates from a harmonic potential function. As the two nuclei approach each other, the increasingly stronger repulsion between their positive charges makes the curve to rise sharply past the equilibrium point, while at larger inter-nuclear distances the curve flattens at an increasingly slower rate because of the long range tail of both electron-nucleus and nuclear-nuclear coulombic interactions. It is approximately harmonic only around the equilibrium geometry of the molecule. For instance, one may expand the diatomic energy curve $E(R)$ around the equilibrium bond length R_{eq} in a Taylor series so as to gain the general analytical expression

$$E(R) = E(R_{eq}) + \left[\frac{dE}{dR}\right]_{R_{eq}} (R - R_{eq}) + \frac{1}{2!}\left[\frac{d^2E}{dR^2}\right]_{R_{eq}} (R - R_{eq})^2 + \frac{1}{3!}\left[\frac{d^3E}{dR^3}\right]_{R_{eq}} (R - R_{eq})^3 + \frac{1}{4!}\left[\frac{d^4E}{dR^4}\right]_{R_{eq}} (R - R_{eq})^4 + \ldots \tag{46.1}$$

The term that is linearly dependent on nuclear displacement vanishes as the first derivative is zero at the minimum of E. The second derivative, that is, the curvature, at the stationary point is known as the force constant k. Thus expression (46.1) reduces to

$$E(x) = E_{min} + \frac{1}{2}kx^2 + \frac{1}{6}gx^3 + \frac{1}{24}hx^4 + \ldots \tag{46.2}$$

where E_{min} is the minimum energy, the variable x is equal to the displacement $(R - R_{eq})$, and g, h, etc. are the third, fourth, etc. derivatives at the stationary point. With $[E(x) - E_{min}]$ as the potential function, Schrödinger equation for the relative movement of the two nuclei is written as

$$(\mathbf{H}^0 + \mathbf{H}')\chi_v(x) = E_{vib}\chi_{vib}(x). \tag{46.3}$$

It describes the vibration of the diatomic molecule. In the above the zero-order Hamiltonian is for the harmonic oscillation,

$$\mathbf{H}^0 = -\frac{\hbar^2}{2\mu}\frac{d^2}{dx^2} + \frac{1}{2}kx^2 \tag{46.4}$$

where μ is the reduced mass of the molecule. Eigenvalues $E_n^{(0)}$'s and the eigenfunctions $\psi_n^{(0)}$'s of $\mathbf{H}^0$ are given by equations (12.10) and (12.12) respectively. The perturbation $\mathbf{H}'$ is given by

$$\mathbf{H}' = \frac{1}{6}gx^3 + \frac{1}{24}hx^4 + \ldots \tag{46.5}$$

Here we would like to calculate E_{vib} correct through first order.

Before the actual calculation starts we want to express $\mathbf{H}'$ in terms of the dimensionless variable ξ. The following expression is achieved:

$$\mathbf{H}' = g'\xi^3 + h'\xi^4 \tag{46.6}$$

where $g' = \frac{1}{6}\left(\frac{\hbar}{\mu\omega_0}\right)^{3/2} g$, $h' = \frac{1}{24}\left(\frac{\hbar}{\mu\omega_0}\right)^2 h$, and $\omega_0 = 2\pi\nu_0 = (k/\mu)^{1/2}$. One may repeatedly use the recurrence relationship of Hermite polynomials from Table 3.1,

$$\xi H_n = \frac{1}{2}(H_{n+1} + 2nH_{n-1}),$$

to obtain

$$\xi^3 H_n = \frac{1}{8}[H_{n+3} + 6(n+1)H_{n+1} + 12n^2 H_{n-1} + 8n(n-1)(n-2)H_{n-3}],$$

$$\xi^4 H_n = \frac{1}{16}[H_{n+4} + 4(2n+3)H_{n+2} + 12(2n^2+2n+1)H_n$$
$$+ 16n(n-1)(2n-1)H_{n-2} + 16n(n-1)(n-2)(n-3)H_{n-4}].$$

Next one may use the orthogonality relationship for Hermite polynomials to calculate the perturbation integral

$$\int_{-\infty}^{\infty} dx\, \psi_m^{(0)*}(x)\, \mathbf{H}'\psi_n^{(0)}(x)$$
$$= 2^{-3/2}g'\left[\{(n+1)(n+2)(n+3)\}^{1/2}\delta_{m,n+3} + 3(n+1)^{3/2}\delta_{m,n+1} + 3n^{3/2}\delta_{m,n-1} + \{n(n-1)(n-2)\}^{1/2}\delta_{m,n-3}\right]$$
$$+ 2^{-2}h'\left[\{(n+1)(n+2)(n+3)(n+4)\}^{1/2}\delta_{m,n+4} + 2(n+1)^{1/2}(n+2)^{1/2}(2n+3)\delta_{m,n+2} + 3(2n^2+2n+1)\delta_{m,n} + 2n^{1/2}(n-1)^{1/2}(2n-1)\delta_{m,n-2} + \{n(n-1)(n-2)(n-3)\}^{1/2}\delta_{m,n-4}\right].$$

Therefore, the first-order correction to energy is given by

$$E_n^{(1)} = \frac{3}{2}\left(n^2+n+\frac{1}{2}\right)h'. \tag{46.7}$$

For transition $n \leftrightarrow n+1$

$$hc\bar{\nu} = E_{vib,n+1} - E_{vib,n} \tag{46.8}$$

so that through first order in perturbation $\mathbf{H}'$, the frequency of transition is given by

$$\bar{\nu} = \bar{\nu}_0 + 2(n+1)\bar{\nu}_0 x_e \tag{46.9}$$

where the anharmonic constant x_e is

$$x_e = \left(\frac{\hbar}{16\mu^2\omega_0^3}\right)h. \tag{46.10}$$

Both $\bar{\nu}_0$ and x_e are measurable from vibrational spectroscopy (infrared spectroscopy and Raman scattering). Quantum chemical calculations generally yield a good measure of both quantities.

Note that Eq. (46.9) is applicable only to the transitions between the low-lying vibrational levels. The second-order correction to energy $E_n^{(2)}$ is small for a small n, but it becomes more prominent for the larger n values. Because of $E_n^{(2)}$ and still higher order corrections, successive levels come close to each other as the vibrational quantum number n increases. But as n increases, one has to additionally consider the higher-order terms in the Taylor expansion. Thus successive energy levels become closer and closer until they become effectively continuous. Also, the population at the higher levels is exceedingly small at any reasonable temperature where the chemical bond is not ruptured. Therefore, the transitions between successive levels of large n values are difficult to detect.

PROBLEMS

1. What Zeeman effect will be observed if the magnetic field is in the x-direction, not along the z-axis?
2. Discuss (the first-order) Stark effect for n=3 levels of hydrogen.
3. Calculate the crystal-field splitting for a square planar complex.
4. Draw the valence energy level diagram for the tetrahedral complex ML_4 when M is a transition metal ion.
5. Give examples of stable 18-electron metal complexes.
6. Find the eigenvectors and eigenfunctions of the Hamiltonian for a particle moving in the potential $V(x) = \frac{1}{2}kx^2 + cx$ (i) by using a suitable transformation, and (ii) from perturbation theory. Compare the results obtained in the two cases.
7. If the oscillator charge is e and the perturbation cx is brought in by applying an electric field of strength E in the positive x-direction, what is the induced dipole moment? The polarizability α is defined as the ratio of the induced moment to the electric field. Show that the perturbation changes the energy of the oscillator by about $-\frac{1}{2}\alpha E^2$.
8. Calculate the second-order correction to energy $E_n^{(2)}$ for the anharmonic oscillator.

CHAPTER 8

MANY-ELECTRON FUNCTION

Molecular Orbital Wave Functions and Methods of Improvement

We have previously seen that a many-electron function should be (i) basically a product of one-electron functions and (ii) antisymmetric under the interchange of the space-spin coordinates of any two electrons. Hence the simplest many-electron fucntion is a single Slater determinant in the nondegenerate case (for closed-shell species) and at most an appropriate linear combination of only a few degenerate or quasi-degenerate determinants for open-shell species. The best wave function of this type is found form the Hartree-Fock theory, and it is called as the wave function representing the electronic configuration or simply as *the electronic configuration.* Orbitals are solutions of the Hatree-Fock equation. These are often determined as linear combinations of chosen one-electron functions just as molecular orbitals are often calculated as the best linear combinations of atomic orbitals (LCAO-MO). The many-electron wave function that represents an electronic configuration is termed a *molecular orbital* (MO) *wave function.*

There are two ways to get a better value, that is, a still lower value for the Hartree-Fock total energy and a better MO wave function. The first method is to vary the parameters such as the exponents which are used to define the basis functions. This results in a considerable improvement in energy as we have discussed in the case of H_2^+. The second method involves increasing the basis size. The increase in the variational degrees of freedom makes the calculated molecular orbitals more accurate. If one chooses good basis functions and increases the basis size sufficiently, one can get the Hartree-Fock total energy *accurate* through a specific number of significant digits. To obtain the Hartree-Fock total energy accurate through a greater number of digits ultimately requires a large time for computation.[1]

[1]In this context accuracy is measured with respect to the total energy that corresponds to an exact solution of the Hartree-Fock equations. Although the Hartree-Fock equations cannot be exactly solved, near-Hartree-Fock solutions can be obtained by numerically solving these equations or by using an extraordinarily extended and optimized basis set as discussed in Chapter 5 so that the calculated total energy converges.

The simple MO wave function is only an approximate solution of the many-electron Schrödinger equation. The exact ground state wave function is vastly different, as we have discussed earlier in the case of the H_2 molecule. The simple MO wave function contains a significant weight of ionic terms, and this is clearly unrealistic for bonding between like atoms such as those in homonuclear diatomic molecules. The situation becomes better for bonds between heteroatoms. An unequal mixing of the orbitals of two different atoms generates polarity. Thus the molecular orbital theory (in fact, LFT) is quite successful in interpreting the electronic structure of those transition metal complexes in which the metal-ligand bonds are considerably ionic. The MO wave function is clearly unsuitable for a description of covalent bonds.

In order to achieve a good description of the electronic structure of covalent molecules one may consider expanding the many-electron wave function in the basis of Slater determinants or configurations. The many-electron basis includes the ground state configuration as well as the excited state ones. The Ritz variation principle indicates that the best solution for the ground as well as excited states are obtained by diagonalizing the corresponding many-electron Hamiltonian matrix, as the configurations are orthogonal to each other. This process of finding the best linear combinations is known as *Configuration Interaction* (CI). As one includes more and more configurations in the treatment, the computed potential energy surface becomes progressively closer to the surface indicated by experimental dissociation energies, vibrational frequencies and anharmonicities, and sometimes the energy of activation. Simultaneously, the calculated many-electron wave function of the lowest energy eigenvalue approaches the true ground state wave function.

Lecture 47: The Valence Bond Theory

The valence bond (VB) method was developed by Heitler and London (1927), and this development was the first successful attempt to describe the covalent bond. From a molecular orbital based treatment, one determines an approximate wave function and then improves it by the CI technique. The valence bond treatment works in the reverse manner. Knowing that the bond is covalent, one writes the corresponding electron-pair wave functions just as Heitler and London wrote the functions

$$\Psi_{\pm}^{VB}(\mathbf{r}_1, \mathbf{r}_2) = [2(1 + S^2)]^{-1/2}\{\varphi_A(\mathbf{r}_1)\varphi_B(\mathbf{r}_2) \pm \varphi_A(\mathbf{r}_2)\varphi_B(\mathbf{r}_2)\} \qquad (47.1)$$

for the H_2 molecule. Note that Ψ_{+}^{VB} is identical with Ψ_{cov} in Eq. (36.16).

The total wave functions must be antisymmetric. Hence a symmetric spatial function like Ψ_{+}^{VB} must be associated with an antisymmetric spin function and an antisymmetric spatial function should be accompanied by a symmetric spin function. Since each electron has two possible spin states represented by α and β, there can be four spin wave functions for two electrons. These are

$$\alpha(s_1)\alpha(s_2),\ \alpha(s_1)\beta(s_2),\ \beta(s_1)\alpha(s_2) \text{ and } \beta(s_1)\beta(s_2). \qquad (47.2)$$

As discussed earlier in connection with the MO function for H_2, the total spin angular momentum operator is written as $\mathbf{S} = \mathbf{S}_1 + \mathbf{S}_2$. All the four products in Eq.(47.2) are eigenfunctions of S_Z with eigenvalues $\hbar$, 0, 0 and $-\hbar$ respectively. The first and the last are eigenfunctions of S^2 with eigenvalues $2\hbar^2$ each, but the second and the third products are not eigenfunctions of this operator. In fact we find that two linear combinations of the second and third products are eigenfunctions of S^2. So the common eigenfunctions of S^2 and S_Z are those shown in Table 8.1. Each of the symmetric spin wave functions multiplicatively combines with Ψ_-^{VB} to yield three total wave functions of same energy. The three-fold degeneracy arises because energy is determined only by the spatial wave function when the Hamiltonian does not depend upon any spin operator. The three degenerate states are referred to as spin triplet. They are characterized by $S = 1$ and M_S = 1, 0, −1. The spin multiplicity is given by $(2S+1)$ that equals 3 in the present circumstances.

The antisymmetric spin wave function in Table 8.1 leads to the total wave function

$$\Phi_+(\mathbf{x}_1, \mathbf{x}_2) = \Psi_+^{VB}(\mathbf{r}_1, \mathbf{r}_2)\frac{1}{\sqrt{2}}\{\alpha(s_1)\beta(s_2) - \alpha(s_2)\beta(s_1)\}. \quad (47.3)$$

The state is nondegenerate. In fact, $S = 0$, $M_S = 0$, and $2S+1 = 1$. This is a singlet state.

The singlet and triplet energies will be explicitly calculated in Part II of these lectures. Nevertheless, it is possible to identify the ground state as the energy is solely determined by the spatial function in the absence of any spin-dependent term in the Hamiltonian. For the lowest-energy states φ_A and φ_B must be chosen as $1s_A$ and $1s_B$, the 1s orbitals of each hydrogen atom. The antisymmetric spatial function Ψ_-^{VB} has a node whenever

$$1s_A(\mathbf{r}_1)1s_B(\mathbf{r}_2) = 1s_A(\mathbf{r}_2)1s_B(\mathbf{r}_1). \quad (47.4)$$

For example, when both the electrons are on the plane that is perpendicular to and bisects the molecular axis, Ψ_-^{VB} becomes zero. Because of the nodes in Ψ_-^{VB}, the triplet state has energy higher than the energy of the singlet state. Thus Φ_+ represents the ground state of the H_2 molecule.

Table 8.1 Two-electron spin eigenfunctions.

Function	Eigenvalues S^2	S_Z	Q. Number S	M_S	Permutation Symmetry
$\alpha(s_1)\alpha(s_2)$	$2\hbar^2$	$\hbar$	1	1	Symmetric
$2^{-1/2}\{\alpha(s_1)\beta(s_2)+\beta(s_1)\alpha(s_2)\}$	$2\hbar^2$	0	1	0	Symmetric
$\beta(s_1)\beta(s_2)$	$2\hbar^2$	$-\hbar$	1	−1	Symmetric
$2^{-1/2}\{\alpha(s_1)\beta(s_2)-\beta(s_1)\alpha(s_2)\}$	0	0	0	0	Antisymmetric

The simple valence bond treatment yields the equilibrium bond length of H_2 molecule as 0.80 Å with dissociation energy 3.14 eV and vibrational frequency 4800 cm^{-1}. This is certainly better than the results from a simple molecular orbital treatment (1.32 Å, 2.69 eV and 4160 cm^{-1}), yet it is bad compared to the experimental values R_{eq} = 0.74 Å, D_{el} = 4.72 eV and $\bar{\nu}$ = 4318 cm^{-1}. The valence bond results can be improved by using better functions for φ_A and φ_B. Table 8.2 contains a list of molecular properties for H_2 which have been computed by various methods. This table gives a clear indication of the general trend of what happens when different strategies are employed for carrying out calculations on other molecules.

Table 8.2 Molecular properties of H_2, computed by different methods.

Method	Effective Nuclear Charge	Dissociation Energy (eV)	Equilibrium Bond length (Å)	Vibrational Frequency (cm^{-1})
MO ($\zeta = 1$)	1.0	2.69	1.32...	4160
VB ($\zeta = 1$) [1]	1.0	3.14	0.80	4800
MO (ζ-optimization)	1.197	3.49	0.73	4580
VB (ζ-optimization) [2]	1.166	3.76	0.76	4900
MO-SCF [3]	0.95	3.64	...	...
ICR [4]	1.193 (C_{ion}= 0.256)	4.00	0.77	4750
MO-CI [5]	(28 configurations)	4.55	...	4385
Series expansion (40 terms) [3]	0.995	4.747	...	...
Series expansion (80 terms) [6]	1.072	4.748	...	4401
Experiment [7]		4.747	0.7395	4401

[1] Sugiura (1927).
[2] Wang (1928).
[3] Kolos and Roothaan (1960).
[4] Ion-covalent resonance: Weinbaum (1933).
[5] McLean, Weiss and Yoshimine (1960).
[6] Kolos and Wolniewicz (1964).
[7] Fisk and Kirtman (1964).

We have discussed earlier that H_2 has about 2 percent ionicity, that is, the ground state wave function has about 14 per cent contribution from ionic structures. This means that in whatever way one may improve φ_A and φ_B, the resulting wave function will never match the experimental ground state. The calculated results can be made better by writing a trial wave function as

$$\Psi(\mathbf{r}_1, \mathbf{r}_2) = N\,(\Psi_{cov} + \lambda\Psi_{ion}) \tag{47.5}$$

and then minimizing the expectation value $\langle \mathbf{H} \rangle$ for variation of λ. The optimized spatial function represents neither the purely covalent structure nor the purely ionic one, rather it stands for a "hybrid" of these two structures. This phenomenon is hence called as the *ion-covalent resonance*, and it results in the achievement of a significant amount of stability, the ion-covalent resonance energy (Δ_{ICR}), relative to the lowest-energy canonical structure, (in this case, the covalent structure), as shown in Figure 8.1. The ion-covalent resonance energy is a measure of the polarity of molecules. It may be estimated for a diatomic species AX by subtracting the covalent bond energy from the actual bond energy while the former is taken as the geometric mean of the bond energies of A_2 and X_2:

$$\Delta = B_{A-X} - B^{cov}_{A-X} \tag{47.6}$$

where

$$B^{cov}_{A-X} = (B_{A-A}B_{X-X})^{1/2}. \tag{47.7}$$

In the above the bond energy is defined as the negative of the binding energy,

$$B_{M-N} = (E_M + E_N) - (E_{M-N})_{min}. \tag{47.8}$$

A list of bond energies is given in Table 8.3. The estimated ion-covalent resonance energies are shown in Table 8.4.

Table 8.3 Bond energies for bonds A—A and X—X in electron Volts.

A_2	Bond energy [a]	AX	Bond energy
H_2	4.47	HF	5.85
Li_2	1.09	HCl	4.45
B_2	3.59	FCl	3.77
C_2	3.58	LiH	2.50
N_2	1.73	NaH	2.26
O_2	1.47	A_2	Double Bond energy
F_2	1.35 [b]	C_2	6.23
Na_2	0.746	N_2	4.33
Mg_2	1.35	O_2	5.12
Si_2	2.30	A_2	Triple Bond energy
P_2	2.08	C_2	8.65
S_2	2.78	N_2	9.76
Cl_2	2.49	AX	Double Bond energy
K_2	0.507	CN	6.37
Ca_2	1.09		
Ge_2	1.95		
As_2	1.51		
Se_2	1.78		
Br_2	1.97		
Rb_2	0.466		
Sr_2	0.870		
Sn_2	1.51		
Sb_2	1.25		
Te_2	1.30		
I_2	1.54		
Cs_2	0.456		

[a] Darwent (1970).
[b] Dibeler et al. (1970).

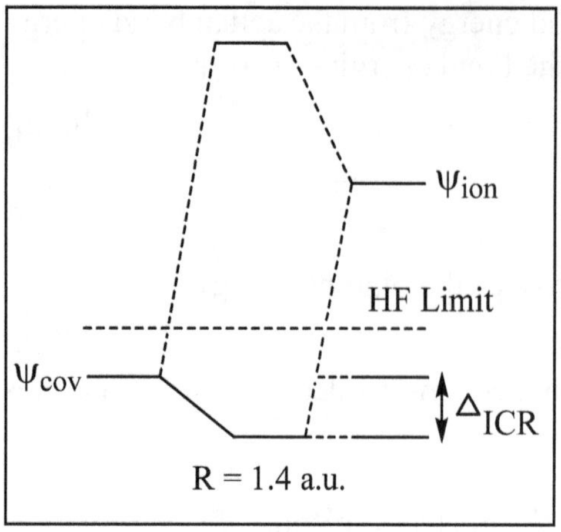

Figure 8.1 Ion-covalent resonance in H_2. The covalent structure is much more stable than the Hartree-Fock structure. The ionic structure is of greater energy. The resonance energy is measured from the energy of the covalent structure.

Table 8.4 Experimental ion-covalent resonance energies in electron Volts.

Species	Bond energy	Covalent bond energy as geometric mean	Δ_{ICR}
HF	5.85	2.46	3.40
HCl	4.45	3.34	1.12
LiH	2.50	2.21	0.29
NaH	2.26	1.83	0.43
FCl	3.77	1.83	1.94

The ion-covalent resonating structure is useful in determining the polarity of a hetero-nuclear diatomic molecule. Suppose that A is less electronegative than B. The ion-covalent resonance is

$$A\text{—}B \leftrightarrow A^{+}B^{-} \tag{47.9}$$

The wave function in Eq. (47.5) yields the fractional ionic character

$$\text{FIC} = N^2\lambda^2 \tag{47.10}$$

with the normalization constant

$$N = (1+2\lambda S+\lambda^2)^{-1/2} \tag{47.11}$$

where S is now the overlap of Ψ_{cov} with Ψ_{ion}. If we approximate the permanent dipole moment of the covalent structure to zero (point charge approximation of the electron density) and neglect the overlap dipole moment, the permanent dipole moment of the molecule will be given by (see Figure 8.2)

$$\mu = |e|R(\text{FIC}) \tag{47.12}$$

where R is the (equilibrium) bond length. Note that the dipole moment depends on both λ and S. This formula may be used to estimate the wave function in Eq. (47.5).

Figure 8.2 Illustration of Eq. (47.12) with q substituting for $|e| \times$ FIC.

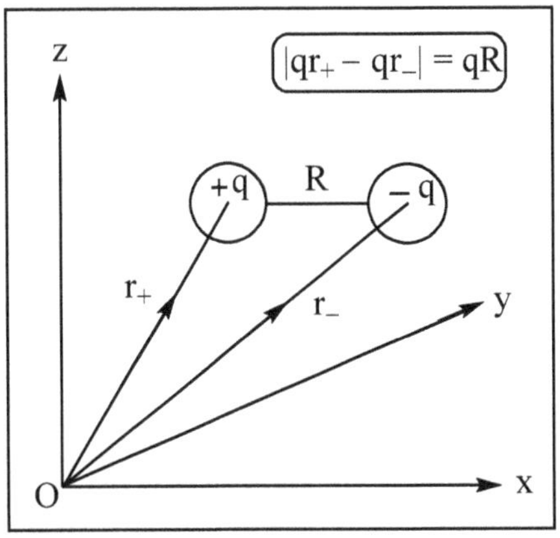

Example 46: To find the ion-covalent wave function

1) $\mu_{HF} = 1.82$ D, $R_{HF} = 0.921$ Å $\Rightarrow$ FIC = 0.412
2) $\mu_{HCl} = 1.03$ D, $R_{HCl} = 1.27$ Å $\Rightarrow$ FIC = 0.168
3) $\mu_{HBr} = 0.83$ D, $R_{HBr} = 1.41$ Å $\Rightarrow$ FIC = 0.123
4) $\mu_{HI} = 0.45$ D, $R_{HI} = 1.61$ Å $\Rightarrow$ FIC = 0.058

If we take $S \approx 0.25$, we get $\lambda \approx 1.03$ for HF, 0.503 for HCl, 0.411 for HBr and 0.264 for HI.

A comparison of the simple valence bond treatment with the simple molecular orbital method is given in Table 8.5.

Table 8.5 Comparison of the simple molecular orbital and the simple valence bond treatments.

M.O.T.	V.B.T.
1) This treatment is entirely based on theoretical principles.	1) This method requires the empirical knowledge of the covalent nature of the bond.
2) The steps are A.O.'s $\rightarrow$ M.O.'s M.O.'s $\rightarrow$ Product function	2) The steps are A.O.'s $\rightarrow$ product functions Products of a.o.'s $\xrightarrow{\text{L.C.}}$ Judiciously chosen combinations.
3) Hybridization is not necessary. A.O.'s $\xrightarrow{\text{L.C.}}$ Symmetry orbitals/M.O.'s S.O.'s $\xrightarrow{\text{L.C.}}$ M.O.'s	3) Hybridization is necessary. Central atom A.O.'s $\rightarrow$ H.O.'s A.O.'s (L) $\rightarrow$ S.O.'s(L) H.O.'s $\otimes$ S.O.'s $\rightarrow$ Product functions
4) The method is successfully applied to the study of polar bonds. Example: LFT	4) Application of this treatment is successful for almost pure covalent bond. Example: Homonuclear diatomics.
5) Results can be improved in two ways: i) by using optimized atomic orbitals, ii) by general configuration interaction technique.	5) Results can be improved in two ways: i) by using optimized atom-centered functions, ii) by ion-covalent resonance.

Lecture 48: Polyatomic Molecules, Resonance and Aromaticity

Although hybridization is not a must for molecular orbital treatments, it is the first step in the valence bond treatment on any nontrivial molecule. The hybrid orbitals are used in writing the electron-pair wave functions representing the covalent bonds. The situation is made clear by the following examples.

Example 47: To find the VB wave function for methane

Methane is known to be primarily covalent and it possesses four equivalent C—H bonds. Let h_1, h_2, h_3 and h_4 be the four sp^3 hybrid orbitals of the central atom. The 1s orbitals of the four ligands (hydrogen atoms) are used to form four C—H valence bond functions of the type

$$\Phi_{C-H_1}(\mathbf{x}_1, \mathbf{x}_2) = [2(1+S^2)]^{-1/2}[h_1(\mathbf{r}_1)1s_1(\mathbf{r}_2) + h_1(\mathbf{r}_2)1s_1(\mathbf{r}_1)] \otimes 2^{-1/2}[\alpha(s_1)\beta(s_2)-\alpha(s_2)\beta(s_1)]. \quad (48.1)$$

etc. The quantity S is now the overlap between the hybrid orbital h_1 directed along the ligand H_1 and the 1s orbital ($1s_1$) on the same ligand. The valence electron wave function for CH_4 can be written as

$$\Phi(\mathbf{x}_1, \mathbf{x}_2, \mathbf{x}_3, \ldots, \mathbf{x}_8) = \Phi_{C-H_1}(\mathbf{x}_1, \mathbf{x}_2)\Phi_{C-H_2}(\mathbf{x}_3, \mathbf{x}_4)\Phi_{C-H_3}(\mathbf{x}_5, \mathbf{x}_6)\Phi_{C-H_4}(\mathbf{x}_7, \mathbf{x}_8). \quad (48.2)$$

The wave function is antisymmetric only for the interchanges of space-spin coordinates $\mathbf{x}_1$ and $\mathbf{x}_2$, $\mathbf{x}_3$ and $\mathbf{x}_4$, etc. Hence an overall antisymmetrization operator **A** may be used. The effect of this operator would be to prepare all distinct pair-permuted functions like

$$\Phi_{C-H_1}(\mathbf{x}_1, \mathbf{x}_4)\,\Phi_{C-H_2}(\mathbf{x}_3, \mathbf{x}_2)\ldots$$

with a negative sign for each permutation and so on, so that the result is a properly antisymmetrized valence-bond wave function

$$\Phi_{VB}(\mathbf{x}_1, \ldots, \mathbf{x}_8) = N\mathbf{A}\{\Phi_{C-H_1}(\mathbf{x}_1, \mathbf{x}_2)\Phi_{C-H_2}(\mathbf{x}_3, \mathbf{x}_4)\Phi_{C-H_3}(\mathbf{x}_5, \mathbf{x}_6)\Phi_{C-H_4}(\mathbf{x}_7, \mathbf{x}_8)\}. \quad (48.3)$$

where N is an overall normalization constant.

The valence bond function in Eq. (48.3) represents four 100% covalent bonds. It can be used to calculate the total valence electron energy as the expectation value of the valence-electronic Hamiltonian. One may extend it by including the wave function for the core electrons of the carbon atom as a multiplicative factor within the product on the right hand side of Eq. (48.3) and then calculate the total electronic energy. One may determine the equilibrium bond length, the electronic dissociation energy, the vibrational frequencies and anharmonic constants. All these results will be close to the observed values but never identical with them. If one wishes to improve upon the results, one should use better functions for the hybrid and ligand orbitals and then one must consider the interaction of different electron-pair wave functions including the possibility of ion-covalent resonance.

Example 48: The π-electron valence bond functions for the benzene molecule.

We write

$$K_1 = \Phi_{C_1-C_2}(\mathbf{x}_1, \mathbf{x}_2)\,\Phi_{C_3-C_4}(\mathbf{x}_3, \mathbf{x}_4)\,\Phi_{C_5-C_6}(\mathbf{x}_5, \mathbf{x}_6). \quad (48.4)$$

for the first canonical structure of Kekule. See Figure 8.3. In the above the "two-π-electron" function $\Phi_{C_1\text{-}C_2}(\mathbf{x}_1, \mathbf{x}_2)$ is given by

$$\Phi_{C_1\text{-}C_2}(\mathbf{x}_1, \mathbf{x}_2) = [2(1+S^2)]^{-1/2}\,[pz1(\mathbf{r}_1)pz2(\mathbf{r}_2) + pz1(\mathbf{r}_2)pz2(\mathbf{r}_1)] \otimes 2^{-1/2}\,[\alpha(s_1)\beta(s_2) - \alpha(s_2)\beta(s_1)]. \qquad (48.5)$$

Figure 8.3 Kekulé and Dewar structures for benzene and the corresponding π-electron energies.

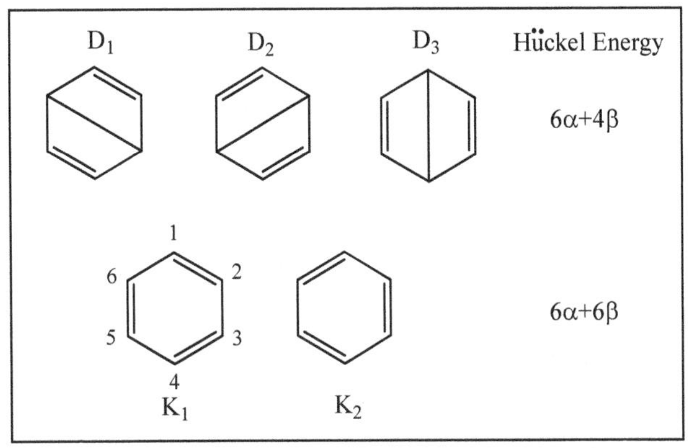

The wave function

$$K_2 = \Phi_{C_2\text{-}C_3}(\mathbf{x}_2, \mathbf{x}_3)\,\Phi_{C_4\text{-}C_5}(\mathbf{x}_4, \mathbf{x}_5)\,\Phi_{C_6\text{-}C_1}(\mathbf{x}_6, \mathbf{x}_1). \qquad (48.6)$$

corresponds to the second Kekule′ structure in Figure 8.3. The three Dewar structures are associated with the functions

$$D_1 = \Phi_{C_1\text{-}C_2}(\mathbf{x}_1, \mathbf{x}_2)\,\Phi_{C_3\text{-}C_6}(\mathbf{x}_3, \mathbf{x}_6)\,\Phi_{C_4\text{-}C_5}(\mathbf{x}_4, \mathbf{x}_5),$$

$$D_2 = \Phi_{C_6\text{-}C_1}(\mathbf{x}_6, \mathbf{x}_1)\,\Phi_{C_2\text{-}C_5}(\mathbf{x}_2, \mathbf{x}_5)\,\Phi_{C_3\text{-}C_4}(\mathbf{x}_3, \mathbf{x}_4), \qquad (48.7)$$

$$D_3 = \Phi_{C_5\text{-}C_6}(\mathbf{x}_5, \mathbf{x}_6)\,\Phi_{C_1\text{-}C_4}(\mathbf{x}_1, \mathbf{x}_4)\,\Phi_{C_2\text{-}C_3}(\mathbf{x}_2, \mathbf{x}_3).$$

Figure 8.4 Resonance in benzene: the valence bond approach is directly based on a chemist's visualization of this phenomenon. The MO interpretation is not so direct.

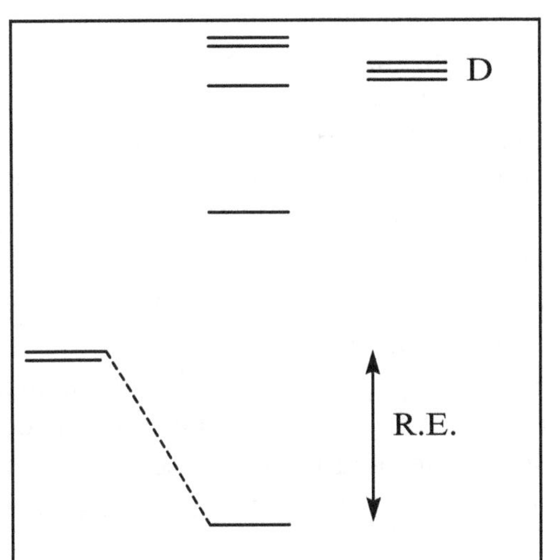

The net π-electron wave function is written as

$$\Phi_{VB}(\mathbf{x}_1, \mathbf{x}_2, \ldots, \mathbf{x}_6) = c_1K_1 + c_2K_2 + c_3D_1 + c_4D_2 + c_5D_3. \quad (48.8)$$

The best values for the coefficients $c_1, c_2, \ldots, c_5$ are then determined. The schematic energy level diagram is shown by Figure 8.4. The actual molecule in its ground state is more stable than any of its cannonical structures.

This discussion clarifies the concept of resonance in chemistry. Resonance is the phenomenon by which we describe reality (the true wave function) in terms of fictions (wave functions corresponding to the canonical structures). The standard structures drawn by a chemist may not exist at all — they reside only in the mind of the chemist — and so they are fictions. The stability gained by the species due to the interaction among the standard structures is called resonance energy. This energy is always measured relative to the energy of the most stable canonical structure. Figure 8.4 shows the resonance energy as the difference of the ground state energy of benzene and the energy of each Kekule′ structure. The unusual stability gained sometimes results in special properties which are collectively termed *aromaticity*. Common examples of aromatic molecules are benzene, napthalene, anthracene, pyrrole, pyridine, etc. and their derivatives. These molecules are known to satisfy Hückel's $(4n+2)$ rule the origin of which is best explained by Hückel's molecular orbital treatment. Dewar has extended a principle originally due to Evans (1939). According to the Evans-Dewar rule, one compares the π-electron energy of a C_n cyclic system with that for the corresponding linear system so as to determine whether the cyclic system is aromatic or antiaromatic. Thus cyclobutadiene is antiaromatic as

$$E_{\text{ring}}(C_4) = 4\alpha + 4\beta > E_{\text{chain}}(C_4) = 4\alpha + 4.47\beta$$

whereas benzene is aromatic as

$$E_{\text{ring}}(C_6) = 6\alpha + 8\beta < E_{\text{chain}}(C_6) = 6\alpha + 6.99\beta$$

Resonance is basically associated with the phenomenon of linear superposition. Thus the molecular orbital is an outcome of resonance involving atomic orbitals, and a MO wave function can be viewed as a product of resonance that involves different covalent and ionic structures. Resonance in chemistry is not very different from the phenomenon of resonance as understood in physics. Two structures of similar symmetries strongly interact only when they are almost equal in energy, that is, the energy difference corresponds to a nearly zero frequency difference. In this sense resonance in chemistry is a mere manifestation of the physical resonance. Suppose that the species is somehow prepared in one of the canonical states. The state of the species will start oscillating very rapidly among all the canonical states until it relaxes by some additional mechanism into the hybrid state that is significantly lower in energy. As an alternative view, the hybrid stationary state is a time-dependent linear combination of the canonical stationary states. The coefficients of linear combination vary periodically in time although the period of oscillation is different for canonical structures of different energy. Thus the name resonance given to the phenomenon is entirely appropriate. In physics, however, transmission resonances such as that occurring for a square well look like long-lived bound states and the resonances in the scattering of elementary particles appear as new fundamental particles with finite lifetimes.

Lecture 49: Electronegativity

The ionization energy or potential is defined as the *minimum* energy required to ionize an *isolated* species in its *ground* state such that one electron is freed from it leaving behind an isolated positively charged ion in its ground state. This is also called the first ionization energy or potential. The second, the third, etc. ionization potentials refer to the successive loss of electrons one at a time. The definition involves the minimum amount of energy to be absorbed as any additional energy will appear as the kinetic energy of the ion-electron system. The term potential was given because earlier experiments involved determining the stopping potential for the ejected photo-electron. The definition can be extended to the ionization process for the loss of one electron from a specific orbital, and then one gets the orbital ionization potentials as discussed earlier in Chapter 5.

The electron affinity is similarly the minimum energy evolved (or the maximum energy absorbed) while one electron attaches itself to an isolated species thereby forming an isolated negatively charged ion in its ground state. The stipulation of the maximum for absorbed energy implies that the free electron has zero kinetic energy and the original species is in its ground state. In practice gaseous atoms or molecules at low pressures may be considered as effectively isolated since these would be in average far away from other atoms or molecules at any particular instant of time.

The important point is that both ionization potential and electron affinity are well-defined and they are quantitatively described in terms of the difference of energy for two states. Both quantities have been conceptualized in physics and measurable from physical experiments. Chemistry, however, involves collective processes. One atom or molecule losing one electron or gaining it is physics. Chemistry is where one atom or molecule loses one electron that is promptly captured by another atom or molecule.

Electronegativity is the relative power of an atom to draw the electron towards itself while it is bonded to another atom. It is essentially a qualitative concept, but the elements can be arranged in an order of increasing electronegativity. While joined by a chemical bond the less electronegative element becomes the positive end and the element of greater electronegativity acquires a partial negative charge. The first quantitative assignment of electronegativity was made by Pauling who introduced a relative scale. Shortly afterwards Mulliken (1934) proposed an absolute scale for electronegativity. Mulliken's definition made electronegativity more versatile as the concept could be extended to define the orbital electronegativity. Recently a new interpretation of electronegativity has been given from a density functional treatment: Parr and his coworkers have shown that the electronegativity of an atom may be approximately defined as $-\mu$ where μ is the chemical potential.

Electronegativity is primarily an empirical concept. It indicates the relative power, but the latter is exactly what determines the course of a reaction. We stress that chemistry is a subject of competition. The formation of positive ions is energetically expensive, yet these ions are formed along with negative ions so long as the total Gibbs free energy decreases. This, of course, indicates the relative power of certain species to draw electrons from other species. This is why electronegativity is so important in chemistry.

Crudely speaking, one may have the luxury of studying noncompetitive phenomena in physics.[2] To exemplify, the determination of ionization energy of a molecule or the work function of a metal is a problem of physics. Solubility is typically a chemical concept, with the guiding principle "like dissolves like" that leads to the maximal decrease in Gibbs free energy.

The Pauling Scale

Pauling's aim was to provide a linear scale for electronegativity (x). Thus if element A is more electronegative than element B which is turn more electronegative than element C, then element A must be more electronegative than element C. Hence Pauling suggested the linear behavior

$$(x_A - x_C) = (x_A - x_B) + (x_B - x_C). \tag{49.1}$$

We have observed that the ion-covalent resonance energy Δ is a qualitative measure of bond. The greater the polarity, the larger Δ is. However, it was found that $\Delta_{AB} + \Delta_{BC} \neq \Delta_{AC}$ though the relation

$$\sqrt{\Delta_{AB}} + \sqrt{\Delta_{BC}} = \sqrt{\Delta_{AC}} \tag{49.2}$$

approximately holds true. Hence Pauling defined $\sqrt{\Delta}$ as a measure of the difference in electronegativity,

$$\sqrt{\Delta_{AB}} \propto |x_A - x_B|. \tag{49.3}$$

He reasoned that for electronegativity to be of daily use to a general chemist, its value should not be very large or very small, but should be of order of unity and preferably positive. The first constraint can be easily met as the ion-covalent resonance energy is known to equal only a few electron volts. The proportionality constant in Eq. (49.3) is chosen to be 1 when Δ is expressed in electron volt:

$$x_A - x_B = \sqrt{\Delta_{AB} \text{ in eV}}\,. \tag{49.4}$$

This yields a relative scale, and x is yet to be uniquely determined. The second constraint is satisfied by fixing x_H at a handy value 2.2 so that elemental electronegativity varies from a minimum of about 0.8 for Cs to a maximum of about 4 for F. The outcome is the Pauling scale of electronegativity (Table 8.6) that has found indeed a robust application to our daily work in chemistry.

As the ion-covalent resonance energy in electron volts is approximately given by

$$\Delta_{AB} \approx \mu_{AB}^2 \tag{49.5}$$

where μ_{AB} is the permanent dipole moment of the neutral species AB in Debye, we get the approximate relationship

$$|x_A - x_B| \approx \mu_{AB} \text{ in D}. \tag{49.6}$$

[2]In physics too one finds collective phenomena such as a collective excitation that accompanies the formation of quasi-particles, and competitive phenomena though the competition appears in a different form. A physicist normally works out the probabilities of competitive transitions, or studies the change of phase. Physicists often deal with the push and pull due to different physical forces.

Table 8.7 illustrates that this rough correlation is valid when $|x_A - x_B|$ is reasonably small.

Essential points about the Pauling scale are as follows:

- The Pauling scale is a relative scale. It pertains to the situation of *atoms in molecules*.
- The numbers are of the *right order* so that they can be memorized and put into use whenever the occasion arises.
- The electronegativity difference gives an approximate value of the permanent dipole moment in Debye unit.

The Mulliken Scale

Mulliken's treatment (1934, 1935) was based on the energetics of the reactions

$$A + B = A^+B^- \qquad \Delta E_1 \approx IP_A - EA_B - e^2/R_{AB}$$

$$A + B = A^-B^+ \qquad \Delta E_2 \approx IP_B - EA_A - e^2/R_{AB} \qquad (49.7)$$

where IP and EA are the ionization potential and the absolute magnitude of the electron affinity. For a strictly nonpolar bond, ΔE_1 and ΔE_2 should be equal to each other, thus yielding

$$IP_A + EA_A = IP_B + EA_B. \qquad (49.8)$$

Also, for $\Delta E_1 > \Delta E_2$,

$$IP_A + EA_A > IP_B + EA_B. \qquad (49.9)$$

But the relation $\Delta E_1 > \Delta E_2$ corresponds to $X_A > X_B$ where X is the electronegativity. Therefore Mulliken defined the *absolute* electronegativity X of an atom as

$$X = \frac{1}{2}(IP + EA) \qquad (49.10)$$

where the IP and EA values are expressed in electron volts.

It is found that the Mulliken electronegativity X varies approximately linearly with the Pauling electronegativity x. One writes

$$x = a(X - b), \qquad (49.11)$$

and finds that the best value of the constant a is given by

$$a = 0.336. \qquad (49.12)$$

Comparing x and X for hydrogen one obtains

$$b = 0.615. \qquad (49.13)$$

A rougher estimate is obtained from the relation

$$X = 2.78x. \qquad (49.14)$$

Mulliken's electronegativity scale is an absolute one. It corresponds to a free atom and not to atoms in a molecule. Mulliken's definition also suffers from the fact that the numbers generated fluctuate more widely and so they are difficult to memorize. The great merit of Mulliken's definition is that the electronegativity values can be assigned to orbitals and to the states of hybridization (Table 8.6). Using Eq. (49.11), Hinze et al. converted (1963) the hybridization-dependent Mulliken-Jaffé electronegativity to the

Pauling scale. A few typical values of the converted electronegativity are given in Table 8.8.

Table 8.6 Electronegativity in the Pauling scale (x) and the Mulliken scale (X).

						Legend
H						Element
2.2						Pauling x
7.176						Mulliken X(s)
...						Mulliken X(p)
...						Mulliken-Jaffé X (hybrid)
...						Mulliken X(d)
Li	Be	B	C	N	O	F
0.98	1.57	2.04	2.55	3.04	3.41	3.98
3.106	5.946	9.594	14.051	19.316	25.390	32.272
1.258	2.563	4.001	5.572	7.275	9.111	11.080
0.837	1.40 (sp)	1.93 (sp^2)	2.48 (sp^3)	3.68 (sp^3)	3.04 (p)	3.90
...	...	...	...	...	...	...
Na	Mg	Al	Si	P	S	Cl
0.93	1.31	1.61	1.90	2.19	2.58	3.16
2.804	5.125	7.771	10.030	14.033	17.650	21.59
1.302	2.052	2.995	4.133	5.464	6.989	8.708
...	...	...	...	...	...	...
0.150	0.162	0.224	0.337	0.500	0.713	0.977
K	Ca		Ge	As	Se	Br
0.82	1.00		2.01	2.18	2.55	2.96
Rb						I
0.82						2.66
Cs						
0.79						

Table 8.7 Electronegativity difference and experimental dipole moment.

Molecule	$\|x_A - x_B\|$	μ (Debye)
HI	0.46	0.45
HBr	0.76	0.83
HCl	0.96	1.03
HF	1.78	1.82

Table 8.8 Hybridization-dependent electronegativity.

	Be	B	C	N	O
sp	1.40		3.29	4.67	
sp^2		1.93	2.75	3.94	
sp^3			2.48	3.68	
p			1.75	2.28	3.04

Lecture 50: Valence-Shell Electron Pair Repulsion (VSEPR)

There are often more valence electrons than what would be required to form bonds in a covalent molecule. The extra electrons remain in nonbonding (and sometimes in antibonding) orbitals. For instance, ammonia has three N—H single bonds and a lone pair of valence electrons. If we take nitrogen to be in the state of sp^3 hybridization in ammonia, then the valence bond wave function is written as

$$\Phi(\mathbf{x}_1, ..., \mathbf{x}_8) = NA\Big[\Phi_{h_1-s_A}(\mathbf{x}_1, \mathbf{x}_2)\,\Phi_{h_2-s_B}(\mathbf{x}_3, \mathbf{x}_4)\,\Phi_{h_3-s_C}(\mathbf{x}_5, \mathbf{x}_6)\times h_4(\mathbf{r}_7)\,h_4(\mathbf{r}_8) \times\frac{1}{\sqrt{2}}\{\alpha(s_7)\beta(s_8)-\beta(s_7)\alpha(s_8)\}\Big] \quad (50.1)$$

where h_1, h_2 and h_3 are the three hybrid orbitals which, along with the 1s orbitals of the three hydrogen atoms H_A, H_B and H_C, respectively, form the three N—H single bonds. Besides, two spin-paired electrons stay in the nonbonding orbital h_4. There are three pairs of electrons (bond pairs) forming the three single bonds and one pair of nonbonding electrons (lone pair).

The VSPER theory was developed by Gillespie and Nyholm (1957, 1970). It is based on a qualitative reasoning that helps us to predict the molecular geometry. The main logic is that once all the bonds are formed, the equilibrium geometry is determined as the configuration with the least amount of coulomb energy of repulsion between the electron pairs. The same configuration is determined in three steps.

Figure 8.5 Disposition of valence shell electron pairs in space and the resulting molecular geometry for acetylene and carbon dioxide.

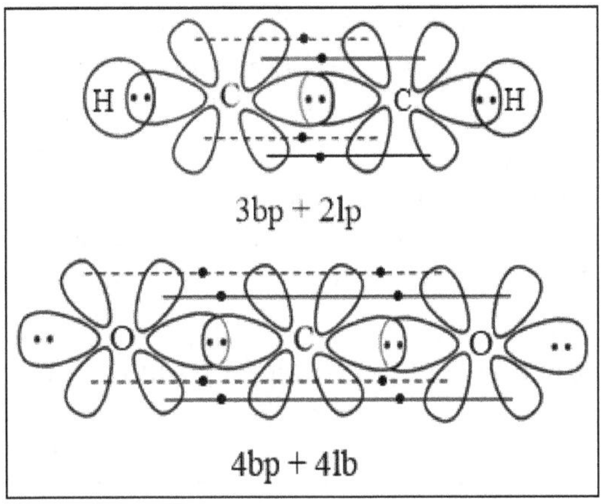

1. The number of bond pairs and that of lone pairs are counted.
2. As the bond pairs and lone pairs are delocalized (or localized) to almost the same extent, the lone pair-lone pair repulsion is almost as strong as the bond pair-lone pair repulsion. Therefore, the molecular geometry in which the bond pairs and the lone pairs are symmetrically distributed such that they remain as far away from each other as possible corresponds to the most stable configuration. This condition also determines the state of hybridization of the central atom in most cases; nevertheless, exceptions are known to occur; (see the following discussion on PF_3).
3. The electrons of a bond pair are really concentrated in the region of the chemical

bond, that is, between the two bonded atoms, whereas the electrons belonging to a lone pair feel less attraction by the atomic core charges. Therefore, the strength of coulombic repulsion really follows the trend

lone pair-lone pair > bond pair-lone pair > bond pair-bond pair. (50.2)

Hence the molecular geometry distorts from the symmetric one to yield the most stable geometry. In the equilibrium structure the bond angles are slightly less than the angles between bonds and the axes of nonbonding orbitals, and the latter angles are in turn less than the angles between nonbonding orbitals.

To clarify these three steps we now consider suitable examples. Let us start with polyatomic molecules which have elements of the second row of the periodic table as central atoms. These elements are stabilized by forming "octets" which fill the four valence orbitals of the atom. So the saturated systems have four electron pairs some of which (at least two) are bond pairs and the rest are lone pairs. Step 2 indicates that these pairs will be directed towards the four corners of a tetrahedron, and so the approximate configuration is tetrahedral. Now we have different possibilities:

(a) If there are 2 bond pairs and 2 lone pairs, step 3 indicates that the unequal coulombic repulsion would lead to a distorted tetrahedral arrangement. Sometimes the distortion is slight as in the H_2O molecule. Water has a bent structure. The H—O—H angle is about 104°, slightly less than the tetrahedral angle 109.47°. In case there is only one lone pair, the distortion is aided by the gain in stability due to the formation of bonds by utilizing all nonbonding orbitals that are not fully occupied, and it results in the linear geometry. This is evidenced in the H—C—C fragment of acetylene. Carbon dioxide is another example. These two cases are illustrated in Figure 8.5.

(b) If there are 3 bond pairs and 1 lone pair, a distorted tetrahedral arrangement is the only outcome of step 3. The resulting molecular geometry is pyramidal. Ammonia has a pyramidal structure with the H—N—H angle a little less than the tetrahedral angle. See Figure 8.6. The molecule NF_3 has a similar structure.

(c) If there are 4 bond pairs (examples: CH_4, CCl_4, CF_4) the molecule has a non-distorted tetrahedral geometry. A very slight distortion from the tetrahedral geometry occurs when the species is partially substituted; (examples: CH_2Cl_2, $CHCl_3$, etc.). Cases (b) and (c) are illustrated by Figure 8.6.

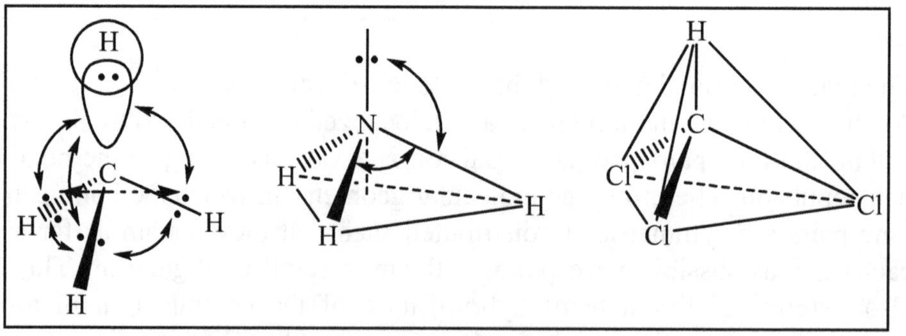

Figure 8.6 Illustration of VSEPR for CH_4, NH_3 and $CHCl_3$.

The situation for electron-deficient species is sometimes simpler and sometimes more complicated. Molecules like BeH_2, $BeCl_2$, etc. have two bond pairs and no lone pair with the central atom. These are necessarily linear. Boron trifluoride (BF_3) has three bond pairs. This molecule is planar with F—B—F angles equal to 120°, that is, the B—F bonds are directed towards the three corners of an equilateral triangle. These structures are trivially deduced from Steps 1 and 2. The molecule BH_3 is not a stable species — it would dimerize easily. Had it been stable, it would have had a planar triangular structure like BF_3. The molecule B_2H_6 is interesting as two of the hydrogen atoms in it are bonded to both the boron atoms. It possesses 4 one-electron bonds between the two boron atoms and the two bridging hydrogen atoms. If we extend the VSPER theory by taking the one-electron bonds into account, each boron atom will be found to be approximately tetrahedrally coordinated. This makes the hydrogen bridge perpendicular to the planes of the coplanar BH_2 fragments (Figure 8.7).

Figure 8.7 VSEPR with one-electron bonds: the B_2H_6 ballad.

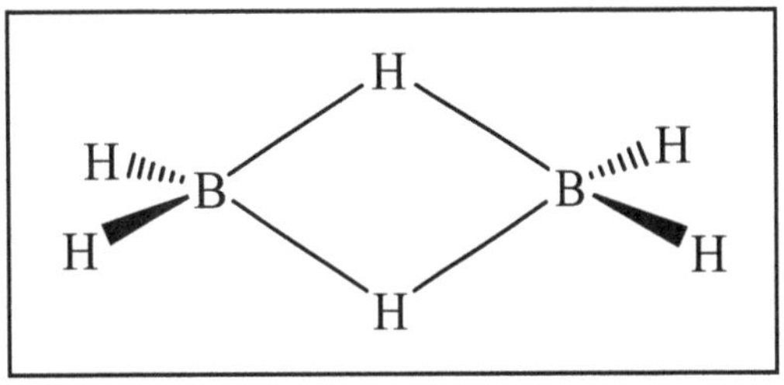

More complex molecules have more complex geometries. Molecules of the type AX_4 with one lone pair have square pyramidal structures, but if there are two lone pairs, the AX_4 species have square planar geometries. The molecule PF_3 has a pyramidal structure, but the angle F—P—F is nearly 90° (Figure 8.8). This happens since the sp^3 hybridization becomes energetically expensive and the p orbitals are involved in the bonding in PF_3, leaving aside the lone pair on the non-bonding s orbital of phosphorous atom. The structure of PF_5 is a trigonal bipyramid. Molecules PCl_5, $TeCl_4$, ClF_3 and ICl_2^- are electronically similar and they have trigonal bipyramidal arrangement of electron pairs which determines the molecular geometry. See Figure 8.8.

Molecules of type AX_6 with no lone pair are octahedral in shape. Examples are SF_6, transition metal complexes of type ML_6^{-n}, etc. Electronically similar molecules, e.g., IF_5, ICl_4^-, etc., have six electron pairs which are approximately octehedrally disposed around the central atom (Figure 8.8). Many cations in solvated state have tetrahedral or octahedral coordination, e.g., NH_4^+ (tetrahedral), H_3O^+ (pyramidal), $[Na(H_2O)_4]^+$ (tetrahedral), $[Na(H_2O)_6]^+$ (octahedral), $[M(H_2O)_6]^{+n}$ (octahedral) where M^{n+} is a transition metal ion, etc.

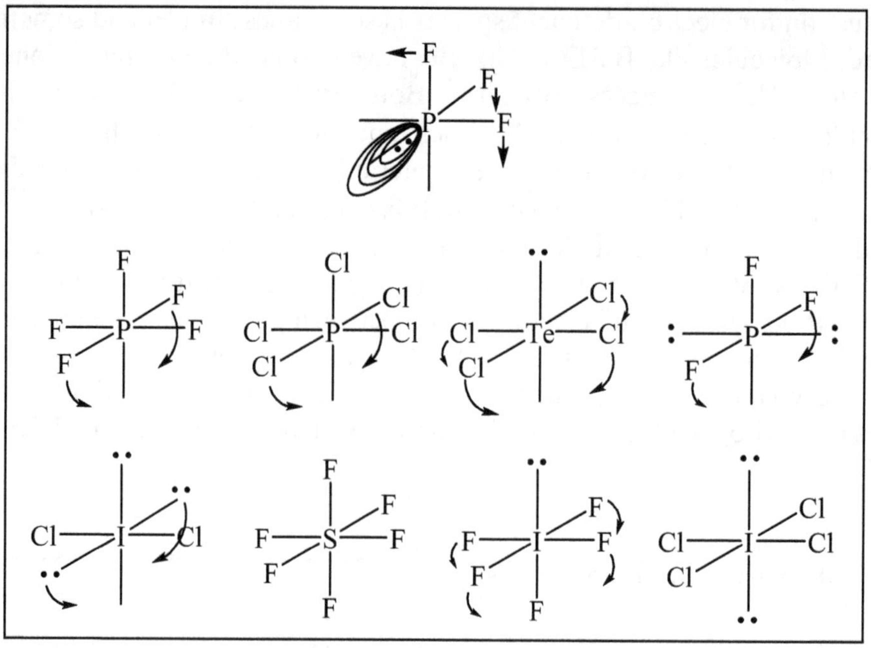

Figure 8.8 Structures predicted by VSEPR theory for (i) PF_3, and molecules isoelectronic with (ii) PF_5 and (iii) SF_6.

Text Suggested for Further Reading

J. Simons, "An Introduction to Theoretical Chemistry", (Cambridge University Press, Cambridge, 2003).

PROBLEMS

1. Write the MO wave function for He_2.
2. Write the VB wave function for O_2 and explain the triplet ground state. Compare your results with the MO wave function for the same molecule.
3. Derive an expression for the electronic dipole moment of a heteronuclear diatomic molecule from a VB description of the wave function.
4. Convert the values of Mulliken electronegativity given in Table 8.6 into the Pauling equivalents and the Pauling electronegativity values into their Mulliken equivalents. What do you observe? Do you find any anomaly?
5. Guess the molecular geometry of $Ni(CO)_4$.

PART II
ELEMENTARY QUANTUM CHEMISTRY

CHAPTER 9

THE MATRIX REPRESENTATIONS

Lecture 51: Transformation Theory

Consider the Schrödinger equation

$$\mathbf{H}\psi_k(\mathbf{r}) = E_k\psi_k(\mathbf{r}). \tag{51.1}$$

Since **H** is hermitian, the functions ψ_k's form an orthonormal set. The index k may stand for a continuous or a discrete variable. The set $\{\psi\}$ is postulated to be complete. To denote the sum over the discrete values of k as well as the continuous values, we use the generalized summation symbol S:

$$\underset{k}{S} = \sum_k \oplus \int dk. \tag{51.2}$$

Let $\mathbf{\Lambda}$ be another hermitian operator with eigenfunctions ϕ_μ's

$$\mathbf{\Lambda}\phi_\mu(\mathbf{r}) = \lambda_\mu\phi_\mu(\mathbf{r}). \tag{51.3}$$

Also, $\{\phi\}$ is another orthonormal complete set. One can write ϕ_μ as

$$\phi_\mu(\mathbf{r}) = \underset{k}{S}\, c_{k\mu}\psi_k(\mathbf{r}). \tag{51.4}$$

In accordance with our discussion in Part I of this book, the coefficient $c_{k\mu}$ will be given by

$$c_{k\mu} = \int d^3r\, \psi_k^*(\mathbf{r})\phi_\mu(\mathbf{r}). \tag{51.5}$$

The complex conjugate of the coefficient is

$$c_{k\mu}^* = \int d^3r\, \phi_\mu^*(\mathbf{r})\, \psi_k(\mathbf{r}). \tag{51.6}$$

Equations (51.4) and (51.5) show that one may also recognize the expansion of ψ_k in the basis $\{\phi\}$ as

$$\psi_k(\mathbf{r}) = \underset{\mu}{S}\, c_{k\mu}^*\phi_\mu(\mathbf{r}). \tag{51.7}$$

We will shortly bring in the matrices. The functions ψ_k's are mutually orthogonal and normalized, that is,

$$\int d^3r\, \psi_k^*(\mathbf{r})\psi_l(\mathbf{r}) = \delta_{k,l}. \tag{51.8}$$

Here $\delta_{k,l}$ is the Kroenecker delta function. Substituting Eq. (51.7) in this equation and exploiting the orthonormality of the functions ϕ's one obtains

$$\underset{\mu}{S}\, c_{k\mu}c_{l\mu}^* = \delta_{k,l}. \tag{51.9}$$

When one arranges the coefficients $c_{k\mu}$'s in an ordered two-dimensional array one obtains the square matrix $\boldsymbol{C}$ (usually of infinite order). The hermitian adjoint matrix is $\boldsymbol{C}^\dagger$, and Eq. (51.9) implies that

$$(\boldsymbol{CC}^\dagger)_{kl} = \delta_{k,l}. \tag{51.10}$$

Similarly one can establish that

$$(\boldsymbol{C}^\dagger\boldsymbol{C})_{\mu\nu} = \delta_{\mu,\nu} \tag{51.11}$$

so that

$$\boldsymbol{CC}^\dagger = \boldsymbol{1} = \boldsymbol{C}^\dagger\boldsymbol{C} \tag{51.12}$$

where $\boldsymbol{1}$ is the unit matrix of the same (infinite) order. We have just found out that the matrix $\boldsymbol{C}$ is unitary:

$$\boldsymbol{C}^\dagger = \boldsymbol{C}^{-1}. \tag{51.13}$$

The closure property may be established in the following way. Write

$$\sum_\mu c_{k\mu} c_{l\mu}^* = \int d^3r\, \psi_k^*(\mathbf{r}) \int d^3r' \left(\sum_\mu \phi_\mu(\mathbf{r})\phi_\mu^*(\mathbf{r}')\right) \psi_l(\mathbf{r}'). \tag{51.14}$$

By comparing this result with Equations (51.8) and (51.9) one finds

$$\sum_\mu \phi_\mu(\mathbf{r})\phi_\mu^*(\mathbf{r}') = \delta^3(\mathbf{r} - \mathbf{r}') \tag{51.15}$$

where the involved δ^3 function is a three-dimensional Dirac delta function.

Next, we want to show that *the solution of the Schrödinger equation is equivalent to diagonalizing a particular Hamiltonian matrix* ($\boldsymbol{H}''$).[1] To do so, we define two Hamiltonian matrices $\boldsymbol{H}'$ and $\boldsymbol{H}''$ by the elements

$$H'_{kl} = \int d^3r\, \psi_k^*(\mathbf{r})\mathbf{H}\psi_l(\mathbf{r}) = E_k\delta_{k,l},$$

$$H''_{\mu\nu} = \int d^3r\, \phi_\mu^*(\mathbf{r})\mathbf{H}\phi_\nu(\mathbf{r}). \tag{51.16}$$

The matrix $\boldsymbol{H}'$ is diagonal with the eigenvalues of operator $\mathbf{H}$ appearing as the diagonal elements, whereas $\boldsymbol{H}''$ is in general a nondiagonal but hermitean matrix. It is easy to establish, by using Eq. (51.7), its complex conjugate version and the definition above, that

$$(\boldsymbol{C}\boldsymbol{H}''\boldsymbol{C}^\dagger)_{kl} = \sum_\mu \sum_\nu c_{k\mu} H''_{\mu\nu} c_{l\nu}^* = H'_{kl}. \tag{51.17}$$

Therefore, one may write

$$\boldsymbol{C}\boldsymbol{H}''\boldsymbol{C}^\dagger = \boldsymbol{H}' \tag{51.18}$$

which establishes our aim.

[1]In principle, this is exact, as the functions form a complete set. In practice, this is generally approximate, as one works with a finite number of standard functions. For instance, in a quantum chemical calculation, one uses a finite basis set. Standard quantum chemical softwares for electronic structure calculations invariably use diagonalization of the Hamiltonian matrix. Therefore, the results one obtains are only approximate solutions of the electronic Schrödinger equation. The results can be made better by using a larger basis set besides adopting a better methodology.

An interesting observation is that the operator $\mathbf{H}(\mathbf{r})$ itself may be recognized as a diagonal element. Let us label the rows and the columns by coordinates $\mathbf{r}$ and $\mathbf{r}'$ respectively. Then the unit matrix will be defined by its elements

$$1_{\mathbf{r}\mathbf{r}'} = \delta^3(\mathbf{r} - \mathbf{r}'). \tag{51.19}$$

The coordinate $\mathbf{r}$ is a continuous three dimensional variable. We write

$$H_{\mathbf{r}\mathbf{r}'} = \mathbf{H}(\mathbf{r})\, \delta^3(\mathbf{r} - \mathbf{r}') \tag{51.20}$$

where $\mathbf{H}(\mathbf{r})$ is the Hamiltonian operator. We now proceed to find the unitary matrices $\boldsymbol{U}$ and $\boldsymbol{V}$ which transform the martrix $\boldsymbol{H}$ into $\boldsymbol{H}'$ and $\boldsymbol{H}''$ respectively.

As $\boldsymbol{U}$ transforms $\boldsymbol{H}$ into $\boldsymbol{H}'$, that is, $\boldsymbol{UHU}^\dagger = \boldsymbol{H}'$, one writes

$$\boldsymbol{HU}^\dagger = \boldsymbol{U}^\dagger\boldsymbol{H}'. \tag{51.21}$$

Equating the $(\mathbf{r}, k)$ elements of the matrix products on both sides we get

$$\sum_{\mathbf{r}'} H_{\mathbf{r}\mathbf{r}'} U^\dagger{}_{\mathbf{r}'k} = \sum_{l} U^\dagger{}_{\mathbf{r}l} H'_{l,k}\,. \tag{51.22}$$

Using Eq. (51.16) for $H'_{k,l}$ and Eq.(51.20) for $H_{\mathbf{r}\mathbf{r}'}$ and by replacing the summation over $\mathbf{r}'$ by an integral over the continuous varible $\mathbf{r}'$ we get

$$\mathbf{H}(\mathbf{r}) U^\dagger{}_{\mathbf{r}k} = E_k U^\dagger{}_{\mathbf{r}k}. \tag{51.23}$$

Thus we get back the Schrödinger equation (51.1). The element $U^\dagger{}_{\mathbf{r}k}$ can be identified with the eigenfunction $\psi_k(\mathbf{r})$. Therefore,

$$U_{k\mathbf{r}} = \psi_k{}^*(\mathbf{r}),\; U^\dagger{}_{k\mathbf{r}} = \psi_k(\mathbf{r}). \tag{51.24}$$

The matrix $\boldsymbol{U}$ is unitary, as

$$(\boldsymbol{UU}^\dagger)_{kl} = \int d^3r\, \psi_k{}^*(\mathbf{r})\psi_l(\mathbf{r}) = \delta_{k,l} = 1_{kl}$$

and (51.25)

$$(\boldsymbol{U}^\dagger\boldsymbol{U})_{\mathbf{r}\mathbf{r}'} = \sum_k \psi_k(\mathbf{r})\psi_k{}^*(\mathbf{r}') = \delta^3(\mathbf{r} - \mathbf{r}') = 1_{\mathbf{r}\mathbf{r}'}.$$

No such simplification results from the transformation using the matrix $\boldsymbol{V}$, but one may deduce a correlation between the unitary matrices $\boldsymbol{U}$, $\boldsymbol{V}$ and $\boldsymbol{C}$. Because

$$\boldsymbol{VHV}^\dagger = \boldsymbol{H}'' \tag{51.26}$$

one may write

$$\boldsymbol{V}^\dagger\boldsymbol{H}''\boldsymbol{V} = \boldsymbol{H} \tag{51.27}$$

By using the unitary nature of $\boldsymbol{V}$. Therefore, from Eq. (51.21),

$$\boldsymbol{H}' = \boldsymbol{UV}^\dagger\boldsymbol{H}''\boldsymbol{VU}^\dagger \tag{51.28}$$

so that

$$\boldsymbol{UV}^\dagger = \boldsymbol{C},\; \boldsymbol{VU}^\dagger = \boldsymbol{C}^\dagger. \tag{51.29}$$

Also, as $\boldsymbol{U}$ is unitary,

$$\boldsymbol{V}^\dagger = \boldsymbol{U}^\dagger\mathbf{C} \tag{51.30}$$

which leads to the identification

$$V^\dagger{}_{\mathbf{r}\mu} = \sum_k U^\dagger{}_{\mathbf{r}k}\, c_{k\mu}$$

$$= \phi_\mu(\mathbf{r}). \qquad \text{[By Equations (51.4) and (51.24)]} \tag{51.31}$$

Lecture 52: Representations

In the last lecture we have discussed three different Hamiltonian matrices, $\boldsymbol{H}$, $\boldsymbol{H}'$ and $\boldsymbol{H}''$. The rows and the columns of these matrices are indexed by the eigenvalues of coordinates, energy and the operator Λ. The three matrices are said to be *representations* of the Hamiltonian operator. Thus $\boldsymbol{H}$, $\boldsymbol{H}'$ and $\boldsymbol{H}''$ are the Hamiltonian in the *coordinate representation*, the Hamiltonian in the *energy representation* and the Hamiltonian in the Λ *representation*, respectively.

The $\boldsymbol{\Lambda}$ matrix in the coordinate representation is defined by

$$\Lambda_{\mathbf{rr}'} = \Lambda(\mathbf{r})\, \delta^3(\mathbf{r} - \mathbf{r}'). \tag{52.1}$$

One can easily verify that

$$(\boldsymbol{U\Lambda U}^\dagger)_{kl} = \int d^3r\, \psi_k^*(\mathbf{r})\Lambda(\mathbf{r})\psi_l(\mathbf{r}), \quad (\boldsymbol{V\Lambda V}^\dagger)_{\mu\nu} = \lambda_\mu \delta_{\mu,\nu}. \tag{52.2}$$

We also have from the definition (52.1)

$$\Lambda(\mathbf{r})\mathrm{f}_2(\mathbf{r}) = \int d^3r'\, \Lambda_{\mathbf{rr}'}\, \mathrm{f}_2(\mathbf{r}') \tag{52.3}$$

where f_2 is any arbitrary function. Hence

$$\int d^3r\, \mathrm{f}_1^*(\mathbf{r})[\Lambda(\mathbf{r})\mathrm{f}_2(\mathbf{r})] = \int d^3r \int d^3r'\, \mathrm{f}_1^*(\mathbf{r})\Lambda_{\mathbf{r}\,\mathbf{r}'}\mathrm{f}_2(\mathbf{r}')$$

$$= \int d^3r'\, [\int d^3r\, \Lambda_{\mathbf{r}'\mathbf{r}}\, \mathrm{f}_1(\mathbf{r})]^*\mathrm{f}_2(\mathbf{r}')$$

where f_1 is another arbitrary function of coordinates, and so

$$\boxed{\int d^3r\, \mathrm{f}_1^*(\mathbf{r})\Lambda(\mathbf{r})\mathrm{f}_2(\mathbf{r}) = \int d^3r\, [\Lambda^\dagger \mathrm{f}_1(\mathbf{r})]^*\mathrm{f}_2(\mathbf{r}).} \tag{52.4}$$

This result is known as the *turn over rule*. Compare Eq. (4.6). This rule is valid for any arbitrary operator Λ.

Any arbitrary function $\mathrm{f}_i(\mathbf{r})$ can be expanded in terms of functions $\psi_k(\mathbf{r})$ or $\phi_\mu(\mathbf{r})$ as

$$\mathrm{f}_i(\mathbf{r}) = \sum_k a_{ki}\psi_k(\mathbf{r}) = \sum_k U^\dagger{}_{\mathbf{r}k} a_{ki}, \tag{52.5}$$

$$\mathrm{f}_i(\mathbf{r}) = \sum_\mu b_{\mu i}\phi_\mu(\mathbf{r}) = \sum_\mu V^\dagger{}_{\mathbf{r}\mu} b_{\mu i}.$$

If we consider $\mathrm{f}_i(\mathbf{r})$ as the $(\mathbf{r}, i)$ element $(f_{\mathbf{r}i})$ of the matrix $\boldsymbol{f}$, we obtain the *ith* column as

$$\boldsymbol{f}_i = \boldsymbol{U}^\dagger \boldsymbol{a}_i = \boldsymbol{V}^\dagger \boldsymbol{b}_i \tag{52.6}$$

which implies

$$\boldsymbol{a}_\mathrm{i} = \boldsymbol{U}\boldsymbol{f}_\mathrm{i}\,; \quad \boldsymbol{b}_\mathrm{i} = \boldsymbol{V}\boldsymbol{f}_\mathrm{i}\,. \tag{52.7}$$

The elements a_{ki} and $b_{\mu i}$ can be found from Eq. (52.5) by using the orthonormality relationships,

$$a_{ki} = \int d^3r\, \psi_k^*(\mathbf{r})\mathrm{f}_i(\mathbf{r}),$$

$$b_{\mu i} = \int d^3r\, \phi_\mu^*(\mathbf{r})\mathrm{f}_i(\mathbf{r}). \tag{52.8}$$

The relationship between the columns $\boldsymbol{a}_i$ and $\boldsymbol{b}_i$ can be established by using Eq. (52.7) and the unitary property of $\boldsymbol{U}$ and $\boldsymbol{V}$. One obtains

$$\boldsymbol{a}_i = \boldsymbol{U}\boldsymbol{V}^{\dagger}\boldsymbol{b}_i = \mathbf{C}\boldsymbol{b}_i. \tag{52.9}$$

The columns $\boldsymbol{f}_i$, $\boldsymbol{a}_i$ and $\boldsymbol{b}_i$ describe the same function $\mathrm{f}_i(\mathbf{r})$ *in the coordinate representation, the energy representation and the* $\boldsymbol{\Lambda}$ *representation respectively.* You may verify that

$$\boldsymbol{f}_i^{\dagger}\boldsymbol{f}_i = \boldsymbol{a}_i^{\dagger}\boldsymbol{U}\boldsymbol{U}^{\dagger}\boldsymbol{a}_i = \boldsymbol{a}_i^{\dagger}\boldsymbol{a}_i\,, \tag{52.10}$$

$$\boldsymbol{f}_i^{\dagger}\boldsymbol{f}_i = \boldsymbol{b}_i^{\dagger}\boldsymbol{V}\boldsymbol{V}^{\dagger}\boldsymbol{b}_i = \boldsymbol{b}_i^{\dagger}\boldsymbol{b}_i$$

where

$$\boldsymbol{f}_i^{\dagger}\boldsymbol{f}_i = \underset{\mathbf{r}}{S}\, f^{\dagger}{}_{i\mathbf{r}} f_{\mathbf{r}i} = \int d^3r\, \mathrm{f}_i^*(\mathbf{r})\mathrm{f}_i(\mathbf{r}). \tag{52.11}$$

Thus the norm of the state function is left unchanged. If the norm is finite, the function $\mathrm{f}_i(\mathbf{r})$ (or the columns $\boldsymbol{f}_i$, $\boldsymbol{a}_i$ and $\boldsymbol{b}_i$) can be normalized to unity by dividing it (or each of the columns) by $(\boldsymbol{f}_i^{\dagger}\boldsymbol{f}_i)^{1/2}$, the square root of the norm.

In matrix representation, the turn over rule (52.4) is trivially satisfied by writing

$$\boldsymbol{f}_1^{\dagger}\boldsymbol{\Lambda}\boldsymbol{f}_2 = (\boldsymbol{\Lambda}^{\dagger}\boldsymbol{f}_1)^{\dagger}\boldsymbol{f}_2. \tag{52.12}$$

The column $\mathbf{f}_i$, $\boldsymbol{a}_i$ or $\boldsymbol{b}_i$ representing the state function $\mathrm{f}_i(\mathbf{r})$ can be regarded as a state vector in an infinite dimensional Hilbert space. Each dimension corresponds to one of the rows of the state vector, and the component of the state vector along the same dimension is given by the corresponding matrix element. Thus a_{ki} in equations (52.8) is the component of the state vector $\boldsymbol{a}_i$ along the dimension k in the energy representation. This is purely a geometrical picture in which different representations correspond to different choices for the orientation of axes in the Hilbert space.[2]

We define the *scalar product* or the *inner product* of two state vectors as

$$(\mathrm{f}_i, \mathrm{f}_j) = \underset{\mathbf{r}}{S}\, \mathrm{f}_{\mathbf{r}i}^*\mathrm{f}_{\mathbf{r}j} = \boldsymbol{f}_i^{\dagger}\boldsymbol{f}_j, \tag{52.13}$$

or

$$(\mathrm{f}_i, \mathrm{f}_j) = \int d^3r\, \mathrm{f}_i^*(\mathbf{r})\mathrm{f}_j(\mathbf{r}). \tag{52.14}$$

---

[2]The Hilbert space is basically a function space with special features. Courant and Hilbert (1962) define the Hilbert space as follows. Let $\varphi_1, \varphi_2, \ldots, \varphi_j, \ldots$ be a sequence of functions with derivatives which are continuous up to order r. The norms of these functions ($\|\varphi_j\|$) and their derivatives ($\|\varphi_j\|_r$) are uniformly bounded. Moreover the functions φ_j converge in the r-norm, that is, if the norm $\|\varphi_i - \varphi_j\|_r \to 0$ for $i \to \infty$ as well as for $j \to \infty$, then an ideal element φ is attributed to the sequence $\{\varphi_j\}$. The limit function has the norm $\|\varphi\|_r = \lim_{j\to\infty} \|\varphi_j\|_r$. Then the Hilbert space $H_r(x)$ or $H_r^*(x)$ is defined as the completion of the space of smooth functions φ with the r-norm $|\varphi(x)|_r$ or $|\varphi(x)|_r^*$.

The inner product is independent of representations and indeed a number. This is also observed by writing

$$f_i^\dagger f_j = a_i^\dagger UU^\dagger a_j = a_i^\dagger a_j\,, \quad f_i^\dagger f_j = b_i^\dagger VV^\dagger b_j = b_i^\dagger b_j\,. \tag{52.15}$$

The transformation caused by the operation of any operator $\mathbf{\Omega}$ on f_j,

$$f_j \to \mathbf{\Omega} f_j = g_j, \tag{52.16}$$

generates a contracted or a stretched state vector g_j that may be aligned along a direction different from the direction of f_j. Hence this transformation is called a generalized rotation of the state vector f_j in Hilbert space. The matrix element

$$\Omega_{ij} = f_i^\dagger \mathbf{\Omega} f_j \tag{52.17}$$

can be viewed as the scalar product $(f_i, \mathbf{\Omega} f_i)$. Note that Ω_{ij} is independent of representation.

Dirac (1930) introduced a compact notation for the state vectors f_i, a_i and b_i. The notation is the *ket* or the *ket vector* $|\mathrm{i}>$. The hermitian adjoint $f_i^\dagger$, $a_i^\dagger$ or $b_i^\dagger$ is represented by the *bra* or the *bra vector* $<\mathrm{i}|$. The inner product of two state vectors f_i and f_j is written as

$$(f_i, f_j) = f_i^\dagger f_j = <\mathrm{i}|\mathrm{j}>. \tag{52.18}$$

The right hand side is in general a complex number, the so-called *c-number* that is formed by the bracket (bra-c-ket). We can write Eq. (52.17) as

$$\Omega_{ij} = <\mathrm{i}|\mathbf{\Omega}|\mathrm{j}> = <\mathrm{i}|\mathbf{\Omega}\mathrm{j}> \tag{52.19}$$

and as

$$\Omega_{ij} = (\mathbf{\Omega}^\dagger f_i)^\dagger f_j = <\mathbf{\Omega}^\dagger\,\mathrm{i}|\mathrm{j}>. \tag{52.20}$$

When $\mathbf{\Omega}$ is the unit operator **1** we retrieve Eq. (52.18). We write, using Eq. (51.5),

$$c_{k\mu} = (\psi_k, \phi_\mu) = <\mathrm{k}|\mu> = <\mu|\mathrm{k}>^*. \tag{52.21}$$

Similarly the consideration of Eq. (52.19) with $\mathbf{\Omega} = \mathbf{H}$ yields from Equations (51.22) and (51.24)

$$U_{k\mathbf{r}} = <\mathrm{k}|\mathbf{r}> = \psi_k^*(\mathbf{r}),$$

$$<\mathbf{r}|\mathrm{k}> = \psi_k(\mathbf{r}), \tag{52.22}$$

and with $\mathbf{\Omega} = \mathbf{\Lambda}$ we obtain from Equations (51.27) and (51.31)

$$V_{\mu\mathbf{r}} = <\mu|\mathbf{r}> = \phi_\mu^*(\mathbf{r}),$$

$$<\mathbf{r}|\mu> = \phi_\mu(\mathbf{r}). \tag{52.23}$$

In a similar way one finds from (52.5) the relation $f_i(\mathbf{r}) = <\mathbf{r}|\mathrm{i}>$, and furthermore from (52.8),

$$a_{ki} = <\mathrm{k}|\mathrm{i}>, \quad b_{\mu i} = <\mu|\mathrm{i}>. \tag{52.24}$$

The expectation value of any arbitrary operator $\mathbf{\Omega}$ in state $|\,\mathrm{i}>$ is now written as

$$<\mathbf{\Omega}> = \frac{<\mathrm{i}|\mathbf{\Omega}|\mathrm{i}>}{<\mathrm{i}|\mathrm{i}>}. \tag{52.25}$$

Lecture 53: Projection Operator, Identity and Equations of Motion

Projection operator

From equations (52.5) and (52.24) one observes that in the bra and ket notation the function $f_i(\mathbf{r})$ and the vector $\boldsymbol{f_i}$ are given by

$$f_i(\mathbf{r}) = <\mathbf{r}|i> = \underset{k}{S} <\mathbf{r}|k><k|i> = \underset{\mu}{S} <\mathbf{r}|\mu><\mu|i> \tag{53.1}$$

and

$$\boldsymbol{f_i} \equiv |i> = \underset{k}{S} |k><k|i> = \underset{\mu}{S} |\mu><\mu|i>. \tag{53.2}$$

One may write

$$|i> = \underset{k}{S} \mathbf{P}_k|i> = \underset{\mu}{S} \mathbf{P}_\mu|i> \tag{53.3}$$

where the operators $\mathbf{P}_k$ and $\mathbf{P}_\mu$ are

$$\mathbf{P}_k = |k><k|,$$

$$\mathbf{P}_\mu = |\mu><\mu|. \tag{53.4}$$

These are called projection operators as the effect of the operation of $\mathbf{P}_k$ or $\mathbf{P}_\mu$ on $\boldsymbol{f_i}$ is to generate the vector component $a_{ki}|k>$ or $b_{\mu i}|\mu>$, that is, to project the vector onto the k-axis or the μ-axis in Hilbert space. This component is the energy eigenstate with eigenvalue E_k or the eigenstate of $\mathbf{\Lambda}$ with eigenvalue λ_μ, but it is in general unnormalized because the coefficient a_{ki} or $b_{\mu i}$ is not necessarily of modulus unity.

The projection operator has the properties

(i) $\mathbf{P}_\mu^\dagger = \mathbf{P}_\mu$	($\mathbf{P}_\mu$ is hermitian),	(53.5)
(ii) $\mathbf{P}_\mu^2 = |\mu><\mu||\mu><\mu| = |\mu><\mu| = \mathbf{P}_\mu$	($\mathbf{P}_\mu$ is idempotent).	(53.6)

Projection operators also satisfy

$$\text{(iii) } \mathbf{P}_\mu\mathbf{P}_\nu = \delta_{\mu,\nu}\mathbf{P}_\mu \tag{53.7}$$

as the left hand side equals $|\mu><\mu|\nu><\nu| = \delta_{\mu,\nu}|\mu><\nu|$.

As the space spanned by the vectors $|\mu>$ is complete,

$$\underset{\mu}{S}\mathbf{P}_\mu = \mathbf{1} \tag{53.8}$$

where $\mathbf{1}$ is the unit operator. Compare Eq. (53.3). We refer to Eq.(53.8) as *resolution of identity (the unit operator)*.

We now derive an analytical expression for the projection operator $\mathbf{P}_\mu$. One may start from the eigenvalue equation

$$\mathbf{\Lambda}|\gamma> = \lambda_\gamma|\gamma> \tag{53.9}$$

for $\gamma = \mu, \nu$, etc. and obtain

$$(\mathbf{\Lambda} - \lambda_\nu)|\mu> = (\lambda_\mu - \lambda_\nu)|\mu> \qquad \text{(for } \gamma = \mu \neq \nu\text{)},$$

and (53.10)

$$(\mathbf{\Lambda} - \lambda_\nu)|\nu> = 0 \qquad \text{(for } \gamma = \nu\text{)}.$$

Therefore,

$$\frac{(\mathbf{\Lambda} - \lambda_\nu)}{(\lambda_\mu - \lambda_\nu)}|\mu> = |\mu> \qquad \text{(for } \mu \neq \nu\text{)},$$

and (53.11)

$$\frac{(\mathbf{\Lambda} - \lambda_\nu)}{(\lambda_\mu - \lambda_\nu)}|\nu> = 0 \qquad \text{(again for } \mu \neq \nu\text{)}.$$

This shows that

$$\prod_{\nu(\neq\mu)} \frac{(\mathbf{\Lambda} - \lambda_\nu)}{(\lambda_\mu - \lambda_\nu)}|\mu> = |\mu>$$

while for any γ not equal to μ,

$$\prod_{\nu(\neq\mu)} \frac{(\mathbf{\Lambda} - \lambda_\nu)}{(\lambda_\mu - \lambda_\nu)}|\gamma> = 0.$$

Hence an analytical expression for $\mathbf{P}_\mu$ is

$$\mathbf{P}_\mu = \prod_{\nu(\neq\mu)} \frac{(\mathbf{\Lambda} - \lambda_\nu)}{(\lambda_\mu - \lambda_\nu)}. \tag{53.12}$$

When this $\mathbf{P}_\mu$ operates on each component of $|i>$, the result is always zero except when the μth component is involved. In the latter case the effect of the operation is to keep the component unchanged: $\mathbf{P}_\mu|\mu><\mu|i> = |\mu><\mu|i>$.

Equations of Motion

We have seen that the state vectors (bra and ket) give a complete representation of the stationary states of the physical system. We are also interested in the dynamical characteristics of the system. The state vectors may be considered to depend on time; in that event the time dependence of these vectors must be precisely described. The evolution of state vectors in time is normally considered by three standard quantum mechanical formulations. These systematics are discussed now.

The Schrödinger Picture

This formulation is based on the choice that *the ket vector* (the wave function) *is explicitly dependent on time*. The time evolution of the ket is given by the Schrödinger equation

$$i\hbar\frac{d}{dt}|\text{is}(t)> = \mathbf{H}|\text{is}(t)>. \tag{53.13}$$

The subscript S refers to the time dependence in the Schrödinger picture. By equating the hermitian adjoints of both sides of Eq. (53.13) we get

$$-i\hbar\frac{d}{dt} < \text{is}(t)| = < \text{is}(t)|\mathbf{H} \tag{53.14}$$

as $\mathbf{H} = \mathbf{H}^\dagger$. The Schrödinger equation is a first order differential equation. It allows us to calculate the state vector at a later (or an earlier) instant of time if we know the state vector at an earlier (or a later) instant. The solutions of equations (53.13) and (53.14) can be written as

$$|\text{is}(t)> = |\text{is}(t_0)> - \frac{i}{\hbar}\int_{t_0}^{t} d\tau\, \mathbf{H}(\tau)\, |\text{is}(\tau)>,$$

and (53.15)

$$< \text{is}(t)| = < \text{is}(t_0)| + \frac{i}{\hbar}\int_{t_0}^{t} d\tau < \text{is}(\tau)|\, \mathbf{H}(\tau)$$

where we have considered $\mathbf{H}$ to be explicitly dependent on time. These equations can be iterated to obtain explicit solutions. For example, the first equation can be iterated as

$$|\text{is}(t)> = |\text{is}(t_0)> - \frac{i}{\hbar}\int_{t_0}^{t} d\tau\, \mathbf{H}(\tau)\, |\text{is}(t_0)>$$
$$+ \left(-\frac{i}{\hbar}\right)^2 \int_{t_0}^{t} d\tau\, \mathbf{H}(\tau) \int_{t_0}^{\tau} d\tau'\, \mathbf{H}(\tau')\, |\text{is}(t_0)> + \ldots \tag{53.16}$$

which can be rearranged as

$$|\text{is}(t)> = |\text{is}(t_0)> - \frac{i}{\hbar}\int_{t_0}^{t} d\tau\, \mathbf{H}(\tau)\, |\text{is}(t_0)>$$
$$+ \frac{1}{2!}\left(-\frac{i}{\hbar}\right)^2 \int_{t_0}^{t} d\tau\, \mathbf{H}(\tau) \int_{t_0}^{t} d\tau'\, \mathbf{H}(\tau')\, |\text{is}(t_0)> + \ldots$$
$$= \exp\left(-\frac{i}{\hbar}\int_{t_0}^{t} d\tau\right) |\text{is}(t_0)>. \tag{53.17}$$

Particularly simple expressions result when the Hamiltonian **H** becomes independent of time. These are as follows:

$$|\text{is}(t)> = e^{-i\mathbf{H}(t-t_0)/\hbar}\, |\text{is}(t_0)>,$$
$$< \text{is}(t)| = < \text{is}(t_0)|\; e^{i\mathbf{H}(t-t_0)/\hbar} \qquad \text{(the adjoint form).} \tag{53.18}$$

Let us determine the rate of change of the expectation values with time. Many of the commonly used operators are *explicitly independent of time* in the Schrödinger picture. A time-dependent expectation value of a general operator is calculated as

$$< \mathbf{O}_S >_t = < is(t)| \mathbf{O}_S |is(t) >$$

$$= < is(t_0)| \exp\left(\frac{i}{\hbar}\int_{t_0}^{t} d\tau \mathbf{H}(\tau)\right) \mathbf{O}_S \exp\left(-\frac{i}{\hbar}\int_{t_0}^{t} d\tau \mathbf{H}(\tau)\right) |is(t_0) >. \qquad (53.19)$$

The last step results from the use of Eq. (53.17). We obtain the rate of change

$$i\hbar\frac{d}{dt} <\mathbf{O}_S>_t$$

$$= < is(t_0)| \exp\left(\frac{i}{\hbar}\int_{t_0}^{t} d\tau \mathbf{H}(\tau)\right) i\hbar\frac{\partial \mathbf{O}_S}{\partial t} \exp\left(-\frac{i}{\hbar}\int_{t_0}^{t} d\tau \mathbf{H}(\tau)\right) |is(t_0) >$$

$$+ < is(t_0)| \exp\left(\frac{i}{\hbar}\int_{t_0}^{t} d\tau \mathbf{H}(\tau)\right) [\mathbf{O}_S, \mathbf{H}] \exp\left(-\frac{i}{\hbar}\int_{t_0}^{t} d\tau \mathbf{H}(\tau)\right) |is(t_0) >,$$

that is,

$$i\hbar\frac{d}{dt} < \mathbf{O}_S >_t = i\hbar\left\langle\frac{\partial \mathbf{O}_S}{\partial t}\right\rangle + < [\mathbf{O}_S, \mathbf{H}] >_t. \qquad (53.20)$$

The expectation value of the partial derivative accounts for the explicit time dependence of operator $\mathbf{O}_S$.

The Heisenberg Picture

In Heisenberg's formulation *the operator is explicitly dependent on time but the state vector remains time-independent.* The Heisenberg operator is

$$\mathbf{O}_H(t) = \exp\left(\frac{i}{\hbar}\int_{t_0}^{t} d\tau \mathbf{H}(\tau)\right) \mathbf{O}_S \exp\left(-\frac{i}{\hbar}\int_{t_0}^{t} d\tau \mathbf{H}(\tau)\right) \qquad (53.21)$$

while the state vector is chosen as $|is(t_0)>$. When the Hamiltonian $\mathbf{H}$ is independent of time the time-dependent operator $\mathbf{O}_H(t)$ is given by

$$\mathbf{O}_H(t) = e^{i\mathbf{H}(t-t_0)/\hbar}\, \mathbf{O}_S\, e^{-i\mathbf{H}(t-t_0)/\hbar}. \qquad (53.22)$$

One customarily chooses $t_0 = 0$ that leads to simplified expressions. Nevertheless, physical results remain independent of any such simplification and the choice of a nonzero value for t_0 will lead to an exactly equivalent treatment. The expectation value is obtained from the expression

$$< \mathbf{O}_H >_t = < is(t_0)| \mathbf{O}_H(t) |is(t_0)> \qquad (53.23)$$

which by use of Eq. (53.21) and subsequently by Eq.(53.19) reduces to the equality

$$< \mathbf{O}_H >_t = < \mathbf{O}_S >_t. \qquad (53.24)$$

Thus the time-dependent expectation values calculated in the two pictures are same. This establishes the equivalence of the two formulations.

The equation of motion for the Heisenberg operator can be obtained from Eq. (53.21),

$$i\hbar\frac{d}{dt}\mathbf{O}_{\mathrm{H}} = i\hbar\frac{\partial \mathbf{O}_{\mathrm{H}}}{\partial t} + [\mathbf{O}_{\mathrm{H}}(t), \mathbf{H}]. \tag{53.25}$$

The partial time derivative takes care of the explicit time dependence of the operator $\mathbf{O}_{\mathrm{H}}$. This leads to the same rate of change as that given by Eq. (53.20). Equation (53.25) was originally adopted by Heisenberg as a suitable extension of the classical equation of motion for observables in terms of Poisson brackets. As discussed in Lecture 1, the observable was replaced by the operator and the Poisson bracket was replaced by the commutator bracket along with the multiplicative factor $(-i/\hbar)$ to obtain Eq. (53.25). The present equation is referred to as Heisenberg's Equation of Motion.

The Interaction Picture

A theoretical formalism carried out either in the Schrödinger picture or in the Heisenberg picture relies on the knowledge of the eigenfunctions and eigenvalues of the operator $\mathbf{H}$. For instance, if one expands an arbitrary ket $|i_{\mathrm{S}}(t_0)>$ in the basis of the eigenkets of the time-independent operator $\mathbf{H}$ to write

$$|i_{\mathrm{S}}(t_0)> = \sum_k d_{ki}|k> \equiv \sum_k d_{ki}|k(t=t_0)>, \tag{53.26}$$

the expectation value of an observable in the *ith* state will be given by

$$<\mathbf{O}_{\mathrm{S}}>_t = <\mathbf{O}_{\mathrm{H}}>_t = \sum_k \sum_l d_{ki}^* d_{li}\, e^{-i(E_k - E_l)(t-t_0)/\hbar} (\mathbf{O}_{\mathrm{S}})_{k,l}. \tag{53.27}$$

The Schrödinger equation can be exactly solved only in a few simple cases. For most of the physical systems, exact solutions of the Schrödinger equation cannot be determined. In the latter cases it is often found to be useful to divide the Hamiltonian into two parts $\mathbf{H}_0$ and $\mathbf{H}'$,

$$\mathbf{H} = \mathbf{H}_0 + \mathbf{H}'. \tag{53.28}$$

The partitioning is carried out in such a way that $\mathbf{H}_0$ is time-independent and its eigenfunctions are easily derivable, whereas $\mathbf{H}'$ contains all the time dependence of $\mathbf{H}$ but it is only a small term representing the interaction of the otherwise unperturbed system, (that is, the system characterized by $\mathbf{H}_0$), with some external agency (such as an internally created or an externally applied electromagnetic field). In these cases, neither the Schrödinger picture nor the Heisenberg picture is convenient, and it is found to be advantageous to work in the interaction picture where *the state vector as well as operator are assigned time dependences*:

$$|i_{\mathrm{I}}(t)> = e^{i\mathbf{H}_0(t-t_0)/\hbar}|i_{\mathrm{S}}(t)>, \tag{53.29}$$

$$\mathbf{O}_{\mathrm{I}}(t) = e^{i\mathbf{H}_0(t-t_0)/\hbar}\,\mathbf{O}_{\mathrm{S}}\, e^{-i\mathbf{H}_0(t-t_0)/\hbar}. \tag{53.30}$$

For $\mathbf{H}_0 = 0$ (s.t. $\mathbf{H} = \mathbf{H}'$), the interaction picture reduces to the Schrödinger picture [with $|i_{\mathrm{I}}(t)> = |i_{\mathrm{S}}(t)>$ and $\mathbf{O}_{\mathrm{I}}(t) = \mathbf{O}_{\mathrm{S}}$], whereas for $\mathbf{H}' = 0$ (s.t. $\mathbf{H} = \mathbf{H}_0$), the interaction picture becomes identical with the Heisenberg picture [with $|i_{\mathrm{I}}(t)> = e^{i\mathbf{H}(t-t_0)/\hbar}|i_{\mathrm{S}}(t)> = |i_{\mathrm{S}}(t_0)>$ and $\mathbf{O}_{\mathrm{I}}(t) = e^{i\mathbf{H}(t-t_0)/\hbar}\,\mathbf{O}_{\mathrm{S}}\, e^{-i\mathbf{H}(t-t_0)/\hbar} = \mathbf{O}_{\mathrm{H}}(t)$].

The equation of motion of the state vector in the interaction picture is obtained by differentiating both sides of Eq. (53.29) with respect to time. One finds

$$i\hbar\frac{d\,|\,i_I(t)>}{dt} = -\mathbf{H}_0|i_I(t)>+ e^{i\mathbf{H}_0(t-t_0)/\hbar}\, i\hbar\frac{d\,|\,i_S(t)>}{dt}$$

$$= -\mathbf{H}_0|i_I(t)>+ e^{i\mathbf{H}_0(t-t_0)/\hbar}\,\mathbf{H}|i_S(t)>$$

$$= -\mathbf{H}_0|i_I(t)>+ e^{i\mathbf{H}_0(t-t_0)/\hbar}\,\mathbf{H}\,e^{-i\mathbf{H}_0(t-t_0)/\hbar}|i_I(t)>$$

$$= e^{i\mathbf{H}_0(t-t_0)/\hbar}(\mathbf{H}-\mathbf{H}_0)\,e^{-i\mathbf{H}_0(t-t_0)/\hbar}|i_I(t)>$$

$$= e^{i\mathbf{H}_0(t-t_0)/\hbar}\,\mathbf{H}'\,e^{-i\mathbf{H}_0(t-t_0)/\hbar}|i_I(t)>.$$

Because of the operator time dependence in Eq. (53.30), the last result can be written in the form

$$i\hbar\frac{d\,|\,i_I(t)>}{dt} = \mathbf{H}_I'(t)\,|i_I(t)>. \qquad (53.31)$$

Thus the time dependence of the state vector is determined by the *time-dependent interaction* operator $\mathbf{H}_I'(t)$, that is, from a wave equation where $\mathbf{H}_I'(t)$ takes the place of the Hamiltonian. The equation of motion for the operator can be found in a similar way. We get

$$\frac{d\mathbf{O}_I}{dt} = \frac{\partial\mathbf{O}_I}{\partial t} - \frac{i}{\hbar}[\mathbf{O}_I, \mathbf{H}_0] = \frac{\partial\mathbf{O}_I}{\partial t} - \frac{i}{\hbar}[\mathbf{O}_I(t), \mathbf{H}_{0I}]. \qquad (53.32)$$

The last step results from the identification of $\mathbf{H}_{0I}$ with $\mathbf{H}_0$ as the latter operator (and in fact every operator) commutes with itself.

The interaction picture has found wide applications in quantum mechanical treatments of interacting systems. It is almost invariably adopted in time-dependent formalisms. For example, almost all many-body theories, Green's function treatments and diagrammatic perturbation theory are based on this picture.

Lecture 54: The Classical Light

A classical light wave propagating in the z-direction is associated with an electric field vector $\boldsymbol{E}$ and a magnetic field vector $\boldsymbol{H}$ which are perpendicular to each other and also perpendicular to the direction of propagation of the light wave. Thus

$$\boldsymbol{E}(\mathbf{r}, t) = \mathbf{i}\,E_x(\mathbf{r}, t) + \mathbf{j}\,E_y(\mathbf{r}, t) \qquad (54.1)$$

and in Gaussian units,

$$\boldsymbol{H}(\mathbf{r}, t) = \mathbf{k}\times\boldsymbol{E}(\mathbf{r}, t) \qquad (54.2)$$

such that $H_x(\mathbf{r}, t) = -E_y(\mathbf{r}, t)$ and $H_y(\mathbf{r}, t) = E_x(\mathbf{r}, t)$. Knowledge of the electric field gives a sufficient description of the classical electromagnetic wave. These fields oscillate in time with frequency ν that is called the frequency of the light wave. The corresponding wavelength and wave vector (magnitude) are

$$\lambda = \frac{c}{\nu}\,, \tag{54.3}$$

$$k = \frac{2\pi}{\lambda} = \frac{2\pi\nu}{c} = \frac{\omega}{c} \tag{54.4}$$

where c is the speed of light and ω is the angular frequency. Components of the electric field can be written in the real form

$$E_x = E_x^0 \cos(kz - \omega t + \delta_x)$$

and (54.5)

$$E_y = E_y^0 \cos(kz - \omega t + \delta_y).$$

The electric field vector determines the state of polarization of the light wave. The wave is *plane polarized* in a specific direction if the component of $\boldsymbol{E}$ in the perpendicular direction vanishes. Thus $E_y = 0$ ($E_y^0 = 0$) for the x-polarized wave (Figure 9.1), $E_x = 0$ ($E_x^0 = 0$) for a y-polarized wave and $E_x = E_y$ for a 45°-polarized wave.

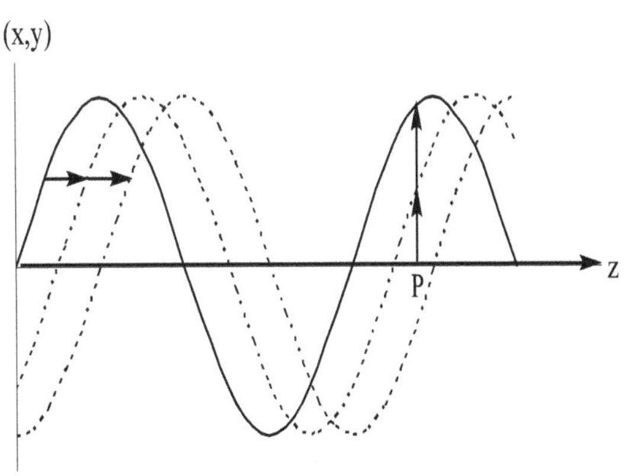

Figure 9.1 The x-polarized (or y-polarized) electric field wave: The solid line refers to an earlier time. The dashed line is obtained after a short period. The wave propagates in the positive direction along the z-axis. The electric field vectors at point P at two time instants have been indicated.

It is also possible to express the components of the electric field in the complex form,

$$E_{xc} = E_c^{\,0}\, e^{i(kz - \omega t + \delta_x)}$$

and (54.6)

$$E_{yc} = E_c^{\,0}\, e^{i(kz - \omega t + \delta_y)}.$$

The wave can be visualized to be *circularly polarized.* If $E_{yc} = iE_{xc}$, the wave is right circularly polarized as the y-component stays 90° behind the x-component, that is, $\delta_y = \delta_x + \pi/2$. Similarly, the wave is left circularly polarized when $E_{yc} = -iE_{xc}$, or $\delta_y = \delta_x - \pi/2$. We use the abbreviations RCP and LCP for the right and the left circular polarizations respectively. The RCP is illustrated in Figure 9.2.

The mixing of two collinear waves of same frequency but opposite circular polarizations generates a plane polarized light wave, the plane of polarization being determined by the phase difference between the two circularly polarized waves. The *unpolarized* light is supposed to be a uniform mixture of plane polarized light waves with 'all' possible planes of polarization, and in practice it turns out to be a random superposition of waves with a priori equal probability for each random orientation of the polarization plane.

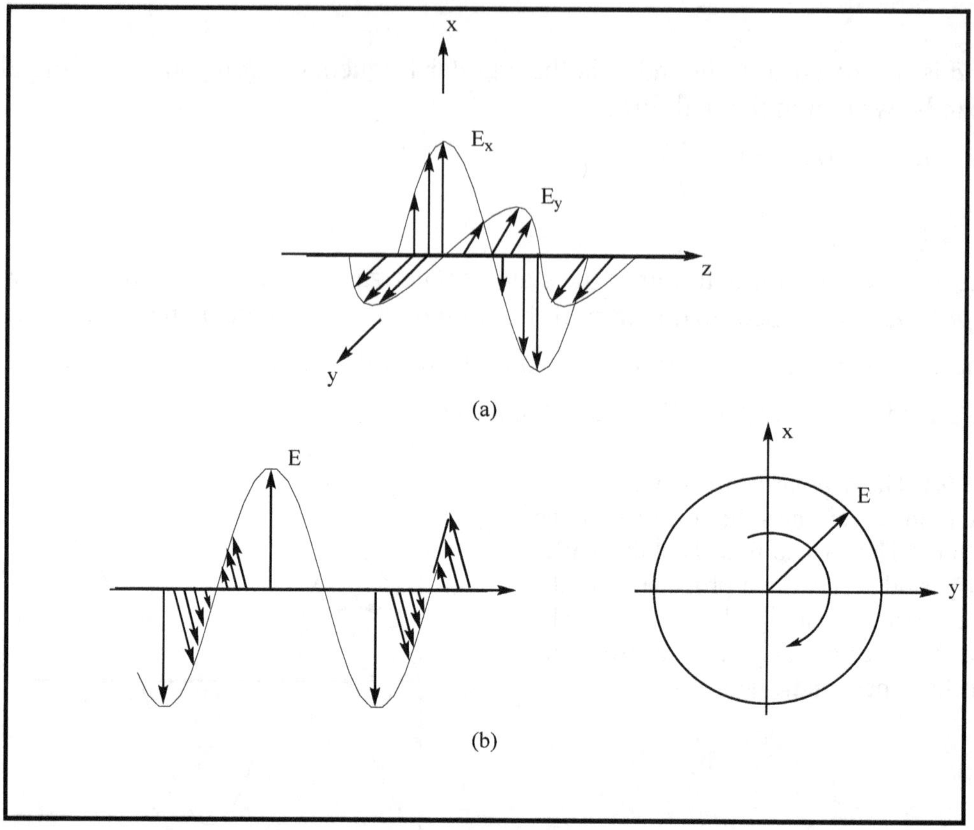

Figure 9.2 The right circularly polarized wave: (a) The y component of the electric field vector lags behind the x component by 90°. The arrows represent the x and y components at each point. The z-axis is the direction of propagation. (b) Left: The arrows represent the vector sum $\boldsymbol{E}$. Right: The vector $\boldsymbol{E}$ precesses around the z axis that points below and away from the plane of paper. The corresponding angular momentum ($\boldsymbol{L}_z$ eigenvalue) is $\hbar$.

The energy density of an electromagnetic field is given by

$$\in(\mathbf{r}, \mathrm{t}) = \frac{1}{8\pi}[\{\mathrm{Re}\, E_x(\mathbf{r}, \mathrm{t})\}^2 + \{\mathrm{Re}\, E_y(\mathbf{r}, \mathrm{t})\}^2 + \{\mathrm{Re}\, H_x(\mathbf{r}, \mathrm{t})\}^2 + \{\mathrm{Re}\, H_y(\mathbf{r}, \mathrm{t})\}^2] \tag{54.7}$$

in Gaussian units. For the plane wave of Eq. (54.2) this expression simplifies to

$$\in(\mathbf{r}, \mathrm{t}) = \frac{1}{4\pi}[(E_x{}^0)^2 \cos^2 (kz - \omega t + \delta_x) + (E_y{}^0)^2 \cos^2 (kz - \omega t + \delta_y)]. \tag{54.8}$$

The average energy per unit volume is

$$\overline{E} = \frac{1}{V}\int d^3r \in(\mathbf{r}, \mathrm{t}) = \frac{1}{8\pi}|\boldsymbol{E}|^2. \tag{54.9}$$

This result can be used to calculate the fraction of light that passes through a polaroid filter or a nicol prism. Suppose that the filter allows only the passage of the x-polarized rays. Classically the fraction of the energy which is passed by the polaroid is

$$\mathrm{f}_1 = \frac{|\boldsymbol{E}_x|^2}{|\boldsymbol{E}|^2}. \tag{54.10}$$

Take the case of a coherent superposition of right circularly polarized light and left circularly polarized light. If the prism allows only the passage of the left circularly polarized light, the fraction of energy passed by the prism will be given by

$$\mathrm{f}_2 = \frac{|\boldsymbol{E}_{\mathrm{LCP}}|^2}{|\boldsymbol{E}_{\mathrm{RCP}}|^2 + |\boldsymbol{E}_{\mathrm{LCP}}|^2}. \tag{54.11}$$

The reason for this review will be clear in a moment. Quantum mechanically, the total energy of radiation is given by

$$E_{total} = N\hbar\,\omega \tag{54.12}$$

where N is the total number of photons. Since all photons are identical and all face identical conditions while passing through the filter or the prism, it is very difficult to understand why only the fraction f_1 or f_2 passes. A possible way to explain the situation is to assign the probability f_1 to each photon for passing through the polaroid filter and the probability f_2 for passing through the nicol prism. This indicates that the state vector representing the state of polarization of the photons must be written in an analogous way. Our task in the next lecture will be to discuss how to write the quantum mechanical polarization state vectors for photons. For the time being, we take a look at the diffraction experiment that was referred to in Lecture 2.

We stated in Lecture 2 that the interference pattern appears *even when* the intensity of light is reduced to such an extent that in average only one photon is in transit between the source and the screen. The meaning of the statement will be clarified here. When two rays 1 and 2 are incident upon the same point P on the screen, the electric field vector at this point is given by the vector sum

$$\boldsymbol{E} = \boldsymbol{E}_1 + \boldsymbol{E}_2 \tag{54.13}$$

so that

$$|\boldsymbol{E}|^2 = |\boldsymbol{E}_1|^2 + |\boldsymbol{E}_2|^2 + 2\boldsymbol{E}_1\cdot\boldsymbol{E}_2. \tag{54.14}$$

The corresponding energy density is

$$\epsilon_{\mathrm{P}}(\mathrm{t}) = \frac{1}{8\pi}|\boldsymbol{E}|^2 = \epsilon_{1\mathrm{P}}(\mathrm{t}) + \epsilon_{2\mathrm{P}}(\mathrm{t}) + \epsilon_{12}(\mathrm{t}) \tag{54.15}$$

where $\epsilon_{12}(\mathrm{t})$ can be written as

$$\begin{aligned}\epsilon_{12}(\mathrm{t}) = \frac{1}{8\pi}\Big[&E_{1x}^0 E_{2x}^0\, [\cos\{(k_1p_1 + k_2p_2) - 2\omega t + (\delta_{1x}+\delta_{2x})\} \\ &\qquad + \cos\{(k_1p_1 - k_2p_2) + (\delta_{1x} - \delta_{2x})\}] \\ &+ E_{1y}^0 E_{2y}^0\, [\cos\{(k_1p_1 + k_2p_2) - 2\omega t + (\delta_{1y} + \delta_{2y})\} \\ &\qquad + \cos\{(k_1p_1 - k_2p_2) + (\delta_{1y}-\delta_{2y})\}]\Big] \cos\theta_{12}.\end{aligned} \tag{54.16}$$

In the above, p_1 and p_2 are the path lengths and θ_{12} is the angle between the two rays incident at point P. Time averages for the first and the third cosine functions equal zero. Assuming that the rays 1 and 2 are monochromatic (see the setup for the diffraction experiment in Lecture 2), $k_1 = k_2 = k$. Furthermore, if we assume that the rays are almost parallel and have equal amplitudes and phases, we get $\theta_{12} = 0$ so that

$$\in_{12}(t) \cong \frac{1}{4\pi}|\boldsymbol{E}_1^0|^2 \cos k(p_1 - p_2) \tag{54.17}$$

which varies from $-2\in$ to $2\in$ where $\in$ is the time-averaged energy density for a single ray. Thus the time-averaged $\in_P$ varies from zero to $4\in$. The important point is that due to the path difference (p_1-p_2) the time average of $\in_{12}$ varies from point to point and gives rise to successively constructive and destructive interferences, thus creating the interference pattern. The intensity is written as

$$I = I_1 + I_2 + I_{12} \tag{54.18}$$

and I is in general different from $(I_1 + I_2) = 2I_1$, the sum of the intensities due to the individual rays.

The intensity I can be varied by varying the number of photons, but the shape of the pattern remains independent of photon number. This phenomenon indicates that if we adopt a corpuscular description of light, we must attach to each photon a probabilistic interpretation so as to account for the fluctuation of the pattern intensity from zero to $4I_1$.

Lecture 55: Photons – The Simplest Quantum Mechnical System

If we compare expression (54.9) for the average energy of photons per unit volume with the total energy expression (54.12) for a single photon, we obtain

$$\left|\sqrt{\frac{V}{8\pi\hbar\omega}}\,\boldsymbol{E}\right|^2 = 1. \tag{55.1}$$

This expression may be used as the norm of the state vector for a single photon. The state vector is then defined by

$$|\Psi> = \sqrt{\frac{V}{8\pi\hbar\omega}}\begin{pmatrix} E_x \\ E_y \end{pmatrix}. \tag{55.2}$$

The corresponding row (or the adjoint) vector is written as

$$<\Psi| = \sqrt{\frac{V}{8\pi\hbar\omega}}\,(E_x{}^* \; E_y{}^*) \tag{55.3}$$

and a scalar product between two states of a photon can be written as

$$<\Psi|\Psi'> = \frac{V}{8\pi\hbar\omega}(E_x{}^*E_x' + E_y{}^*E_y') = <\Psi'|\Psi>^*. \tag{55.4}$$

The bases we have chosen for photon polarization are in reality the state vectors for x- and y- polarizations:

$$|\Psi_x> = \begin{pmatrix} 1 \\ 0 \end{pmatrix} \qquad (x\text{-polarization}),$$

and (55.5)

$$|\Psi_y> = \begin{pmatrix} 0 \\ 1 \end{pmatrix} \qquad (y\text{-polarization}).$$

The state vector for θ° polarization is

$$|\Psi_\theta> = \cos\theta|\Psi_x> + \sin\theta|\Psi_y> = \begin{pmatrix} \cos\theta \\ \sin\theta \end{pmatrix}. \tag{55.6}$$

Alternatively, the state vectors for the circular polarizations may be chosen as bases. The state vectors for RCP and LCP will be written as

$$|\Psi_{RCP}> = 2^{-1/2}\begin{pmatrix}1\\ i\end{pmatrix},$$

and (55.7)

$$|\Psi_{LCP}> = 2^{-1/2}\begin{pmatrix}1\\ -i\end{pmatrix}.$$

Any arbitrary one-photon state of the form given by Eq. (55.2) can be expanded in the basis set given in Eq. (55.5). We get

$$|\Psi> = |\Psi_x>< \Psi_x|\Psi> + |\Psi_y>< \Psi_y|\Psi>$$

which yields the superposition

$$|\Psi> = \sqrt{\frac{V}{8\pi\hbar\omega}}\ (E_x|\Psi_x> + E_y|\Psi_x>). \tag{55.8}$$

The same state may be expressed in the basis of circularly polarized states by writing

$$\begin{aligned}|\Psi> &= |\Psi_{RCP}>< \Psi_{RCP}|\Psi> + |\Psi_{LCP}>< \Psi_{LCP}|\Psi> \\ &= \sqrt{\frac{V}{16\pi\hbar\omega}}\ [(E_{xc} - iE_{yc})|\Psi_{RCP}> + (E_{xc} + iE_{yc})|\Psi_{LCP}>] \\ &=. \sqrt{\frac{V}{8\pi\hbar\omega}}\ (E_{RCP}|\Psi_{RCP}> + E_{LCP}|\ \Psi_{LCP}>).\end{aligned} \tag{55.9}$$

With E_{yc}=$+iE_{xc}$ we get pure RCP while E_{yc}=$-iE_{xc}$ gives pure LCP.

Therefore, the probability that a photon will pass through a polaroid filter that allows only the passage of the x-polarized classical wave is

$$P_1 = \frac{V}{8\pi\hbar\omega}|E_x|^2 \tag{55.10}$$

which, because of the normalization condition in Eq. (55.1), reduces to the precise result

$$P_1 = \frac{|E_x|^2}{|E_x|^2 + |E_y|^2} = f_1. \tag{55.11}$$

Similarly the probability that each photon in state $|\Psi>$ will pass through the nicol prism is

$$P_2 = \frac{V}{16\pi\hbar\omega}\ |E_{xc} - iE_{yc}|^2 = f_2. \tag{55.12}$$

The photon has an angular momentum. The right circularly polarized light has $L_z = +\hbar$ and for the left circularly polarized light, $L_z = -\hbar$. When a photon propagating along the *z*-direction is absorbed by (or emitted from) an atom, (the *z*-component of) the total orbital angular momentum of the atom changes by either $+\hbar$ or $-\hbar$. The Lapporté rule is $\Delta l = \pm 1$. The average value of the photon angular momentum in the polarization state $|\Psi>$ is

$$\begin{aligned}<\boldsymbol{L}_z> &= \hbar\,(|<\Psi_{RCP}|\Psi>|^2 - |<\Psi_{LCP}|\Psi>|^2) \\ &= \frac{V}{8\pi\omega}\ (|E_{RCP}|^2 - |E_{LCP}|^2).\end{aligned} \tag{55.13}$$

This $< L_Z >$ appears as the angular momentum transfer averaged over many experiments each of which involves the absorption (or emission) of a photon in the polarization state $|\Psi >$.

Texts Suggested for Further Study

G. Baym, "Lectures on Quantum Mechanics", (W. A. Benjamin, London, 1969) Ch. 1.
R. Courant and D. Hilbert, "Methods of Mathematical Physics", (Interscience, New York, 1962) p. 668.
P. A. M. Dirac, "The Principles of Quantum Mechanics", 4th edition, (Oxford University Press, Oxford, 1976) Ch. 5.
L. I. Schiff, "Quantum Mechanics", 3rd edition, (McGraw-Hill, New York, 1968) Ch. 6.

PROBLEMS

1. Using the linear expansion $\psi_k(\mathbf{r}) = \underset{\mu}{\mathbf{S}} c_{k\mu}\phi_\mu(\mathbf{r})$ show that

$$\underset{\mu}{\mathbf{S}} c_{k\mu}{}^* c_{l\mu} = \delta_{kl}\,,$$

$$\underset{\mu}{\mathbf{S}} \phi_\mu(\mathbf{r})\phi_\mu{}^*(\mathbf{r}') = \delta^3(\mathbf{r}-\mathbf{r}')\,.$$

2. Write three unitary transformations to relate the Hamiltonian matrix in three different representations. Identify the unitary operators used by you.
3. Write three unitary transformations to relate any other operator matrix in three different representations.
4. Justify the statement: Solution of the time-independent form of Schrödinger equation is equivalent to the diagonalization of the Hamiltonian matrix. Do you follow this statement in your work, analytical (Hückel treatment) or quantum chemical calculations (using Gaussian, HyperChem, etc. software)? Explain.
5. Derive the turn over rule for a hermitean operator $\mathbf{\Lambda}$ using the coordinate representation.

EXERCISES

1. Obtain the turn over rule in a complete matrix representation.
2. Verify that if $\mathbf{P}$ is a projection operator, $\mathbf{Q} = (\mathbf{1} - \mathbf{P})$ is another projection operator.
3. Show that the operator $\mathbf{P}_\mu$ is hermitean, $\mathbf{P}_\mu{}^2 = \mathbf{P}_\mu$ and $\sum_\mu \mathbf{P}_\mu = 1$.
4. Write the state vectors for 45° and 135° polarizations of a single photon. Calculate the matrix for transformation from the (x, y) basis to the (RCP, LCP) basis.
5. Verify that the photon polarization bases form orthonormal sets.

CHAPTER 10

BASIC OPERATOR FORMALISM

The bra and ket vectors have been introduced in the previous chapter. It would be appropriate to discuss a few aspects of the operator formalism in the new notation. The formulation is not only theoretically exact but also extremely popular for its notational convenience, and it can be regarded as an abbreviated form of the treatment in coordinate representation that requires operators and integrals to be explicitly written as practiced in Chapters 2-8.

Lecture 56: Definitions and Basic Relationships

In Eq. (52.4) we have taken the definition of the adjoint operator to be synonymous with the definition of the adjoint matrix,

$$(\mathbf{\Omega}^\dagger)_{ij} = \mathbf{\Omega}_{ji}^* \Leftrightarrow \langle i|\mathbf{\Omega}^\dagger|j\rangle = \langle j|\mathbf{\Omega}|i\rangle^*. \qquad (56.1)$$

For $|i\rangle = |j\rangle$ we have the average value

$$\langle \mathbf{\Omega}^\dagger \rangle_i = \langle \mathbf{\Omega} \rangle_i^*. \qquad (56.2)$$

When the operator is self-adjoint or hermitian, that is, $\mathbf{\Omega}^\dagger = \mathbf{\Omega}$, *the average value is real:*

$$\langle \mathbf{\Omega}^\dagger \rangle_i = \langle \mathbf{\Omega} \rangle_i^* = \langle \mathbf{\Omega} \rangle_i = \text{a real quantity}. \qquad (56.3)$$

In particular, *the eigenvalues of hermitian operators are real.* Equation (56.1) is equivalent to the *turn-over rule*

$$\langle i|\mathbf{\Omega}^\dagger|j\rangle = \langle \mathbf{\Omega} i|j\rangle. \qquad (56.4)$$

Equation (52.4) is a complex-conjugated version.

In fact, Eq. (56.2) *suffices as the definition of the adjoint operator*, and Eq. (56.1) is derivable from this definition. To show this we write, for Eq. (56.2),

$$\langle \phi|\mathbf{\Omega}^\dagger|\phi\rangle = \langle \phi|\mathbf{\Omega}|\phi\rangle^* \qquad \text{(definition of } \Omega^\dagger)$$

and then get, for $|\phi\rangle = |i\rangle + |j\rangle$ and using Eq. (56.3),

$$\langle i|\mathbf{\Omega}^\dagger|j\rangle + \langle j|\mathbf{\Omega}^\dagger|i\rangle = \langle i|\Omega|j\rangle^* + \langle j|\Omega|i\rangle^* \qquad (56.5)$$

while for $|\phi\rangle = |i\rangle + i|j\rangle$, again by making use of Eq. (56.3),

$$\langle i|\mathbf{\Omega}^\dagger|j\rangle - \langle j|\Omega^\dagger|i\rangle = -\langle i|\Omega|j\rangle^* + \langle j|\Omega|i\rangle^*. \qquad (56.6)$$

These two equations, (56.5) and (56.6), immediately lead to Eq. (56.1).

The importance of boundary conditions in determining the hermitian property of operators has been discussed earlier in Chapter 2. For the bra and ket notations it has been necessary to introduce an infinite-dimensional Hilbert space which has axes representing well-behaved functions. It is almost trivial to show from Eq. (56.4) that the non-degenerate eigenfunctions of hermitian operators are orthogonal to each other. If $|i>$ and $|j>$ are eigenkets with eigenvalues ω_i and ω_j, then Eq. (56.4) is translated as

$$(\omega_i^* - \omega_j) < i|j > = 0 \tag{56.7}$$

which indicates that for $|i> = |j>$, $\omega_i = \omega_i^*$ = real and for $\omega_i \neq \omega_j$, $< i|j > = 0$.

One useful representation is the *momentum representation.* This is obtained by considering the operator $\mathbf{\Lambda}$ of the previous chapter as the operator for linear momentum. Wave functions in coordinate and momentum representations are related to each other merely by a Fourier transformation (and the inverse transformation):

$$\hat{\mathbf{p}} = -i\hbar\nabla \qquad \psi(\mathbf{r}) = \hbar^{-3/2} \int d^3p\, e^{i\mathbf{p}.\mathbf{r}/\hbar}\, \varphi(\mathbf{p}),$$

$$\hat{\mathbf{r}} = i\hbar\nabla_p \qquad \varphi(\mathbf{p}) = \hbar^{-3/2} \int d^3r\, e^{-i\mathbf{p}.\mathbf{r}/\hbar}\, \psi(\mathbf{r}). \tag{56.8}$$

We now use the turn-over rule to establish a few operator relationships.

Theorem 14. $(\mathbf{A} + \mathbf{B})^\dagger = \mathbf{A}^\dagger + \mathbf{B}^\dagger$

Proof: Take $\mathbf{\Omega} = \mathbf{A} + \mathbf{B}$. Then by the turn-over rule

$$< i|(\mathbf{A} + \mathbf{B})^\dagger|j > = < (\mathbf{A} + \mathbf{B})i|j > = < \mathbf{A}i|j > + < \mathbf{B}i|j >$$
$$= < i|\mathbf{A}^\dagger|j > + < i|\mathbf{B}^\dagger|j > = < i\,|\mathbf{A}^\dagger + \mathbf{B}^\dagger|\, j>.$$

Multiplying with $|i>$ from left and $<j|$ from the right and by summing over all states i and j one obtains

$$\sum_i \sum_j |i>< i|\,(\mathbf{A} + \mathbf{B})^\dagger|j >< j| = \sum_i |i >< i|\,(\mathbf{A}^\dagger + \mathbf{B}^\dagger) \sum_j |j><j|.$$

Use of the resolution of identity yields

$$(\mathbf{A} + \mathbf{B})^\dagger = \mathbf{A}^\dagger + \mathbf{B}^\dagger. \text{ QED} \tag{56.9}$$

Theorem 15. $(\mathbf{AB})^\dagger = \mathbf{B}^\dagger\mathbf{A}^\dagger$.

Proof: Take $\mathbf{\Omega} = \mathbf{AB}$ and use the turn-over rule successively to write

$$< i|(\mathbf{AB})^\dagger|j > = < \mathbf{AB}i|j > = < \mathbf{A}(\mathbf{B}i)|j >$$
$$= < \mathbf{B}i|\mathbf{A}^\dagger|j > = < \mathbf{B}i|\mathbf{A}^\dagger j >$$
$$= < i|\mathbf{B}^\dagger|\mathbf{A}^\dagger j > = < i|\mathbf{B}^\dagger\mathbf{A}^\dagger|j >.$$

Repeating the second and third steps in the proof of the previous theorem one gets

$$(\mathbf{AB})^\dagger = \mathbf{B}^\dagger\mathbf{A}^\dagger. \text{ QED} \tag{56.10}$$

There is nothing surprizing about these relationships as they hold for matrices and the turn-over rule establishes the correspondence between matrices (and their adjoints) and operators (and their adjoints).

Theorem 16. $< \mathbf{A}^\dagger\mathbf{A} >_\varphi \geq 0$, with the equality valid *iff* $\mathbf{A}|\varphi > = 0$ always.

Proof: $< \varphi|\mathbf{A}^\dagger\mathbf{A}|\varphi > = < \varphi|\mathbf{A}^\dagger|\mathbf{A}\varphi > = < \mathbf{A}\varphi|\mathbf{A}\varphi >$, the last step being an outcome of the turn-over rule. Now, $\mathbf{A}|\varphi> = |\mathrm{f}>$ which is a contracted or an expanded state vector. Hence

$$< \mathbf{A}^{\dagger}\mathbf{A} > = < \mathrm{f}|\mathrm{f} >.$$

But $< \mathrm{f}|\mathrm{f} >$ is the integral $\int dv\,|\,\mathrm{f}\,|^2$ that may be viewed as a continuous sum of square modulus, and so

$$< \mathrm{f}|\mathrm{f} > \geq 0. \tag{56.11}$$

The equlity holds *iff* $\mathrm{f}(\mathbf{r}) = 0$ at every point $\mathbf{r}$ in space. **QED**

Corollary: $\sigma_A^2 = < \varphi|(\mathbf{A} - \overline{\mathrm{A}})^{\dagger}(\mathbf{A} - \overline{\mathrm{A}})|\varphi> \; \geq \; 0$ (56.12)

where $\overline{\mathrm{A}}$ is the average value of observable A in state $|\varphi>$.

If $|\varphi>$ is an eigenket, $\sigma_A^2 = 0$. For a general ket $|\varphi>$ one obtains

$$\sigma_A^2 = < \mathbf{A}^{\dagger}\mathbf{A} > - \; |\overline{\mathrm{A}}\,|^2. \tag{56.13}$$

We have already seen in Chapter 2 that Equations (56.11) and (56.12) can be used to derive the uncertainty theorem (Theorem 8),

$$\sigma_A\sigma_B \geq \frac{1}{2}|< i[\mathbf{A}, \mathbf{B}] >|. \tag{7.5}$$

Lecture 57: The Virial Theorem

The Schrödinger equation is a first-order differential equation in time. Therefore, it gives the rate of change of any arbitrary wave function with time. If we know the wave function at an earlier moment, using this equation we can in principle obtain the wave function at a later instant. The square modulus of the wave function is physically interpreted as the probability density of finding the constituent particles in the system at the respective coordinates at the given instant of time. The probabilistic interpretation is a central theme of quantum theory.

Why is the Schrödinger equation so good an equation? The answer is that it allows the assignment of two important physical relationships to the probability cloud (the charge cloud or the electron cloud). The probability fluid satisfies two necessary and sufficient criteria of fluid dynamics. These relationships are derived in the following.

Let us deal with a one-particle system for the sake of simplicity. The system is characterized by an external potential $\mathrm{V}(\mathbf{r})$ that is a real function of the space coordinates. The wave function is written as

$$\psi(\mathbf{r},t) = R(\mathbf{r},t)e^{-\frac{i}{\hbar}S(\mathbf{r},t)} \tag{57.1}$$

where R and S are real functions of coordinates and time. If this expression is inserted back in the Schrödinger equation (5.6), and then the real parts of the two sides of the resulting expression are equated to each other and similarly the imaginary parts are equated, one finds

$$R\dot{S} = \frac{\hbar^2}{2m}\nabla^2 R - \frac{1}{2m} R\nabla S\bullet\nabla S - VR, \tag{57.2}$$

and

$$\dot{R} = -\frac{1}{2m}(2\nabla R\bullet\nabla S + R\nabla^2 S). \tag{57.3}$$

The wave function in (57.1) is associated within the probability density

$$\rho(\mathbf{r}, t) = |\psi|^2 = R^2(\mathbf{r}, t) \tag{57.4}$$

and the *current density* (probability current)

$$\boldsymbol{J}(\mathbf{r}, t) = \frac{1}{2m}[\psi^*\mathbf{p}\psi + (\mathbf{p}\psi)^*\psi] = \frac{1}{m}R^2\nabla S. \tag{57.5}$$

The rate of change of the probability density with time is $\dot{\rho} = 2R\dot{R}$ which by Eq. (57.3) is found as

$$\dot{\rho} = -\nabla\bullet\boldsymbol{J}. \tag{57.6}$$

This result is known as the *equation of continuity*. It holds for an incompressible fluid.

The quantity $m\boldsymbol{J}$ is analogous to the momentum density, and its rate of change with time must be analogous to force density. In fact we find

$$\frac{d}{dt}\left(\frac{m\boldsymbol{J}}{\rho}\right) = -\nabla V_Q - \nabla V \tag{57.7}$$

where the quantum potential V_Q is given by

$$V_Q = Re\left(-\frac{\hbar^2}{2m}\frac{\nabla^2\psi}{\psi}\right). \tag{57.8}$$

Equation (57.7) is analogous to the equations of stresses in hydrodynamics, the latter being fluid-dynamical analogues of Newton's equation (1.3).

Virial Theorem

The virial theorem is known to hold for periodic motions in classical mechanics.[1] It is also found to be valid for periodic quantum mechanical systems, that is, for bound states. It is relatively easy to derive this theorem for an energy eigenstate. Hence this derivation is discussed at first. The derivation for a general bound state is more involved and so it will be given next.

---

[1]The classical equations are obtained for a periodic system by setting the constraint $\oint pdx$ is minimum.

The Heisenberg equation of motion may be used to write the rate of change of the average value of **r•p** in a particular state. In case the state is an eigenstate of the Hamiltonian, the rate equals zero:[2]

$$\frac{d}{dt} < \mathbf{r \cdot p} > = -\frac{i}{\hbar} < [\mathbf{r \cdot p}, \mathbf{H}] > = -\frac{i}{\hbar} \{<\mathbf{r \cdot p}>E - E<\mathbf{r \cdot p}>\} = 0. \tag{57.9}$$

Let us calculate the commutator for a one-particle system:

$$[\mathbf{r \cdot p}, \mathbf{H}] = 2i\hbar \mathbf{T} - i\hbar \mathbf{r} \cdot \nabla V \tag{57.10}$$

Therefore, one obtains for an eigenstate

Theorem 17. General Virial Theorem for eigenstates: $2 < \mathbf{T} > = < \mathbf{r} \cdot \nabla V >$ (57.11)

This is the general statement of the virial theorem. When the potential V is a homogeneous function of degree n in coordinates, the virial theorem is found as

Theorem 18. Special Virial Theorem for eigenstates: $2 < \mathbf{T} > = n < V >$. (57.12)

The degree n is 2 for a one-dimensional harmonic oscillator and –1 for the electron in hydrogen atom. Hence for the harmonic oscillator eigenfunctions $< \mathbf{T} >_v = < V >_v = \frac{1}{2} E_v$, v being the vibrational quantum number, whereas for the hydrogen atomic orbitals $< \mathbf{T} >_{nlm} = -\frac{1}{2} < V >_{nlm} = -E_n$.

The virial theorem for a general bound state (not necessarily an eigenstate) was discussed by Fock (1978) in terms of a variation of scale. A scaling operation is defined by the change of the position vector as

$$\mathbf{r} \to \mathbf{r}' = \mathbf{r}/\lambda. \tag{57.13}$$

Under scaling, the normalized wavefunction $\psi(\mathbf{r})$ transforms as

$$\psi(\mathbf{r}) \to \psi'(\mathbf{r}) = \lambda^{3/2}\psi(\lambda\mathbf{r}) \tag{57.14}$$

such that the norm of the wave function remains preserved:

$$\int d^3r' \psi'^*(\mathbf{r}')\psi'(\mathbf{r}') \equiv \lambda^3 \int d^3r' \psi^*(\lambda\mathbf{r}')\psi(\lambda\mathbf{r}') = \int d^3r\, \psi^*(\mathbf{r})\psi(\mathbf{r}) = 1. \tag{57.15}$$

For example, look at the 1s wavefunction for a hydrogen-like system. In arbitrary unit,

$$\psi_{1s}(\boldsymbol{r}) = \frac{Z^{3/2}}{\pi^{1/2} a_0^{3/2}} e^{-\frac{Zr}{a_0}}$$

whereas in atomic unit ($\boldsymbol{r}' = \frac{\boldsymbol{r}}{a_0}$),

$$\psi_{1s}'(\boldsymbol{r}') = \frac{Z^{3/2}}{\pi^{1/2}} e^{-Zr} = a_0^{3/2} \psi_{1s}(a_0 \boldsymbol{r}).$$

[2]This implies that the space-average <**r.p**> is a constant of motion.

By a change of scale, the kinetic energy operator varies as

$$\mathbf{T} \to \mathbf{T}' = \lambda^2\mathbf{T} \tag{57.16}$$

and if V is a homogeneous function of degree n in coordinates, it varies as

$$\mathrm{V} \to \mathrm{V}' = \lambda^{-n}\mathrm{V}. \tag{57.17}$$

Then the expectation values transform as

$$<\mathbf{T}> \to <\mathbf{T}>' = \int d^3r'\psi'^*(\mathbf{r}')\mathbf{T}'\psi'(\mathbf{r}') = \lambda^3\int d^3r'\psi^*(\lambda\mathbf{r}')\mathbf{T}'\psi(\lambda\mathbf{r}')$$

$$= \lambda^2\int d^3r\,\psi^*(\mathbf{r})\mathbf{T}\psi(\mathbf{r}) = \lambda^2 <\mathbf{T}>$$

$$<\mathrm{V}> \to <\mathrm{V}>' = \int d^3r'\psi'^*(\mathbf{r}')\mathrm{V}(\mathbf{r}')\psi'(\mathbf{r}') = \lambda^3\int d^3r'\psi^*(\lambda\mathbf{r}')\mathrm{V}(\mathbf{r}')\psi(\lambda\mathbf{r}')$$

$$= \int d^3r\,\psi^*(\mathbf{r})\mathrm{V}\left(\frac{\mathbf{r}}{\lambda}\right)\psi(\mathbf{r})$$

$$= \lambda^{-n}\int d^3r\,\psi^*(\mathbf{r})\mathrm{V}(\mathbf{r})\psi(\mathbf{r})$$

$$= \lambda^{-n} <\mathrm{V}>$$

so that

$$<\mathbf{H}> \to <\mathbf{H}>' = \lambda^2 <\mathbf{T}> + \lambda^{-n} <\mathrm{V}>.$$

The time-dependent Schrödinger equation can be derived from the variation of the integral of action per unit time,

$$A = \int d^3r\,\Psi^*(\mathbf{r}, t)\,(\mathbf{H} - i\hbar\frac{\partial}{\partial t})\,\Psi(\mathbf{r}, t). \tag{57.18}$$

The time-dependent Schrödinger equation (5.6) results when A is stationary for all arbitrary variation of $\Psi(\mathbf{r}, t)$. The time-independent version of Eq. (57.18) is obtained by assigning a stationary state-like time dependence to $\Psi(\mathbf{r}, t)$,

$$\Psi(\mathbf{r}, t) = e^{-i\varepsilon t/\hbar}\,\psi(\mathbf{r}), \tag{57.19}$$

whence

$$A = \int d^3r\,\psi^*(\mathbf{r})\,(\mathbf{H} - \varepsilon)\,\psi(\mathbf{r}). \tag{57.20}$$

For an arbitrary state of the form (57.19) the quantity ε is determined if we put $A = 0$, that is, $|A|$ is minimum, and then ε is seen as the expectation value. Thus *the time-independent Schrödinger equation is obtained by considering that A or $|A|$ is minimum subject to the constraint that ε is stationary for all arbitrary variations of* $\psi(\mathbf{r})$. This observation is the starting point for the derivation of virial theorem for a general state.

The generalized action integral in Eq. (57.20) transforms under scaling as

$$A(\lambda) = <\mathbf{H}>' - \varepsilon(\lambda) = (\lambda^2 <\mathbf{T}> + \lambda^{-n} <\mathrm{V}>) - \varepsilon(\lambda). \tag{57.21}$$

When $|A|$ is minimum for the variation of the scale parameter λ subject to the constraint that $\varepsilon(\lambda)$ remains stationary, one obtains

$$2\lambda <\mathbf{T}> - n\lambda^{-(n+1)} <\mathrm{V}> = 0.$$

The result of the variation of scale is sought for the unchanged scale, that is, for $\lambda = 1$. The desired result is the familiar virial theorem that holds for *bound state wave functions for which the energy expectation value has been made stationary for variation of the scale factor,*

Theorem 19. The Generalized Virial Theorem: $2 < \mathbf{T} > = n < \mathrm{V} >$.	(57.22)

Eigenfunctions are extremely special cases of these wave functions. Generally the interactions in atoms and molecules are coulombic in nature ($n = -1$) and this is a great help. Equations similar to (57.22) hold for the *best* Slater determinant that has also been optimized by a scale variation. In the molecular case, optimization of the electronic energy with respect to the scale parameter while the intermolecular distances are kept fixed leads to a modified relationship known as the hypervirial theorem.

Lecture 58: Normal Operators

Let two operators **A** and **B** be hermitian. Then

$$(\mathbf{AB})^\dagger = \mathbf{B}^\dagger\mathbf{A}^\dagger = \mathbf{BA} \tag{58.1}$$

but $(\mathbf{AB})^\dagger \neq \mathbf{AB}$ unless **A** and **B** commute with each other. The sum $(\mathbf{AB} + \mathbf{BA})$ is always hermitian and the difference $(\mathbf{AB} - \mathbf{BA})$ is *anti-hermitian*, as

$$(\mathbf{AB} + \mathbf{BA})^\dagger = \mathbf{BA} + \mathbf{AB} = (\mathbf{AB} + \mathbf{BA}),$$

$$(\mathbf{AB} - \mathbf{BA})^\dagger = \mathbf{BA} - \mathbf{AB} = -(\mathbf{AB} - \mathbf{BA}). \tag{58.2}$$

Therefore, $\mathbf{AB} = \frac{1}{2}(\mathbf{AB}+\mathbf{BA}) + \frac{1}{2}(\mathbf{AB} - \mathbf{BA})$ is the sum of a hermitian part and an anti-hermitian one.

A general operator $\mathbf{\Omega}$ can be always written as a sum of a hermitian and an antihermitian parts,

$$\mathbf{\Omega} = \mathbf{C} + i\mathbf{D}. \tag{58.3}$$

The operator **C** is hermitian and **D** is also hermitian such that $i\mathbf{D}$ is antihermitian. We note

$$\mathbf{C} = \frac{1}{2}(\mathbf{\Omega} + \mathbf{\Omega}^\dagger) = \mathbf{C}^\dagger,$$

$$\mathbf{D} = \frac{-i}{2}(\mathbf{\Omega} - \mathbf{\Omega}^\dagger) = \mathbf{D}^\dagger. \tag{58.4}$$

The commutator $[\mathbf{\Omega}, \mathbf{\Omega}^\dagger]$ turns out as

$$[\mathbf{\Omega}, \mathbf{\Omega}^\dagger] = -2i\,[\mathbf{C}, \mathbf{D}] \tag{58.5}$$

so that $\mathbf{\Omega}$ and $\mathbf{\Omega}^\dagger$ mutually commute when the hermitian and the antihermitian parts of $\mathbf{\Omega}$ commute with each other.

The operator $\mathbf{\Omega}$ is called *normal* if it commutes with its hermitian adjoint. We see that the necessary and sufficient condition for $\mathbf{\Omega}$ to be normal is

$$[\mathbf{C}, \mathbf{D}] = 0. \tag{58.6}$$

If in particular either **C** or **D** is zero, the commutator vanishes. However, neither **C** nor **D** may be zero, yet their commutator may equal zero. Therefore *all normal operators are not Hermitian or anti-hermitean, althogh all hermitian or anti-hermitean operators are normal.*

We now prove the *eigenvalue theorem*,

Theorem 20. If $\boldsymbol{\Omega}$ is normal, $\boldsymbol{\Omega}$ and $\boldsymbol{\Omega}^\dagger$ have common eigenfunctions ψ_k with eigenvalues that are complex conjugates of each other.

To prove this theorem we write

$$< (\boldsymbol{\Omega}^\dagger - \omega_k^*)\psi_k|(\boldsymbol{\Omega}^\dagger - \omega_k^*)\psi_k > = < \psi_k|(\boldsymbol{\Omega} - \omega_k)(\boldsymbol{\Omega}^\dagger - \omega_k^*)|\psi_k >.$$

Since $\boldsymbol{\Omega}$ is normal, $[\boldsymbol{\Omega},\boldsymbol{\Omega}^\dagger] = 0$ so that

$$\text{R.H.S.} = < \psi_k|(\boldsymbol{\Omega}^\dagger - \omega_k^*)(\boldsymbol{\Omega} - \omega_k)|\psi_k > = < (\boldsymbol{\Omega} - \omega_k)\psi_k|(\boldsymbol{\Omega} - \omega_k)\psi_k >.$$

Therefore, if ψ_k is an eigenfunction of $\boldsymbol{\Omega}$ with eigenvalue ω_k,

$$< (\boldsymbol{\Omega}^\dagger - \omega_k^*)\psi_k|(\boldsymbol{\Omega}^\dagger - \omega_k^*)\psi_k > = 0$$

and by Theorem 16,

$$\boldsymbol{\Omega}^\dagger|\psi_k > = \omega_k^*|\psi_k >. \tag{58.7}$$

Hence the theorem is proved.

Corollary: If an operator $\boldsymbol{\Omega}$ is hermitian, its eigenvalues are real.

For a hermitian operator, $\boldsymbol{\Omega}^\dagger = \boldsymbol{\Omega}$ so that $\omega_k^* = \omega_k$. This proves the corollary.

Corollary: If two hermitian operators **A** and **B** commute with each other, they *can* have common eigenfunctions.

To prove this Corollary, write $\boldsymbol{\Omega} = \mathbf{A} + i\mathbf{B}$ such that $\boldsymbol{\Omega}^\dagger = \mathbf{A} - i\mathbf{B}$. Now $\boldsymbol{\Omega}$ is normal as $[\mathbf{A}, \mathbf{B}] = 0 \Rightarrow [\boldsymbol{\Omega}, \boldsymbol{\Omega}^\dagger] = 0$. So $\boldsymbol{\Omega}$ and $\boldsymbol{\Omega}^\dagger$ have common eigenfunctions such that if

$$\boldsymbol{\Omega}|\psi_k > = \omega_k|\psi_k >$$

then

$$\boldsymbol{\Omega}^\dagger|\psi_k > = \omega_k^*|\psi_k >.$$

Therefore,

$$\mathbf{A}|\psi_k > = \frac{1}{2}(\boldsymbol{\Omega} + \boldsymbol{\Omega}^\dagger)\ |\psi_k > = \frac{1}{2}(\omega_k + \omega_k^*)|\psi_k > = Re\ (\omega_k)\ |\psi_k >,$$

$$\mathbf{B}|\psi_k > = \frac{-i}{2}(\boldsymbol{\Omega} - \boldsymbol{\Omega}^\dagger)\ |\psi_k > = \frac{-i}{2}(\omega_k - \omega_k^*)|\psi_k > = Im\ (\omega_k)\ |\psi_k >$$

This shows that if $[\mathbf{A}, \mathbf{B}] = 0$, the hermitian operators **A** and **B** can have common eigenfunctions.

Corollary: If two hermitian operators **A** and **B** have common eigenfucntions ψ_k's which form a complete set, the operators commute with each other.

To establish this corollary assume that the eigenvalues associated with $|\psi_k >$ are a_k and b_k respectively for the two operators. Then

$$\mathbf{AB}|\psi_k > = a_k b_k|\psi_k >$$

which has been achieved by successive application of the operators **B** and **A** on $|\psi_k >$. Similarly

$$\mathbf{BA}|\psi_k > = b_k a_k|\psi_k >$$

But a_k and b_k are numbers (eigenvalues), and they commute with each other. Therefore,

$$[\mathbf{A}, \mathbf{B}]\,|\psi_k\rangle = 0$$

that gives

$$[\mathbf{A}, \mathbf{B}] \sum_k |\psi_k\rangle\langle\psi_k| = 0.$$

This yields the desired result

$$[\mathbf{A}, \mathbf{B}] = 0. \tag{58.8}$$

We now prove the *orthogonality theorem*,

> **Theorem 21.** If an operator $\boldsymbol{\Omega}$ is normal and $|\psi_k\rangle$ and $|\psi_l\rangle$ are its eigenfunctions with two different eigenvalues ω_k and ω_l ($\omega_k \neq \omega_l$), then $\langle\psi_k|\psi_l\rangle = 0$.

To prove this, we write

$$\langle\boldsymbol{\Omega}\psi_k|\psi_l\rangle = \langle\psi_k|\boldsymbol{\Omega}^\dagger|\psi_l\rangle$$

The left hand side is evaluated as

$$\text{L.H.S.} = \omega_k^* \langle\psi_k|\psi_l\rangle$$

whereas the right hand side is calculated by using Theorem 20 to obtain

$$\text{R.H.S.} = \omega_l^* \langle\psi_k|\psi_l\rangle.$$

Hence $(\omega_k^* - \omega_l^*)\langle\psi_k|\psi_l\rangle = 0$.

Since $\omega_k \neq \omega_l$, $\langle\psi_k|\psi_l\rangle = 0$. **QED**

A very useful theorem is now encountered:

> **Theorem 22.** If an operator $\boldsymbol{\Omega}$ is normal with two non-degenerate eigenfunctions ψ_k and ψ_l, and if it commutes with another operator $\mathbf{A}$, then $\langle\psi_k|\mathbf{A}|\psi_l\rangle = 0$.

We write $\langle\psi_k|\boldsymbol{\Omega}\mathbf{A}|\psi_l\rangle = \langle\psi_k|\mathbf{A}\boldsymbol{\Omega}|\psi_l\rangle$ so that

$$(\omega_k - \omega_l)\langle\psi_k|\mathbf{A}|\psi_l\rangle = 0.$$

Since $\omega_k \neq \omega_l$, $\langle\psi_k|\mathbf{A}|\psi_l\rangle = 0$. **QED**

As an application of this theorem we give the following example:

Example 49: The hydrogen atom Hamiltonian commutes with $\boldsymbol{L}^2$ and $\boldsymbol{L}_z$. Therefore,

$$\langle\psi_{nlm}|\boldsymbol{L}^2|\psi_{n'l'm'}\rangle = \hbar^2\, l(l+1)\, \boldsymbol{\delta}_{\mathbf{n,n'}}\, \delta_{l,l'}\delta_{m,m'},$$

$$\langle\psi_{nlm}|\boldsymbol{L}_z|\psi_{n'l'm'}\rangle = \hbar m\, \boldsymbol{\delta}_{\mathbf{n,n'}}\, \delta_{l,l'}\delta_{m,m'}. \tag{58.9}$$

Similarly all the off-diagonal matrix elements corresponding to operators $\boldsymbol{L}_x$ and $\boldsymbol{L}_y$ vanish. Thus magnetic dipole transitions are forbidden for isolated atoms and atomic ions in the absence of a magnetic field that may be externally applied, or created internally through the spin angular momentum. The magnetic-dipole mechanism for electronic transitions is found to be operative for atoms or ions in a molecular or crystal-field environment.

Lecture 59: Linear Harmonic Oscillator — Solution by Ladder Operator Technique

The Hamiltonian for the linear harmonic oscillator is

$$\mathbf{H} = \frac{p^2}{2m} + \frac{1}{2}m\omega^2x^2$$

from which a dimensionless operator can be prepared,

$$\frac{\mathbf{H}}{\hbar\omega} = \frac{p^2}{2m\hbar\omega} + \frac{m\omega}{2\hbar}x^2 \tag{59.1}$$

such that the eigenvalue of the latter is a pure number. This operator is of the form of a sum of two dimensionless squared operators,

$$\frac{\mathbf{H}}{\hbar\omega} = \mathbf{P}^2 + \mathbf{Q}^2 \tag{59.2}$$

where

$$\mathbf{P} = \frac{1}{(2m\hbar\omega)^{1/2}}p,$$

$$\mathbf{Q} = \left(\frac{m\omega}{2\hbar}\right)^{1/2}x.$$

Operators **P** and **Q** are hermitian. The operator $\mathbf{H}/\hbar\omega$ is now converted into the form of a hermitian product by writing

$$\frac{\mathbf{H}}{\hbar\omega} = (\mathbf{Q} - i\mathbf{P})(\mathbf{Q} + i\mathbf{P}) - i[\mathbf{Q}, \mathbf{P}] = \mathbf{a}^\dagger\mathbf{a} + \frac{1}{2} \tag{59.3}$$

since $[\mathbf{Q}, \mathbf{P}] = \frac{i}{2}$. The two new dimensionless operators **a** and $\mathbf{a}^\dagger$ are defined by

$$\mathbf{a} = \mathbf{Q} + i\mathbf{P} = \left(\frac{m\omega}{2\hbar}\right)^{1/2}x + i\frac{1}{(2m\hbar\omega)^{1/2}}p, \tag{59.4}$$

$$\mathbf{a}^\dagger = \mathbf{Q} - i\mathbf{P} = \left(\frac{m\omega}{2\hbar}\right)^{1/2}x - i\frac{1}{(2m\hbar\omega)^{1/2}}p.$$

Let us write

$$\mathbf{N} = \mathbf{a}^\dagger\mathbf{a} \tag{59.5}$$

Then

$$\mathbf{H} = \hbar\omega\,(\mathbf{N} + \frac{1}{2}) \tag{59.6}$$

so that if $|n>$ is an eigenket of the dimensionless operator **N** with eigenvalue n,

$$\mathbf{N}\,|n> = n\,|n>, \tag{59.7}$$

then $|n>$ is also an eigenket of **H** with eigenvalue $(n+\frac{1}{2})\hbar\omega$:

$$\mathbf{H}\,|n> = E_n\,|n>, \tag{59.8}$$

and

$$E_n = (n+\frac{1}{2})\,\hbar\omega. \tag{59.9}$$

The eigenvalues of the operator $\mathbf{N}$ can be systematically derived. Since $\mathbf{N}$ is dimensionless, n must be a pure number. As $\mathbf{N} = \mathbf{N}^\dagger$, n must be real. Again,

$$n = < n|\mathbf{N}|n > = < n|\mathbf{a}^\dagger \mathbf{a}|n >$$

and by Theorem 16, $n \geq 0$. It is easy to check that

$$[\mathbf{a}, \mathbf{a}^\dagger] = 1. \tag{59.10}$$

Hence

$$[\mathbf{a}, \mathbf{N}] = \mathbf{a}, \tag{59.11}$$

$$[\mathbf{a}^\dagger, \mathbf{N}] = -\mathbf{a}^\dagger. \tag{59.12}$$

From Eq.(59.11) one observes

$$\mathbf{Na}|n > = \mathbf{a}(\mathbf{N} - 1)|n > = (n - 1)\mathbf{a}|n >.$$

Thus $\mathbf{a}|n >$ is an eigenket of $\mathbf{N}$ with eigenvalue $(n - 1)$ such that

$$\mathbf{a}|n > = c|n - 1 >$$

where c is a multiplicative factor. The quantity c can be evaluated from the expectation value of $\mathbf{N}$. Since

$$< n|\mathbf{N}|n > = c^2 < n - 1|n - 1 >$$

one obtains $c = \sqrt{n}$ because the eigenket $|n - 1 >$ is normalized, and correspondingly

$$\mathbf{a}|n > = \sqrt{n}|n - 1 >. \tag{59.13}$$

By successive application of the operator $\mathbf{a}$ one would generate the states $|n - 2 >$, $|n - 3 >$, etc., corresponding to the successively lower eigenvalues $(n - 2)$, $(n - 3)$, etc., of $\mathbf{N}$. Therefore, the operator $\mathbf{a}$ is a *step-down* operator. But we have already noticed that these eigenvalues must be greater than or equal to zero. Hence the downward series must terminate and it would do so in a natural way if one gets $\sqrt{n}=0$, that is,

$$\ldots$$

$$\mathbf{a}|1 > = |0 >,$$

$$\mathbf{a}|0 > = 0.$$

Therefore, n must be equal to zero or a positive integer,

$$n = 0, 1, 2, \ldots \tag{59.14}$$

Instead of starting from Eq. (59.11) if one starts from Eq. (59.12) one obtains

$$\mathbf{Na}^\dagger|n > = \mathbf{a}^\dagger(\mathbf{N} + 1)|n > = (n + 1)\mathbf{a}^\dagger|n >$$

and recognizes $\mathbf{a}^\dagger|n >$ as an eigenket of $\mathbf{N}$ with eigenvalue $(n + 1)$ such that

$$\mathbf{a}^\dagger|n > = c'|n + 1 >.$$

Hence $\mathbf{a}^\dagger$ is called a *step-up* operator. By forming the inner product one finds

$$< n|\mathbf{aa}^\dagger|n > = c'^2 < n + 1|n + 1 >$$

The commutator in Eq. (59.10) leads to the result $c' = \sqrt{n+1}$ so that

$$\mathbf{a}^\dagger|n > = \sqrt{n+1}\,|n + 1 >. \tag{59.15}$$

The equation can be utilized in forming the eigenkets from $|0 >$ by repeated applications of the operator $\mathbf{a}^\dagger$:

$$|1 > = \mathbf{a}^\dagger|0 >$$

$$|2 > = \frac{1}{\sqrt{2}}\mathbf{a}^\dagger|1 > = \frac{1}{\sqrt{2}}\mathbf{a}^{\dagger\,2}|0 >$$

and in general

$$|n> = \frac{1}{\sqrt{n!}}\, \mathbf{a}^{\dagger n}|0>. \tag{59.16}$$

This fantastic derivation is due to Dirac. Both $\mathbf{a}$ and $\mathbf{a}^\dagger$ are called *ladder operators*. When they are applied on the state vector $|n>$ with eigenvalue n, the result is essentially a state vector $|n'>$ with eigenvalue $n' = n - 1$ or $n' = n + 1$. The most important aspect of this derivation is that a wholly new interpretation arises from it.

The eigenvalues of $\mathbf{H}$ are given by Eq. (59.9) along with (59.14). Hence the successive eigenvalues differ from each other by the amount $\hbar\omega$. This amount is recognized as the elementary excitation energy. The state $|n>$ is associated with n quanta of vibrational excitation. In other words, the eigenvalue of the operator $\mathbf{N}$ represents the number of excitation quanta. The operator $\mathbf{N}$ is hence *the number operator*, and the quantum of excitation with energy $\hbar\omega$ stands for a "particle". The state $|n>$ is a number state, that is, it is an eigenstate of the number operator. As eigenfunctions of $\mathbf{H}$, they may be viewed as state vectors in *the number representation.* The effect of the operator $\mathbf{a}$ (or $\mathbf{a}^\dagger$) is to create a state with one less (or one more) "particle". See Equations (59.13) and (59.15). Therefore, we refer to $\mathbf{a}$ and $\mathbf{a}^\dagger$ as the particle *annihilation (or destruction) operator and the particle creation operator*. "Particles" (excitation quanta) can be created or destroyed! The state $|0>$ corresponds to the vacuum state (excitation vacuum).

If we adopt this physical picture, we do not have to refer to the conventional Schrödinger equation at all. The operator eigenvalue problem (59.8) is exactly equivalent to the wave equation in coordinate space. The definitions of $\mathbf{N}$ in terms of $\mathbf{a}$ and $\mathbf{a}^\dagger$ in Eq. (59.5) and of $\mathbf{H}$ in terms of $\mathbf{N}$ in Eq. (59.6), and the commutation property in Eq. (59.10) are enough to derive the requisite solutions. This treatment is usually called as the second quantum formalism as the number of "particles" is quantized. But the name should not generate a misconception – there is no additional quantization and the formalism is exactly equivalent to the "first quantum treatment" discussed in Lecture 12. The second quantum treatment here is based on the commutator algebra for the creation and destruction operators, and it offers solution to the problem in a concise step while avoiding complicated differential equations. Particles for which creation and annihilation operators obey the commutation relationship in Eq. (59.10) are known as bosons.

Finally, one cannot escape the question: how does one find the wave functions in coordinate representation? The answer to this question is really simple: these functions can be easily derived from the number states. From our discussion on transformation theory we know that the wave function in coordinate representation will be given by

$$\psi_n(x) = <x|n>. \tag{59.17}$$

Since $\mathbf{a}|0> = 0$, we write

$$\mathbf{a} < \mathrm{x}|0> = < \mathrm{x}|\mathbf{a}|0> = 0$$

so that $\psi_0(\mathrm{x}) = < \mathrm{x}|0>$ has to be found from the first-order differntial equation

$$\left[\left(\frac{m\omega}{2\hbar}\right)^{1/2} x + \left(\frac{\hbar}{2m\omega}\right)^{1/2} \frac{d}{dx}\right]\psi_0(x) = 0. \tag{59.18}$$

The solution is

$$\psi_0(x) = \left(\frac{m\omega}{\pi\hbar}\right)^{1/4} e^{-m\omega x^2/2\hbar} \tag{59.19}$$

where we have also considered ψ_0 to be normalized. The function $\psi_1(x)$ is found from

$\psi_1(x) = <x|1> = <x|\mathbf{a}^\dagger|0> = \mathbf{a}^\dagger<x|0>$

so that

$$\psi_1(x) = \left[\left(\frac{m\omega}{2\hbar}\right)^{1/2} x - \left(\frac{\hbar}{2m\omega}\right)^{1/2} \frac{d}{dx}\right]\psi_0(x) = \left(\frac{m\omega}{\pi\hbar}\right)^{1/4}\left(\frac{2m\omega}{\hbar}\right)^{1/2} x\psi_0(x). \tag{59.20}$$

All $\psi_n(x)$, ($n = 2, 3$, etc.), can be found in a similar way. In general

$$\psi_n(x) = <\mathrm{x}|\mathrm{n}> = \left(\frac{m\omega}{\pi\hbar}\right)^{1/4} (2^n n!)^{-1/2} H_n[x] e^{-m\omega x^2/2\hbar} \tag{12.12}$$

where H_n is the Hermite polynomial of degree n. The orthonormality of the wave-function in coordinate representation is assured by the matrix formulation:

$$\int_{-\infty}^{\infty} dx\ \psi_n(x)\psi_m(x) = \underset{x}{\mathrm{S}} < n|x >< x|m > = < n|\mathbf{1}|m > = \delta_{n,m}. \tag{59.21}$$

Texts Suggested for Further Reading

P. A. M. Dirac, "The Principles of Quantum Mechanics", 4th edition, (Oxford University Press, Oxford, 1976).

F. L. Pilar, "Elementary Quantum Chemistry", International ed. (McGraw-Hill, New York, 1990).

EXERCISES

1. Show that the average value of a self-adjoint operator is real.
2. Prove the turn over rule. Hence write the turn over rule for a hermitean operator.
3. Discuss that the boundary conditions are important in deciding whether an operator is hermitean or not.
4. Prove that hermitean operators have real eigenvalues and orthogonal eigenfunctions.
5. Write the operator x in terms of the momentum coordinate.
6. Relate the wave function in coordinate representation to that in momentum representation, and vice versa. Establish that the two forms are equivalent to each other.
7. Prove the general inequality $< a^\dagger a > \ \geq 0$.
8. Show that the Schrödinger equation leads to the equation of continuity for probability waves.
9. Derive Virial theorem for a stationary state.

PROBLEMS

1. Show that $[x, p_x] = i\hbar$ in momentum representation.
2. (a) Prove that $\langle \varphi|\mathbf{G}^\dagger|\varphi \rangle = \langle \mathbf{G}\varphi|\varphi \rangle$

   (b) Use Equations (56.11) and (56.12) to derive the result

$$\langle \mathbf{F}^\dagger\mathbf{F}\rangle \langle \mathbf{G}^\dagger\mathbf{G}\rangle \geq \frac{1}{4}\langle(\mathbf{F}^\dagger\mathbf{G} + \mathbf{G}^\dagger\mathbf{F})\rangle$$

   (c) Show that $\sigma_A{}^2\sigma_B{}^2 \geq \frac{1}{4}\langle i[\mathbf{A}, \mathbf{B}]\rangle^2$

   (d) Will it be possible to find a common eigenfunction of $\boldsymbol{L}^2$ and $\boldsymbol{L}_z$? Discuss in the light of the relation (c).
3. (a) Express the following operators in terms of creation and annhilation operators for an excitation quantum of the linear harmonic oscillator: x, p, x^2 and p^2.

   (b) Use the number representation to evaluate $<x>$, $<p>$, $<x^2>$, $<p^2>$, σ_x and σ_p.

   (c) Show that the excitation vacuum corresponds to the minimum uncertainty product.
4. (a) Consider an electron oscillating in a one-dimensional cavity. Show that the dipole moment operator is given by

$$\mu_x = e\left(\frac{\hbar}{2m\omega}\right)^{1/2}(\mathbf{a}+\mathbf{a}^\dagger).$$

   (b) Deduce the selection rule $\Delta n = \pm 1$ for electric dipole transitions.
5. Form the spherical harmonics for $l = 4$. (Write $\boldsymbol{L}_+$, and set $\boldsymbol{L}_+|l,l\rangle = 0$. This gives the function for $|l,l\rangle$. Use the effect of $\boldsymbol{L}_-$ on $|l,m\rangle$ to generate other functions.)
6. Solve the oscillator problem in second quantization when the potential is written as $V(x) = \varpi x + \frac{1}{2}m\omega^2 x^2$.
7. Find the perturbed energy values and wave functions in the second quantized notation for an anharmonic oscillator with $V(x) = \frac{1}{2}m\omega^2 x^2 + \frac{1}{6}gx^3 + \frac{1}{24}hx^4$.

CHAPTER 11

ANGULAR MOMENTUM REVISITED

Lecture 60: Ladder Operator Technique for Angular Momentum

A general angular momentum is represented by the operator $\boldsymbol{M}$ with Cartesian components $\boldsymbol{M}_x$, $\boldsymbol{M}_y$ and $\boldsymbol{M}_z$. Each component is hermitean so that $\boldsymbol{M}$ as a whole is hermitean:

$$\boldsymbol{M} = \boldsymbol{M}^\dagger = \mathbf{i}\boldsymbol{M}_x + \mathbf{j}\boldsymbol{M}_y + \mathbf{k}\boldsymbol{M}_z. \qquad (60.1)$$

By definition the components satisfy the special commutation relationships[1]

$$[\boldsymbol{M}_x, \boldsymbol{M}_y] = i\hbar\boldsymbol{M}_z, \qquad (x, y, z \text{ cyclically varying}). \qquad (60.2)$$

The orbital angular momentum $\boldsymbol{L}$ and the spin angular momentum $\boldsymbol{S}$ are two simple examples of $\boldsymbol{M}$. The square of the angular momentum corresponds to the operator

$$\boldsymbol{M}^2 = \boldsymbol{M}\bullet\boldsymbol{M} = \boldsymbol{M}_x^2 + \boldsymbol{M}_y^2 + \boldsymbol{M}_z^2 \qquad (60.3)$$

that commutes with every component,

$$[\boldsymbol{M}^2, \boldsymbol{M}_i] = 0, \qquad (i = x, y, z). \qquad (60.4)$$

Therefore it is possible to find a set of common eigenfunctions of $\boldsymbol{M}^2$ and $\boldsymbol{M}_x$, another set of common eigenfunctions of $\boldsymbol{M}^2$ and $\boldsymbol{M}_y$, and yet another set for $\boldsymbol{M}^2$ and $\boldsymbol{M}_z$. Functions belonging to each set are generally distinct from those belonging to another set since the components do not commute with each other. Therefore, one may choose any specific set. Conventionally, one derives common eigenfunctions of $\boldsymbol{M}^2$ and $\boldsymbol{M}_z$, and the corresponding eigenvalues, and uses these results in a theoretical treatment. It is our intention to show that the common eigenfunctions of $\boldsymbol{M}^2$ and $\boldsymbol{M}_z$ and the corresponding eigenvalue spectra can be derived simply from the definition given by Equations (60.1) and (60.2).

Two new operators can be defined by

---

[1] If we number components as $\boldsymbol{M}_1$, $\boldsymbol{M}_2$ and $\boldsymbol{M}_3$, Eq. (60.2) will be written as

$$[\boldsymbol{M}_i, \boldsymbol{M}_j] = i\hbar\varepsilon_{ijk}\boldsymbol{M}_k$$

where the tensor ε is defined by

$\varepsilon_{ijk} = 1$ for i, j, k varying cyclically as (x, y, z), (y, z, x) and (z, x, y);

$\varepsilon_{ijk} = -1$ for i, j, k varying cyclically as (x, z, y), (y, x, z) and (z, y, x);

$\varepsilon_{ijk} = 0$ in all other cases.

$$\boldsymbol{M}_+ = \boldsymbol{M}_x + i\boldsymbol{M}_y, \tag{60.5a}$$

and

$$\boldsymbol{M}_- = \boldsymbol{M}_x - i\boldsymbol{M}_y. \tag{60.5b}$$

Since $\boldsymbol{M}_x$ and $\boldsymbol{M}_y$ are hermitean, $\boldsymbol{M}_+$ and $\boldsymbol{M}_-$ are hermitean adjoints of each other. This allows us to calculate two new hermitean products:

$$\boldsymbol{M}_+\boldsymbol{M}_- = \boldsymbol{M}^2 - \boldsymbol{M}_z^2 + \hbar\boldsymbol{M}_z, \tag{60.6a}$$

and

$$\boldsymbol{M}_-\boldsymbol{M}_+ = \boldsymbol{M}^2 - \boldsymbol{M}_z^2 - \hbar\boldsymbol{M}_z. \tag{60.6b}$$

One may also observe the relations

$$[\boldsymbol{M}_z, \boldsymbol{M}_+] = \hbar\boldsymbol{M}_+, \tag{60.7a}$$

and

$$[\boldsymbol{M}_z, \boldsymbol{M}_-] = -\hbar\boldsymbol{M}_-. \tag{60.7b}$$

These results will be used in our derivation.

Let there be a common eigenket $|\lambda, \mu>$ of operators $\boldsymbol{M}^2$ and $\boldsymbol{M}_z$ such that the corresponding eigenvalues are $\hbar^2\Lambda(\lambda)$ and $\hbar m(\mu)$ respectively,

$$\boldsymbol{M}^2|\lambda, \mu> = \hbar^2\Lambda(\lambda)|\lambda,\mu> \tag{60.8a}$$

and

$$\boldsymbol{M}_z|\lambda, \mu> = \hbar m(\mu)|\lambda,\mu>. \tag{60.8b}$$

Using Equations (60.7a) and (60.7b) one can write

$$\boldsymbol{M}_z\boldsymbol{M}_+|\lambda, \mu> = \boldsymbol{M}_+(\boldsymbol{M}_z + \hbar)|\lambda, \mu> = \hbar[m(\mu) + 1]\,\boldsymbol{M}_+|\lambda, \mu>, \tag{60.9a}$$

and

$$\boldsymbol{M}_z\boldsymbol{M}_-|\lambda, \mu> = \boldsymbol{M}_-(\boldsymbol{M}_z - \hbar)|\lambda, \mu> = \hbar[m(\mu) - 1]\boldsymbol{M}_-|\lambda, \mu>. \tag{60.9b}$$

Thus $\boldsymbol{M}_+|\lambda, \mu>$ and $\boldsymbol{M}_-|\lambda, \mu>$ are eigenkets of $\boldsymbol{M}_z$ with eigenvalues respectively raised and lowered by $\hbar$. In other words $\boldsymbol{M}_+$ and $\boldsymbol{M}_-$ are step-up and step-down or ladder operators. One may write

$$\boldsymbol{M}_+|\lambda, \mu> = c_+|\lambda, \mu'>$$

where (60.10)

$$m(\mu') = m(\mu) + 1,$$

and

$$\boldsymbol{M}_-|\lambda, \mu> = c_-|\lambda, \mu''>$$

where (60.11)

$$m(\mu'') = m(\mu) - 1.$$

The quantities c_+ and c_- are as yet undetermined constant multipliers. They can be determined by requiring that the kets are normalized. This requirement leads to the simplifications

$$<\lambda, \mu|\,\boldsymbol{M}_-\boldsymbol{M}_+|\lambda, \mu> = c_+^2 <\lambda, \mu'|\lambda, \mu'> = c_+^2$$

and

$$<\lambda, \mu|\,\boldsymbol{M}_+\boldsymbol{M}_-|\lambda, \mu> = c_-^2 <\lambda, \mu''|\lambda, \mu''> = c_-^2$$

where we have used $\boldsymbol{M}_+^\dagger = \boldsymbol{M}_-$ and $\boldsymbol{M}_-^\dagger = \boldsymbol{M}_+$. Use of equations (60.6a) and (60.6b) along with Equations (60.8a) and (60.8b) leads to the results

$$c_+ = \hbar[\Lambda(\lambda) - m^2(\mu) - m(\mu)]^{1/2} \tag{60.12a}$$

and

$$c_- = \hbar[\Lambda(\lambda) - m^2(\mu) + m(\mu)]^{1/2}. \tag{60.12b}$$

Repeated applications of M_+ and M_- will continue to generate new eigenkets with increasing and decreasing eigenvalues of M_z. However, as $M^2 \geq M_z^2$, we must have

$$\Lambda(\lambda) \geq m^2(\mu) \tag{60.13}$$

and so the series of eigenvalues cannot extend in any direction indefinitely. The series must terminate, and the natural way for it to do so is to require $c_+ = 0$ for, say, the combination (λ, μ_1) and similarly $c_- = 0$ for the combination (λ, μ_2). Consequently, $m(\mu_1) \geq m(\mu_2)$. The expressions for c_+ and c_- in Equations (60.12a) and (60.12b) suggest the relationships

$$\Lambda(\lambda) = m^2(\mu_1) + m(\mu_1) = m^2(\mu_2) - m(\mu_2) \tag{60.14}$$

that yields

$$[m(\mu_1) + m(\mu_2)][m(\mu_1) - m(\mu_2) + 1] = 0.$$

Therefore, either

$$m(\mu_1) = -\,m(\mu_2) \tag{60.15}$$

or $m(\mu_1) = m(\mu_2) - 1$. But the latter is impossible as $m(\mu_1) \geq m(\mu_2)$. Hence Eq. (60.15) is the only acceptable solution. As the successive $m(\mu)$ values differ by 1, the quantity $[m(\mu_1) - m(\mu_2)]$ must be equal to zero or a positive integer, so that $m(\mu_1)$ must be zero, a half integer greater than zero or a positive integer. Without any loss of generality one can identify λ with the maximum value $m(\mu_1)$ and write

$\lambda = 0, 1, 2, \ldots$	(for $2\lambda = 0$ or an even integer)	
and		(60.16)
$\lambda = 1/2, 3/2, 5/2, \ldots$	(for $2\lambda =$ an odd integer).	

The number $m(\mu)$ varies as

$$m(\mu) = -\lambda, -\lambda + 1, \ldots, \lambda - 1, \lambda.$$

There is no constraint to guide the relationship between $m(\mu)$ and μ except that successive $m(\mu)$ values differ by unity. Hence the use of m has been superfluous, that is, without any loss of generality we can write μ in place of $m(\mu)$. Therefore, we find

$$\mu = -\lambda, -\lambda + 1, \ldots, \lambda - 1, \lambda \tag{60.17}$$

and from Eq. (60.14),

$$\Lambda(\lambda) = \lambda(\lambda + 1). \tag{60.18}$$

With these results one may rewrite the eigenvalue equations (60.8a) and (60.8b). The revised equations are

$$M^2\,|\lambda, \mu\rangle = \hbar^2\lambda(\lambda + 1)\,|\lambda, \mu\rangle$$

and

$$M_z\,|\lambda, \mu\rangle = \hbar\mu\,|\lambda, \mu\rangle. \tag{60.19}$$

For the orbital angular momentum of one electron ($M = L$), λ is written as l and μ as m_l or simply as m. The quantum numbers l and m are zero or integers. For the electron spin angular momentum ($\boldsymbol{M} = \boldsymbol{S}$), we write λ as s and μ as m_s. We know that $s = ½$ and m_s has two possible values, $+½$ and $-½$.

Furthermore, if one combines Equations (60.10), (60.11), (60.12), (60.17) and (60.18) together one can explicitly describe the effect of the application of the ladder operators on the state vector $|\lambda, \mu>$ as

$$\boldsymbol{M}_+|\lambda, \mu> = \hbar[(\lambda - \mu)(\lambda + \mu + 1)]^{1/2}\,|\lambda, \mu+1> \tag{60.20a}$$

and

$$\boldsymbol{M}_-|\lambda, \mu> = \hbar[(\lambda + \mu)(\lambda - \mu + 1)]^{1/2}\,|\lambda, \mu-1>. \tag{60.20b}$$

The general treatment of angular momentum is useful in three specific ways. In the case of an angular momentum for which a classical description is possible, such as the orbital angular momentum, the general treatment shows that the structure of the eigenvalue spectrum and the relationships among the eigenkets are strictly the outcome of the commutation rules which have been used to define the angular momentum. In the case of angular momenta with no classical correspondence, such as the spin or the isospin, no pictorial interpretation or geometrical description can be given and the general treatment is the only way of handling the problem. The general treatment also represents a systematic procedure for dealing with the theory of the addition of more than one angular momentum. Two or more types of angular momenta of a specific particle can be added together to form the total angular momentum for the same species. Angular momenta of the same kind but for different electrons may be combined to form the total angular momentum of the specific type but for the total system. Total angular momenta of different types (total orbital angular momentum, total spin angular momentum, etc.) for N electrons can be added to form the sum total. These aspects and their relevance for the atomic structure will be discussed in the following lectures.

Lecture 61: Ladder Operator Method of Solving the Orbital Angular Momentum Problem

We have seen in Lecture 13 that the orbital angular momentum is defined by

$$\boldsymbol{L} = \mathbf{r}\times\mathbf{p} = \mathbf{i}\boldsymbol{L}_x + \mathbf{j}\boldsymbol{L}_y + \mathbf{k}\boldsymbol{L}_z, \tag{13.1}$$

it is hermitean, that is,

$$\boldsymbol{L} = \boldsymbol{L}^\dagger,$$

and its components satisfy commutation rules of the type

$$[\boldsymbol{L}_x, \boldsymbol{L}_y] = i\hbar\boldsymbol{L}_z, \qquad \text{(x, y, z cyclically varying).} \tag{13.6}$$

The commutation relations can be cast into the form

$$[\boldsymbol{L}_i, \boldsymbol{L}_j] = i\hbar\varepsilon_{ijk}\boldsymbol{L}_k.$$

The components $\boldsymbol{L}_x$, $\boldsymbol{L}_y$ and $\boldsymbol{L}_z$ and the squared angular momentum operator $\boldsymbol{L}^2$ are written in terms of the spherical polar coordinates as (see Appendix II)

$$L_x = i\hbar\left(\cot\theta\ \cos\varphi\frac{\partial}{\partial\theta} + \sin\varphi\frac{\partial}{\partial\varphi}\right),$$

$$L_y = i\hbar\left(\cot\theta\ \sin\varphi\frac{\partial}{\partial\theta} - \cos\varphi\frac{\partial}{\partial\varphi}\right), \tag{13.13}$$

$$L_z = -i\hbar\frac{\partial}{\partial\varphi},$$

and

$$L^2 = -\hbar^2\left[\frac{1}{\sin\theta}\frac{\partial}{\partial\theta}\left(\sin\theta\frac{\partial}{\partial\theta}\right) + \frac{1}{\sin^2\theta}\frac{\partial^2}{\partial\varphi^2}\right] \tag{13.14}$$

We have also found the eigenfunctions of L_z in the coordinate representation. These are

$$\Phi_m(\varphi) = \frac{1}{\sqrt{2\pi}}e^{im\varphi}, \qquad (m = 0, \pm1, \pm2, \dots). \tag{13.19}$$

The corresponding eigenvalues are $\hbar m$. Comparing with the general treatment we find $\mu = m$. Since μ (= m) is an integer, λ (written as l in the present context) is also an integer. Thus we have

$$l\ (=\lambda) = 0, 1, 2, \dots$$
$$m\ (=\mu) = -l, -l+1, \dots, 0, \dots, l-1, l \tag{61.1}$$

with the eigenvalue relations

$$L^2|l, m> = \hbar^2 l(l+1)\,|l, m>$$

and (61.2)

$$L_z|l, m> = \hbar m\,|l, m>$$

in vector space. The same relations appear in the coordinate representation as

$$L^2\,Y_{l,m}(\theta, \varphi) = \hbar^2 l(l+1)\,Y_{l,m}(\theta, \varphi),$$

and (13.29)

$$L_z\,Y_{l,m}(\theta, \varphi) = \hbar m\,Y_{l,m}(\theta, \varphi).$$

There are $(2l + 1)$ values of m. We have also noted that $Y_{l,m}(\theta, \varphi)$ can be separated,

$$Y_{l,m}(\theta, \varphi) = \Theta_{l,m}(\theta)\,\Phi_m(\varphi). \tag{13.27}$$

Our aim here is to use the step-up and step-down operators to determine the function $\Theta_{l,m}(\theta)$.

The ladder operators are obtained from Equations (60.5a) and (60.5b):

$$L_+ = L_x + iL_y, \tag{61.3a}$$

$$L_- = L_x - iL_y. \tag{61.3b}$$

Using Equations (13.13) we write these operators as

$$L_+ = \hbar\,e^{i\varphi}\left(\frac{\partial}{\partial\theta} + i\cot\theta\frac{\partial}{\partial\varphi}\right), \tag{61.4a}$$

$$L_- = -\hbar\,e^{-i\varphi}\left(\frac{\partial}{\partial\theta} - i\cot\theta\frac{\partial}{\partial\varphi}\right). \tag{61.4b}$$

To check that these two operators are Hermitian conjugates of each other, remember that the derivative operators are antihermitean.

One can generate the functions $Y_{l,m+1}$, $Y_{l,m+2}$, etc. (or $Y_{l,m-1}$, $Y_{l,m-2}$, etc.) by successively applying the operator $\boldsymbol{L}_+$ (or $\boldsymbol{L}_-$) on $Y_{l,m}$. The series must terminate on both the sides as $l \geq |m|$. From the general treatment we know that the series terminates in the positive side after $m = l$ because of the operation

$$\boldsymbol{L}_+ Y_{l,l} = 0. \tag{61.5}$$

This yields the first order differential equation

$$\frac{\partial}{\partial\theta}\Theta_{l,l}(\theta) = l \cot\theta\, \Theta_{l,l}(\theta) \tag{61.6}$$

with solution

$$\Theta_{l,l}(\theta) = N \sin^l\theta. \tag{61.7}$$

If we normalize this function we get

$$N^2 \int_0^\pi d\theta \sin^{2l+1}\theta = 1$$

so that

$$\Theta_{l,l}(\theta) = \left[\frac{(2l+1)!}{2(2^l l!)^2}\right]^{\frac{1}{2}} \sin^l\theta. \tag{61.8}$$

If we investigate the termination of the negative side of m by

$$\boldsymbol{L}_- Y_{l,-l} = 0 \tag{61.9}$$

we obtain

$$\Theta_{l,-l}(\theta) = \Theta_{l,l}(\theta). \tag{61.10}$$

This is not surprizing at all. In fact for all valid $|m|$,

$$\Theta_{l,m}(\theta) = \Theta_{l,-m}(\theta) \tag{61.11}$$

as the Θ function satisfies the differential equation (13.22) which depends upon m^2 and hence on $|m|$ but not on the sign of m.

The analysis above leads to the expression

$$Y_{l,\pm l}(\theta, \phi) = \frac{1}{2(2^l l!)}\left[\frac{(2l+1)!}{\pi}\right]^{\frac{1}{2}} \sin^l\theta\, e^{\pm il\varphi}. \tag{61.12}$$

For convenience we attach a factor of $(-1)^l$ to $Y_{l,l}$. All other $\Theta_{l,|m|}$ functions and hence all other $Y_{l,\pm m}(\theta, \phi)$ can be found from equations (60.20a) and (60.20b):

$$Y_{l,m+1}(\theta, \phi) = \hbar^{-1}\,[(l-m)(l+m+1)]^{-1/2}\, \boldsymbol{L}_+ Y_{l,m}(\theta, \phi), \tag{61.13a}$$

and

$$Y_{l,m-1}(\theta, \phi) = \hbar^{-1}\,[(l+m)(l-m+1)]^{-1/2}\, \boldsymbol{L}_- Y_{l,m}(\theta, \phi). \tag{61.13b}$$

For $m = -l$, Eq. (61.13a) yields

$$Y_{l,-(l-1)}(\theta, \phi) = \frac{1}{2(2^l\, l!)}\left[\frac{(2l+1)!2l}{\pi}\right]^{\frac{1}{2}} \sin^{l-1}\theta\cos\theta e^{-i(l-1)\varphi}, \tag{61.14a}$$

and for $m = l$, Eq. (61.13b) leads to the spherical harmonics

$$Y_{l,(l-1)}(\theta, \phi) = \frac{(-1)^{l-1}}{2(2^l\, l!)}\left[\frac{(2l+1)!2l}{\pi}\right]^{\frac{1}{2}} \sin^{l-1}\theta\cos\theta\, e^{i(l-1)\varphi}. \tag{61.14b}$$

All the spherical harmonics can be determined by this procedure. See Table 3.2 for a list of these functions up to $l = 4$.

Lecture 62: Differential Equation Method of Solving the Orbital Angular Momentum Problem

We have just reviewed the orbital angular momentum problem. The Φ-equation can be easily solved. In Lecture 61 we have discussed how the Θ-equation may be solved in an abstract way by using the ladder operators $\boldsymbol{L}_+$ and $\boldsymbol{L}_-$. Here we discuss the traditional way of solving the Θ-equation,

$$\frac{1}{\sin\theta}\frac{d}{d\theta}\left(\sin\theta\frac{d\Theta(\theta)}{d\theta}\right)+\left(\frac{\Lambda}{\hbar^2}-\frac{m^2}{\sin^2\theta}\right)\Theta(\theta) = 0. \tag{13.22}$$

Let us put $\cos\theta = z$ and $\Theta(\theta) = P^m(z) \equiv P^m(\cos\theta)$ in this equation. The result is

$$(1-z^2)\frac{d^2P^m(z)}{dz^2} - 2z\frac{dP^m(z)}{dz} + \left(\frac{\Lambda}{\hbar^2}-\frac{m^2}{1-z^2}\right)P^m(z) = 0 \tag{62.1}$$

which needs to be solved within the range $-1 \le z \le 1$.

A function $P_l(z)$ is introduced here. It is defined by the *generating function* $G(z, s)$:

$$G(z, s) = (1-2zs+s^2)^{-1/2} = \sum_{l=0}^{\infty} P_l(z)\, s^l \tag{62.2}$$

such that

$$P_l(z) = \frac{1}{l!}\left[\frac{\partial^l G(z,s)}{\partial s^l}\right]_{s=0}. \tag{62.3}$$

The function G is always real for a real s if z lies in the range (−1, 1). Differentiation of $G(z, s)$ with z leads to the *differential relation*

$$(1-z^2)\frac{dP_l(z)}{dz} = -\,lzP_l + lP_{l-1} \tag{62.4}$$

and differentiation with respect to s yields the *recurrence relation*

$$(l+1)P_{l+1} = (2l+1)\, zP_l - lP_{l-1}. \tag{62.5}$$

By differentiating both sides of Eq. (62.4) with respect to z and using the recurrence relation Eq. (62.5) to simplify the resulting equation, one obtains the *differential equation for the polynomials* $P_l(z)$:

$$(1-z^2)\,\frac{d^2P_l(z)}{dz^2} - 2z\frac{dP_l(z)}{dz} + l(l+1)\,P_l(z) = 0 \qquad (62.6)$$

For $l = 0, 1, 2, ...$ This equation was studied by Legendre and the polynomial $P_l(z)$ is known as Legendre polynomial of degree l.

Mathematical background

Legendre's equation can be solved by the power series method. This task is simple enough, and it is left to the reader. Also, the polynomial $P_l(z)$ can be directly found from the generating relationship (62.2). Rodrigues derived the formula

$$P_l(z) = \frac{1}{2^l\,l!}\frac{d^l(z^2-1)^l}{dz^l} \qquad (62.7)$$

by applying the binomial theorem to $(z^2-1)^l$ and differentiating l times term by term. One special value is

$$P_l(z) = (-1)^l\,P_l(-z) \qquad (62.8)$$

which is easily derivable from Rodrigues' formula (62.7). Another is

$$P_l(1) = 1 \qquad (62.9)$$

that can be obtained directly by inserting $z = 1$ in the generating function in Eq.(62.2). In particular, $P_0(z)$ is always equal to 1. The first few polynomials are

$$\begin{aligned} P_0 &= 1 & P_1 &= z \\ P_2 &= \frac{1}{2}(3z^2-1) & P_3 &= \frac{1}{2}(5z^3-3z) \\ P_4 &= \frac{1}{8}(35z^4-30z^2+3) & P_5 &= \frac{1}{8}(63z^5-70z^3+15z) \end{aligned} \qquad (62.10)$$

Legendre's equation (62.6) is in reality a Sturm-Liouville equation and the orthonormality of Legendre polynomials can be worked out accordingly. Normalization and orthogonality are given without proof in the following:

$$\int_{-1}^{+1} dz\;P_l(z)P_{l'}(z) = \frac{2}{2l+1}\,\delta_{l,l'}. \qquad (62.11)$$

Now consider Eq. (62.1) and put $\Lambda = \hbar^2 l(l+1)$. Substituting

$$P^m(z) = (1-z^2)^{m/2}\,u(z) \qquad (62.12)$$

in the resulting expression, one may show that $u(z)$ satisfies the differential equation

$$(1-z^2)\frac{d^2u}{dz^2} - 2(m+1)z\frac{du}{dz} + [l(l+1) - m(m+1)]u = 0. \qquad (62.13)$$

By differentiating the left side of Legendre's equation (62.6) m times one recovers essentially Eq. (62.13) so that a nontrivial solution for $u(z)$ is

$$u(z) = \frac{d^mP_l(z)}{dz^m}, \qquad \text{(for } 0 \le m \le l\text{)}. \qquad (62.14)$$

Hence we write

$$P_{l,m}(z) \equiv P^m(z) = (-1)^m(1-z^2)^{m/2}\frac{d^mP_l(z)}{dz^m} \qquad (62.15)$$

where an additional factor of $(-1)^m$ has been incorporated. The polynomial $P_{l,m}(z)$ is known as the *associated Legendre polynomial*. One can also show

$$P_{l,-m}(z) = (-1)^m \frac{(l-m)!}{(l+m)!} P_{l,m}(z), \qquad \text{(for } 0 \le m \le l). \tag{62.16}$$

Associated Legendre polynomials satisfy the *differential equation*

$$(1-z^2)\frac{d^2P_{l,m}(z)}{dz^2} - 2z\frac{dP_{l,m}(z)}{dz} + \left[l(l+1) - \frac{m^2}{1-z^2}\right]P_{l,m}(z) = 0. \tag{62.17}$$

They obey the *recurrence relations*

$$(2l+1)\, zP_{l,m} = (l-m+1)\, P_{l+1,m} + (l+m)\, P_{l-1,m}$$

$$(2l+1)(1-z^2)^{1/2}\, P_{l,m} = P_{l-1,m+1} - P_{l+1,m+1}$$

$$(1-z^2)\,\frac{dP_{l,m}(z)}{dz} = (l+1)zP_{l,m} - (l-m+1)P_{l+1,m} = -lz\, P_{l,m} + (l+m)\, P_{l-1,m}$$

$$(1-z^2)^{1/2}\, P_{l,m+1} = (l-m)\, zP_{l,m} - (l+m)\, P_{l-1,m}$$

$$= -(l+m+1)\, P_{l,m} + (l-m+1)\, P_{l+1,m}\,. \tag{62.18}$$

The normalization and orthogonality relationships for the associated Legendre polynomials are written as

$$\int_{-1}^{+1} dz\, P_{l,m}(z)\, P_{l',m}(z) = \frac{2}{2l+1}\frac{(l+m)!}{(l-m)!}\,\delta_{l,l'}. \tag{62.19}$$

Therefore, the normalized angular function Θ is written as

$$\Theta_{l,m}(\theta) = \left[\frac{2l+1}{2}\frac{(l-m)!}{(l+m)!}\right]^{1/2} P_{l,m}(\cos\theta) \tag{62.20}$$

with the spherical harmonics

$$Y_{l,m}(\theta,\varphi) = \left[\frac{2l+1}{4\pi}\frac{(l-m)!}{(l+m)!}\right]^{1/2} P_{l,m}(\cos\theta)\, e^{im\varphi} \tag{62.21}$$

for $-l \le m \le l$.

Lecture 63: Spin Angular Momentum; Reconciliation of Spin with Spatial Orbitals

Spin angular momentum is expressed as the vector operator $\boldsymbol{S} = (\boldsymbol{S}_x, \boldsymbol{S}_y, \boldsymbol{S}_z)$. The components are hermitean. They satisfy the requisite commutation relationships $[\boldsymbol{S}_i, \boldsymbol{S}_j] = i\hbar\varepsilon_{ijk}\boldsymbol{S}_k$. Many fundamental particles have spin number ½. Electrons, protons, neutrons, μ mesons, neutrinos, Λ, Σ and Ξ particles are all examples of half-spin particles. When one measures the spin angular momentum of any of these particles along the direction of an externally applied magnetic field, one finds a component that equals either $\frac{\hbar}{2}$ or $-\frac{\hbar}{2}$. This direction is conventionally taken as the z-axis. Thus there are two states, $|\alpha>$ and $|\beta>$, such that $m_s \equiv \mu = +½$ or $-½$. Therefore, the spin quantum number s ($\equiv \lambda$) is ½ and the state vectors are written as

$$|\alpha> = |½, ½\rangle$$

and (63.1)

$$|\beta> = |½, -½\rangle.$$

We rewrite the eigenvalue equation (60.8a) as

$$S^2|½, ½\rangle = \tfrac{3}{4}\hbar^2\,|½, ½\rangle$$

and (63.2)

$$S^2|½, -½\rangle = \tfrac{3}{4}\hbar^2\,|½, -½\rangle$$

respectively for the two state vectors, and similarly, Eq.(60.8b) turns out as

$$S_Z\,|½, ½\rangle = \frac{\hbar}{2}\,|½, ½\rangle$$

and (63.3)

$$S_Z\,|½, -½\rangle = -\frac{\hbar}{2}\,|½, -½\rangle$$

in the two cases. The operators S_+ and S_- are given by

$$S_+ = S_X + iS_Y$$

and (63.4)

$$S_- = S_X - iS_Y$$

and the results of their operation on the state vectors can be obtained from Equations (60.20). We find

$$S_+|\alpha> \equiv S_+|½, ½\rangle = 0$$

$$S_+|\beta> \equiv S_+|½, -½\rangle = \hbar|½, ½\rangle \equiv \hbar|\alpha> \qquad (63.5)$$

$$S_-|\alpha> \equiv S_-|½, ½\rangle = \hbar|½, -½\rangle \equiv \hbar|\beta>$$

$$S_-|\beta> \equiv S_-|½, -½\rangle = 0.$$

A matrix representation for the spin operators and the state vectors can be constructed with ease. There are only two spin states for the spin-½ particles, and the two orthonormal state vectors can be represented by column vectors of dimension two or more than two. The simplest representative column vectors are as follows:

$$|\alpha> \equiv |½, ½\rangle = \begin{pmatrix}1\\0\end{pmatrix},$$

and (63.6)

$$|\beta> \equiv |½, -½\rangle = \begin{pmatrix}0\\1\end{pmatrix}.$$

Note the orthonormality

$$<\alpha|\alpha> = (1\ \ 0)\begin{pmatrix}1\\0\end{pmatrix} = 1; \qquad <\alpha|\beta> = (1\ \ 0)\begin{pmatrix}0\\1\end{pmatrix} = 0;$$

$$<\beta|\alpha> = (0\ \ 1)\begin{pmatrix}1\\0\end{pmatrix} = 0; \qquad <\beta|\beta> = (0\ \ 1)\begin{pmatrix}0\\1\end{pmatrix} = 1.$$

With this choice of state vectors, S_Z, S_+ and S_- are square matrix operators of rank 2 which can be determined by inspecting the effects in equations (63.3) and (63.5). The representation matrices calculated in this way are

$$S_z = \frac{\hbar}{2}\begin{pmatrix} 1 & 0 \\ 0 & -1 \end{pmatrix} \; ; S_+ = \hbar\begin{pmatrix} 0 & 1 \\ 0 & 0 \end{pmatrix} \; ; S_- = \hbar\begin{pmatrix} 0 & 0 \\ 1 & 0 \end{pmatrix} . \tag{63.7}$$

Matrices representing the operators S_x and S_y are found from equations (63.4) and the last two relations in (63.7):

$$S_x = \frac{\hbar}{2}\begin{pmatrix} 0 & 1 \\ 1 & 0 \end{pmatrix} ; \quad S_y = \frac{\hbar}{2}\begin{pmatrix} 0 & -i \\ i & 0 \end{pmatrix} . \tag{63.8}$$

Hence the representation for S^2 is

$$S^2 = \frac{3\hbar^2}{4}\begin{pmatrix} 1 & 0 \\ 0 & 1 \end{pmatrix} \tag{63.9}$$

which is found while the operator products in S^2 are replaced by the products of matrices.

These representation matrices are related to the Pauli spin matrices by

$$S_i = \frac{\hbar}{2}\sigma_i . \tag{63.10}$$

The Pauli spin matrices are

$$\sigma_x = \begin{pmatrix} 0 & 1 \\ 1 & 0 \end{pmatrix} ; \sigma_y = \begin{pmatrix} 0 & -i \\ i & 0 \end{pmatrix} ; \sigma_z = \begin{pmatrix} 1 & 0 \\ 0 & -1 \end{pmatrix} . \tag{63.11}$$

They follow the commutation relations

$$[\sigma_i, \sigma_j] = 2i\varepsilon_{ijk}\sigma_k . \tag{63.12}$$

These matrices were used by Pauli (1927) to construct a relativistic wave equation for a spin-½ particle, notably, the electron.

The spin orbital can be written in terms of the column vectors which stand for |α> and |β>. A general spin orbital that includes the effects of the spin-orbit coupling is written as the two-component spinor

$$\psi(\mathrm{x}) \;\Leftrightarrow\; \psi(\mathbf{r}) = \begin{pmatrix} \phi_1(\mathbf{r}) \\ \phi_2(\mathbf{r}) \end{pmatrix} = \phi_1(\mathbf{r})|\alpha> + \phi_2(\mathbf{r})|\beta> . \tag{63.13}$$

The one-electron Hamiltonian operator $\boldsymbol{H}(\mathrm{x})$ is now written as the matrix operator

$$\boldsymbol{H} = \begin{pmatrix} h_{\alpha\alpha} & h_{\alpha\beta} \\ h_{\beta\alpha} & h_{\beta\beta} \end{pmatrix} \tag{63.14}$$

with the elements given by

$$h_{\alpha\alpha} = <\alpha|\mathbf{h}|\alpha>,$$

$$h_{\alpha\beta} = <\alpha|\mathbf{h}|\beta>, \tag{63.15}$$

etc. The off-diagonal elements $h_{\alpha\beta}$ and $h_{\beta\alpha}$ arise from the spin-orbit interaction.

Under space-spin separation the spin orbitals are found as spinors with only one non-zero component,

$$\psi_\uparrow(\mathrm{x}) = \phi(\mathbf{r})\alpha(\mathrm{s}) \;\Leftrightarrow\; \psi_\uparrow = \begin{pmatrix} \phi(\mathbf{r}) \\ 0 \end{pmatrix},$$

and (63.16)

$$\psi_\downarrow(\mathrm{x}) = \phi(\mathbf{r})\beta(\mathrm{s}) \;\Leftrightarrow\; \psi_\downarrow = \begin{pmatrix} 0 \\ \phi(\mathbf{r}) \end{pmatrix}.$$

This separation is achieved when the Hamiltonian operator is replaced by the diagonal matrix

$$\boldsymbol{H} = \begin{pmatrix} \boldsymbol{h} & 0 \\ 0 & \boldsymbol{h} \end{pmatrix}. \tag{63.17}$$

To clarify the situation we will write the Schrödinger equation for one electron in a hydrogen-like system. Two equations can be written, one for states with $m_S = ½$ and the other for states with $m_S = -½$. These are

$$\begin{pmatrix} \boldsymbol{h} & 0 \\ 0 & \boldsymbol{h} \end{pmatrix} \begin{pmatrix} \phi_{nlm}(\mathbf{r}) \\ 0 \end{pmatrix} = E_n \begin{pmatrix} \phi_{nlm}(\mathbf{r}) \\ 0 \end{pmatrix} \qquad (\text{for } m_S = ½) \tag{63.18a}$$

and

$$\begin{pmatrix} \boldsymbol{h} & 0 \\ 0 & \boldsymbol{h} \end{pmatrix} \begin{pmatrix} 0 \\ \phi_{nlm}(\mathbf{r}) \end{pmatrix} = E_n \begin{pmatrix} 0 \\ \phi_{nlm}(\mathbf{r}) \end{pmatrix} \qquad (\text{for } m_S = -½). \tag{63.18b}$$

In the above the operator $\boldsymbol{h}$ is the spin-free Hamiltonian

$$\boldsymbol{h} = -\frac{1}{2}\nabla^2 - \frac{Z}{r} \tag{63.19}$$

and each of the two equations (63.18a) and (63.18b) is equivalent to the spin-free Schrödinger equation

$$\boldsymbol{h}\phi_{nlm}(\mathbf{r}) = E_n\phi_{nlm}(\mathbf{r}). \tag{63.20}$$

Thus the spin-free Schrödinger equation is in reality the wave equation for a particle in a specific spin state, and not for an electron with zero spin. This statement remains somewhat ambiguous here, but the relativistic electron theory which will not be discussed in these lectures makes this point abundantly clear. The spinor formulation is a perfectly general methodology. The same technique may be extended to the treatment of any other atom and also to any formalism involving molecular electrons.

Lecture 64. How to Treat Two Angular Momenta Together?

The sum of two independent angular momenta $\boldsymbol{M}_1$ and $\boldsymbol{M}_2$ is given by

(i) $\boldsymbol{M} = \boldsymbol{M}_1 + \boldsymbol{M}_2$ (64.1)

s.t. $M_x = M_{1x} + M_{2x}$, $M_y = M_{1y} + M_{2y}$, $M_z = M_{1z} + M_{2z}$; (64.2)

(ii) $\boldsymbol{M} = \boldsymbol{M}^\dagger$ (64.3)

since both $\boldsymbol{M}_1$ and $\boldsymbol{M}_2$ are hermitean; and

(iii) $[M_x, M_y] = i\hbar M_z$, (x, y, z cyclically varying). (64.4)

The last result can be established from the independence of the angular momenta $\boldsymbol{M}_1$ and $\boldsymbol{M}_2$ which is given by the nine possible commutations

$$[M_{1x,\, y \text{ or } z}, M_{2x,\, y \text{ or } z}] = 0. \tag{64.5}$$

The square of the total angular momentum is easily calculated. One finds

$$\boldsymbol{M}^2 = \boldsymbol{M}_1^2 + \boldsymbol{M}_2^2 + 2(M_{1x}M_{2x} + M_{1y}M_{2y}) + 2M_{1z}M_{2z}. \tag{64.6}$$

We know that the ladder operators $\boldsymbol{M}_{1\pm}$ and $\boldsymbol{M}_{2\pm}$ are

$$\boldsymbol{M}_{1\pm} = \boldsymbol{M}_{1x} \pm i\boldsymbol{M}_{1y} \qquad (64.7)$$

$$\boldsymbol{M}_{2\pm} = \boldsymbol{M}_{2x} \pm i\boldsymbol{M}_{2y}$$

so that $\boldsymbol{M}_{1x}$, $\boldsymbol{M}_{1y}$, $\boldsymbol{M}_{2x}$ and $\boldsymbol{M}_{2y}$ can be expressed as

$$\boldsymbol{M}_{1x} = \frac{1}{2}(\boldsymbol{M}_{1+} + \boldsymbol{M}_{1-});\ \boldsymbol{M}_{2x} = \frac{1}{2}(\boldsymbol{M}_{2+} + \boldsymbol{M}_{2-}); \qquad (64.8)$$

$$\boldsymbol{M}_{1y} = -\frac{i}{2}(\boldsymbol{M}_{1+} - \boldsymbol{M}_{1-});\ \boldsymbol{M}_{2y} = -\frac{i}{2}(\boldsymbol{M}_{2+} - \boldsymbol{M}_{2-}).$$

These expressions can be used to rewrite $\boldsymbol{M}^2$ in terms of the ladder operators. The final form for $\boldsymbol{M}^2$ is

$$\boldsymbol{M}^2 = \boldsymbol{M}_1^2 + \boldsymbol{M}_2^2 + (\boldsymbol{M}_{1+}\boldsymbol{M}_{2-} + \boldsymbol{M}_{1-}\boldsymbol{M}_{2+}) + 2\boldsymbol{M}_{1z}\boldsymbol{M}_{2z}. \qquad (64.9)$$

Our objective is to find the eigenvalues and the common eigenfunctions or eigenkets of the operators $\boldsymbol{M}^2$ and $\boldsymbol{M}_z$, that is, to find $\mathbf{Y}_{\lambda,m}$ or $|\lambda, m>$ of the eigenvalue problems

$$\boldsymbol{M}^2\mathbf{Y}_{\lambda,m} = \hbar^2\lambda(\lambda+1)\mathbf{Y}_{\lambda,m} \quad \Leftrightarrow \quad \boldsymbol{M}^2|\lambda, m> = \hbar^2\lambda(\lambda+1)|\lambda, \mathrm{m}>$$

$$(\lambda = 0, 1, 2, \ldots \text{ or } \frac{1}{2}, \frac{3}{2}, \frac{5}{2}, \ldots) \qquad (64.10)$$

$$\boldsymbol{M}_z\mathbf{Y}_{\lambda,m} = \hbar m\mathbf{Y}_{\lambda,m} \quad \Leftrightarrow \quad \boldsymbol{M}_z|\lambda, m> = \hbar m|\lambda, m>$$

$$(m = -\lambda, -\lambda+1, \ \ldots\ , \lambda-1, \lambda)$$

in terms of the common eigenfunctions (or eigenkets) and the corresponding eigenvalues of $(\boldsymbol{M}_1^2, \boldsymbol{M}_{1z})$ and $(\boldsymbol{M}_2^2, \boldsymbol{M}_{2z})$.

Let $\mathrm{Y}_{\lambda_1\mu_1}(1) \equiv \mathrm{Y}_{\lambda_1\mu_1}(\theta_1, \varphi_1)$ be the common eigenfunctions of $\boldsymbol{M}_1^2$ and $\boldsymbol{M}_{1z}$, and $\mathrm{Y}_{\lambda_2\mu_2}(2) \equiv \mathrm{Y}_{\lambda_2\mu_2}(\theta_2, \varphi_2)$ be the common eigenfunctions of $\boldsymbol{M}_2^2$ and $\boldsymbol{M}_{2z}$. The corresponding kets are $|\lambda_1, \mu_1>^{(1)}$ and $|\lambda_2, \mu_2>^{(2)}$. The product functions

$$\mathbf{Y}_{\lambda_1\mu_1,\lambda_2\mu_2} = \mathrm{Y}_{\lambda_1\mu_1}(1)\ \mathrm{Y}_{\lambda_2\mu_2}(2) \qquad (64.11)$$

corresponding to the ket

$$|\lambda_1\lambda_2\mu_1\mu_2> = |\lambda_1, \mu_1>^{(1)}|\lambda_2, \mu_2>^{(2)} \qquad (64.12)$$

form a basis for the functions $\mathbf{Y}_{\lambda,m}$.

It is easy to check the commutation of $\boldsymbol{M}^2$ with the individual angular momentum operators. One finds

$$[\boldsymbol{M}^2, \boldsymbol{M}_1^2] = 0;\ [\boldsymbol{M}^2, \boldsymbol{M}_2^2] = 0 \qquad (64.13)$$

but

$$[\boldsymbol{M}^2, \boldsymbol{M}_{1z}] = -\hbar(\boldsymbol{M}_{1+}\boldsymbol{M}_{2-} - \boldsymbol{M}_{1-}\boldsymbol{M}_{2+}),$$

$$[\boldsymbol{M}^2, \boldsymbol{M}_{2z}] = -\hbar(\boldsymbol{M}_{2+}\boldsymbol{M}_{1-} - \boldsymbol{M}_{2-}\boldsymbol{M}_{1+}),$$

s.t.

$$[\boldsymbol{M}^2, \boldsymbol{M}_z] = [\boldsymbol{M}^2, \boldsymbol{M}_{1z}] + [\boldsymbol{M}^2, \boldsymbol{M}_{2z}] = 0. \qquad (64.14)$$

The commutation properties of $\boldsymbol{M}_z$ can be investigated in a similar way. It is easy to show that

$$[\boldsymbol{M}_z, \boldsymbol{M}_1^2] = 0; \ [\boldsymbol{M}_z, \boldsymbol{M}_2^2] = 0. \tag{64.15}$$

But in this case the operator commutes with both $\boldsymbol{M}_{1z}$ and $\boldsymbol{M}_{2z}$,

$$[\boldsymbol{M}_z, \boldsymbol{M}_{1z}] = 0; \ [\boldsymbol{M}_z, \boldsymbol{M}_{2z}] = 0. \tag{64.16}$$

Thus $\boldsymbol{M}^2$ can have common eigenfunctions with $\boldsymbol{M}_1^2$, $\boldsymbol{M}_2^2$ and $\boldsymbol{M}_z$ but not with $\boldsymbol{M}_{1z}$ and $\boldsymbol{M}_{2z}$, whereas $\boldsymbol{M}_z$ can have common eigenfunctions with all the four operators $\boldsymbol{M}_1^2$, $\boldsymbol{M}_2^2$, $\boldsymbol{M}_{1z}$ and $\boldsymbol{M}_{2z}$. Therefore, the functions $\mathbf{Y}_{\lambda,m}$ may be chosen to be eigenfunctions of $\boldsymbol{M}_1^2$ and $\boldsymbol{M}_2^2$ as well. This choice is written as

$$\mathbf{Y}_{\lambda m;\lambda_1\lambda_2} \Leftrightarrow |\lambda_1\lambda_2\lambda m>. \tag{64.17}$$

One may expand it or the corresponding ket vector in the basis $\{\mathbf{Y}_{\lambda_1\mu_1,\lambda_2\mu_2}\}$ or $\{|\lambda_1\lambda_2\mu_1\mu_2>\}$:

$$\mathbf{Y}_{\lambda m;\lambda_1\lambda_2} = \sum_{\lambda_1'=0}^{\infty}\sum_{\mu_1'=-\lambda_1'}^{\lambda_1'}\sum_{\lambda_2'=0}^{\infty}\sum_{\mu_2'=-\lambda_2'}^{\lambda_2'} c_{\lambda_1'\mu_1'\lambda_2'\mu_2';\lambda m\lambda_1\lambda_2}\,\mathbf{Y}_{\lambda_1'\mu_1';\lambda_2'\mu_2'} \tag{64.18}$$

$\Leftrightarrow$

$$|\lambda_1\lambda_2\lambda m> = \sum_{\lambda_1'=0}^{\infty}\sum_{\mu_1'=-\lambda_1'}^{\lambda_1'}\sum_{\lambda_2'=0}^{\infty}\sum_{\mu_2'=-\lambda_2'}^{\lambda_2'} c_{\lambda_1'\mu_1'\lambda_2'\mu_2';\lambda m\lambda_1\lambda_2}\,|\lambda_1'\lambda_2'\mu_1'\mu_2'>. \tag{64.19}$$

As the kets of type $|\lambda_1\lambda_2\mu_1\mu_2>$ are orthonormal, the coefficients c's are easily found from a calculation of the inner products

$$c_{\lambda_1'\mu_1'\lambda_2'\mu_2';\lambda m\lambda_1\lambda_2} = <\lambda_1'\lambda_2'\mu_1'\mu_2'|\lambda_1\lambda_2\lambda m>. \tag{64.20}$$

These coefficients are known as *Clebsch-Gordan coefficients*.

The Clebsch-Gordan coefficients are simplified a lot by manipulations in two steps. The first step starts with the observation

(i) $<\lambda_1'\lambda_2'\mu_1'\mu_2'|\boldsymbol{M}_1^2|\lambda_1\lambda_2\lambda m> = \hbar^2\lambda_1'(\lambda_1'+1) <\lambda_1'\lambda_2'\mu_1'\mu_2'|\lambda_1\lambda_2\lambda m>$

$$= <\lambda_1'\lambda_2'\mu_1'\mu_2'|\lambda_1\lambda_2\lambda m>\hbar^2\lambda_1(\lambda_1+1)$$

that implies

$$(\lambda_1-\lambda_1')(\lambda_1+\lambda_1'+1)<\lambda_1'\lambda_2'\mu_1'\mu_2'|\lambda_1\lambda_2\lambda m> = 0.$$

For $\lambda_1 \neq \lambda_1'$, neither the first factor nor the second one is zero. Hence the coefficient (the inner product) is zero. Repeating this argument in the case where we use $\boldsymbol{M}_2^2$ in place of $\boldsymbol{M}_1^2$, we can write

$$<\lambda_1'\lambda_2'\mu_1'\mu_2'|\lambda_1\lambda_2\lambda m> = \delta_{\lambda_1,\lambda_1'}\delta_{\lambda_2,\lambda_2'}<\lambda_1\lambda_2\mu_1\mu_2|\lambda_1\lambda_2\lambda m>. \tag{64.21}$$

In the second step one uses $\boldsymbol{M}_z = \boldsymbol{M}_{1z} + \boldsymbol{M}_{2z}$ in place of $\boldsymbol{M}_1^2$ in (i) and obtains

$$\text{(ii)} < \lambda_1\lambda_2\mu_1\mu_2|\boldsymbol{M}_z|\lambda_1\lambda_2\lambda m> = \hbar(\mu_1+\mu_2) < \lambda_1\lambda_2\mu_1\mu_2|\lambda_1\lambda_2\lambda m>$$
$$= < \lambda_1\lambda_2\mu_1\mu_2|\lambda_1\lambda_2\lambda m>\hbar m$$

that indicates the coefficient to be zero unless $m = \mu_1+\mu_2$. This allows one to simplify Eq. (64.21) and write

$$< \lambda_1'\lambda_2'\mu_1'\mu_2'|\lambda_1\lambda_2\lambda m> = \delta_{\lambda_1,\lambda_1'}\delta_{\lambda_2,\lambda_2'}\delta_{\mu_1+\mu_2,m} < \lambda_1\lambda_2\mu_1\mu_2|\lambda_1\lambda_2\lambda m>. \tag{64.22}$$

With this result expression (64.19) can be recast into a vastly simplified form,

$$|\lambda_1\lambda_2\lambda m> = \sum_{\mu_1=-\lambda_1}^{\lambda_1}\sum_{\substack{\mu_2=-\lambda_2 \\ (\mu_1+\mu_2=m)}}^{\lambda_2} |\lambda_1\lambda_2\mu_1\mu_2>< \lambda_1\lambda_2\mu_1\mu_2|\lambda_1\lambda_2\lambda m>. \tag{64.23}$$

The Clebsch-Gordan coefficients can be evaluated by using three basic properties: orthonormality, that these coefficients can be chosen as real, and recursion relationships. The orthonormality prescriptions are

$$\sum_\lambda\sum_m < \lambda_1\lambda_2\mu_1\mu_2|\lambda_1\lambda_2\lambda m>< \lambda_1\lambda_2\lambda m\,|\lambda_1\lambda_2\mu_1'\mu_2'> = \delta_{\mu_1,\mu_1'}\delta_{\mu_2,\mu_2'} \tag{64.24}$$

and

$$\sum_{\mu_1}\sum_{\mu_2} < \lambda_1\lambda_2\lambda m\,|\lambda_1\lambda_2\mu_1\mu_2>< \lambda_1\lambda_2\mu_1\mu_2|\lambda_1\lambda_2\lambda' m'> = \delta_{\lambda,\lambda'}\delta_{m,m'}. \tag{64.25}$$

For real coefficients,

$$< \lambda_1\lambda_2\lambda m|\lambda_1\lambda_2\mu_1\mu_2> = < \lambda_1\lambda_2\mu_1\mu_2|\lambda_1\lambda_2\lambda m>. \tag{64.26}$$

The third property, the recursion relation, is obtained by using the operators $\boldsymbol{M}_\pm$ in place of $\boldsymbol{M}_1^2$ in (i):

$$\begin{aligned}&[\lambda(\lambda+1) - m(m\pm1)]^{1/2} < \lambda_1\lambda_2\mu_1\mu_2|\lambda_1\lambda_2\lambda(m\pm1)> \\ &= [\lambda_1(\lambda_1+1) - \mu_1(\mu_1\mp1)]^{1/2} < \lambda_1\lambda_2(\mu_1\mp1)\mu_2|\lambda_1\lambda_2\lambda m> \\ &+ [\lambda_2(\lambda_2+1) - \mu_2(\mu_2\mp1)]^{1/2} < \lambda_1\lambda_2\mu_1(\mu_2\mp1)|\lambda_1\lambda_2\lambda m>.\end{aligned} \tag{64.27}$$

Furthermore, there are $(2\lambda_1+1)(2\lambda_2+1)$ states in a specific (λ_1, λ_2) multiplet out of which only one state corresponds to $m = (\lambda_1+\lambda_2)$, one corresponds to $m = -(\lambda_1+\lambda_2)$, two states are for $m = (\lambda_1+\lambda_2-1)$ and two more for $m = -(\lambda_1+\lambda_2-1)$, and so on till we get p states for $m = (\lambda_1+\lambda_2-p+1)$ and p more for $m = -(\lambda_1+\lambda_2-p+1)$. Finally there are $(p+1)$ states for each m lying in the range $-(\lambda_1+\lambda_2-p) \leq m \leq (\lambda_1+\lambda_2-p)$. Therefore,

$$2\sum_{i=1}^{p} i + [2(\lambda_1+\lambda_2-p)+1](p+1) = (2\lambda_1+1)(2\lambda_2+1)$$

which gives

$$p = \inf(2\lambda_1, 2\lambda_2).$$

These m values correspond to $\lambda = (\lambda_1+\lambda_2)$, $\lambda = (\lambda_1+\lambda_2-1)$, ... , $\lambda = (\lambda_1+\lambda_2-p)$ so that

$$|\lambda_1-\lambda_2| \leq \lambda \leq (\lambda_1+\lambda_2). \tag{64.28}$$

These five equations, from (64.24) to (64.28), can be used to determine unique values for the Clebsch-Gordan coefficients.

Lecture 65: Addition of Orbital and Spin Angular Momenta

The electron has the spin angular momentum ($\boldsymbol{S}$) – and the orbital angular momentum ($\boldsymbol{L}$) in a central field. The total angular momentum is a vector sum of $\boldsymbol{L}$ and $\boldsymbol{S}$:

$$\boldsymbol{J} = \boldsymbol{L} + \boldsymbol{S}. \tag{65.1}$$

We seek the solutions of the eigenvalue equations

$$\boldsymbol{J}^2\,|\chi> = \hbar^2 j(j+1)|\chi> \tag{65.2}$$

and

$$J_z|\chi> = \hbar m_j|\chi> \tag{65.3}$$

where $\boldsymbol{J}^2$ is given by [see Eq.(64.9)]

$$\boldsymbol{J}^2 = \boldsymbol{L}^2 + \boldsymbol{S}^2 + (\boldsymbol{L}_+\boldsymbol{S}_- + \boldsymbol{L}_-\boldsymbol{S}_+) + 2\boldsymbol{L}_z\boldsymbol{S}_z, \tag{65.4}$$

and, of course,

$$\boldsymbol{J}_z = \boldsymbol{L}_z + \boldsymbol{S}_z. \tag{65.5}$$

As $s = ½$ for the electron, one can write from Eq. (64.28)

$$j = l - ½ \qquad \text{with } m_j = -(l-½), \dots, (l-½)$$

and (65.6)

$$j = l + ½ \qquad \text{with } m_j = -(l+½), \dots, (l+½).$$

This accounts for the total multiplicity $[2(l-½)+1]+[2(l+½)+1] = 2(2l+1) = (2s+1)(2l+1)$. There is only one choice for j corresponding to $l=0$, viz., $j=½$. This j value is accompanied by two values of m_j, $-½$ and $½$.

The states are given by

$$|l\,½\,jm> = \sum_{\substack{m_l=-l \\ (m_l+m_s=m)}}^{+l}\ \sum_{m_s=-1/2}^{+1/2} |l\,½\,m_l m_s>< l\,½\,m_l m_s|l\,½\,jm>. \tag{65.7}$$

For specific j values we get

$$\begin{aligned}|l\,½\,(l-½)\,m> &= |l\,½\,(m+½)\,-½>< l\,½\,(m+½)\,-½|l\,½\,(l-½)\,m> \\ &+ |l\,½\,(m-½)\,½>< l\,½\,(m-½)\,½|l\,½\,(l-½)\,m>\end{aligned} \tag{65.8a}$$

and

$$\begin{aligned}|l\,½\,(l+½)\,m> &= \eta(l-m-½)\,|l\,½\,(m+½)\,-½>< l\,½\,(m+½)\,-½|l\,½\,(l+½)\,m> \\ &+ \eta(l+m-½)\,|l\,½\,(m-½)\,½>< l\,½\,(m-½)\,½|l\,½\,(l+½)\,m>\end{aligned} \tag{65.8b}$$

where the step function η is defined as

$$\eta(i) = \begin{cases} 0 & \text{for } i < 0, \\ 1 & \text{for } i \geq 0. \end{cases} \tag{65.9}$$

Thus in this case there are only four types of Clebsch-Gordan coefficients. These can be evaluated by using the conditions (64.24) - (64.27). The results of such a calculation are

$$< l\,½\,(m\pm½)\,\mp½|\,l\,½\,(l-½)\,m> = \pm\left(\frac{l\pm m+1/2}{2l+1}\right)^{1/2}, \tag{65.10}$$

$$< l\,½\,(m\pm½)\,\mp½|\,l\,½\,(l+½)\,m> = \left(\frac{l\mp m+1/2}{2l+1}\right)^{1/2}.$$

If we adopt matrix representation for the spin vectors, the basis functions corresponding to the ket $|l\ \tfrac{1}{2}\ m_l m_s>$ are written as the spinors

$$\begin{pmatrix} Y_{l,m_l}(\theta,\varphi) \\ 0 \end{pmatrix} \qquad \text{(for } m_s = \tfrac{1}{2}\text{)}, \qquad (65.11)$$

$$\begin{pmatrix} 0 \\ Y_{l,m_l}(\theta,\varphi) \end{pmatrix} \qquad \text{(for } m_s = -\tfrac{1}{2}\text{)}.$$

When these are multiplied with the radial components we obtain the spinors which appear in equations (63.18a) and (63.18b). The kets $|lsjm>$ correspond to linear combinations of the spinors in Equations (65.11).

The spin-orbit interaction is approximately proportional to $\boldsymbol{L}\bullet\boldsymbol{S}$. Because

$$\boldsymbol{L}\bullet\boldsymbol{S} = \tfrac{1}{2}(\boldsymbol{J}^2-\boldsymbol{L}^2-\boldsymbol{S}^2), \qquad (65.12)$$

the kets $|lsjm>$ are eigenkets of the same operator:

$$\begin{aligned} \boldsymbol{L}\bullet\boldsymbol{S}\,|lsjm> &= (l/2)\,|lsjm> && \text{(for } j = l+\tfrac{1}{2}\text{)};\\ &= -\,[(l+1)/2]|lsjm> && \text{(for } j = l-\tfrac{1}{2}\text{)}. \end{aligned} \qquad (65.13)$$

Therefore, the spin-orbit interaction vanishes for $l = 0$. Indeed we have found from the second equation in (65.10) that

$$|l=0, s=\tfrac{1}{2}, j=\tfrac{1}{2}, m=\tfrac{1}{2}> \;=|\, l=0, s=\tfrac{1}{2}, m_l=0, m_s=\tfrac{1}{2}> \Leftrightarrow \begin{pmatrix} Y_{00} \\ 0 \end{pmatrix}, \qquad (65.14)$$

$$|l=0, s=\tfrac{1}{2}, j=\tfrac{1}{2}, m=-\tfrac{1}{2}> \;=|\, l=0, s=\tfrac{1}{2}, m_l=0, m_s=-\tfrac{1}{2}> \Leftrightarrow \begin{pmatrix} 0 \\ Y_{00} \end{pmatrix}.$$

Notice from Eq. (65.13) that the spin-orbit coupling term $<\boldsymbol{L}\bullet\boldsymbol{S}>$ is positive when $\boldsymbol{L}$ and $\boldsymbol{S}$ are parallel, and negative when the two angular momenta oppose each other.

Figure 7.2 illustrates the splitting of the atomic energy levels in the *absence* of an externally applied magnetic field as well as the splitting pattern that evolves when an external field is switched on.

Lecture 66: Addition of Two Spins

The helium atom and the diatomic hydrogen molecule have two electrons each. Most of the atoms, ions and molecules have two or more valence electrons. If the spin-orbit coupling and the higher-order terms are neglected, the electronic Hamiltonian of a two-electron or a two-valence-electron species commutes with the operators $\boldsymbol{S}^2$ and $\boldsymbol{S}_z$ where $\boldsymbol{S}$ is the vector sum of the spins of the two electrons,

$$\boldsymbol{S} = \boldsymbol{S}_1 + \boldsymbol{S}_2. \qquad (66.1)$$

The species can be found in a definite state of total spin. This state is represented by the spin eigenfunction $\eta(s_1, s_2)$ that satisfies

$$\begin{aligned} \boldsymbol{S}^2\eta(s_1, s_2) &= \hbar^2 S(S+1)\eta(s_1, s_2), \\ \boldsymbol{S}_z\eta(s_1, s_2) &= \hbar M_S\eta(s_1, s_2). \end{aligned} \qquad (66.2)$$

The coordinates s_1 and s_2 are the "spin coordinates" of electrons 1 and 2. The nature of these coordinates may not be known, but their existence is taken for granted here. In fact their nature can be understood only from a fully relativistic treatment. The total spin operators can be written in terms of the operators for the spins of individual electrons. Thus

$$\boldsymbol{S}^2 = \boldsymbol{S}_1^2 + \boldsymbol{S}_2^2 + (\boldsymbol{S}_{1+}\boldsymbol{S}_{2-} + \boldsymbol{S}_{1-}\boldsymbol{S}_{2+}) + 2\boldsymbol{S}_{1z}\boldsymbol{S}_{2z} \qquad (66.3)$$

and

$$\boldsymbol{S}_z = \boldsymbol{S}_{1z} + \boldsymbol{S}_{2z}. \qquad (66.4)$$

Following Eq. (64.28) we find four spin states corresponding to the quantum numbers

$$S = s_1 - s_2 = \tfrac{1}{2} - \tfrac{1}{2} = 0 \quad \text{with} \quad M_S = 0$$

and (66.5)

$$S = s_1 + s_2 = 1 \quad \text{with} \quad M_S = -1, 0, +1.$$

For $S = 0$ there is $2S+1 = 1$ state, that is, only one state. This is called the *singlet* spin state. The $S = 1$ case has $2S+1 = 3$ spin states which together are referred to as *triplet*. The three states belonging to the triplet are degenerate as long as the Hamiltonian does not have a spin-dependent operator. The total number of spin states is four, and this is expected as each electron can be in any of the two independent states of spin, α and β.

The Clebsch-Gordan technique is vastly simplified here. The basis functions are $\alpha(s_1)\alpha(s_2)$, $\alpha(s_1)\beta(s_2)$, $\beta(s_1)\alpha(s_2)$, and $\beta(s_1)\beta(s_2)$. Let us investigate the effect of the application of the operators $\boldsymbol{S}_z$ and $\boldsymbol{S}^2$ on these basis functions. For the first operator we find

$$\begin{aligned}
&\boldsymbol{S}_z\alpha(s_1)\alpha(s_2) = \hbar\alpha(s_1)\alpha(s_2) && \Leftrightarrow && M_S = 1,\\
&\boldsymbol{S}_z\alpha(s_1)\beta(s_2) = 0 && \Leftrightarrow && M_S = 0,\\
&\boldsymbol{S}_z\beta(s_1)\alpha(s_2) = 0 && \Leftrightarrow && M_S = 0,\\
&\boldsymbol{S}_z\beta(s_1)\beta(s_2) = -\hbar\beta(s_1)\beta(s_2) && \Leftrightarrow && M_S = -1.
\end{aligned} \qquad (66.6)$$

All the four products are eigenfunctions of $\boldsymbol{S}_z$, but the functions $\alpha(s_1)\alpha(s_2)$ and $\beta(s_1)\beta(s_2)$ are nondegenerate with distinct eigenvalues whereas the products $\alpha(s_1)\beta(s_2)$ and $\beta(s_1)\alpha(s_2)$ have the same eigenvalue and are degenerate. The nondegenerate eigenvalues ($M_S = \pm 1$) ensure that $\alpha(s_1)\alpha(s_2)$ and $\beta(s_1)\beta(s_2)$ are eigenfunctions of $\boldsymbol{S}^2$ with eigenvalue $1(1+1)\hbar^2 = 2\hbar^2$. The degenerate products $\alpha(s_1)\beta(s_2)$ and $\beta(s_1)\alpha(s_2)$ may or may not be eigenfunctions of the same operator. Indeed a brief calculation using Eq. (66.3) shows

$$\begin{aligned}
&\boldsymbol{S}^2\,\alpha(s_1)\alpha(s_2) = 2\hbar^2\,\alpha(s_1)\alpha(s_2) && \Leftrightarrow && S = 1,\\
&\boldsymbol{S}^2\,\beta(s_1)\beta(s_2) = 2\hbar^2\,\beta(s_1)\beta(s_2) && \Leftrightarrow && S = 1,
\end{aligned} \qquad (66.7)$$

but

$$\begin{aligned}
&\boldsymbol{S}^2\,\alpha(s_1)\beta(s_2) = \hbar^2\,[\alpha(s_1)\beta(s_2) + \beta(s_1)\alpha(s_2)],\\
&\boldsymbol{S}^2\,\beta(s_1)\alpha(s_2) = \hbar^2\,[\alpha(s_1)\beta(s_2) + \beta(s_1)\alpha(s_2)].
\end{aligned} \qquad (66.8)$$

The last pair of equations indicates that orthonormal combinations of the form

$$\eta = c_1\alpha(s_1)\beta(s_2) + c_2\beta(s_1)\alpha(s_2) \qquad (66.9)$$

where $c_1^2+c_2^2 = 1$ can be eigenfunctions of $\boldsymbol{S}^2$ and indeed common eigenfunctions of $\boldsymbol{S}^2$ and $\boldsymbol{S}_Z$. This argument leads to an algebraic equation that may be solved very easily, and one obtains

$$
\begin{aligned}
c_1 = c_2 = \frac{1}{\sqrt{2}} \quad & \Leftrightarrow \eta_1 = \frac{1}{\sqrt{2}}[\alpha(s_1)\beta(s_2) + \beta(s_1)\alpha(s_2)] \\
\boldsymbol{S}^2\eta_1 = 2\hbar^2\eta_1 \quad & \Leftrightarrow S = 1 \\
\boldsymbol{S}_Z\eta_1 = 0 \quad & \Leftrightarrow M_S = 0
\end{aligned} \tag{66.10}
$$

and

$$
\begin{aligned}
c_1 = -c_2 = \frac{1}{\sqrt{2}} \quad & \Leftrightarrow \eta_2 = \frac{1}{\sqrt{2}}[\alpha(s_1)\beta(s_2) - \beta(s_1)\alpha(s_2)] \\
\boldsymbol{S}^2\eta_2 = 0 \quad & \Leftrightarrow \mathrm{S} = 0 \\
\boldsymbol{S}_Z\eta_2 = 0 \quad & \Leftrightarrow M_S = 0.
\end{aligned} \tag{66.11}
$$

Alternatively, these coefficients can be determined just by the Clebsch-Gordan technique.

The two-electron spin states have been compiled in Table 11.1.

Table 11.1 Two-electron spin states.

State function Eigenket $\lvert S, M_S\rangle$ **expressed in the basis** $\lvert s_1, s_2; m_{s1}, m_{s2}\rangle$	**Eigenvalue of** $\boldsymbol{S}^2$	$\boldsymbol{S}_Z$	**Quantum Numbers** S	M_S
Singlet: *antisymmetric under spin interchange*				
$2^{-1/2}[\alpha(s_1)\beta(s_2) - \beta(s_1)\alpha(s_2)]$	0	0	0	0
$\lvert 0,0\rangle = \frac{1}{\sqrt{2}}[\lvert \frac{1}{2},\frac{1}{2};\frac{1}{2},-\frac{1}{2}\rangle - \lvert \frac{1}{2},\frac{1}{2};-\frac{1}{2},\frac{1}{2}\rangle]$				
Triplet: *symmetric under spin exchange*				
$\alpha(s_1)\alpha(s_2)$	$2\hbar^2$	$\hbar$	1	1
$\lvert 1,1\rangle = \lvert \frac{1}{2},\frac{1}{2};\frac{1}{2},\frac{1}{2}\rangle$				
$2^{-1/2}[\alpha(s_1)\beta(s_2) + \beta(s_1)\alpha(s_2)]$	$2\hbar^2$	0	1	0
$\lvert 1,0\rangle = \frac{1}{\sqrt{2}}[\lvert \frac{1}{2},\frac{1}{2};\frac{1}{2},-\frac{1}{2}\rangle + \lvert \frac{1}{2},\frac{1}{2};-\frac{1}{2},\frac{1}{2}\rangle]$				
$\beta(s_1)\,\beta(s_2)$	$2\hbar^2$	$-\hbar$	1	−1
$\lvert 1,-1\rangle = \lvert \frac{1}{2},\frac{1}{2};-\frac{1}{2},-\frac{1}{2}\rangle$				

Note that the singlet spin function is antisymmetric and the triplet function is symmetric under the interchange of the spin coordinates. The total wave function must be antisymmetric under the exchange of the space-spin coordinates of the two electrons. Thus the spatial part of the singlet state must be symmetric and that of the triplet states must be antisymmetric under the interchange of the spatial coordinates. This analysis leads to the interesting result that while nondegenerate spatial orbitals are involved in the bonding, the two-electron ground state must be a spin singlet as in the case of H_2. It corresponds to less nodes of the spatial wave function such that the energy is lower for the singlet. When degenerate orbitals are involved, the two-electron ground state can be either a singlet or even a spin triplet as it is found for the diatomic oxygen molecule. In such cases the ground state is often determined by the sign of the exchange integral over the degenerate orbitals, and a positive exchange integral stabilizes the triplet state.

The two-electron spinors are four-component column vectors since there are four possible orthonormal products. These have been shown against the ket vectors in Table 11.2.

Table 11.2 Two-electron spinors.

Basis Ket $\vert s_1, s_2; m_{s1}, m_{s2}>$	**Spinor**	**Eigenket**	**Eigenspinor**
$\vert \frac{1}{2}, \frac{1}{2}; \frac{1}{2}, \frac{1}{2} \rangle$	$\begin{pmatrix}1\\0\\0\\0\end{pmatrix}$	$\vert 1, 1 >$	$\begin{pmatrix}1\\0\\0\\0\end{pmatrix}$
$\vert \frac{1}{2}, \frac{1}{2}; \frac{1}{2}, -\frac{1}{2} \rangle$	$\begin{pmatrix}0\\1\\0\\0\end{pmatrix}$	$\vert 0, 0 >$	$2^{-1/2}\begin{pmatrix}0\\1\\-1\\0\end{pmatrix}$
$\vert \frac{1}{2}, \frac{1}{2}; -\frac{1}{2}, \frac{1}{2} \rangle$	$\begin{pmatrix}0\\0\\1\\0\end{pmatrix}$	$\vert 1, 0 >$	$2^{-1/2}\begin{pmatrix}0\\1\\1\\0\end{pmatrix}$
$\vert \frac{1}{2}, \frac{1}{2}; -\frac{1}{2}, -\frac{1}{2} \rangle$	$\begin{pmatrix}0\\0\\0\\1\end{pmatrix}$	$\vert 1, -1 >$	$\begin{pmatrix}0\\0\\0\\1\end{pmatrix}$

If we identify the unit operator as

$$\mathbf{1} = |S{=}0, M_s{=}0 >< 0, 0| + |1, 1 >< 1, 1| + |1, 0 >< 1, 0| + |1, -1 >< 1, -1| \qquad (66.12)$$

then the operator $\boldsymbol{S}_1 \bullet \boldsymbol{S}_2$ can be formulated as

$$\boldsymbol{S}_1 \bullet \boldsymbol{S}_2 = \frac{1}{2}(\boldsymbol{S}^2 - \boldsymbol{S}_1^2 - \boldsymbol{S}_2^2)\,\mathbf{1} = \frac{1}{2}(\boldsymbol{S}^2 - \frac{3}{2}\hbar^2)\,\mathbf{1} = \frac{1}{2}\boldsymbol{S}^2 - \frac{3}{4}\hbar^2\mathbf{1}.$$

The eigenvalue of $\frac{1}{2}\boldsymbol{S}^2$ is 0 for the singlet state and $\hbar^2$ for the triplet state. Therefore, the eigenvalue of $\hbar^{-2}\boldsymbol{S}_1 \bullet \boldsymbol{S}_2$ is $-\frac{3}{4}\mathbf{1}$ for the singlet state and $\frac{1}{4}\mathbf{1}$ for the triplet state. This allows us to write the projection operator for the singlet spin state as

$$\mathbf{P}_0 = \frac{1}{4}\mathbf{1} - \frac{1}{\hbar^2}\boldsymbol{S}_1 \bullet \boldsymbol{S}_2 \equiv |0, 0 >< 0, 0| \qquad (66.13)$$

since it destroys the triplet state and leaves the singlet state unchanged while operating on it. Similarly

$$\mathbf{P}_1 = \frac{3}{4}\mathbf{1} + \frac{1}{\hbar^2}\,\boldsymbol{S}_1\boldsymbol{\cdot}\boldsymbol{S}_2 \equiv \sum_{M_S=-1}^{1} |1, M_S><1, M_S| \qquad (66.14)$$

is the projection operator for the triplet spin state. The sum of $\mathbf{P}_0$ and $\mathbf{P}_1$ is the unit operator **1**.

Lecture 67: Angular Momentum for More Than One Electron in a Centrally Symmetric Field

In Lecture 65 we found for one electron moving in a central field

$$[\boldsymbol{L}^2, \boldsymbol{J}^2] = 0 \text{ and } [\boldsymbol{S}^2, \boldsymbol{J}^2] = 0$$

but

$$[L_z, \boldsymbol{J}^2] \neq 0 \text{ and } [S_z, \boldsymbol{J}^2] \neq 0 \text{ although } [L_z + S_z, \boldsymbol{J}^2] = 0.$$

This is why we could choose common eigenfunctions *either* for the operators $\boldsymbol{L}^2$, $\boldsymbol{S}^2$, L_z, S_z and J_z *or* for $\boldsymbol{L}^2$, $\boldsymbol{S}^2$, $\boldsymbol{J}^2$ and J_z. In fact the second choice was made in Lecture 65, but the situation radically changes for more than one electron.

The total angular momenta for a multiple number of electrons are given by the vector sums of the corresponding operators for individual electrons. Thus the total orbital angular momentum, the total spin angular momentum and the total angular momentum operators are

$$\boldsymbol{L} = \boldsymbol{L}_1 + \boldsymbol{L}_2 + \ldots,$$
$$\boldsymbol{S} = \boldsymbol{S}_1 + \boldsymbol{S}_2 + \ldots,$$

and (67.1)

$$\boldsymbol{J} = \boldsymbol{J}_1 + \boldsymbol{J}_2 + \ldots = \boldsymbol{L} + \boldsymbol{S}.$$

These definitions, the vector sums, lead to the expressions

$$\boldsymbol{L}^2 = (\boldsymbol{L}_1^2 + \boldsymbol{L}_2^2 + \ldots) + 2(L_{1z}L_{2z} + L_{1z}L_{3z} + \ldots) + (L_{1+}L_{2-} + L_{1-}L_{2+} + \ldots),$$

$$L_z = L_{1z} + L_{2z} + \ldots,$$

$$\boldsymbol{S}^2 = (\boldsymbol{S}_1^2 + \boldsymbol{S}_2^2 + \ldots) + 2(S_{1z}S_{2z} + S_{1z}S_{3z} + \ldots) + (S_{1+}S_{2-} + S_{1-}S_{2+} + \ldots),$$

$$S_z = S_{1z} + S_{2z} + \ldots,$$

$$\boldsymbol{J}^2 = \boldsymbol{L}^2 + \boldsymbol{S}^2 + 2L_zS_z + (L_{1+}S_{2-} + L_{1-}S_{2+})$$
$$= (\boldsymbol{J}_1^2 + \boldsymbol{J}_2^2 + \ldots) + 2(J_{1z}J_{2z} + J_{1z}J_{3z} + \ldots) + (J_{1+}J_{2-} + J_{1-}J_{2+} + \ldots),$$

and (67.2)

$$J_z = L_z + S_z = J_{1z} + J_{2z} + \ldots$$

The following commutations hold:

i) $[\boldsymbol{L}^2, \boldsymbol{L}_1^2 + \boldsymbol{L}_2^2 + \ldots] = 0.$

ii) $[\boldsymbol{L}^2, L_z] = 0$ but $[\boldsymbol{L}^2, L_{1z}] \neq 0$ etc.

iii) $[\boldsymbol{L}_1^2 + \boldsymbol{L}_2^2 + \ldots, L_z] = 0.$

iv) $[\boldsymbol{S}^2, \boldsymbol{S}_1^2 + \boldsymbol{S}_2^2 + \ldots] = 0.$ (67.3a)

v) $[S^2, S_z] = 0$ but $[S^2, S_{1z}] \neq 0$ etc.

vi) $[S_1^2 + S_2^2 + ..., S_z] = 0$.

Also,

vii) Every orbital angular momentum operator commutes with all spin angular momentum operators.

viii) $[J^2, J_1^2 + J_2^2 + ...] = 0$.

ix) $[J^2, J_z] = 0$ but $[J^2, J_{1z}] \neq 0$ etc. (67.3b)

x) $[J_1^2 + J_2^2 + ..., J_z] = 0$.

xi) $[J^2, L^2 + S^2] = 0$.

xii) $[J^2, L_z] = -[J^2, S_z] \neq 0$.

Therefore, one may seek common eigenfunctions of the folowing groups of operators:

I. $\sum_{i=1}^{N} L_i^2, L_z, \sum_{i=1}^{N} S_i^2, S_z,$ (and J_z);

II. $L^2, L_z, \sum_{i=1}^{N} S_i^2, S_z,$ (and J_z);

III. $L^2, L_z, S^2, S_z,$ (and J_z);

IV. $L^2, S^2, J^2, J_z,$ (and $\sum_{i=1}^{N} J_i^2$).

Group I eigenfunctions are the easiest to prepare. The formation of Group II functions is easier than the formation of the common eigenfunctions of the operators belonging to Group III. Preparation of the Group IV eigenfunctions is the most difficult task. For physical reasons we are interested in Group III and Group IV eigenfunctions. These two groups of functions are generated by following the *L-S* (or *Russell-Saunders*) *coupling scheme* and the *j-j coupling scheme* respectively.

The Hamiltonian for a many-electron atom can be written as

$$\mathbf{H} = \mathbf{H}^{(1)} + \mathbf{H}^{(2)} + \mathbf{H}_{s.o.}. \tag{67.4}$$

The operator $\mathbf{H}^{(1)}$ is the sum of the one-electron Hamiltonian operators

$$\mathbf{H}^{(1)} = -\frac{1}{2}\sum_{i=1}^{N} \frac{1}{r_i^2}(\mathbf{D}_i - L_i^2) - \sum_{i=1}^{N} \frac{Z}{r_i}. \tag{67.5}$$

It commutes with $\sum_{i=1}^{N} L_i^2$ as well as L^2, and also with L_z, S^2, $\sum_{i=1}^{N} S_i^2$, S_z, J^2 and J_z.

The operator $\mathbf{H}^{(2)}$ consists of all distinct instantaneous Coulomb interactions among the electrons,

$$\mathbf{H}^{(2)} = \frac{1}{2}\sum_{i\neq j} \frac{1}{|\mathbf{r}_i - \mathbf{r}_j|}. \tag{67.6}$$

It is possible to show that $\mathbf{H}^{(2)}$ commutes with L^2 and L_z but not with the individual *z*-components L_{iz} for $i = 1, 2, ..., N$.

The operator $\mathbf{H}_{s.o.}$ can be approximately described by the spin-orbit interaction

$$\mathbf{H}_{s.o.} = \sum_{i=1}^{N} A_i \boldsymbol{L}_i \bullet \boldsymbol{S}_i. \tag{67.7}$$

This interaction is actually an outcome of the relativistic nature of the dynamics of electrons. The term A_i is proportional to $m^{-2}c^{-2} <r^{-3}>_i$ when the electron occupies the *ith* orbital. For the low-Z atoms the A_i's are small enough and the effect of $\mathbf{H}_{s.o.}$ can be evaluated by first-order perturbation theory while the zeroth-order functions, the eigenfunctions of $\mathbf{H}^{(1)} + \mathbf{H}^{(2)}$, are chosen to be common eigenfunctions of $\boldsymbol{L}^2$, $\boldsymbol{L}_z$, $\boldsymbol{S}^2$, $\boldsymbol{S}_z$ and $\boldsymbol{J}_z$. This is the L-S or Russell-Saunders scheme. But neither $\boldsymbol{L}_z$ nor $\boldsymbol{S}_z$ commutes with $\mathbf{H}_{s.o.}$ although $\boldsymbol{J}_z$ commutes with this operator:

$$[\boldsymbol{L}_z, \mathbf{H}_{s.o.}] = i\hbar \sum_{i=1}^{N} A_i(\boldsymbol{L}_{iy}\boldsymbol{S}_{ix} - \boldsymbol{L}_{ix}\boldsymbol{S}_{iy}) = -[\boldsymbol{S}_z, \mathbf{H}_{s.o.}]. \tag{67.8}$$

Therefore, for small A_i's (for a low-Z atom)

$$[\boldsymbol{L}_z, \mathbf{H}] \approx 0$$

and (67.9)

$$[\boldsymbol{S}_z, \mathbf{H}] \approx 0,$$

that is, M_L and M_S are approximately good quantum numbers. It is usual in Russell-Saunders coupling scheme to treat $\mathbf{H}_{s.o.}$ through first order in perturbation theory and so for qualitative discussions the effect of the spin-orbit coupling can be approximately evaluated by using $\mathbf{H}_{s.o.}'$ instead of $\mathbf{H}_{s.o.}$ where the primed operator is of the form

$$\mathbf{H}_{s.o.}' = A\boldsymbol{L} \bullet \boldsymbol{S}. \tag{67.10}$$

Since $\boldsymbol{J}^2$ commutes with $\mathbf{H}^{(1)}$ and $\mathbf{H}^{(2)}$ as well as with $\mathbf{H}_{s.o.}$ and $\mathbf{H}_{s.o.}'$, the quantum number J may be used to characterize the state. There is a (2J + 1)-fold degeneracy in the absence of an external electric or magnetic field. In the L-S scheme, the energy levels are characterized by the quantum numbers L, S and J. They are represented by the *term symbols*

$$^{2S+1}L_J. \tag{67.11}$$

It is vastly easier to write down the common eigenfunctions of the operators belonging to Group I. For closed shells, these functions are the single determinants which can be written in terms of the orthonormal spin orbitals $\phi\alpha$'s and $\phi\beta$'s as

$$\Psi(\mathbf{x}_1, \mathbf{x}_2, \dots, \mathbf{x}_N) = (N!)^{-1/2}\, \mathrm{Det}\{\phi_a(\mathbf{r}_1)\alpha(s_1)\ \phi_a(\mathbf{r}_2)\beta(s_2) \dots \phi_v(\mathbf{r}_N)\beta(s_N)\}. \tag{67.12}$$

One can easily establish that the following results hold:

$$\boldsymbol{L}^2 \{Y_{l,-l}(\theta_1, \varphi_1)\, Y_{l,-l+1}(\theta_2, \varphi_2) \dots Y_{l,l}(\theta_{2l+1}, \varphi_{2l+1})\} = 0,$$

$$\boldsymbol{L}_z \{Y_{l,-l}(\theta_1, \varphi_1)\, Y_{l,-l+1}(\theta_2, \varphi_2) \dots Y_{l,l}(\theta_{2l+1}, \varphi_{2l+1})\} = 0,$$

$$\boldsymbol{S}^2 \{\alpha(s_1)\beta(s_2) \dots \beta(s_N)\} = 0,$$

and (67.13)

$$\boldsymbol{S}_z \{\alpha(s_1)\beta(s_2) \dots \beta(s_N)\} = 0.$$

The implication is that an atomic closed shell has zero angular momentum: $L = 0$, $S = 0$ and $J = 0$. Thus any non-vanishing angular momentum is contributed by the open shells. In other words, the term is determined by the open shells and, in the case of ground state atoms, by the valence shell. Ground state term symbols for the very-low-Z atoms are given in Table 11.3.

Table 11.3 Ground states of low-Z atoms.

Atom	Configuration	L	S	J	Terms
H	1s	0	½	½	$^2S_{1/2}$
He	$(1s^2)$	0	0	0	$^1S_0$
Li	$(1s^2)2s$	0	½	½	$^2S_{1/2}$
Be	$(1s^22s^2)$	0	0	0	$^1S_0$
B	$(1s^22s^2)2p$	1	½	½, 3/2	$^2P_{1/2}$ (g), $^2P_{3/2}$
C	$(1s^22s^2)2p^2$	0	0	0	$^1S_0$
		1	1	0, 1, 2	$^3P_0$ (g), $^3P_1$, $^3P_2$
		2	0	2	$^1D_2$
N	$(1s^22s^2)2p^3$	0	3/2	3/2	$^4S_{3/2}$ (g)
		1	½	½, 3/2	$^2P_{1/2}$, $^2P_{3/2}$
		2	½	3/2, 5/2	
O	$(1s^22s^2)2p^4$	0	0	0	$^1S_0$
		1	1	0, 1, 2	$^3P_0$, $^3P_1$, $^3P_2$ (g)
		2	0	2	$^1D_2$
F	$(1s^22s^2)2p^5$	1	½	½, 3/2	$^2P_{1/2}$, $^2P_{3/2}$ (g)
Ne	$(1s^22s^22p^6)$	0	0	0	$^1S_0$

The *j-j* coupling scheme becomes important for high-Z atoms which must be relativistically treated. The motion of the valence electrons in neutral atoms is essentially in the non-relativistic limit. So for a large nuclear charge ($Z > 50$), the neutral atom ground state is still given in the *L-S* notation although the overall electronic structure has to be determined by the *j-j* coupling scheme. Relativistic effects like mass-velocity corrections, Darwin interaction and spin-orbit interaction leave their mark on the energy and the probability distribution of the core electrons and this feature in turn influences the energy and the probability distribution of the valence electrons. X-ray spectroscopy involves excitations from the core levels of an atom to one of the levels unoccupied in the ground state configuration. The corresponding excitations must be described in the *j-j* scheme whenever the atom has a medium or larger nuclear charge ($Z \geq 50$).

Texts Suggested for Further Reading

S. N. Datta, C. O. Trindle and F. Illas, "Theoretical and Computational Aspects of Magnetic Organic Molecules", (Imperial College Press, London, 2014).

EXERCISES

1. For $\boldsymbol{M} = \boldsymbol{M}_1 + \boldsymbol{M}_2$, prove the range for Λ, $|\Lambda_1 - \Lambda_2| \leq \Lambda \leq \Lambda_1 + \Lambda_2$.
2. Why are the common eigenkets of $\boldsymbol{M}^2$ and $\boldsymbol{M}_z$ selected as $|\Lambda_1, \Lambda_2; \Lambda, m\rangle$ and not as $|\Lambda_1, \Lambda_2; m_1, m_2\rangle$?
3. Verify that the matrix representation for $\boldsymbol{S}^2$ as given in Eq. (63.9) indeed satisfies the eigenvalue relations (63.2).
4. Verify the commutation relationship $[\boldsymbol{S}_i, \boldsymbol{S}_j] = i\hbar\varepsilon_{ijk}\boldsymbol{S}_k$ in matrix representation.
5. Find the effect of $\boldsymbol{L.S}$ operating on $|l, s; j, m\rangle$.
6. What are Clebsch-Gordan coefficients? Derive two properties of the Clebsch-Gordan coefficients and simplify the expression for $|\Lambda_1, \Lambda_2; \Lambda, m\rangle$.
7. Compile the rules for determining the Clebsch-Gordan coefficients.
8. Prove the four equations (64.24)-(64.27).
9. Use the same equations to prove the relation (65.10).
10. Write down the formal expressions for two-electron spin projection operators.
11. Derive analytical expressions for two-electron spin projection operators.
12. Discuss the applicability of L-S and j-j coupling schemes.

PROBLEMS

1. What are the wave functions and kets corresponding to the spinors

$$\begin{pmatrix} 1/\sqrt{2} \\ 1/\sqrt{2} \end{pmatrix}; \quad \begin{pmatrix} \phi(\mathbf{r})Y_{11} \\ 0 \end{pmatrix}; \quad \frac{1}{\sqrt{3}}\begin{pmatrix} \sqrt{2}\phi(\mathbf{r})Y_{10} \\ \phi(\mathbf{r})Y_{11} \end{pmatrix}; \quad \begin{pmatrix} 1/2 \\ 1/2 \\ 1/2 \\ 1/2 \end{pmatrix}.$$

2. Show that $\boldsymbol{H}^{(2)}$ in Eq. (67.6) commutes with $\boldsymbol{L}^2$ and $\boldsymbol{L}_z$ but not with $\boldsymbol{L}_{iz}$.

   Hint: Use $\frac{1}{r_{12}}$ and $\boldsymbol{L} = \boldsymbol{L}_1 + \boldsymbol{L}_2$ only. Show that $[\frac{1}{r_{12}}, \boldsymbol{L}_{1z}] = -[\frac{1}{r_{12}}, \boldsymbol{L}_{2z}] \neq 0$.
3. Construct the spherical harmonics for $l = 4$.
4. Calculate the $|lsjm\rangle$ states for a p electron ($l = 1$). Derive the spinor notations for these states.
5. What is the use of the kets $|\Lambda_1, \Lambda_2; m_1, m_2\rangle$? Are these common eigenkets of some operators?
6. Determine the Clebsch-Gordan coefficients when $\boldsymbol{M}_1$ is $\boldsymbol{L}$ and $\boldsymbol{M}_2$ is $\boldsymbol{S}$.

CHAPTER 12

EXPLICIT TREATMENT OF MANY-ELECTRON ATOMS

Lectures 68–70: Explicit Derivations of Hartree and Hartree-Fock Equations

Lecture 68: General Aspects

In these lectures we will explicitly derive Hartree and Hartree-Fock equations in the general way so that these may be equally applicable to many-electron atoms and many-electron molecules. The electronic Hamiltonian is written as

$$\mathbf{H}_{\mathrm{el}} = \sum_{i=1}^{N} \mathbf{h}(i) + \frac{1}{2}\sum_{i\neq j}^{N}\frac{1}{r_{ij}} \tag{68.1}$$

where

$$\mathbf{h}(i) = -\frac{1}{2}\nabla_i^2 + V_{ext}(\mathbf{r}_i) \tag{68.2}$$

in atomic units. For simplicity we leave out the spin-orbit and other interactions. The external scalar potential is

$$V_{ext}(\mathbf{r}_i) = -\sum_{v=1}^{n}\frac{Z_v}{|\mathbf{r}_i - \mathbf{R}_v|} \tag{68.3}$$

where the sum is over all the nuclei. The total number of nuclei is n, and $n = 1$ for atoms. In the atomic case there is only one nucleus that is conventionally placed at the origin. This is merely a matter of convenience: the results should remain unaltered wherever the nucleus might be placed.

Trial Functions

Hartree (1928, 1938) treated the trial function as a single product of one-electron functions,

$$\Psi_{\mathrm{H}}(\mathbf{r}_1, \mathbf{r}_2, \ldots, \mathbf{r}_N) = \phi_{\mathrm{p}}(\mathbf{r}_1)\phi_{\mathrm{q}}(\mathbf{r}_2)\ldots\phi_{\mathrm{t}}(\mathbf{r}_N), \tag{68.4}$$

and found the best wave function of this form while each spatial orbital is filled with at most two electrons. The term "best" implies the total energy to be stationary, and the normal expectation is that the stationary point is also a minimum. The minimum point is often realized. This happens because of the way the basis functions ϕ's are chosen, namely, that they are representative of the lowest lying orbitals of the lowest-energy configuration.

Fock (1930) modified this treatment by including the indistinguishability of electrons in it. So in his treatment the trial function is a single anti-symmetrized product or a determinant of the form

$$\Psi_{HF}(\mathbf{x}_1, \mathbf{x}_2, \ldots, \mathbf{x}_N) = (N!)^{-1/2} \|\psi_p(\mathbf{x}_1)\psi_q(\mathbf{x}_2)\ldots\psi_t(\mathbf{x}_N)\| \tag{68.5}$$

where $\mathbf{x}$ represents both space and spin coordinates and ψ is a spin orbital basis function. An important point may be noted here: in a rigorous sense Fock's wave function too corresponds to a closed-shell species.

The open-shell wave function must be an eigenfunction of $\boldsymbol{L}^2$, $\boldsymbol{S}^2$, $\boldsymbol{L}_Z$, $\boldsymbol{S}_Z$, $\boldsymbol{J}^2$ and $\boldsymbol{J}_Z$, and hence it is generally given by a linear combination of a few almost degenerate Slater determinants. The treatment of open-shell species was established by Roothaan (1960).

Pople and Nesbet (1954) developed an alternative treatment for the open-shells, which account for a part of the spin-orbit coupling although the resulting many-electron wave function is not strictly an eigenfunction of $\boldsymbol{S}^2$. In this development the wave function for an open-shell species is basically written in the form of Eq. (68.5), the space-spin separation is invoked yet one allows the one-electron spatial functions for the two different spin states to be different.

Energy expectation value

We start with the calculation of the expectation values of a one-electron operator $\theta^{(1)}$ and a two-electron operator $\theta^{(2)}$ when the many-electron wave function is given by a single product, say, by Ψ_H in Eq. (68.4). By separating integrals as factors we get

$$\begin{aligned}\langle\theta^{(1)}(\mathbf{r}_i)\rangle_{\Psi_H} &= \int dv_1 \int dv_2 \ldots \int dv_N\, \Psi_H^*(\mathbf{r}_1, \ldots, \mathbf{r}_N)\, \theta^{(1)}(\mathbf{r}_i)\, \Psi_H(\mathbf{r}_1, \ldots, \mathbf{r}_N)\\ &= \int dv_1\, \phi_p^*(\mathbf{r}_1)\phi_p(\mathbf{r}_1) \int dv_2\, \phi_q^*(\mathbf{r}_2)\phi_q(\mathbf{r}_2) \ldots \int dv_i\, \phi_s^*(\mathbf{r}_i)\, \theta^{(1)}(\mathbf{r}_i)\, \phi_s(\mathbf{r}_i)\\ &\quad \ldots \int dv_N\, \phi_t^*(\mathbf{r}_N)\phi_t(\mathbf{r}_N); \end{aligned}\tag{68.6}$$

$$\begin{aligned}\langle\theta^{(2)}(\mathbf{r}_i, \mathbf{r}_j)\rangle_{\Psi_H} &= \int dv_1 \int dv_2 \ldots \int dv_N\, \Psi_H^*(\mathbf{r}_1, \ldots, \mathbf{r}_N)\, \theta^{(2)}(\mathbf{r}_i, \mathbf{r}_j)\, \Psi_H(\mathbf{r}_1, \ldots, \mathbf{r}_N)\\ &= \int dv_1\, \phi_p^*(\mathbf{r}_1)\phi_p(\mathbf{r}_1) \int dv_2\, \phi_q^*(\mathbf{r}_2)\phi_q(\mathbf{r}_2) \ldots\\ &\quad \int dv_i \int dv_j\, \phi_s^*(\mathbf{r}_i)\phi_{s'}^*(\mathbf{r}_j)\, \theta^{(2)}(\mathbf{r}_i, \mathbf{r}_j)\, \phi_s(\mathbf{r}_i)\phi_{s'}(\mathbf{r}_j) \ldots \int dv_N\, \phi_t^*(\mathbf{r}_N)\phi_t(\mathbf{r}_N).\end{aligned}$$

Since the one-electron functions are normalized,

$$\langle\theta^{(1)}(\mathbf{r}_i)\rangle_{\Psi_H} = \int dv_i\, \phi_s^*(\mathbf{r}_i)\, \theta^{(1)}(\mathbf{r}_i)\, \phi_s(\mathbf{r}_i) = \langle\theta^{(1)}\rangle_{\phi_S}$$

and (68.7)

$$\langle\theta^{(2)}(\mathbf{r}_i, \mathbf{r}_j)\rangle_{\Psi_H} = \int dv_i \int dv_j\, \phi_s^*(\mathbf{r}_i)\phi_{s'}^*(\mathbf{r}_j)\, \theta^{(2)}(\mathbf{r}_i, \mathbf{r}_j)\, \phi_s(\mathbf{r}_i)\phi_{s'}(\mathbf{r}_j) = \langle\theta^{(2)}\rangle_{\phi_S,\phi_{S'}}.$$

These values are independent of the coordinates $\mathbf{r}_i$ and $\mathbf{r}_j$, but they depend on the functions ϕ_s and $\phi_{s'}$. Similar results are obtained when the function Ψ_H is replaced by the product of spin-orbitals

$$\Phi(\mathbf{x}_1, \mathbf{x}_2, \ldots, \mathbf{x}_N) = \psi_p(\mathbf{x}_1)\psi_q(\mathbf{x}_2)\ldots\psi_t(\mathbf{x}_N). \tag{68.8}$$

The results are

$$\langle \theta^{(1)}(\mathbf{r}_i)\rangle_\Phi = \theta^{(1)}_{s,s} \equiv \int dx\, \psi_s^*(\mathbf{x})\theta^{(1)}(\mathbf{r})\psi_s(\mathbf{x})$$

and (68.9)

$$\langle \theta^{(2)}(\mathbf{r}_i, \mathbf{r}_j)\rangle_\Phi = \theta^{(2)}_{s,s'} \equiv \int dx \int dx'\, \psi_s^*(\mathbf{x})\, \psi_{s'}^*(\mathbf{x}')\, \theta^{(2)}(\mathbf{r},\mathbf{r}')\, \psi_s(\mathbf{x})\, \psi_{s'}(\mathbf{x}').$$

Equations (68.7) are used to calculate the expectation value of energy with the wave function Ψ_H. The one-electron part is evaluated as

$$\langle \sum_{i=1}^{N} \mathbf{h}(i)\rangle_{\Psi_H} = h_{pp} + h_{qq} + \ldots + h_{tt} = \sum_s n_s h_{ss} \tag{68.10}$$

where s is an orbital index and n_s is the number of electrons in orbital ϕ_s. Since n_s equals 0, 1 or 2, the sum is necessarily over the occupied orbitals. The term h_{ss} is the integral

$$h_{ss} = \int dv\, \phi_s^*(\mathbf{r})\, \mathbf{h}(\mathbf{r})\, \phi_s(\mathbf{r}). \tag{68.11}$$

The two-electron contribution is obtained in a similar way. One writes

$$\frac{1}{2}\langle \sum_{i\neq}\sum_j \frac{1}{r_{ij}}\rangle_{\Psi_H} = (J_{pq} + \ldots + J_{ps} + \ldots + J_{pt}) + (J_{qp} + \ldots + J_{qs} + \ldots + J_{qt})$$

$$+ \ldots + (J_{tp} + J_{tq} + \ldots + J_{ts} + \ldots)$$

$$= \sum_s \delta_{n_s,2} J_{ss} + \frac{1}{2}\sum_{s\neq s'} n_s n_{s'} J_{ss'} \tag{68.12}$$

where $J_{ss'}$ is the direct coulomb repulsion integral

$$J_{ss'} = \int dv_1 \int dv_2\, \phi_s^*(\mathbf{r}_1)\phi_{s'}^*(\mathbf{r}_2)\frac{1}{r_{12}}\phi_s(\mathbf{r}_1)\phi_{s'}(\mathbf{r}_2). \tag{68.13}$$

Thus for a closed-shell species with two electrons in each of the occupied orbitals ϕ_1, ϕ_2, ... , $\phi_{N/2}$, the energy expectation value is

$$E_{\Psi_H} = 2\sum_{s=1}^{N/2} h_{ss} + \sum_{s=1}^{N/2} J_{ss} + 2\sum_{s\neq s'}^{N/2} J_{ss'}. \tag{68.14}$$

Equations (68.9) are to be used in evaluating the energy expectation value when the trial function is the Slater determinant Ψ_{HF}. Complications arise from the fact that Ψ_{HF} is in fact a linear combination of $N!$ product functions of the type Φ given in (68.8). We first write the determinant as

$$\Psi_{HF} = (N!)^{-1/2}\sum_{\mathbf{P}} (-1)^P\, \mathbf{P}\Phi(\mathbf{x}_1, \mathbf{x}_2, \ldots, \mathbf{x}_N) \tag{68.15}$$

where the operator **P** causes all distinct permutations of the coordinates of electrons, and P is the number of pair permutations required to achieve the result of the application of **P** on Φ. Next we use the orthogonality of the one-electron functions in a direct way. For instance,

$$\int dx_1 \ldots \int dx_N \ \psi_p{}^*(\mathbf{x}_1)\ldots\psi_s{}^*(\mathbf{x}_i)\ldots\psi_t{}^*(\mathbf{x}_N)\ \theta^{(1)}(\mathbf{r}_i)\ \psi_{p'}(\mathbf{x}_1)\ldots\psi_{s'}(\mathbf{x}_i)\ldots\psi_{t'}(\mathbf{x}_N)$$

$$= \left(\int dx_1\ \psi_p{}^*(\mathbf{x}_1)\psi_{p'}(\mathbf{x}_1)\right)\left(\int dx_i\ \psi_s{}^*(\mathbf{x}_i)\theta^{(1)}(\mathbf{r}_i)\psi_{s'}(\mathbf{x}_i)\right)\left(\int dx_N\ \psi_t{}^*(\mathbf{x}_N)\psi_{t'}(\mathbf{x}_N)\right)$$

$$= \delta_{p,p'}\delta_{q,q'}\ldots\theta^{(1)}_{s,s'}\ldots\delta_{t,t'}$$

$$= \delta_{p,p'}\delta_{q,q'}\delta_{s,s'}\ldots\delta_{t,t'}\ \theta^{(1)}_{s,s}. \tag{68.16}$$

Hence

$$\int dx_1 \ldots \int dx_N\ \Psi_{HF}{}^*(\mathbf{x}_1, \ldots, \mathbf{x}_N)\ \boldsymbol{\theta}^{(1)}(\mathbf{r}_i)\ \Psi_{HF}(\mathbf{x}_1, \ldots, \mathbf{x}_N)$$

$$= (N!)^{-1} \sum_{\mathbf{P},\mathbf{P}'} (-1)^{P+P'} \int dx_1 \ldots \int dx_N\ [\mathbf{P}\Phi(\mathbf{x}_1, \ldots, \mathbf{x}_N)]^*\ \boldsymbol{\theta}^{(1)}(\mathbf{r}_i)\ [\mathbf{P}'\Phi(\mathbf{x}_1, \ldots, \mathbf{x}_N)]$$

$$= (N!)^{-1} \sum_{\mathbf{P},\mathbf{P}'} \delta_{\mathrm{P,P'}} (-1)^{P+P'} \int dx_1 \ldots \int dx_N\ [\mathbf{P}\Phi(\mathbf{x}_1, \ldots, \mathbf{x}_N)]^*\ \boldsymbol{\theta}^{(1)}(\mathbf{r}_i)\ [\mathbf{P}'\Phi(\mathbf{x}_1, \ldots, \mathbf{x}_N)]$$

$$= (N!)^{-1} \sum_{\mathbf{P}} \int dx_1 \ldots \int dx_N\ [\mathbf{P}\Phi(\mathbf{x}_1, \ldots, \mathbf{x}_N)]^*\ \boldsymbol{\theta}^{(1)}(\mathbf{r}_i)\ [\mathbf{P}\Phi(\mathbf{x}_1, \ldots, \mathbf{x}_N)]. \tag{68.17}$$

Suppose that the function $\mathbf{P}\Phi(\mathbf{x}_1, \ldots, \mathbf{x}_N)$ identifies the *ith* electron is in the spin-orbital ψ_u. Then the integral in the last line of equations (68.17) yields the result $\theta^{(1)}_{u,u}$. Now the number of distinct permutations which place the *ith* electron in ψ_u is $(N-1)!$. This is so because there are $(N-1)!$ ways in which the rest $(N-1)$ electrons can be distributed in the rest $(N-1)$ spin orbitals. Hence the number of times we get $\theta^{(1)}_{u,u}$ in the sum of Eq. (68.17) is $(N-1)!$. Similarly, $\theta^{(1)}_{v,v}$ is obtained $(N-1)!$ times, and so on. This leads to the result

$$\langle \boldsymbol{\theta}^{(1)}(\mathbf{r}_i)\rangle_{\Psi_{HF}} = (N!)^{-1}\ (N-1)! \sum_{s}^{occ} \theta^{(1)}_{s,s} = N^{-1} \sum_{s}^{occ} \theta^{(1)}_{s,s} \tag{68.18}$$

where the sum is over all the occupied spin-orbitals. This result is obtained for each electron i. Since the result is independent of the electron index, we have

$$\left\langle \sum_{i=1}^{N} \theta^{(1)}(\mathbf{r}_i)\right\rangle_{\Psi_{HF}} = \sum_{s}^{occ} \theta^{(1)}_{s,s} \tag{68.19}$$

and in particular,

$$\left\langle \sum_{i=1}^{N} \mathbf{h}(\mathbf{r}_i)\right\rangle_{\Psi_{HF}} = \sum_{s}^{occ} h_{ss}. \tag{68.20}$$

The calculation of the expectation value of a two-electron operator is similar but much more tedious. Also, one obtains two possible integrals, direct and exchange, from this calculation:

$$\int dx_1 \ldots \int dx_N\ \psi_p^*(\mathbf{x}_1)\ldots\psi_s^*(\mathbf{x}_i)\ldots\psi_{s''}^*(\mathbf{x}_j)\ldots\psi_t^*(\mathbf{x}_N)\ \theta^{(2)}(\mathbf{r}_i, \mathbf{r}_j)$$

$$\times\ \psi_{p'}(\mathbf{x}_1)\ldots\psi_{s'}(\mathbf{x}_i)\ldots\psi_{s'''}(\mathbf{x}_j)\ldots\psi_{t'}(\mathbf{x}_N)$$

$$= \delta_{p,p'}\delta_{q,q'}\ldots\delta_{t,t'}\ [\delta_{s,s'}\delta_{s'',s'''}\int dx_i \int dx_j\ \psi_s^*(\mathbf{x}_i)\psi_{s''}^*(\mathbf{x}_j)\ \theta^{(2)}(\mathbf{r}_i, \mathbf{r}_j)\ \psi_{s'}(\mathbf{x}_i)\psi_{s'''}(\mathbf{x}_j)$$

$$+\ \delta_{s,s'''}\delta_{s'',s'}\int dx_i \int dx_j\ \psi_s^*(\mathbf{x}_i)\psi_{s''}^*(\mathbf{x}_j)\ \theta^{(2)}(\mathbf{r}_i, \mathbf{r}_j)\ \psi_{s'}(\mathbf{x}_i)\psi_{s'''}(\mathbf{x}_j)].$$

$$= \delta_{p,p'}\delta_{q,q'}\ldots\delta_{t,t'}\ [\delta_{s,s'}\delta_{s'',s'''}(ss'|s''s''') + \delta_{s,s'''}\delta_{s'',s'}(ss'|s''s''')]. \tag{68.21}$$

We have used the notation

$$(ab|cd) = \int dx_1 \int dx_2\ \psi_a^*(\mathbf{x}_1)\psi_c^*(\mathbf{x}_2)\ \theta^{(2)}(\mathbf{r}_1, \mathbf{r}_2)\ \psi_b(\mathbf{x}_1)\psi_d(\mathbf{x}_2). \tag{68.22}$$

After a somewhat lengthy calculation one obtains

$$\langle\theta^{(2)}(\mathbf{r}_i, \mathbf{r}_j)\rangle_{\Psi_{HF}} = \int dx_1 \ldots \int dx_N\ \Psi_{HF}(\mathbf{x}_1, \ldots, \mathbf{x}_N)\ \theta^{(2)}(\mathbf{r}_i, \mathbf{r}_j)\ \Psi_{HF}(\mathbf{x}_1, \ldots, \mathbf{x}_N)$$

$$= (N!)^{-1}(N-2)! \sum_{s\neq s'}^{occ} [(ss|s's') - (ss'|s's)]$$

$$= \frac{1}{N(N-1)} \sum_{s\neq s'}^{occ} [(ss|s's') - (ss'|s's)]. \tag{68.23}$$

The minus sign arises with the exchange integral (ss′|s′s). The two product functions involved in this integration differ by a single pair-permutation which results in the factor $(-1)^{P+P'} = -1$. The result in Eq. (68.23) is independent of the electron indices i and j. Hence the same result is obtained from each two-particle interaction, and the average value of the sum of these interaction terms is given by

$$\langle\frac{1}{2}\sum_{i\neq j}\ \theta^{(2)}(\mathbf{r}_i, \mathbf{r}_j)\rangle_{\Psi_{HF}} = \frac{1}{2}\sum_{s\neq s'}\ [(ss|s's') - (ss'|s's)]. \tag{68.24}$$

In the special case of coulomb interaction the direct integral (ss|s's') will be written as $J_{ss'}$ and the exchange integral (ss'|s's) will be denoted by $K_{ss'}$ so that

$$\langle\frac{1}{2}\sum_{i\neq j}\frac{1}{r_{ij}}\rangle = \frac{1}{2}\sum_{s\neq s'}\ (J_{ss'} - K_{ss'}). \tag{68.25}$$

Thus the energy of a single determinant wave function is

$$E_{\Psi_{HF}} = \sum_{s}^{occ} h_{ss} + \frac{1}{2}\sum_{s\neq s'}\ (J_{ss'} - K_{ss'}). \tag{68.26}$$

The indices s and s′ vary from 1 to N when Ψ_{HF} represents the ground state configuration.

Lecture 69: Aspects of Variation

Space-spin separation

When the dependences on the space and spin coordinates are separable, the spin orbital is the product of a spatial function and the spin function. In the restricted formalism, one writes the spin orbitals of two types of spin in terms of the same set of spatial functions,

$$\psi_{2u-1}(\mathbf{x}) = \phi_u(\mathbf{r})\alpha(s),$$

$$\psi_{2u}(\mathbf{x}) = \phi_u(\mathbf{r})\beta(s). \tag{69.1}$$

The total energy of the ground state configuration can be calculated from Eq. (68.26). Let ψ_s and $\psi_{s'}$ correspond to $\phi_u\chi$ and $\phi_v\chi'$ respectively, where χ and χ' are the conventional spin states of an electron. Then the one-electron integral h_{ss} converts as

$$h_{ss} = \int dv \int ds\, \phi_u^*(\mathbf{r})\chi^*(s)h(\mathbf{r})\phi_u(\mathbf{r})\chi(s)$$

$$= \left(\int ds\, \chi^*(s)\chi(s)\right) \int ds\; \phi_u^*(\mathbf{r})h(\mathbf{r})\phi_u(\mathbf{r}) = h_{uu} \tag{69.2}$$

as the spin function χ is normalized. The two-electron integrals also reduce to the form of the two-electron integrals over spatial functions. The reduction of the direct operator is carried out as follows:

$$J_{ss'} = \int dv_1 ds_1 \int dv_2 ds_2\; \phi_u^*(\mathbf{r}_1)\chi^*(s_1)\phi_v^*(\mathbf{r}_2)\chi'^*(s_2)\frac{1}{r_{12}}\phi_u(\mathbf{r}_1)\chi(s_1)\phi_v(\mathbf{r}_2)\chi'(s_2)$$

$$= \left(\int ds_1\, |\chi(s_1)|^2\right)\left(\int ds_2\, |\chi'(s_2)|^2\right)\int dv_1 \int dv_2\;\; \phi_u^*(\mathbf{r}_1)\phi_v^*(\mathbf{r}_2)\frac{1}{r_{12}}\phi_u(\mathbf{r}_1)\phi_v(\mathbf{r}_2)$$

$$= J_{uv}, \tag{69.3}$$

again because both χ and χ' are normalized. This result is obtained regardless of whether χ is α or it is β and χ' is α or β. The exchange integral is more sensitive to the integration over spin coordinates. In fact

$$K_{ss'} = \int dv_1 ds_1 \int dv_2 ds_2\; \phi_u^*(\mathbf{r}_1)\chi^*(s_1)\phi_v^*(\mathbf{r}_2)\chi'^*(s_2)\frac{1}{r_{12}}\phi_v(\mathbf{r}_1)\chi'(s_1)\phi_u(\mathbf{r}_2)\chi(s_2)$$

$$= \left(\int ds_1\, \chi^*(s_1)\chi'(s_1)\right)\left(\int ds_2\;\; \chi'^*(s_2)\chi(s_2)\right)$$

$$\times \int dv_1 \int dv_2\; \phi_u^*(\mathbf{r}_1)\, \phi_v^*\, (\mathbf{r}_2)\frac{1}{r_{12}}\phi_v(\mathbf{r}_1)\phi_u(\mathbf{r}_2). \tag{69.4}$$

Hence $K_{ss'}=0$ by spin orthogonality whenever χ and χ' are different spin functions. When both χ and χ' are α or both are β,

$$K_{ss'} = \int dv_1 \int dv_2\; \phi_u^*(\mathbf{r}_1)\, \phi_v^*\, (\mathbf{r}_2)\frac{1}{r_{12}}\phi_v(\mathbf{r}_1)\phi_u(\mathbf{r}_2) = K_{uv}. \tag{69.5}$$

Using the reductions (69.2), (69.3) and (69.5), we get the total electronic energy of a closed-shell species,

$$E = 2\sum_{u=1}^{N/2} h_{uu} + \sum_{u=1}^{N/2}\sum_{v=1}^{N/2} (2J_{uv} - K_{uv}). \tag{69.6}$$

To obtain this energy one must recognize that $J_{uv} = J_{vu}$ and $K_{uv} = K_{vu}$. Furthermore, $J_{uu} = K_{uu}$ so that the energy E can be recast as

$$E = 2\sum_{u=1}^{N/2} h_{uu} + \sum_{u=1}^{N/2} J_{uu} + \sum_{u\neq v}^{N/2} (2J_{uv} - K_{uv}). \tag{69.7}$$

The Hartree total energy (68.14) can be obtained from this expression by neglecting the exchange integrals. Putting all K_{uv} equal to zero is tantamount to the use of a single product instead of a determinant.

The space-spin separation as defined by Eq. (69.1) is really valid for a closed-shell for which $L = 0$ and the spin-orbit coupling vanishes. Otherwise, orbitals of type α have spatial functions different from the spatial parts of orbitals of type β. Besides, if the spin-orbit interaction (the $\boldsymbol{L}\cdot\boldsymbol{S}$ term) is included in the calculation, equations (69.1) become certainly unrealistic for the open shells.

Variation subject to constraints – the Hartree case

The total energy expressions (68.14), (68.26) and (69.7) are now to be used for finding a stationary point for variation of the trial function. The trial function is varied by changing the one-electron functions. The stationarity should be subject to the constraint that the one-electron functions remain normalized and mutually orthogonal.

In Hartree theory (and in the closed-shell Hartree-Fock treatment) one considers small changes in the spatial orbitals

$$\phi_u(\mathbf{r}) \rightarrow \phi_u(\mathbf{r}) + \delta\phi_u(\mathbf{r})$$

$$\phi_u^*(\mathbf{r}) \rightarrow \phi_u^*(\mathbf{r}) + \delta\phi_u^*(\mathbf{r}) \tag{69.8}$$

which result in the change

$$E \rightarrow E + \delta E + \delta^2 E. \tag{69.9}$$

The Hartree δE is found from Eq. (68.14):

$$\delta E = 2\int dv\ [\delta\phi_u^*(\mathbf{r})\,\mathbf{h}(\mathbf{r})\,\phi_u(\mathbf{r}) + \phi_u^*(\mathbf{r})\,\mathbf{h}(\mathbf{r})\,\delta\phi_u(\mathbf{r})]$$

$$+2\int dv_1 \int dv_2\ [\delta\phi_u^*(\mathbf{r}_1)\phi_u^*(\mathbf{r}_2)\frac{1}{r_{12}}\phi_u(\mathbf{r}_1)\phi_u(\mathbf{r}_2) + \phi_u^*(\mathbf{r}_1)\phi_u^*(\mathbf{r}_2)\frac{1}{r_{12}}\delta\phi_u(\mathbf{r}_1)\phi_u(\mathbf{r}_2)$$

$$+4\sum_{s\neq u}^{N/2} \int dv_1 \int dv_2\ [\phi_s^*(\mathbf{r}_1)\delta\phi_u^*(\mathbf{r}_2)\frac{1}{r_{12}}\phi_s(\mathbf{r}_1)\phi_u(\mathbf{r}_2)$$

$$+\phi_s^*(\mathbf{r}_1)\phi_u^*(\mathbf{r}_2)\frac{1}{r_{12}}\phi_s(\mathbf{r}_1)\delta\phi_u(\mathbf{r}_2)]. \tag{69.10}$$

At the stationary point

$$\delta E = 0 \tag{69.11}$$

which implies

$$\int dv\ \delta\phi_u^*(\mathbf{r})\ [\mathbf{h}(\mathbf{r}) + \mathbf{J}_u(\mathbf{r}) + 2\sum_{s(\neq u)}^{N/2} \mathbf{J}_s(\mathbf{r})]\ \phi_u(\mathbf{r})$$

$$+ \int dv\ \phi_u^*(\mathbf{r})\ [\mathbf{h}(\mathbf{r}) + \mathbf{J}_u(\mathbf{r}) + 2\sum_{s(\neq u)}^{N/2} \mathbf{J}_s(\mathbf{r})]\ \delta\phi_u(\mathbf{r}) = 0 \tag{69.12}$$

where the operator $\mathbf{J}_u(\mathbf{r})$ is the direct coulomb repulsion operator defined in terms of the spatial orbital ϕ_u,

$$\mathbf{J}_u(\mathbf{r}) = \int dv'\ \phi_u^*(\mathbf{r}') \frac{1}{|\mathbf{r} - \mathbf{r}'|} \phi_u(\mathbf{r}'). \tag{69.13}$$

The stationary point should be found under the orthogonality constraints

$$\int dv\ \phi_i^*\phi_u = \delta_{i,u} \rightarrow \int dv\ \phi_i^*(\phi_u + \delta\phi_u) = 0$$

and (69.14)

$$\int dv\ \phi_u^*\phi_i = \delta_{u,i} \rightarrow \int dv\ (\phi_u^* + \delta\phi_u^*)\phi_i = 0$$

for i = 1, 2, etc. These imply

$$\int dv\ (\delta\phi_u^*)\phi_i = 0,$$

$$\int dv\ \phi_i^*(\delta\phi_u) = 0. \qquad \text{(for i = 1, 2, ...)} \tag{69.15}$$

Thus Eq. (69.12) must hold subject to equations (69.15), and they can be combined by using Lagrange's undetermined multipliers of types λ_{iu} and λ_{ui} to obtain

$$\int dv\ \ \delta\phi_u^*(\mathbf{r})\ [\{\mathbf{h}(\mathbf{r}) + \mathbf{J}_u(\mathbf{r}) + 2\sum_{s(\neq u)}^{N/2} \mathbf{J}_s(\mathbf{r})\}\phi_u(\mathbf{r}) - \sum_i \lambda_{iu}\phi_i(\mathbf{r})]$$

$$+ \int dv\ [\phi_u^*(\mathbf{r})\{\mathbf{h}(\mathbf{r}) + \mathbf{J}_u(\mathbf{r}) + 2\sum_{s(\neq u)}^{N/2} \mathbf{J}_s(\mathbf{r})\} - \sum_i \phi_i^*(\mathbf{r})\lambda_{ui}]\ \delta\phi_u(\mathbf{r}) = 0. \tag{69.16}$$

The changes $\delta\phi_u$ and $\delta\phi_u^*$ can be independent of each other. Hence each integral in the left side in Eq. (69.16) must be equal to zero. Thus

$$\int dv\ \ \delta\phi_u^*(\mathbf{r})\ [\{\mathbf{h}(\mathbf{r}) + \mathbf{J}_u(\mathbf{r}) + 2\sum_{s(\neq u)}^{N/2} \mathbf{J}_s(\mathbf{r})\}\phi_u(\mathbf{r}) - \sum_i \lambda_{iu}\phi_i(\mathbf{r})] = 0$$

and (69.17)

$$\int dv\ \ [\phi_u^*(\mathbf{r})\{\mathbf{h}(\mathbf{r}) + \mathbf{J}_u(\mathbf{r}) + 2\sum_{s(\neq u)}^{N/2} \mathbf{J}_s(\mathbf{r})\} - \sum_i \phi_i^*(\mathbf{r})\lambda_{ui}]\ \delta\phi_u(\mathbf{r}) = 0.$$

Since $\delta E = 0$ for completely arbitrary $\delta\phi_u$ and $\delta\phi_u^*$, the integrand in each of the above integrals must vanish, which leave aside

$$[\mathbf{h}(\mathbf{r}) + \mathbf{J}_u(\mathbf{r}) + 2\sum_{s(\neq u)}^{N/2} \mathbf{J}_s(\mathbf{r})]\,\phi_u(\mathbf{r}) = \sum_i \lambda_{iu}\phi_i(\mathbf{r}) \tag{69.18}$$

and

$$\phi_u^*(\mathbf{r})\,[\mathbf{h}(\mathbf{r}) + \mathbf{J}_u(\mathbf{r}) + 2\sum_{s(\neq u)}^{N/2} \mathbf{J}_s(\mathbf{r})] = \sum_i \phi_i^*(\mathbf{r})\,\lambda_{ui}. \tag{69.19}$$

Since the operators **h** and **J** are hermitian, the operator $[\mathbf{h}(\mathbf{r}) + \mathbf{J}_u(\mathbf{r}) + 2\sum_{s(\neq u)}^{N/2} \mathbf{J}_s(\mathbf{r})]$ is hermitian also. Multiplying both sides of Eq. (69.18) by $\phi_j^*(\mathbf{r})$ from the left and integrating over all space, and similarly multiplying both sides of Eq.(69.19) by $\phi_j(r)$ from the right and again integrating over all space, and finally using the turn over rule one finds

$$\sum_i \lambda_{iu}\int dv\ \phi_j^*(\mathbf{r})\phi_i(\mathbf{r}) = \sum_i \left(\int dv\ \phi_i^*(\mathbf{r})\phi_j(\mathbf{r})\right)^* \lambda_{ui}^*$$

so that

$$\lambda_{ju} = \lambda_{uj}^*. \tag{69.20}$$

This holds for all j. Hence the λ matrix is hermitian. Equation (69.19) can now be recognized as merely the hermitian adjoint of Eq. (69.18). Therefore, we will henceforth deal only with only one equation, viz., (69.18).

Unitary Transformation to Hartree equation

Convert Eq. (69.18) into

$$h_{iu} + (\mathbf{J}_u)_{iu} + 2\sum_{s(\neq u)}^{N/2} (\mathbf{J}_s)_{iu} = \lambda_{iu} \tag{69.21}$$

by multiplying both sides with ϕ_i^* and integrating over all space. The matrix correspondence is

$$\mathbf{h} + \mathbf{J_u} + 2\sum_{s(\neq u)}^{N/2} \mathbf{J_s} = \mathbf{\Lambda}\,. \tag{69.22}$$

Since $\mathbf{\Lambda}$ is hermitian, it can be diagonalized by an orthogonal transformation. The transformation is in particular unitary as the elements of $\mathbf{\Lambda}$ are formed by using orthonormal functions. Consider the unitary transformation

$$\boldsymbol{\phi} \to \boldsymbol{\phi}' = \boldsymbol{\phi}\,\mathbf{U} \tag{69.23}$$

under which $\boldsymbol{\Lambda}$ converts into the diagonal matrix $\boldsymbol{\varepsilon}$,

$$\boldsymbol{\Lambda} \rightarrow \boldsymbol{\Lambda}' = \mathbf{U}^{\dagger}\boldsymbol{\Lambda}\mathbf{U} = \boldsymbol{\varepsilon}\,. \tag{69.24}$$

The matrices $\mathbf{h}$ and $\mathbf{J}_{\mathbf{i}}$ are simultaneously transformed as

$$\mathbf{h} \rightarrow \mathbf{h}' = \mathbf{U}^{\dagger}\mathbf{h}\mathbf{U} \tag{69.25}$$

and

$$\mathbf{J}_{\mathbf{i}} \rightarrow \mathbf{J}_{\mathbf{i}'}{}' = \mathbf{U}^{\dagger}\mathbf{J}_{\mathbf{i}'}\mathbf{U} \tag{69.26}$$

where the $\mathbf{J}_{\mathrm{i}'}$ operator is formed with the functions $\phi_{\mathrm{i}}{}'(\mathbf{r})$, etc. Hence the whole of Eq. (69.22) transforms into

$$\mathbf{h}' + \mathbf{J}_{\mathbf{u}'}{}' + 2\sum_{\mathrm{s}'(\neq \mathrm{u}')}^{N/2} \mathbf{J}_{\mathbf{s}'}{}' = \boldsymbol{\varepsilon} \tag{69.27}$$

that corresponds to the operator equation

$$\boxed{[\mathbf{h}(\mathbf{r}) + \mathbf{J}_{\mathrm{u}'}(\mathbf{r}) + 2\sum_{\mathrm{s}'(\neq \mathrm{u}')}^{N/2} \mathbf{J}_{\mathrm{s}'}(\mathbf{r})]\phi_{\mathrm{u}'}(\mathbf{r}) = \varepsilon_{\mathrm{u}}\phi_{\mathrm{u}'}(\mathbf{r})} \tag{69.28}$$

where

$$\mathbf{J}_{\mathrm{s}'}(\mathbf{r}) = \int dv'\ \phi_{\mathrm{s}'}{}^*(\mathbf{r}')\frac{1}{|\mathbf{r} - \mathbf{r}'|}\phi_{\mathrm{s}'}(\mathbf{r}').$$

Equation (69.28) is in fact the Hartree equation. The Hartree operator is

$$\mathrm{h}_{\mathrm{Hartree}}(\mathbf{r}) = \mathrm{h}(\mathbf{r}) + \mathbf{J}_{\mathrm{u}'}(\mathbf{r}) + 2\sum_{\mathrm{s}'(\neq \mathrm{u}')}^{N/2} \mathbf{J}_{\mathrm{s}'}(\mathbf{r}), \tag{69.29}$$

defined in terms of direct coulomb repulsion operators which are in turn written in terms of the eigenfuntions of $\mathrm{h}_{\mathrm{Hartree}}$. Thus the Hartree equation represents a pseudo-eigenvalue problem that must be solved in a self-consistent fashion, and henceforth we will consider this wave equation using indexes without primes.

Lecture 70: Aspects of Variation — The Hartree-Fock Treatment

If we repeat the treatment that leads to the Hartree equation but starts from the energy expression (68.26) instead of (68.14), we arrive at the following results. When the spin orbitals ψ_{s} and $\psi_{\mathrm{s}}{}^*$ are varied as

$$\psi_{\mathrm{s}}(\mathbf{x}) \rightarrow \psi_{\mathrm{s}}(\mathbf{x}) + \delta\psi_{\mathrm{s}}(\mathbf{x}), \quad \psi_{\mathrm{s}}{}^*(\mathbf{x}) \rightarrow \psi_{\mathrm{s}}{}^*(\mathbf{x}) + \delta\psi_{\mathrm{s}}{}^*(\mathbf{x}), \tag{70.1}$$

the total energy E in Eq. (68.26) changes as

$$E \rightarrow E + \delta E + \delta^2 E \tag{70.2}$$

where

$$\delta E = \int dx\, \{\delta\psi_s^*(\mathbf{x})\, \mathbf{h}(\mathbf{r})\, \psi_s(\mathbf{x}) + \psi_s^*(\mathbf{x})\, \mathbf{h}(\mathbf{r})\, \delta\psi_s(\mathbf{x})\}$$

$$+ \sum_{p(\neq s)}^{occ} \{\int dx_1 \int dx_2\, \delta\psi_s^*(\mathbf{x}_1)\psi_p^*(\mathbf{x}_2)\frac{1}{r_{12}}\psi_s(\mathbf{x}_1)\psi_p(\mathbf{x}_2)$$

$$-\int dx_1 \int dx_2\, \psi_s^*(\mathbf{x}_1)\psi_p^*(\mathbf{x}_2)\frac{1}{r_{12}}\delta\psi_s(\mathbf{x}_1)\psi_p(\mathbf{x}_2)\}. \tag{70.3}$$

The condition of stationarity now yields

$$\int dx\, \delta\psi_s^*(\mathbf{x})[\mathbf{h}(\mathbf{r}) + \sum_{p(\neq s)}^{occ} (\boldsymbol{J}_p(\mathbf{x}) - \boldsymbol{K}_p(\mathbf{x}))]\psi_s(\mathbf{x})$$

$$+ \int dx\, \psi_s^*(\mathbf{x})[\mathbf{h}(\mathbf{r}) + \sum_{p(\neq s)}^{occ} (\boldsymbol{J}_p(\mathbf{x}) - \boldsymbol{K}_p(\mathbf{x}))]\delta\psi_s(\mathbf{x}) = 0 \tag{70.4}$$

where the direct operator is

$$\boldsymbol{J}_p(\mathbf{x}) = \int dx'\, \psi_p^*(\mathbf{x}')\frac{1}{|\mathbf{r}-\mathbf{r}'|}\psi_p(\mathbf{x}') \tag{70.5}$$

and the exchange operator is defined as

$$\boldsymbol{K}_p(\mathbf{x})\psi_s(\mathbf{x}) = \int dx'\, \psi_p^*(\mathbf{x}')\frac{1}{|\mathbf{r}-\mathbf{r}'|}\psi_s(\mathbf{x}')\psi_p(\mathbf{x}). \tag{70.6}$$

The condition (70.4) is to be subject to the constraints of orthonormality:

$$\int dx\, \psi_q^*(\mathbf{x})\psi_s(\mathbf{x}) = \delta_{q,s} \rightarrow \int dx\, \psi_q^*(\mathbf{x})[\psi_s(\mathbf{x}) + \delta\psi_s(\mathbf{x})] = \delta_{q,s}$$

and (70.7)

$$\int dx\, \psi_s^*(\mathbf{x})\psi_q(\mathbf{x}) = \delta_{s,q} \rightarrow \int dx\, [\psi_s^*(\mathbf{x}) + \delta\psi_s^*(\mathbf{x})]\psi_q(\mathbf{x}) = \delta_{s,q}.$$

These yield

$$\int dx\, \psi_q^*(\mathbf{x})\delta\psi_s(\mathbf{x}) = 0$$

and (70.8)

$$\int dx\, \delta\psi_s^*(\mathbf{x})\psi_q(\mathbf{x}) = 0$$

for $q = 1, 2, \ldots$. Combining Eqs. (70.4) and (70.8) by Langrange's technique one obtains

$$\int dx\, \delta\psi_s^*(\mathbf{x})[\{\mathbf{h}(\mathbf{r}) + \sum_{p(\neq s)}^{occ} (\boldsymbol{J}_p(\mathbf{x}) - \boldsymbol{K}_p(\mathbf{x}))\}\psi_s(\mathbf{x}) - \sum_q \lambda_{qs}\psi_q(\mathbf{x})] \tag{70.9}$$

$$+\int dx\, [\psi_s^*(\mathbf{x})\{\mathbf{h}(\mathbf{r}) + \sum_{p(\neq s)}^{occ} (\boldsymbol{J}_p(\mathbf{x}) - \boldsymbol{K}_p(\mathbf{x}))\} - \sum_q \psi_q^*(\mathbf{x})\lambda_{sq}]\, \delta\psi_s(\mathbf{x}) = 0.$$

Because $\delta\psi_s^*(\mathbf{x})$ and $\delta\psi_s(\mathbf{x})$ can be independent of each other, the two integrals in Eq. (70.9) are individually equal to zero. Furthermore, $\delta\psi^*$ and $\delta\psi$ are completely arbitrary so that one can write

$$[\mathbf{h}(\mathbf{r}) + \sum_{p(\neq s)}^{occ} (\boldsymbol{J}_p(\mathbf{x}) - \boldsymbol{K}_p(\mathbf{x}))]\psi_s(\mathbf{x}) = \sum_q \lambda_{qs}\psi_q(\mathbf{x}), \tag{70.10}$$

and

$$\psi_s^*(\mathbf{x})[\mathbf{h}(\mathbf{r}) + \sum_{p(\neq s)}^{occ} (J_p(\mathbf{x}) - K_p(\mathbf{x}))] = \sum_q \psi_q^*(\mathbf{x})\lambda_{sq}. \tag{70.11}$$

Again it is easy to prove that $\lambda_{sq} = \lambda_{qs}^*$, and Eq. (70.11) is merely the hermitian adjoint form of Eq. (70.10). Hence the matrix $\mathbf{\Lambda}$ can be diagonalized to matrix $\boldsymbol{\varepsilon}$ by a unitary transformation

$$\psi \to \psi' = \psi\mathbf{V},$$
$$\Lambda \to \Lambda' = \mathbf{V}^\dagger\Lambda\mathbf{V} = \boldsymbol{\varepsilon}. \tag{70.12}$$

Parallel to the development in our previous lecture, we find that Eq. (70.10) transforms into the Hartree-Fock equation

$$\boxed{\mathbf{F}(\mathbf{r})\psi_s'(\mathbf{x}) = \varepsilon_s\psi_s'(\mathbf{x})} \tag{70.13}$$

where the effective one-electron Hamiltonian is given by

$$\mathbf{F}(\mathbf{r}) = \mathbf{h}(\mathbf{r}) + \sum_{p(\neq s)}^{occ} [J_{p'}(\mathbf{x}) - K_{p'}(\mathbf{x})]. \tag{70.14}$$

The operators $J_{p'}$ and $K_{p'}$ are defined in terms of the primed (transformed) orbitals. For future use we will drop the primes and the unprimed functions will be taken as those for which the multiplier matrix is diagonal.

Since

$$J_s(\mathbf{x})\psi_s(\mathbf{x}) = K_s(\mathbf{x})\psi_s(\mathbf{x})$$

one may also write the Fock operator $\mathbf{F}(\mathbf{r})$ in terms of both occupied and unoccupied orbitals in a uniform way:

$$\boxed{\mathbf{F}(\mathbf{r}) = \mathbf{h}(\mathbf{r}) + \sum_{p}^{occ} [J_p(\mathbf{x}) - K_p(\mathbf{x})].} \tag{70.15}$$

Under the space-spin separation (69.1) one derives for a closed-shell system the *restricted Hartree-Fock* (RHF) equations

$$\boxed{\mathbf{F}(\mathbf{r})\phi_u(\mathbf{r}) = \varepsilon_u\phi_u(\mathbf{r})} \tag{70.16}$$

with the restricted closed-shell Fock operator given by

$$\mathbf{F}(\mathbf{r}) = \mathbf{h}(\mathbf{r}) + \sum_{p=1}^{N/2} [2\mathbf{J}_p(\mathbf{r}) - \mathbf{K}_p(\mathbf{r})]. \tag{70.17}$$

These results for the closed shell may be derived directly from Eq. (69.7).

Open shell Hartree-Fock – Unrestricted form

Pople and Nesbet (1954) considered that the spatial component of a spin orbital can be different for α and β spins while the shell is open:

$$\psi_{2u-1}(\mathbf{x}) = \phi_u^{\alpha}(\mathbf{r})\alpha(s),$$
$$\psi_{2u}(\mathbf{x}) = \phi_u^{\beta}(\mathbf{r})\beta(s). \tag{70.18}$$

The energy expression (68.26), the Hartree-Fock equation (70.13) and the definition of the Fock operator (70.15) are taken valid for an open-shell system with as many open-shell electrons of spin α as it is necessary to have the maximum possible M_S values. See Figure 12.1. This formalism is known as the *Unrestricted Hartree-Fock* (UHF) treatment. This procedure yields, for N_α electrons with spin α and N_β electrons with spin β ($N = N_\alpha + N_\beta$ and $N_\alpha \neq N_\beta$), the total energy

$$E = [\sum_{u=1}^{N_\alpha} h_{u\alpha,u\alpha} + \sum_{v=1}^{N_\beta} h_{v\beta,v\beta}] + [\frac{1}{2}\sum_{u\neq u'}^{N_\alpha} (J_{u\alpha,u'\alpha} - K_{u\alpha,u'\alpha})$$
$$+ \frac{1}{2}\sum_{v\neq v'}^{N_\beta} (J_{v\beta,v'\beta} - K_{v\beta,v'\beta}) + \sum_{u=1}^{N_\alpha}\sum_{v=1}^{N_\beta} J_{u\alpha,v\beta}] \qquad (70.19)$$

as $K_{u\alpha,v\beta} = 0$.

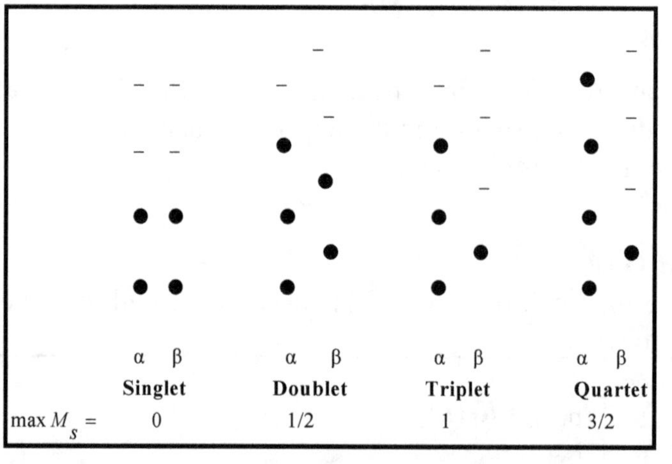

Figure 12.1 Orbital occupancy in UHF treatment of open-shell species: electron occupying each spin orbital is shown by a black circle, and alpha and beta orbital energies differ for open shells.

One obtains two distinct Hartree-Fock equations with two distinct Fock operators for the different spins:

(i) $$\mathbf{F}^\alpha(\mathbf{r})\phi_u^\alpha(\mathbf{r}) = \varepsilon_u^\alpha \phi_u^\alpha(\mathbf{r}), \qquad (70.20)$$

$$\mathbf{F}^\alpha(\mathbf{r}) = \mathrm{h}(\mathbf{r}) + \sum_{p=1}^{N_\alpha} [\boldsymbol{J}_{p\alpha}(\mathbf{r}) - \boldsymbol{K}_{p\alpha}(\mathbf{r})] + \sum_{q=1}^{N_\beta} \boldsymbol{J}_{q\beta}(\mathbf{r}) \qquad (70.21)$$

and

(ii) $$\mathbf{F}^\beta(\mathbf{r})\phi_v^\beta(\mathbf{r}) = \varepsilon_v^\beta \phi_v^\beta(\mathbf{r}) \qquad (70.22)$$

$$\mathbf{F}^\beta(\mathbf{r}) = \mathrm{h}(\mathbf{r}) + \sum_{p=1}^{N_\alpha} \boldsymbol{J}_{p\alpha}(\mathbf{r}) + \sum_{q=1}^{N_\beta} [\boldsymbol{J}_{q\beta}(\mathbf{r}) - \boldsymbol{K}_{q\beta}(\mathbf{r})]. \qquad (70.23)$$

Because the number of α electrons differs from that of β electrons, electrons of different spins face different amount of exchange repulsion. The set of eigenvalues ε_u^{α}'s are necessarily different from the set ε_v^{β}'s, and usually found to be lower because there are more α electrons as shown in Figure 12.1. The spatial functions ϕ_u^{α}'s are also different from ϕ_v^{β}'s. Both α and β spatial orbitals are normalized. The α spatial orbitals are mutually orthogonal as are the β spatial orbitals, but the α spatial orbitals *are not necessarily orthogonal* to the β spatial orbitals. Nevertheless, the α spin orbitals are orthogonal to the β spin orbitals as the spin function parts are orthogonal to each other.

In general, the UHF method does not give pure spin states.

Restricted open shell Hartree-Fock

Roothaan (1960) presented an open-shell formalism within the framework of the RHF treatments. In Roothaan's formulation, the total wave function is written as a weighted sum of determinants such that it becomes an eigenfunction of the total angular momentum operators $\boldsymbol{L}^2$ and $\boldsymbol{S}^2$ as well as $\boldsymbol{L}_Z$ and $\boldsymbol{S}_Z$ for an atom or a simple molecule. Each determinant contains a set of doubly occupied orbitals $\{\phi_c\}$ belonging to a closed shell, and an open shell with the occupied orbitals chosen from a set $\{\phi_o\}$. Only a single open shell of a specific angular momentum symmetry is considered. The expectation value of energy is written in RHF formulation as

$$E = 2\sum_c h_{cc} + \sum_{c,c'} (2J_{cc'} - K_{cc'})$$
$$+ f[2\sum_o h_{oo} + f\sum_{o,o'} (2aJ_{oo'} - bK_{oo'})] + 2f\sum_c \sum_o (2J_{co} - K_{co}). \qquad (70.24)$$

In the above f is the fractional occupancy of the open shell, and a and b are numerical constants which depend on the number of electrons in the open shell and on the angular momenta associated with the total wave function. An ordinary treatment starting from this point leads to two different Hartree-Fock equations for the closed and open shells. After some complicated analysis Roothaan established that *both closed-shell and open-shell orbitals as well as the virtual orbitals* can be taken as eigenfunctions of a *unique* Fock operator:

$$\mathbf{F}\phi_i = \varepsilon_i\phi_i \qquad (70.25)$$

where

$$\mathbf{F(r)} = \mathbf{h(r)} + [\sum_c (2\mathbf{J}_c - \mathbf{K}_c) + f\sum_o (2\mathbf{J}_o - \mathbf{K}_o)] + 2\alpha[\sum_c \mathbf{L}_c + f\sum_o (\mathbf{L}_o - \mathbf{J}_o)]$$
$$- \beta[\sum_c \mathbf{M}_c + f\sum_o (\mathbf{M}_o - \mathbf{K}_o)]. \qquad (70.26)$$

The Roothaan α and β coefficients are given by

$$\alpha = \frac{(1-a)}{(1-f)}, \qquad \beta = \frac{(1-b)}{(1-f)}. \tag{70.27}$$

Roothaan's *coulomb and exchange coupling operators* $\mathbf{L}_i$ and $\mathbf{M}_i$ are defined by

$$\mathbf{L}_i\phi = f\sum_o [\langle \phi_i|\mathbf{J}_o|\phi\rangle\phi_i + \langle \phi_i|\phi\rangle\mathbf{J}_o\phi_i],$$

$$\mathbf{M}_i\phi = f\sum_o [\langle \phi_i|\mathbf{K}_o|\phi\rangle\phi_i + \langle \phi_i|\phi\rangle\mathbf{K}_o\phi_i]. \tag{70.28}$$

A list of the Roothaan's a, b, α and β coefficients for atoms is given in Table 12.1. In the extreme case of a closed-shell atom or molecule (with term necessarily $^1S_0$), a = 1, b = 1, α = 0 and β = 0. In the interesting case of the ground state of a half-closed shell ($f = ½$), the wave functions for the two components with all spin functions same are single determinants so that a = 1 and b = 2 whence α = 0 and β = −2. This situation is found with H, $^2S_{1/2}$; Li, $(1s^2)2s$, $^2S_{1/2}$; N, $(1s^22s^2)2p^3$, $^4S_{3/2}$; etc.

Table 12.1 Roothaan a, b, α and β coefficients for a few atoms.

	N	Electronic Configuration	N_c	N_o	*f*	Term Symbol	Roothaan coefficients a	b	α	β
H	1	1s	0	1	½	$^2S_{1/2}$	1	2	0	−2
He	2	$1s^2$	2	0	0	$^1S_0$	1	1	0	0
Li	3	$(1s^2)2s$	2	1	½	$^2S_{1/2}$	1	2	0	−2
Be	4	$1s^22s^2$	4	0	0	$^1S_0$	1	1	0	0
B	5	$(1s^22s^2)2p$	4	1	⅙	$^2P_{1/2,3/2}$	0	0	6/5	6/5
C	6	$(1s^22s^2)2p^2$	4	2	⅓	$^3P_{0,1,2}$	¾	3/2	⅜	−¾
						$^1D_2$	9/20	−3/10	33/40	39/20
						$^1S_0$	0	−3	3/2	6
N	7	$(1s^22s^2)2p^3$	4	3	½	$^4S_{3/2}$	1	2	0	−2
						$^2D_{3/2,5/2}$	4/5	4/5	2/5	2/5
						$^2P_{1/2,3/2}$	⅔	0	⅔	2
O	8	$(1s^22s^2)2p^4$	4	4	⅔	$^3P_{0,1,2}$	15/16	9/8	3/16	−⅜
						$^1D_2$	69/80	27/40	33/80	39/40
						$^1S_0$	¾	0	¾	3
F	9	$(1s^22s^2)2p^5$	4	5	⅚	$^2P_{1/2,3/2}$	24/25	24/25	6/25	6/25

Roothaan's open-shell treatment is obviously carried out in the *L-S* scheme for many-electron atoms. With suitable modifications the method can be extended to the *j-j* scheme. The methodology also applies to linear molecules for which the a, b, α and β coefficients are given in Table 12.2. The main triumph of Roothaan's method is that it gives rise to a correct orbital-cum-spin angular momentum. Therefore, in one form or another, the treatment is often taken as the first step of rigorous ab-initio methodologies for solving the many-electron Schrödinger equation pertaining to open shell atoms, molecules or clusters.

Both Hartree and Hartree-Fock equations must be solved in a self-consistent way. This leads to the *self-consistent field* treatment which has been discussed in Chapter 5.

Table 12.2 Roothaan a, b, α and β coefficients for a linear molecule.

Open shell configuration	f	Term	a	b	α	β
σ^1	½	$^2\Sigma$	1	2	0	−2
π^1	¼	$^2\Pi$	0	0	4/3	4/3
π^2	½	$^3\Sigma^-$	1	2	0	−2
		$^1\Delta$	½	0	1	2
		$^1\Sigma^+$	−2	−2	2	6
π^3	¾	$^2\Pi$	8/9	8/9	4/9	4/9
δ^1	¼	$^2\Delta$	0	0	4/3	4/3
δ^2	½	$^3\Sigma^-$	1	2	0	−2
		$^1\Gamma$	½	0	1	2
		$^1\Sigma^+$	0	−2	2	6
δ^3	¾	$^2\Delta$	0	0	4/3	4/3

Lecture 71: Interpretations

Orbital energy

One may use Eq. (70.13) to write

$$\varepsilon_s = \langle \psi_s | \mathbf{F} | \psi_s \rangle \quad (71.1)$$

which can be calculated further by using the expression for the Fock operator in Eq. (70.15). Thus the general URHF (UHF) orbital energy is found out:

$$\varepsilon_s = h_{ss} + \sum_{p}^{occ} (J_{ps} - K_{ps}). \quad (71.2)$$

The RHF orbital energy can be calculated either from this expression or from Eq. (70.17). We find

$$\varepsilon_u = h_{uu} + \sum_{p=1}^{N/2} (2J_{pu} - K_{pu}) = h_{uu} + J_{uu} + \sum_{\substack{p=1 \\ (p \neq u)}}^{N/2} (2J_{pu} - K_{pu}). \quad (71.3)$$

The Hartree orbital energy can be obtained by equating the exchange integral in this equation to zero or it can be derived straightforwardly from the Hartree equation. One finds

$$\varepsilon_u^{\text{Hartree}} = h_{uu} + J_{uu} + 2 \sum_{\substack{p=1 \\ (p \neq u)}}^{N/2} J_{pu}. \tag{71.4}$$

The URHF orbital energies for an open-shell species can be directly derived from equations (70.20) and (70.22) or from Eq. (71.2). One obtains

$$\varepsilon_u^{\alpha} = h_{u\alpha,u\alpha} + \sum_{p=1}^{N_\alpha} (J_{p\alpha,u\alpha} - K_{p\alpha,u\alpha}) + \sum_{q=1}^{N_\beta} J_{q\beta,u\alpha}, \tag{71.5}$$

$$\varepsilon_v^{\beta} = h_{v\beta,v\beta} + \sum_{p=1}^{N_\alpha} J_{p\alpha,v\beta} + \sum_{q=1}^{N_\beta} (J_{q\beta,v\beta} - K_{q\beta,v\beta}).$$

Total energy

Using the expressions for orbital energy in Eqs. (71.2), (71.3), (71.4) and (71.5) one may rewrite the Hartree and Hartree-Fock total energies in the following way.

a) For a closed-shell system Eqs. (68.14) and (71.4) lead to

$$E_{\text{Hartree}} = \sum_{u=1}^{N/2} (h_{uu} + \varepsilon_u). \tag{71.6}$$

b) The general URHF total energy expression that uses the spin-orbitals as the one-electron functions is obtained from equations (68.26) and (71.2) as

$$E_{\text{URHF}} = \frac{1}{2} \sum_{s=1}^{N} (h_{ss} + \varepsilon_s). \tag{71.7}$$

c) The RHF total energy for a closed shell is calculated from equations (69.7) and (71.3):

$$E_{\text{RHF}} = \sum_{u=1}^{N/2} (h_{uu} + \varepsilon_u). \tag{71.8}$$

d) The Pople-Nesbet URHF total energy for an open-shell species results from the combination of Eq. (70.19) with Eq. (71.5),

$$E_{\text{URHF}}^{\text{PN}} = \frac{1}{2} \sum_{u=1}^{N_\alpha} (h_{u\alpha,u\alpha} + \varepsilon_{u\alpha}) + \frac{1}{2} \sum_{v=1}^{N_\beta} (h_{v\beta,v\beta} + \varepsilon_{v\beta}). \tag{71.9}$$

e) The total energy for an open-shell system in Roothaan's treatment is given by

$$E^{OS}_{RHF} = \sum_{c} (h_{cc} + \varepsilon_c) + f\sum_{o} (h_{oo} + \varepsilon_o)$$

$$-f \sum_{c,\, o} (2\alpha J_{co} - \beta K_{co}) - f^3 \sum_{o,\, o'} (2\alpha J_{oo'} - \beta K_{oo'}). \tag{71.10}$$

where α, β, a and b are Roothaan parameters.

We now discuss the importance of these energy values.

Let us take the energy expression (68.26), and consider two configurations differing by only one orbital. For the sake of simplicity let one configuration represent the Hartree-Fock ground state. The excited configuration yields a different Fock operator that should be associated with orbitals different from the orbitals in the ground state. An excited state may be prepared by using the orbitals in the Hartree-Fock ground state, but in a time-dependent formalism these orbitals are seen to relax to the orbitals typical of the excited state. This process is called orbital relaxation. Here, for the sake of simplicity, we assume that the orbital relaxation is negligible and the excited state configuration is adequately described by the occupied and unoccupied orbitals of the ground state. Thus we write as usual

$$E_g = \sum_{s=1}^{N} h_{ss} + \frac{1}{2} \sum_{s=1}^{N} \sum_{s'=1}^{N} (J_{ss'} - K_{ss'}). \tag{71.11}$$

where we have omitted the condition $s \neq s'$ as $J_{ss} = K_{ss}$, and obtain for the excited state energy

$$E^{\gamma \to m}_{ex} = \sum_{\substack{s=1 \\ (\neq \gamma)}}^{N} h_{ss} + h_{mm} + \frac{1}{2} \sum_{\substack{s=1 \\ (\neq \gamma)}}^{N} \sum_{\substack{s'=1 \\ (\neq \gamma)}}^{N} (J_{ss'} - K_{ss'}) + \sum_{\substack{s=1 \\ (\neq \gamma)}}^{N} (J_{sm} - K_{sm}). \tag{71.12}$$

Therefore,

$$E^{\gamma \to m}_{ex} - E_g = \Big[h_{mm} + \sum_{\substack{s=1 \\ (\neq \gamma)}}^{N} (J_{sm} - K_{sm})\Big] - \Big[h_{\gamma\gamma} + \sum_{\substack{s=1 \\ (\neq \gamma)}}^{N} (J_{s\gamma} - K_{s\gamma})\Big]$$

$$= (\varepsilon_m - \varepsilon_\gamma) - (J_{m\gamma} - K_{m\gamma}). \tag{71.13}$$

The last step of simplification results from the substitution of Eq. (71.2). This shows that when orbital relaxation is neglected, the spectroscopic transition energy is given by the difference of the orbital energies minus the coulombic repulsion energy for two electrons in the involved orbitals. This is an overstatement, as the Hartree-Fock theory yields only an approximate solution. It works, although qualitatively, for transitions from one of the HOMOs to one of the LUMOs.

We have already made a few critical observations on the Hartree and Hartree-Fock treatments in Chapter 5. For example, we have noted the following important points:

a) Atomic and molecular orbitals are (pseudo) eigenfunctions of the atomic Fock operator and the molecular Fock operator respectively.

b) Unoccupied orbitals are virtual orbitals with unrealistically high energies because an electron in any of these orbitals faces a greater repulsion.

c) Orbital energy is the negative of ionization potential when orbital relaxation effects are neglected (Koopmans' theorem – Theorem 11).

It is very easy to derive Koopmans' theorem, Eq. (25.1). If one neglects the change of orbitals while the N-electron species is ionized to form a positive ion with (N–1) electrons, the URHF total energy for the ion is obtained as

$$E^{N-1} = \sum_{\substack{s=1 \\ (\neq p)}}^{N} h_{ss} + \frac{1}{2} \sum_{\substack{s=1 \\ (\neq p)}}^{N} \sum_{\substack{s=1 \\ (\neq p)}}^{N} (J_{ss'} - K_{ss'}) \tag{71.14}$$

where the missing electron has gone out of the *pth* spin orbital. Therefore,

$$E^{N-1} - E^{N} = -\left[h_{pp} + \sum_{s=1}^{N} (J_{sp} - K_{sp})\right] \tag{71.15}$$

which shows that

$$I_p = -\varepsilon_p. \tag{71.16}$$

At this point we make one more observation. The Hartree and the Hartree-Fock equations are achieved from the stationarity of the energy expectation value for variation of the one-electron functions. The stationary point is not necessarily a global minimum for a single determinant. Normally one chooses the trial function in a specific way so as to achieve a global minimum from a specific calculation. It is equally possible to calculate a saddle point where energy is maximum for variation of certain orbitals and minimum for variation of other orbitals, although one rarely attempts this task. The nature of the stationary point can be established by studying the second differential $\delta^2 E$. The stability of the Hartree-Fock states was discussed by Thouless (1960). In the general URHF scheme the second derivative of total energy is determined by

$$\delta^2 E = \frac{1}{2} \sum_{s} \{<\delta\psi_s|\, \mathbf{h} + \mathbf{F}\, |\delta\psi_s> + <\delta\psi_s|\, \delta\mathbf{F}\, |\psi_s>$$

$$+ <\psi_s|\, \delta\mathbf{F}\, |\delta\psi_s> + <\psi_s|\, \delta^2\mathbf{F}\, |\psi_s>\} \tag{71.17}$$

where $\delta\mathbf{F}$ and $\delta^2\mathbf{F}$ are given by

$$\delta\mathbf{F}\psi_i(\mathbf{x}) = \sum_{s}^{occ} \int dx' \left\{\delta\psi_s^*(\mathbf{x}') \frac{1}{r_{12}} \psi_s(\mathbf{x}')\psi_i(\mathbf{x}) + \psi_s^*(\mathbf{x}') \frac{1}{r_{12}} \delta\psi_s(\mathbf{x}')\psi_i(\mathbf{x}) \right.$$

$$\left. - \delta\psi_s^*(\mathbf{x}') \frac{1}{r_{12}} \psi_i(\mathbf{x}')\psi_s(\mathbf{x}) - \psi_s^*(\mathbf{x}') \frac{1}{r_{12}} \psi_i(\mathbf{x}')\delta\psi_s(\mathbf{x})\right\}$$

and (71.18)

$$\delta^2\mathbf{F}\psi_i(\mathbf{x}) = \sum_{s}^{occ} \int dx' \left\{\delta\psi_s^*(\mathbf{x}')\frac{1}{r_{12}}\delta\psi_s(\mathbf{x}')\psi_i(\mathbf{x}) - \delta\psi_s^*(\mathbf{x}')\frac{1}{r_{12}}\psi_i(\mathbf{x}')\delta\psi_s(\mathbf{x})\right\}.$$

The saddle point character may be established by calculating $\delta^2 E$.

Lecture 72: Examples of Hartree-Fock Atoms

Example 50: Helium atom

The helium atom is the most trivial many-electron atomic system. The ground state electronic configuration $1s^2$ corresponds to the determinant

$$\Psi(\mathbf{x}_1, \mathbf{x}_2) = 2^{-1/2} \| 1s(\mathbf{r}_1)\alpha(s_1) \quad 1s(\mathbf{r}_2)\beta(s_2) \|$$

$$= 1s(\mathbf{r}_1)1s(\mathbf{r}_2) \frac{1}{\sqrt{2}} \left[\alpha(s_1)\beta(s_2) - \beta(s_1)\alpha(s_2)\right]. \tag{72.1}$$

The corresponding total energy is

$$E = 2h_{1s,1s} + J_{1s,1s} \tag{72.2}$$

where

$$h_{1s,1s} = \int dv\ 1s(\mathbf{r})[-\frac{1}{2}\nabla^2 - \frac{2}{r}]1s(\mathbf{r}) \tag{72.3}$$

and

$$J_{1s,1s} = \int dv_1 \int dv_2\ 1s(\mathbf{r}_1)1s(\mathbf{r}_2)\, r_{12}^{-1}\, 1s(\mathbf{r}_1)1s(\mathbf{r}_2)$$

$$= \int dv_1 \int dv_2\ 1s(\mathbf{r}_1)^2 1s(\mathbf{r}_2)^2\, r_{12}^{-1}. \tag{72.4}$$

This leads to the closed-shell Fock operator

$$\mathbf{F}(\mathbf{r}) = \mathbf{h}(\mathbf{r}) + \mathbf{J}_{1s}(\mathbf{r}) \tag{72.5}$$

so that the 1s orbital energy is

$$\varepsilon_{1s} = h_{1s,1s} + J_{1s,1s} \tag{72.6}$$

whereas the virtual orbital energies are given by

$$\varepsilon_{nl,nl} = h_{nl,nl} + (2J_{1s,nl} - K_{1s,nl}). \tag{72.7}$$

These equations give a complete description of the Hartree-Fock treatment of helium atom except for the actual steps involved in solving the Hartree-Fock equation.

Example 51: Lithium atom

It would be instructive to work out a simple example of an open shell and to inspect the differences between the Pople-Nesbet approach and Roothaan's method. Take the example of lithium atom. The URHF method prescribes the wave function

$$\Psi(\mathbf{x}_1, \mathbf{x}_2, \mathbf{x}_3) = \frac{1}{\sqrt{6}} \|\phi_{1s\alpha}(\mathbf{r}_1)\alpha(s_1) \quad \phi_{1s\beta}(\mathbf{r}_2)\beta(s_2) \quad \phi_{2s\alpha}(\mathbf{r}_3)\alpha(s_3)| \tag{72.8a}$$

that gives the total energy

$$E = (h_{1s\alpha,1s\alpha} + h_{1s\beta,1s\beta} + h_{2s\alpha,2s\alpha})$$

$$+ [J_{1s\alpha,1s\beta} + (J_{1s\alpha,2s\alpha} - K_{1s\alpha,2s\alpha}) + J_{1s\beta,2s\alpha}] \qquad (72.9a)$$

with the Fock operators

$$\mathbf{F}^{\alpha} = [-\frac{1}{2}\nabla^2 - \frac{3}{r}] + [(\boldsymbol{J}_{1s\alpha} - \boldsymbol{K}_{1s\alpha}) + \boldsymbol{J}_{1s\beta} + (\boldsymbol{J}_{2s\alpha} - \boldsymbol{K}_{2s\alpha})], \qquad (72.10a)$$

$$\mathbf{F}^{\beta} = [-\frac{1}{2}\nabla^2 - \frac{3}{r}] + [\boldsymbol{J}_{1s\alpha} + (\boldsymbol{J}_{1s\beta} - \boldsymbol{K}_{1s\beta}) + \boldsymbol{J}_{2s\alpha}], \qquad (72.10b)$$

satisfying the Hartree-Fock equations (70.21) and (70.23) respectively. The orbital energies are

$$\varepsilon_{1s\alpha} = h_{1s\alpha,1s\alpha} + J_{1s\alpha,1s\beta} + (J_{1s\alpha,2s\alpha} - K_{1s\alpha,2s\alpha});$$

$$\varepsilon_{1s\beta} = h_{1s\beta,1s\beta} + J_{1s\alpha,1s\beta} + J_{1s\beta,2s\alpha};$$

$$\varepsilon_{2s\alpha} = h_{2s\alpha,2s\alpha} + (J_{1s\alpha,2s\alpha} - K_{1s\alpha,2s\alpha}) + J_{1s\beta,2s\alpha}; \qquad (72.11a)$$

$$\varepsilon_{2s\beta} = h_{2s\beta,2s\beta} + J_{1s\alpha,2s\beta} + (J_{1s\beta,2s\beta} - K_{1s\beta,2s\beta}) + J_{2s\alpha,2s\beta}; \quad \text{(virtual)}.$$

In the URHF method, $\phi_{1s\alpha}(\mathbf{r})$ is different from $\phi_{1s\beta}(\mathbf{r})$, and $\phi_{2s\alpha}(\mathbf{r})$ is different from $\phi_{2s\beta}(\mathbf{r})$. The last orbital $\phi_{2s\beta}$ is a virtual orbital.

Alternatively, if we follow Roothaan's procedure with $f = ½$, and a = 1 and b = 2 for the half-filled open shell we get the wave functions

$$\Phi(M_S = ½) = \frac{1}{\sqrt{6}}\,||1s(\mathbf{r}_1)\alpha(s_1)\ \ 1s(\mathbf{r}_2)\beta(s_2)\ \ 2s(\mathbf{r}_3)\alpha(s_3)||, \qquad (72.8b)$$

$$\Phi(M_S = -½) = \frac{1}{\sqrt{6}}\,||1s(\mathbf{r}_1)\alpha(s_1)\ \ 1s(\mathbf{r}_2)\beta(s_2)\ \ 2s(\mathbf{r}_3)\beta(s_3)||.$$

The total energy is calculated from (70.24). One finds

$$E = (2h_{1s,1s} + h_{2s,2s}) + [J_{1s,1s} + (2J_{1s,2s} - K_{1s,2s})] \qquad (72.9b)$$

with the common Fock operator (for $\alpha = 0$ and $\beta = -2$)

$$\mathbf{F} = [-\frac{1}{2}\nabla^2 - \frac{3}{r}] + [(2\mathbf{J}_{1s} - \mathbf{K}_{1s}) + (\mathbf{J}_{2s} - \frac{1}{2}\mathbf{K}_{2s})] + 2[\mathbf{M}_{1s} + \frac{1}{2}(\mathbf{M}_{2s} - \mathbf{K}_{2s})], \qquad (72.10c)$$

so that the orbital energies are

$$\varepsilon_{1s} = h_{1s,1s} + J_{1s,1s} + (J_{1s,2s} + ½K_{1s,2s});$$

$$\varepsilon_{2s} = h_{2s,2s} + (2J_{1s,2s} - K_{1s,2s}) + ½J_{2s,2s}. \qquad (72.11b)$$

The RHF open-shell method is based on identical spatial functions for α and β orbitals.

Example 52: Carbon atom

As an excellent demonstration of the open-shell theory we consider the multiplet structure of carbon atom. The open shell configuration is $2p^2$. In the following we will not explicitly write the closed shell part of the wave function, but it will be understood that the closed shell part is present. Thus the determinant for the configuration $1s^2 2s^2 2p_1\alpha 2p_1\beta$ will be denoted by $\{(211\tfrac{1}{2})(211\tfrac{\bar{1}}{2})\}$:

$\{(211\tfrac{1}{2})(211\tfrac{\bar{1}}{2})\}$

$$\equiv \frac{1}{\sqrt{6}}\|1s(\mathbf{r}_1)\alpha(s_1)\ \ 1s(\mathbf{r}_2)\beta(s_2)\ \ 2s(\mathbf{r}_3)\alpha(s_3)\ \ 2s(\mathbf{r}_4)\beta(s_4)\ \ 2p_1(\mathbf{r}_5)\alpha(s_5)\ \ 2p_1(\mathbf{r}_6)\beta(s_6)\|. \qquad (72.12a)$$

Since there are two open-shell electrons in six open-shell orbitals, the total number of open-shell determinant is ${}^6C_2 = 15$. These are all eigenfunctions of $\boldsymbol{L}_z$ and $\boldsymbol{S}_z$:

Determinant	M_L	M_S	Determinant	M_L	M_S
$D_1 = \{(211\frac{1}{2})\,(211\frac{\bar{1}}{2})\}$	2	0	$D_9 = \{(210\frac{1}{2})\,(210\frac{\bar{1}}{2})\}$	0	0
$D_2 = \{(211\frac{1}{2})\,(210\frac{1}{2})\}$	1	1	$D_{10} = \{(211\frac{\bar{1}}{2})\,(21\bar{1}\frac{\bar{1}}{2})\}$	0	-1
$D_3 = \{(211\frac{1}{2})\,(210\frac{\bar{1}}{2})\}$	1	0	$D_{11} = \{(210\frac{1}{2})\,(21\bar{1}\frac{1}{2})\}$	-1	1
$D_4 = \{(211\frac{\bar{1}}{2})\,(210\frac{1}{2})\}$	1	0	$D_{12} = \{(210\frac{1}{2})\,(21\bar{1}\frac{\bar{1}}{2})\}$	-1	0
$D_5 = \{(211\frac{\bar{1}}{2})\,(210\frac{\bar{1}}{2})\}$	1	-1	$D_{13} = \{(210\frac{\bar{1}}{2})\,(21\bar{1}\frac{1}{2})\}$	-1	0
$D_6 = \{(211\frac{1}{2})\,(21\bar{1}\frac{1}{2})\}$	0	1	$D_{14} = \{(210\frac{\bar{1}}{2})\,(21\bar{1}\frac{\bar{1}}{2})\}$	-1	-1
$D_7 = \{(211\frac{1}{2})\,(21\bar{1}\frac{\bar{1}}{2})\}$	0	0	$D_{15} = \{(21\bar{1}\frac{1}{2})\,(21\bar{1}\frac{\bar{1}}{2})\}$	-2	0
$D_8 = \{(211\frac{\bar{1}}{2})\,(21\bar{1}\frac{1}{2})\}$	0	0			

(72.12)

Since both l_1 and l_2 are equal to 1, the total orbital angular momentum quantum number L will be 0, 1 or 2. By applying $\boldsymbol{L}^2$ and $\boldsymbol{S}^2$ on these determinants we find that eight of them are eigenfunctions of these operators:

$D_1 \Leftrightarrow L=2, S=0$ Term: ${}^1D$		
$D_2 \Leftrightarrow L=1, S=1$ Term: ${}^3P$	$D_5 \Leftrightarrow L=1, S=1$: ${}^3P$	$D_6 \Leftrightarrow L=1, S=1$: ${}^3P$
$D_{10} \Leftrightarrow L=1, S=1$ Term: ${}^3P$	$D_{11} \Leftrightarrow L=1, S=1$: ${}^3P$	$D_{14} \Leftrightarrow L=1, S=1$: ${}^3P$
$D_{15} \Leftrightarrow L=2, S=0$ Term: ${}^1D$		(72.13)

That D_1 and D_{15} are ${}^1D$ states can be guessed directly from the eigenvalues M_L and M_S. As $M_L = \pm 2$, L should be at least 2, and $L = 2$ is the only possibility here. With M_L either equal to 2 or equal to -2, the only M_S is zero. Therefore, S must be zero.

The application of $\boldsymbol{L}^2$ on D_7, D_8 and D_9 generates linear combinations of these functions:

$$L^2D_7 = 2\hbar^2D_7 + 2\hbar^2D_9,$$

$$L^2D_8 = 2\hbar^2D_8 - 2\hbar^2D_9,$$

$$L^2D_9 = 2\hbar^2D_7 - 2\hbar^2D_8 + 4\hbar^2D_9.$$

All the three determinants correspond to $M_L = 0$ and $M_S = 0$. Therefore, the eigenfunctions will be prepared by linearly combining these determinants as $A = c_1D_7 + c_2D_8 + c_3D_9$ which leads to the algebraic equation

$$\begin{vmatrix} 2-L(L+1) & 0 & 2 \\ 0 & 2-L(L+1) & -2 \\ 2 & -2 & 4-L(L+1) \end{vmatrix} = 0\,.$$

The roots are $L = 0$, 1 and 2. For $L = 0$, the linear combination coefficients satisfy

$c_1 + c_3 = 0,$

$c_2 - c_3 = 0,$

and

$c_1 - c_2 + 2c_3 = 0$

so that one finds the normalized solution

$A_7 = 3^{-1/2}(-D_7 + D_8 + D_9).$ $\qquad (L=0)$

One can similarly obtain

$A_8 = 2^{-1/2}(D_7 + D_8),$ $\qquad (L=1),$

$A_9 = 6^{-1/2}(D_7 - D_8 + 2D_9),$ $\qquad (L=2).$

By applying the $\boldsymbol{S}^2$ operator on D_7, D_8 and D_9 and then on A_7, A_8 and A_9 one finds $\boldsymbol{S}^2A_7 = 0$, $\boldsymbol{S}^2A_8 = 2\hbar^2A_8$ and $\boldsymbol{S}^2A_9 = 0$. Hence one has the following functions

$$\begin{array}{lll} A_7 = 3^{-1/2}(-D_7 + D_8 + D_9) & \Leftrightarrow L=0, S=0 & \text{Term: } {}^1\text{S} \\ A_8 = 2^{-1/2}(D_7 + D_8), & \Leftrightarrow L=1, S=1 & \text{Term: } {}^3\text{P} \\ A_9 = 6^{-1/2}(D_7 - D_8 + 2D_9) & \Leftrightarrow L=2, S=0 & \text{Term: } {}^1\text{D}. \end{array} \qquad (72.14)$$

It is possible to derive the remaining four eigenfunctions of $\boldsymbol{L}^2$, $\boldsymbol{S}^2$, $\boldsymbol{L}_Z$ and $\boldsymbol{S}_Z$ in the same way:

$$\begin{array}{lll} A_3 = 2^{-1/2}(D_3 + D_4) & \Leftrightarrow L=1, S=1 & \text{Term: } {}^3\text{P} \\ A_4 = 2^{-1/2}(D_3 - D_4) & \Leftrightarrow L=2, S=0 & \text{Term: } {}^1\text{D} \\ A_{12} = 2^{-1/2}(D_{12} + D_{13}) & \Leftrightarrow L=1, S=1 & \text{Term: } {}^3\text{P} \\ A_{13} = 2^{-1/2}(D_{12} - D_{13}) & \Leftrightarrow L=2, S=0 & \text{Term: } {}^1\text{D}. \end{array} \qquad (72.15)$$

Let us consider the ground state ^{3}P for the carbon atom. The total energy is obtained from Eq. (70.24) using $f = 1/3$, $a = 3/4$ and $b = 3/2$:

$$\begin{aligned} E = {} & 2[\text{h1s,1s}+\text{h2s,2s}+\frac{1}{3}(\text{h2p}_1\text{,2p}_1+\text{h2p}_0\text{,2p}_0+\text{h2p}_{-1}\text{,2p}_{-1})] \\ & +[\text{J1s,1s}+\text{J2s,2s}+2(2\text{J1s,2s}-\text{K1s,2s}) \\ & +\frac{1}{3}(\text{J2p}_1\text{,2p}_0-\text{K2p}_1\text{,2p}_0+\text{J2p}_1\text{,2p}_{-1}-\text{K2p}_1\text{,2p}_{-1}+\text{J2p}_0\text{,2p}_{-1}-\text{K2p}_0\text{,2p}_{-1})] \\ & +\frac{2}{3}[(2\text{J1s,2p}_1-\text{K1s,2p}_1)+(2\text{J2s,2p}_1-\text{K2s,2p}_1)+(2\text{J1s,2p}_0-\text{K1s,2p}_0) \\ & +(2\text{J2s,2p}_0-\text{K2s,2p}_0)+(2\text{J1s,2p}_{-1}-\text{K1s,2p}_{-1})+(2\text{J2s,2p}_{-1}-\text{K2s,2p}_{-1})]. \end{aligned} \qquad (72.16)$$

This expression correspond to each of the wave functions

$$\varphi(M_s=-1)=3^{-1/2}[D_2+D_{11}-D_6];$$
$$\varphi(M_s=0)=3^{-1/2}[A_3+A_{12}-A_8]; \qquad (72.17)$$
$$\varphi(M_s=-1)=3^{-1/2}[D_5+D_{14}-D_{10}].$$

Note the following equalities:

$$h_{2p_1,2p_1}=h_{2p_0,2p_0}=h_{2p_{-1},2p_{-1}}$$
$$J_{1s,2p_1}=J_{1s,2p_0}=J_{1s,2p_{-1}}$$
$$J_{2s,2p_1}=J_{2s,2p_0}=J_{2s,2p_{-1}}$$
$$K_{1s,2p_1}=K_{1s,2p_0}=K_{1s,2p_{-1}} \qquad (72.18)$$
$$K_{2s,2p_1}=K_{2s,2p_0}=K_{2s,2p_{-1}}$$
$$J_{2p_1,2p_0}=J_{2p_1,2p_{-1}}=J_{2p_0,2p_{-1}}$$
$$K_{2p_1,2p_0}=K_{2p_1,2p_{-1}}=K_{2p_0,2p_{-1}}$$

These relations can be used to simplify the Hartree-Fock calculation. The orbital energies are calculated from Eq. (70.26) with $\alpha = 3/8$ and $\beta = -3/4$. We get

$$\varepsilon_{1s}=h_{1s,1s}+[J_{1s,1s}+(2J_{1s,2s}-K_{1s,2s})+\frac{11}{12}(J_{1s,2p_1}+J_{1s,2p_0}+J_{1s,2p_{-1}})$$
$$-\frac{1}{12}(K_{1s,2p_1}-K_{1s,2p_0}+K_{1s,2p_{-1}})]$$

$$\varepsilon_{2s}=h_{2s,2s}+[(2J_{1s,2s}-K_{1s,2s})+2J_{2s,2s}+\frac{11}{12}(J_{2s,2p_1}+J_{2s,2p_0}+J_{2s,2p_{-1}})$$
$$-\frac{1}{12}(K_{2s,2p_1}+K_{2s,2p_0}+K_{2s,2p_{-1}})] \qquad (72.19)$$

$$\varepsilon_{2p}=h_{2p_0,2p_0}+[(2J_{1s,2p_0}-K_{1s,2p_0})+(2J_{2s,2p_0}-K_{2s,2p_0})$$
$$+\frac{7}{12}(J_{2p_1,2p_0}+J_{2p_0,2p_0}+J_{2p_0,2p_{-1}})$$
$$-\frac{5}{12}(K_{2p_1,2p_0}+K_{2p_0,2p_0}+K_{2p_0,2p_{-1}})]$$

The reader is urged to write similar equations for Be atom which is a closed-shell species in the ground state and the N atom which has a ^4S ground state.

Lecture 73: The Roothaan Equations

We have already seen that the Hartree-Fock equation is of the general form

$$\mathbf{F(r)}\ \phi_i(\mathbf{r}) = \varepsilon_i\phi_i(\mathbf{r}). \tag{70.16}$$

The exact expression of the Fock operator depends on the nature of the system and the nature of the treatment. Thus Pople-Nesbet open-shell equations (70.20) and (70.22) involve different Fock operators, and the RHF open-shell equations (70.25) – (70.28) involve yet another Fock operator. The Fock operator is basically of the type

$$\mathbf{F(r)} = -\frac{1}{2}\nabla^2 - \sum_{n}^{\text{Nuclei}} Z_n r^{-1} + \mathrm{U}(\mathbf{r}) \tag{73.1}$$

where U(**r**) is some sort of an average two-electron interaction operator consisting of both local (direct) and nonlocal (exchange) interactions. A nonlocal operator is defined as an operator which, when operating on a function at a specific point, yields a result that depends on the value of the function at some other point. In any case the form (73.1) makes it explicit that like Schrödinger equation the Hartree-Fock equation is a second-order differential equation. Therefore, in principle Eq. (70.16) can be solved by numerical techniques.

The numerical method of solution becomes practicable for atoms and small linear molecules. For these systems the angular behavior of the orbitals are exactly known, leaving aside a differential equation involving only a few variables (such as radial distances). For atoms, one needs to solve the second-order differential equation for the radial function for each (n, l) subshell. This implies only a 3-variable problem for atoms B-Ne. Such equations have been very accurately solved by Froese-Fischer (1972, 1973, 1977). The energy values are correct through six significant digits. Nevertheless, they are still not exact values, and the calculated values of orbitals are not correct at different points. These solutions are almost universally taken as standard and often referred to as solutions in the *Hartree-Fock limit*. Unfortunately for a general molecule not much simplification is attained since the angular movement of the electrons is extraordinarily complicated. Hence the numerical procedure for a general molecule will involve the solution of a many-dimensional integro-differential equation. This task is enormously complicated and it would be computationally taxing if at all it becomes tractable.

Hence the general procedure for solving the Hartree-Fock equations is to employ a set of known or basis functions and find the best linear expressions in this basis as the required solutions. A set of known functions $\{\chi_\mu(\mathbf{r}) \mid \mu = 1, 2,..., K\}$ is introduced. The unknown atomic or molecular orbitals are linearly expanded in this basis as

$$\phi_i = \sum_{\mu=1}^{K} \chi_\mu C_{\mu i} \qquad \text{for i} = 1, 2, \ldots, \mathrm{K}. \tag{73.2}$$

These expressions can be exact if the set $\{\chi_\mu\}$ is complete. In practice one uses a finite set of basis functions. So the orbitals calculated are not exactly eigenfunctions of **F**. They represent approximate atomic or molecular orbitals. It is necessary to employ a basis that can yield reasonably accurate expansions for the orbitals occupied in the ground state. If we put the expansion (73.2) in the restricted Hartree-Fock equation (70.16), multiply both sides of the resulting expression with χ_μ^* and integrate over all space, we get the *Roothaan equations*

$$\sum_\nu F_{\mu\nu}C_{\nu i} = \sum_\nu S_{\mu\nu}C_{\nu i}\,\varepsilon_i \qquad \text{for } i = 1, 2, \ldots, K. \tag{73.3}$$

The two quantities $F_{\mu\nu}$ and $S_{\mu\nu}$ represent the integrals

$$F_{\mu\nu} = \int dv_1\, \chi_\mu^*(\mathbf{r}_1)\, \mathbf{F}(\mathbf{r}_1)\, \chi_\nu(\mathbf{r}_1)$$

and (73.4)

$$S_{\mu\nu} = \int dv_1\, \chi_\mu^*(\mathbf{r}_1)\, \chi_\nu(\mathbf{r}_1).$$

These can be taken as (μ, ν) elements of the Fock operator matrix **F** and the overlap matrix **S**. Both the matrices are square of rank *K*. Equations (73.3) for i = 1, 2, ... , *K* together are then realized as the matrix equation

$$\mathbf{FC} = \mathbf{SC\varepsilon} \tag{73.5}$$

which is often termed as the *Hartree-Fock-Roothaan equation.* The eigenvalue matrix **ε** is also square of rank *K*, and in fact diagonal. Each column of the matrix **C** is an eigenvector. Elements of a particular column, say, the *ith* one, are the coefficients for the particular (*ith*) orbital.

The way to solve the Hartree-Fock-Roothaan equation involves the definition of the square-root matrix $\mathbf{S}^{1/2}$ and its inverse $\mathbf{S}^{-1/2}$ such that

$$\mathbf{S}^{1/2}\,\mathbf{S}^{1/2} = \mathbf{S},$$

and (73.6)

$$\mathbf{S}^{1/2}\,\mathbf{S}^{-1/2} = \mathbf{1}.$$

One now rewrites the Hartree-Fock-Roothaan equation as

$$\tilde{F}\tilde{C} = \tilde{C}\boldsymbol{\varepsilon} \tag{73.7}$$

where

$$\tilde{F} = \mathbf{S}^{-1/2}\mathbf{F}\mathbf{S}^{-1/2}$$

and (73.8)

$$\tilde{C} = \mathbf{S}^{1/2}\mathbf{C}.$$

Equation (73.7) represents a simple matrix diagonalization problem. The $\tilde{F}$ matrix is diagonalized to form the orbital energy matrix **ε**. The eigenvector matrix **C** is found from the eigenvector matrix $\tilde{C}$ of matrix $\tilde{F}$:

$$\mathbf{C} = \mathbf{S}^{-1/2}\,\tilde{C}. \tag{73.9}$$

Hence given a Fock matrix, one can solve Eq. (73.5) provided that one has a suitable computer program. Everything now depends on the formation of the Fock matrix.

The Fock-Dirac density for a closed-shell species in its ground state configuration is written as

$$\rho^{(1)}(\mathbf{r}) = 2\sum_{u=1}^{N/2} |\phi_u(\mathbf{r})|^2. \tag{73.10}$$

It represents the probability density of finding an electron at the specific point indicated by the position vector **r**. Using expansion (73.2) $\rho^{(1)}(\mathbf{r})$ can be found as

$$\rho^{(1)}(\mathbf{r}) = \sum_{\mu=1}^{K}\sum_{\nu=1}^{K} P_{\mu\nu}\chi_\mu(\mathbf{r})\chi_\nu^*(\mathbf{r}) \qquad (73.11)$$

where $P_{\mu\nu}$ is a *density matrix*, or as it is often called, the *Mulliken population matrix*. It is given by

$$P_{\mu\nu} = 2\sum_{u=1}^{N/2} C_{\mu u}C_{\nu u}^*. \qquad (73.12)$$

The quantity $P_{\mu\nu}S_{\nu\mu}$ is interpreted as the electronic population of the overlap distribution $\chi_\mu\chi_\nu^*$, and $P_{\mu\mu}S_{\mu\mu} = P_{\mu\mu}$ represents the net electronic charge occupying the (atomic) orbital ϕ_μ. The population matrix P can be found using the expansion coefficients. One can show

$$F_{\mu\nu} = h_{\mu\nu} + \sum_{\lambda\sigma} P_{\lambda\sigma}[(\mu\nu|\sigma\lambda) - \frac{1}{2}(\mu\lambda|\sigma\nu)]. \qquad (73.13)$$

Hence the population matrix is needed to form the Fock matrix.

Similar population matrices can be defined in the open shell case. There we come across two different Fock operators for α and β spins, and similarly two different population matrices. In every case, the Fock matrix (or matrices) will be formed with the population matrix (or matrices).

Therefore, the systematics of the HFSCF procedure for the basis set (or analytical) method can be described as follows:

i. The system is specified (by the number of nuclei, nuclear coordinates, atomic numbers, and the number of electrons). The basis set is also specified. These are the input data.

ii. All integrals such as $S_{\mu\nu}$, $h_{\mu\nu}$ and $(\mu\nu|\lambda\sigma)$ are calculated.

iii. The overlap matrix **S** is diagonalized to form $\mathbf{d} = \mathbf{U}^{-1}\mathbf{S}\mathbf{U}$ where **U** is unitary, then $\mathbf{d}^{1/2}$ is obtained with square roots of diagonal elements, $\mathbf{d}^{-1/2}$ is obtained with inverse of the square roots, and finally $\mathbf{S}^{1/2}$ and $\mathbf{S}^{-1/2}$ are prepared as $\mathbf{S}^{1/2} = \mathbf{U}\mathbf{d}^{1/2}\mathbf{U}^{-1}$ and $\mathbf{S}^{-1/2} = \mathbf{U}\mathbf{d}^{-1/2}\mathbf{U}^{-1}$.

iv. The initial population matrix **P** is guessed directly as input, or indirectly from the input matrix **C**, or formed from **C** which may be obtained by diagonlizing a suitable matrix such as the matrix **h**.

v. The Fock matrix **F** is formed by using Eq. (73.13) or similar equations for open-shell systems.

vi. Matrix $\tilde{F}$ is formed.

vii. Matrix $\tilde{F}$ is diagonalized to obtain the matrices $\tilde{C}$ and $\boldsymbol{\varepsilon}$.

viii. Matrix **C** is formed from $\tilde{C}$ and then a new population matrix is prepared.

ix. Convergence of the procedure is checked by comparing a suitable quantity such as the total energy. If the procedure has not converged beyond a tolerance, steps (v)–(ix) are repeated.

x. If the convergence has been achieved, E, ε_i's, **C** and **P** are printed. One may calculate molecular properties such as the electric dipole moment, quadrupole moments, etc.

These steps have been illustrated by the charts 12.2(a) and 12.2(b).

Chart 12.2(a) Schematic representation of the HFSCF treatment for a general molecule: formation of the transformed Fock operator for diagonalization.

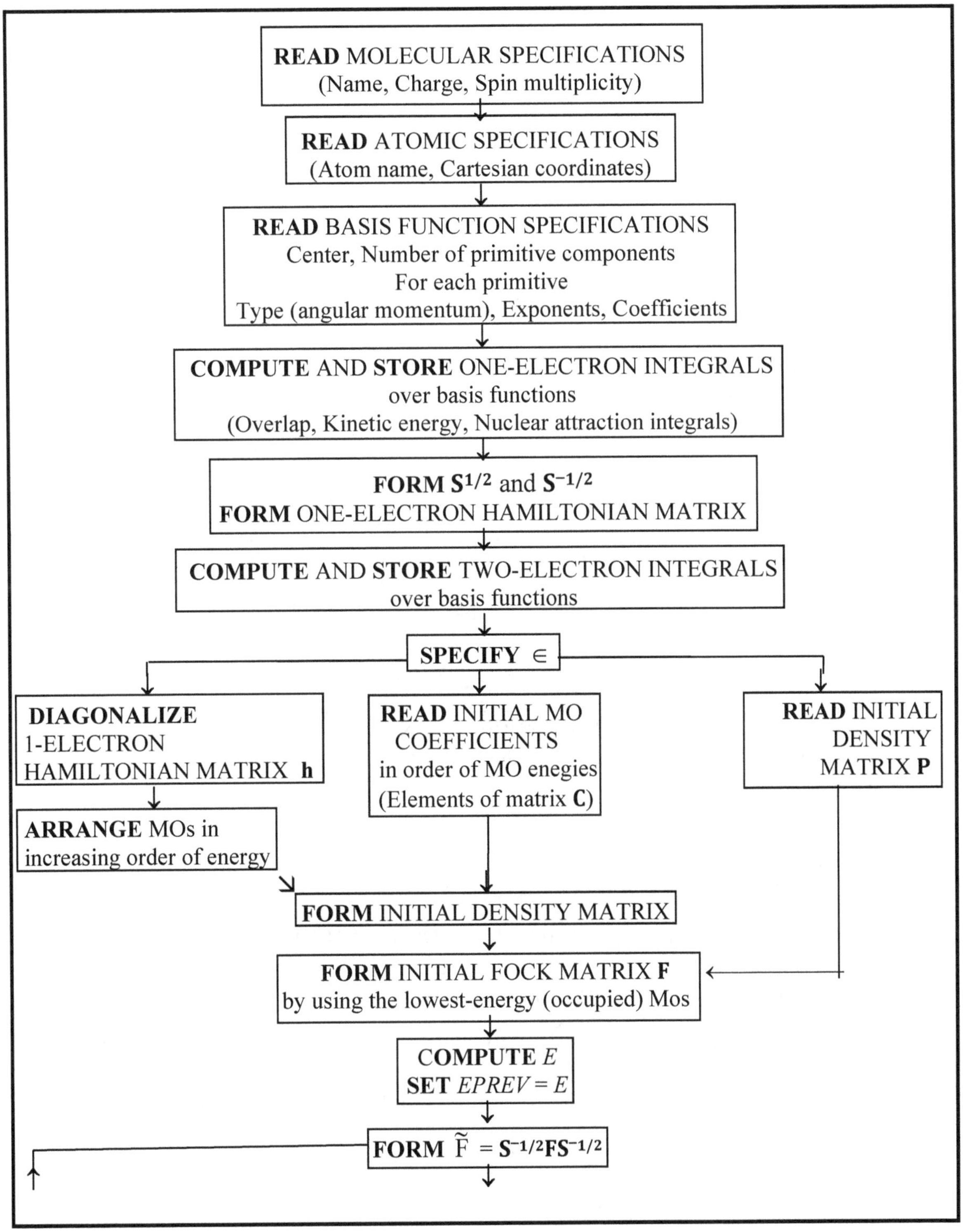

Chart 12.2(b) Schematic representation of the HFSCF treatment … (Continued)

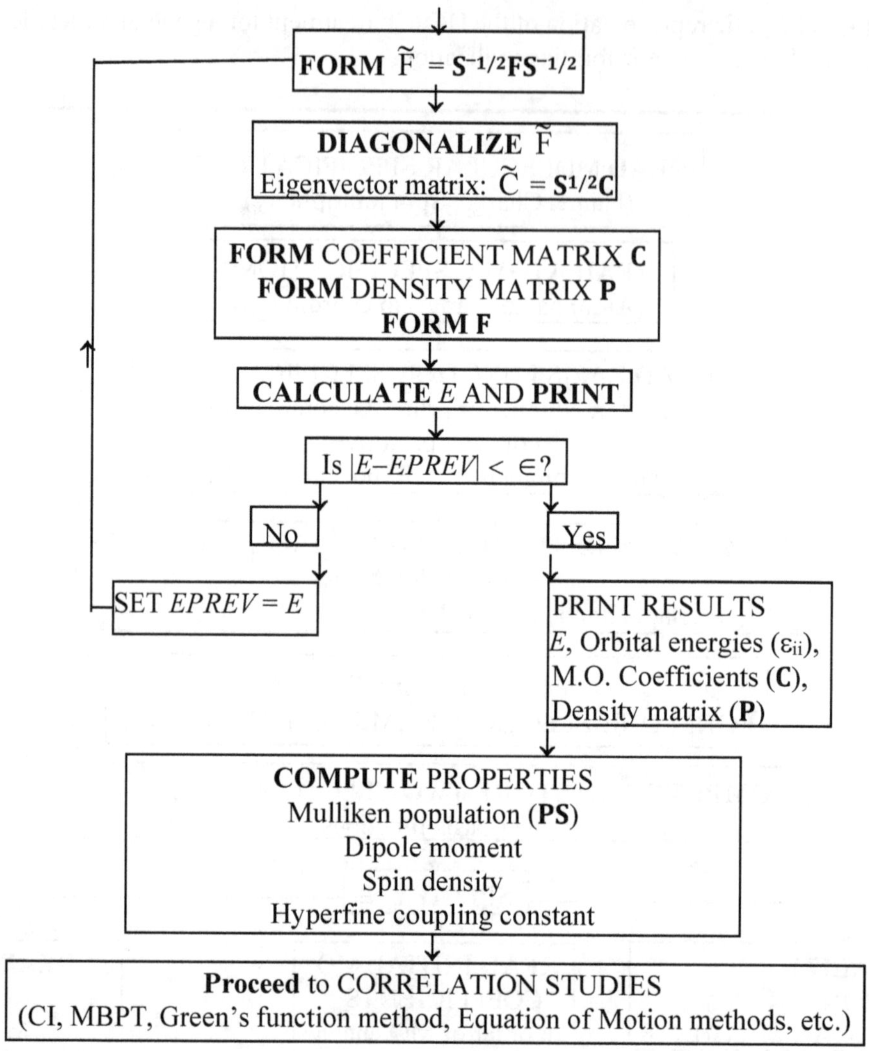

Population Analysis

The number of electrons associated with a given atom or a molecule can be defined in many ways. It is useful to know the atomic or orbital population, as this information can help in interpreting the properties of a molecule. Observe that

$$\text{Trace (PS)} = \sum_{\mu} (\text{PS})_{\mu\mu} = \sum_{\mu} \sum_{\nu} P_{\mu\nu} S_{\nu\mu}. \tag{73.14}$$

When the expression (73.12) is substituted for the population matrix element one obtains

$$\begin{aligned} \text{Trace (PS)} &= \sum_{\mu} (\text{PS})_{\mu\mu} = 2 \sum_{u=1}^{N/2} \int dv\, |\phi_u(\mathbf{r})|^2 \\ &= N. \end{aligned} \tag{73.15}$$

Hence one may also define $(PS)_{\mu\mu}$ as the number of electrons associated with the basis function χ_μ. The quantity $(PS)_{\mu\mu}$ is known as *Mulliken population*. The net atomic charge is written as

$$q_A = Z_A - \sum_{\mu \in A} (PS)_{AA}. \qquad (73.16)$$

We can also write Eq. (73.15) as

$$N = \sum_{\mu} (S^{\alpha}PS^{1-\alpha})_{\mu\mu} = \sum_{\mu} (P')_{\mu\mu} \qquad (73.17)$$

as Trace (AB) = Trace (BA). One may choose any arbitrary α. In Löwdin's (1950) *population analysis*, $\alpha = ½$ and the net charge on atom A is defined by

$$q_A = Z_A - \sum_{\mu \in A} (S^{1/2}PS^{1/2})_{\mu\mu}. \qquad (73.18)$$

This makes it clear that the population analysis scheme is not unique. Therefore, adequate care must be taken in attaching physical significance to the net atomic populations calculated by any scheme. If we continue to think along this line, it will eventually bring us to the definition of natural orbitals, a concept that will be discussed in Lecture 76.

Lecture 74: Basis Sets

STOs

We have already discussed in Chapter 5 that Slater type orbitals (STOs) are good choices for basis functions, as they constitute an approximate representation of atomic orbitals. These functions are given by

$$\phi_{nlm}(\mathbf{r}) = \frac{(2\varsigma_{nl})^{n+1/2}}{(2n!)^{1/2}} r^{n-1} e^{-\varsigma_{nl} r} Y_{lm}(\theta,\varphi). \qquad (26.2)$$

These are very suitable for an atomic calculation since the overlap integrals, the expectation values of kinetic and nuclear potential energies and even the two-electron integrals can be easily calculated with them. Calculation of these integrals for a molecule is not so easy. Nevertheless, expressions for most of the integrals can be derived. The problem arises with 3-centre and especially with 4-centre electron repulsion integrals. Exact analytical expressions for electron repulsion over orbitals three or all four of which are centered on different nuclei are not always available. For these integrals one has to rely on numerical integration techniques, and since the number of such integrals increases very rapidly with the number of basis functions, the STO's are not considered to be very suitable for a general molecular calculation.

Gaussian functions

Boys (1950) proposed the use of Gaussian functions of the form

$$g_{nlm}(\alpha, \mathbf{r}) = N\, r^{n-1}\, e^{-\alpha r^2}\, Y_{lm}(\theta, \phi) \qquad (74.1a)$$

as atomic orbitals. Gaussian wave functions became very rapidly popular in atomic and molecular calculations. Normalized cartesian gaussian primitive functions of type 1s, 2p, 3d, etc. are written as

$$\mathrm{g_s}(\alpha, \mathbf{r}) = \left(\frac{8\alpha^3}{\pi^3}\right)^{1/4} e^{-\alpha r^2},$$

$$\mathrm{g_{p}}_x(\alpha, \mathbf{r}) = \left(\frac{128\alpha^5}{\pi^3}\right)^{1/4} x e^{-\alpha r^2}, \tag{74.1b}$$

$$\mathrm{g_{d}}_{xy}(\alpha, \mathbf{r}) = \left(\frac{2048\alpha^7}{\pi^3}\right)^{1/4} xy e^{-\alpha r^2},$$

etc. The Gaussian functions have two advantages. The *first* advantage is that often a three-dimensional integral over the Gaussian functions can be reorganized as a product of three integrals involving x, y and z variables separately. The calculation of a one-dimensional integral is certainly easier than that of a three-dimensional integral. The *second* advantage is more intrinsic: the product of two Gaussian functions is another Gaussian function. In fact

$$e^{-\alpha_1 |\mathbf{r}-\mathbf{R}_A|^2} e^{-\alpha_2 |\mathbf{r}-\mathbf{R}_B|^2} = \mathrm{C_{AB}}\, e^{-(\alpha_1+\alpha_2)|\mathbf{r}-\mathbf{R}_P|^2} \tag{74.2}$$

where

$$\mathbf{R}_P = (\alpha_1\mathbf{R}_A + \alpha_2\mathbf{R}_B)/(\alpha_1+\alpha_2),$$

$$\mathrm{C_{AB}} = e^{-\gamma |\mathbf{R}_A - \mathbf{R}_B|^2},$$

$$\gamma = \alpha_1\alpha_2/(\alpha_1+\alpha_2).$$

Gaussian functions of the type shown in equations (74.1a) and (74.1b) can be directly chosen as basis functions for a Hartree-Fock calculation. The resulting atomic orbitals would be linear combinations of Gaussian functions of same symmetries. The disadvantage of this procedure is that a large number of Gaussian functions are required to describe all relevant (occupied) atomic orbitals. The number of two-electron integrals is of order $K^4/8$ where K is the basis size, and it increases very rapidly as K increases. For $K = 20$ the number of two-electron integrals is of order 10^4 whereas for $K = 60$, it is of the order of 10^6! This in turn needs the calculation of too many two-electron integrals and the use of these integrals in forming the Fock matrix in every loop of the SCF process.

Contractions

To get rid of the dilemma one may use approximations to STO's as basis functions. These approximations are prepared as fixed linear combinations of the primitive Gaussian functions. These combinations are called contractions. The *contracted Gaussian function* (CGF) is generally written as

$$\chi_\mu^{\mathrm{CGF}}(\mathbf{r}-\mathbf{R}_A) = \sum_{\mathrm{p}=1}^{L} d_{\mathrm{p}\mu}\, \mathrm{g_p}(\alpha_{\mathrm{p}\mu}, \mathbf{r}-\mathbf{R}_A). \tag{74.3}$$

The number of primitives L is called the *length of contraction* and $d_{\mathrm{p}\mu}$ is called a *contraction coefficient*. All the primitives used in such a contraction have to be preferably of the same symmetry. Primitives of two or more symmetries may be combined together when the angular dependence is fairly complicated. For example, a d_{x2-y2} orbital may be described as an appropriate linear combination of d_{x2} primitive Gaussians and d_{y2} primitive Gaussians.

Gaussian-type orbitals (GTOs)

A STO may be fitted to a CGF. The CGF is then called a *Gaussian-type orbital* or in short, a *GTO*. If the contraction length is n, the expansion (74.3) is called a STO-nG basis function. The STO-nG basis implies that each STO in the set has been expanded as a linear combination of n primitive Gaussian functions. The best linear combination may be determined from a least-square analysis. Table 12.3 contains a list of optimum exponents and contraction coefficients for 1G, 2G and 3G expansions of the 1s STO with $\zeta_{10} = 1$.

Table 12.3 Optimum Gaussian exponents and contraction coefficients for STO-nG expansions for the 1s orbital of hydrogen atom (the 1s STO).

n	Exponent	Contraction coefficient
1	0.270950	1.0
2	0.151623	0.678914
	0.851819	0.430129
3	0.109818	0.444635
	0.405771	0.535328
	2.22766	0.154329

O-Ohata, Taketa and Huzinaga (1966) gave a list of such Gaussian exponents and coefficients for STO-nG ($n \geq 4$) expansions of 1s, 2s, 2p, 3s, 3p and 3d orbitals each with $\zeta = 1$. It is very easy to prepare STO-nG expansions for other values of the STO exponent. The exponent ζ can be viewed as a scale factor that multiplies the radial distance r. Let

$$\chi_\mu(\zeta; \mathbf{r}) = \sum_{p=1}^{n} d_{p\mu}\, g_p(\alpha_{p\mu}; \mathbf{r}) \tag{74.4}$$

be a STO-nG expansion that has been optimized by the least-square analysis, that is, the integral

$$\mathcal{I} = \int dv\, [\phi_\mu(\zeta; \mathbf{r}) - \sum_{p=1}^{n} d_{p\mu}\, g_p(\alpha_{p\mu}; \mathbf{r})]^2$$

is stationary for variation of the $d_{p\mu}$'s and the $\alpha_{p\mu}$'s. See that for normalized functions one obtains from (25.2), $\phi_\mu(\zeta; \mathbf{r}) = \zeta^{3/2}\phi_\mu(1; \mathbf{r}')$, and similarly from (74.1a), $g_p(\alpha_{p\mu}; \mathbf{r}) = \zeta^{3/2}$ $g_p(\zeta^{-2}\alpha_{p\mu}; \mathbf{r}')$ where $\mathbf{r}' = \zeta\mathbf{r}$. Therefore, the same integral can be written as

$$\mathcal{I} = \int dv'\, [\phi_\mu(1; \mathbf{r}') - \sum_{p=1}^{n} d_{p\mu}\, g_p(\alpha_{p\mu}/\zeta^2; \mathbf{r}')]^2$$

as $dv' = \zeta^3 dv$. In turn, for $\mathbf{r}' = \zeta'\mathbf{r}''$, the integral can be written as

$$\mathcal{I} = \int dv'' \,[\phi_\mu(\zeta'; \mathbf{r}'') - \sum_{p=1}^{n} d_{p\mu}\, g_p(\alpha_{p\mu}'; \mathbf{r}'')]^2$$

where we use the substitution

$$\alpha_{p\mu}' = \alpha_{p\mu}(\zeta'/\zeta)^2. \qquad (74.5)$$

This shows that the least fit for a STO with a new exponent ζ' can be obtained with unchanged contraction coefficients $\{d_{p\mu}\}$ but with a new set of Gaussian exponents $\{\alpha_{p\mu}'\}$.

Minimal basis

A minimal basis set is a set in which the number of basis functions per atom is the least number with the proper angular momentum symmetries required to describe all the occupied orbitals in the atomic ground state. A minimal basis calculation involves a drastically restricted number of variational degrees of freedom. So the results of a minimal basis calculation are less accurate, particularly when the exponents are not optimized for the specific atom or molecule. The minimal basis set of STO's for an all-electron calculation on a compound containing only an element of the second row in the periodic table and hydrogen is denoted by (1s2s2p/1s) where the slash separates the functions for the second row atom and hydrogen. The minimal basis set for a valence-electron calculation on the same molecule would be designated as (2s2p/1s).

More exotic basis sets

A *double-zeta* (DZ) basis is obtained by replacing each primitive function (STO or GTO) of (often) a minimal basis set (but in general any basis set) by two primitive functions which differ in their orbital exponents.

A *split-valence* (SV) basis set uses two primitives for each valence atomic orbital but only one function for each atomic orbital belonging to the inner shells of each atom. The valence orbitals become generally distorted by the formation of molecular orbitals, that is, the linear combinations require *polarization functions* with higher l quantum numbers. For instance, the *double zeta plus polarization* (DZ+P or DZP) basis involves the DZ basis plus the 2p function for the hydrogen atom or five 3d functions for a second or a third-row atom. The *extended basis set* involves more than two primitive functions for each atomic orbital.

The STO-nG series developed by Pople and coworkers (1969) is one of the most widely used series for molecular calculations. The basis sets are small in size yet very effective in predicting molecular geometries and other properties. The STO-nG* and STO-nG** series are obtained by adding to the basis set one or two polarization functions for each STO belonging to the STO-nG series. Pople and coworkers (1971, 1972, 1980) also prepared a series of Gaussian basis sets through a simultaneous optimization of the exponents and the contraction coefficients so as to achieve the best atomic SCF energies. The n-ijG basis sets fall under this category. The contraction length is n for the inner shell atomic orbital (the 1s orbital). The valence shell is split into a DZ basis, the single zeta components having contraction lengths i and j respectively. Examples are 4-31G, 5-31G, 6-31G etc. Sometimes the split-valence basis is a triple zeta one, like 6-311G. The basis sets 6-31G and 6-311G for carbon consists of 9 and 13 basis functions (CGF) respectively. Calculated results are improved by adding one or two types of polarization functions like the 3d functions to form 6-311G* basis or the 3d and 4f functions to form the set 6-311G**. Lists of optimized exponents and coefficients for STO-nG, STO-nG*, STO-nG** and n-ijG, n-ijkG, n-ijkG* and n-ijkG** series are available from *Ab-initio molecular orbital theory* by W. J. Hehre, L. Radom, P. V. R. Schleyer and J. A. Pople (Wiley, NewYork, 1986). The optimized parameters (such as Gaussian exponents and contraction coefficients) for molecular calculations are found to slightly vary from those for the atomic calculations. The optimized parameters for thermochemical calculations often vary from the optimized parameters for spectroscopic calculations.

Lecture 75: The Hartree-Fock Hamiltonian; Configuration Interaction

The Hartree-Fock Hamiltonian

Because the orbitals are eigenfunctions of the Fock operator, a product of orbitals such as $\psi_p(\mathbf{x}_1)\psi_q(\mathbf{x}_2)...\psi_t(\mathbf{x}_N)$ will be an eigenfunction of the sum of the Fock operators for all the N electrons, and the corresponding eigenvalue will be given by the sum of the corresponding orbital energies:

$$\sum_{i=1}^{N} \mathbf{F}(\mathbf{x}_i)\psi_p(\mathbf{x}_1)\psi_q(\mathbf{x}_2)...\psi_t(\mathbf{x}_N)$$

$$= (\varepsilon_p+\varepsilon_q+ ... +\varepsilon_t)\ \psi_p(\mathbf{x}_1)\psi_q(\mathbf{x}_2)...\psi_t(\mathbf{x}_N). \tag{75.1}$$

Any other product that can be derived from $\psi_p(\mathbf{x}_1)\psi_q(\mathbf{x}_2)...\psi_t(\mathbf{x}_N)$ through the permutation of the coordinates of electrons is also an eigenfunction with the same eigenvalue. Hence the normalized determinant $(N!)^{-1/2}||\psi_p(\mathbf{x}_1)...\psi_t(\mathbf{x}_N)||$ is also an eigenfunction with the same eigenvalue,

$$\sum_{i=1}^{N} \mathbf{F}(\mathbf{x}_i)\ (N!)^{-1/2}||\psi_p(\mathbf{x}_1)...\psi_t(\mathbf{x}_N)||$$

$$= (\varepsilon_p+\varepsilon_q+ ... +\varepsilon_t)\ (N!)^{-1/2}||\psi_p(\mathbf{x}_1)...\psi_t(\mathbf{x}_N)||. \tag{75.2}$$

In the above we have only assumed that ψ_p, ψ_q, etc. are eigenfunctions of operator $\mathbf{F}$. Therefore, the eigenvalue relation (75.2) holds for *any* determinant prepared by using the Hartree-Fock orbitals, that is, it holds not only for the ground state determinant (where ψ_p, ψ_q, etc. are the lowest-lying orbitals) but also for *every* configuration. The operator sum is called the *Hartree-Fock Hamiltonian*,

$$\mathbf{H}_{HF} = \sum_{i=1}^{N} \mathbf{F}(\mathbf{x}_i). \tag{75.3}$$

Its eigenvalues are the sums of the corresponding orbital energies.

The many-electron Hamiltonian can be written as

$$\mathbf{H}_{el} = \mathbf{H}_{\mathrm{HF}} + \mathbf{H}' \tag{75.4}$$

where $\mathbf{H}' = \mathbf{H}_{el} - \mathbf{H}_{\mathrm{HF}}$ is known as the many-electron perturbation. The latter can be determined by using the explicit form of the Fock operator. The one-electron operators in $\mathbf{H}_{el}$ exactly cancel those in $\mathbf{H}_{\mathrm{HF}}$, which leaves

$$\mathbf{H}' = \sum_{\substack{i<j \\ (elec)}}^{N} \frac{1}{r_{ij}} - \sum_{\substack{i=1 \\ (elec)}}^{N} \sum_{\substack{s=1 \\ (occ)}}^{N} [J_s(\mathbf{x}_i) - K_s(\mathbf{x}_i)]. \tag{75.5}$$

Slater determinants written in terms of normalized Hartree-Fock orbitals are normalized many-electron functions. Because of the orthogonality of the Hartree-Fock orbitals, they are orthogonal to each other. They are also eigenfunctions of Hartree-Fock Hamiltonian. They interact with each other only through the interaction operator $\mathbf{H}'$. In principle, the ket vector representing an exact state of the many-electron system can be expanded as the linear combination

$$|\Phi\rangle = C_0|\Psi_0\rangle + \sum_{am} C_a^m |\Psi_a^m\rangle + \sum_{\substack{a<b \\ m<n}} C_{ab}^{mn} |\Psi_{ab}^{mn}\rangle + \sum_{\substack{a<b<c \\ m<n<p}} C_{abc}^{mnp} |\Psi_{abc}^{mnp}\rangle + \ldots \tag{75.6}$$

where $|\Psi_0\rangle$ is the ket vector for the Hartree-Fock ground state configuration; $|\Psi_a^m\rangle$ represents a singly-excited determinant with one electron excited from orbital ψ_a that is occupied in the Hartree-Fock ground state configuration $|\Psi_0\rangle$ to an orbital ψ_m that is unoccupied in $|\Psi_0\rangle$. Singles are represented by the orbital population in Figure 12.3.

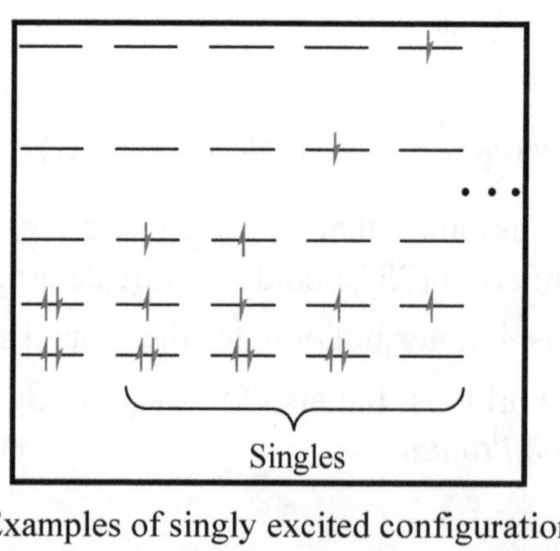

Figure 12.3 Examples of singly excited configurations (Singles).

Similarly, $|\Psi_{ab}^{mn}>$ represents a doubly-excited determinant, where two electrons have been displaced from orbitals ψ_a and ψ_b to orbitals ψ_m and ψ_n. A few doubles are shown in Figure 12.4. A triply excited determinant is written as $|\Psi_{abc}^{mnp}>$, etc.

Figure 12.4 Examples of doubly excited configurations (Doubles).

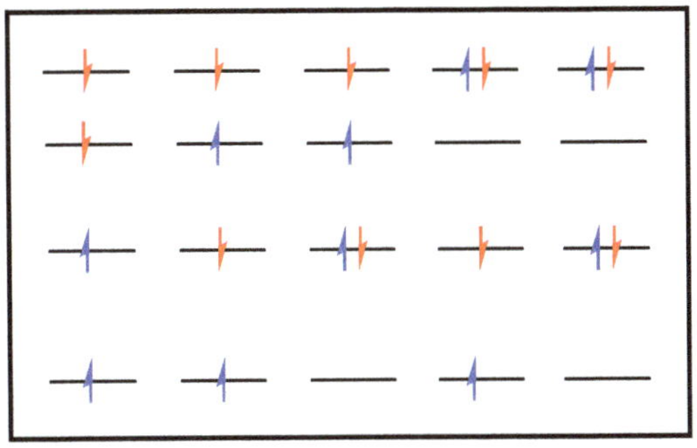

The excited determinants $|\Psi_a^m>$, $|\Psi_{ab}^{mn}>$, $|\Psi_{abc}^{mnp}>$, etc. are in short called as singles, doubles, triples, etc. respectively. As mentioned earlier, these are eigenfunctions of the Hartree-Fock Hamiltonian and the eigenvalue relations are given below.

$$\mathbf{H}_{HF} |\Psi_0> = E_0^{(0)} |\Psi_0> \quad \text{where} \quad E_0^{(0)} = \sum_{\alpha}^{occ.} \varepsilon_\alpha;$$

$$\mathbf{H}_{HF} |\Psi_a^m> = (E_0^{(0)} - \varepsilon_a + \varepsilon_m) |\Psi_a^m>;$$

$$\mathbf{H}_{HF} |\Psi_{ab}^{mn}> = (E_0^{(0)} - \varepsilon_a - \varepsilon_b + \varepsilon_m + \varepsilon_n) |\Psi_{ab}^{mn}>;$$

$$\mathbf{H}_{HF} |\Psi_{abc}^{mnp}> = (E_0^{(0)} - \varepsilon_a - \varepsilon_b - \varepsilon_c + \varepsilon_m + \varepsilon_n + \varepsilon_p) |\Psi_{abc}^{mnp}>; \text{ etc.}$$

Integrals of one-electron operators over determinants are given in Table 12.3, and two-electron integrals are given in Table 12.4.

There are two main ways of determining the coefficients C_0, C_a^m, C_{ab}^{mn}, C_{abc}^{mnp}, etc. The first method involves the use of $|\Phi>$ of Eq. (75.6) as a trial function in a linear variational procedure. The Slater determinants are treated as basis functions. This technique leads to the algebraic equations, and the existence of non-trivial solutions requires the secular determinant to vanish,

$$||(H_{el})_{ij} - E\delta_{ij}|| = 0 \tag{75.7}$$

where $(H_{el})_{ij}$ is the Hamiltonian integral over two reference determinants D_i and D_j, and the corresponding overlap integral is given by the Kroenecker delta δ_{ij}.

Table 12.3 Integrals of one-electron operators over determinants that are written in terms of spin orbitals: the indices m, n, p, etc. are for general orbitals, whereas i is an electron index.

$\|K> = \|...mnp...>$	$\mathbf{O}^{(1)} = \sum_{i=1}^{N} \mathbf{h}(i)$
$\|L> = \|...mnp...>$	$<K\|\mathbf{O}^{(1)}\|K> = \sum_{s\in(...m,n,p,...)} <s\|\mathbf{h}\|s> = \sum_{s} h_{ss}$
$\|L> = \|...mnp'...>\ (p' \neq p)$	$<K\|\mathbf{O}^{(1)}\|L> = <p\|h\|p'> = h_{pp'}$
$\|L> = \|...mn'p'...>, \|...m'n'p'...>$, etc. $(m \neq m', n \neq n', p \neq p', ...)$	$<K\|\mathbf{O}^{(1)}\|L> = 0$

Table 12.4 Integrals of two-electron operators over determinants which are expressed in terms of spin orbitals: the indices m, n, p, etc. are general orbital indices while i and j are electron indices.

$\|K\rangle = \|...mnp...\rangle$	$\mathbf{O}^{(2)} = \frac{1}{2} \sum_{i \neq j}^{N} r_{ij}^{-1}$
$\|L\rangle = \|...mnp...\rangle$	$\langle K\|\mathbf{O}^{(2)}\|K\rangle = \frac{1}{2} \sum_{s \neq s'} [(ss\|s's') - (ss'\|s's)] = \frac{1}{2} \sum_{s,s'} \langle ss'\|\|ss'\rangle$
$\|L\rangle = \|...mnp'...\rangle$ $(p \neq p')$	$\langle K\|\mathbf{O}^{(2)}\|L\rangle = \sum_{s\in(...m,n,p,...)} \langle sp\|\|sp'\rangle$
$\|L\rangle = \|...mn'p'...\rangle$ $(n \neq n', p \neq p')$	$\langle K\|\mathbf{O}^{(2)}\|L\rangle = \langle np\|\|n'p'\rangle$
$\|L\rangle = \|...m'n'p'...\rangle$, etc. $(m \neq m', n \neq n', p \neq p', ...)$	$\langle K\|\mathbf{O}^{(2)}\|L\rangle = 0$

The integral $(H_{el})_{ij}$ is calculated as follows. Since

$$\langle D_i|\mathbf{H}_{HF}|D_j\rangle = \delta_{ij}\, E_i^{(0)} = \delta_{ij} \sum_{s \in D_i} \varepsilon_s, \tag{75.8}$$

the integral $(H_{el})_{ij}$ is obtained as

$$(H_{el})_{ii} = E_i^{(0)} + \langle D_i|\mathbf{H}'|D_i\rangle = E_i^{(0)} + H'_{ii}$$

$$(H_{el})_{ij} = \langle D_i|\mathbf{H}'|D_j\rangle = H'_{ij}. \tag{75.9}$$

The matrix elements H'_{ij} can be easily calculated. In particular, the element H'_{00} is given by

$$H'_{00} = -\frac{1}{2} \sum_{\substack{s \in D_0 \equiv |\Psi_0> \\ s' \in D_0}} (J_{ss'} - K_{ss'}). \tag{75.10}$$

This is in support of the fact that $(H_{el})_{00}$ is indeed the Hartree-Fock total energy of the ground state configuration. The solution of Eq. (75.7) yields energy values, and these may be used in turn to calculate the linear expansion coefficients. This process is known as *Configuration Interaction* (CI) method.

The second methodology depends upon the use of $\mathbf{H}'$ as perturbation. Configurations (Slater determinants for closed shells and linear combination of determinants for open shells) are taken as the zeroth-order functions, and corrections to energy and wave functions are calculated by Rayleigh-Schrödinger perturbation theory. This approach is named as *Many-Body Perturbation Theory* (MBPT). Corrections to various orders are interpreted in terms of successive interactions which can be diagrammatically represented. Because of this a very lucid physical interpretation can be given to these corrections. The correction terms can also be enumerated and evaluated rather easily from the diagrams. *Diagrammatic perturbation techniques* have become standard methods for the investigation of many-body effects.

The Hartree-Fock theory is in reality a *mean-field* theory. It is also referred to as the *one-electron approximation*, as in this theory one electron is considered to move in the average electric field created by the probability distribution of all other electrons. However, the term r_{ij}^{-1} stands for an instantaneous repulsion between two electrons which makes the dynamics of two electrons explicitly correlated. The explicit correlation is known as the many-body effect. The difference between the exact ground state energy E_0 and the Hartree-Fock energy E_0^{HF} is called the *correlation energy*,

$$E_{correl} = E_0 - E_0^{HF}. \tag{75.11}$$

Configuration Interaction (CI)

If one uses K basis functions, $2K$ spin orbitals are obtained from the Hartree-Fock calculation. Then one can prepare a total of $\binom{2K}{N}$ determinants and exactly as many configurations. There is only one ground state configuration (a determinant) for a closed-shell species, and a handful of configurations for the ground state of an open-shell system.

Some of the configurations are singles, some are doubles, some are triples, etc. The total number of n-tuply excited determinants is $\binom{N}{n} \times \binom{2K-N}{n}$. If all the $\binom{2K}{N}$ determinants are involved in the linear variation treatment, the method is referred to as the *full CI* (FCI). This statement assumes that the electronic Hamiltonian has the spin-orbit interaction that connects one spin configuration with another. In case the spin-orbit coupling is absent, FCI for different spin states are separated and for each spin configuration, the FCI involves a considerably smaller number of determinants.

Example 53: Calculation on Be with $K = 5$ (1s, 2s and 2p orbitals) and $N = 4$

Closed-shell species with Hartree-Fock ground state configuration $1s^2 2s^2$ and a single determinant stands for $|\Psi_0\rangle$. The total number of possible determinants is $\binom{10}{4} = 210$. There are $\binom{4}{1} \times \binom{6}{1} = 24$ singles, $\binom{4}{2} \times \binom{6}{2} = 90$ doubles, $\binom{4}{3} \times \binom{6}{3} = 80$ triples and $\binom{4}{4} \times \binom{6}{4} = 15$ quadruples. The full CI involves all these 210 determinants while the *singles-doubles CI* (SDCI) involves only 1+24+90= 115 determinants. In both cases we get singlet, triplet and quintet states.

Example 54: Calculation on Ne atom with $K = 9$ (1s, 2s, 2p, 3s, 3p) and $N = 10$. The FCI involves $\binom{18}{10} = 78336$ configurations whereas the SDCI would involve only $1+10\times 8 + \binom{10}{2} \times \binom{8}{2} = 1341$ configurations. The FCI gives states with spin number varying from 0 to 4. As in the previous case, the SDCI gives only singlet, triplet and quintet states.

We notice that the number of configurations increases extremely rapidly with the number of basis functions. Besides, in the restricted orbital scheme it is possible to select a smaller number of configurations for each type of spin state. Configurations of two different spins do not interact with each other when the spin-orbit interaction is absent. This helps in considerably reducing the size of the CI matrix. The matrix elements H'_{ij} can be calculated by using the integrals in Tables 12.4 and 12.5. Using these integrals one comes across a remarkable theorem.

Theorem 23. Brillouin theorem: Singly excited determinants do not interact directly with a reference Hartree-Fock determinant.

This theorem is proved as follows. When the integral for interaction of $|\Psi_a^m\rangle$ with $|\Psi_0\rangle$ is calculated, one finds

$$< \Psi_0|\mathbf{H}_{el}|\Psi_a^m> = < a|\mathbf{h}|m> + \sum_b^{occ} < ab||mb > \tag{75.12}$$

where

$$< ab||mb > = (am|bb) - (ab|bm). \quad (75.13)$$

But the right side of (75.12) also equals the Fock matrix element between orbitals a and m,

$$<a|\mathbf{F}|m> = <a|\mathbf{h}|m> + \sum_{b}^{occ} < ab||mb >.$$

Therefore,

$$< \Psi_0|\mathbf{H}_{el}|\Psi_a^m > = < a|\mathbf{F}|m > = \varepsilon_m < a|m > = 0 \quad (75.14)$$

which is the desired result.

The impact of Brillouin's theorem is that singles do not directly mix with the Hartree-Fock reference determinant $|\Psi_0>$, since the overlap $<\Psi_0|\Psi_a^m>$ is also equal to zero. The true ground state may contain singly excited determinants which mix indirectly through their interaction with doubly excited determinants, as the latter directly interact with $|\Psi_0>$.

A list of interaction matrix elements calculated using different determinants is given in Table 12.5. Note the following points about the non-vanishing interaction. Non-zero interaction integrals are obtained only between the states of same spin symmetry. The Hartee-Fock ground (or reference state) does not interact with single excitations (or single deexcitations). There is no direct coupling between $|\Psi_0>$ and triples and higher *n*-tuples. Single excitations and the triples contribute to the ground state only through their interaction with the doubles which mix directly with $|\Psi_0>$. Thus the doubles determine the dominant chunk of the ground state correlation energy, both in CI treatments and in MBPT.

Lecture 76: Intermediate Normalization, Natural Orbitals, and Configuration Interaction Treatment of Helium

Intermediate Normalization

In MBPT an unnormalized perturbed wave function is written in the form of Eq. (75.6) with one of the coefficients of configurations (and not necessarily determinants) equal to one. For instance the unnormalized ground state vector is of the form

$$|\Phi_0> = |\Psi_0> + \sum_{am} C_a^m |\Psi_a^m> + \sum_{a<b} \sum_{m<n} C_{ab}^{mn} |\Psi_{ab}^{mn}> + \ldots \quad (76.1)$$

This is also the form of the unnormalized CI wave function. Note the result

$$< \Phi_0|\Phi_0 > = 1 + \sum_{am} (C_a^m)^2 + \sum_{a<b} \sum_{m<n} (C_{ab}^{mn})^2 + \ldots \quad (76.2)$$

However, the state vector is said to be intermediately normalized as

Table 12.5. Hamiltonian matrix elements between singlet configurations constructed from real orbitals in the RHF scheme. The following notations have been used: α, β, etc. are occupied spatial orbitals and m, n, etc. are spatial orbitals unoccupied in the HF ground state.

Ground-Singles

$$\langle \Psi_0 | \mathbf{H}_{el} | {}^{1,3}\Psi_{\alpha}^{m} \rangle = 0$$

Singles-Singles

$$\langle {}^{1}\Psi_{\alpha}^{m} | \mathbf{H}_{el} | {}^{1}\Psi_{\beta}^{n} \rangle = \delta_{\alpha,\beta}\delta_{m,n}[E_0^{HF} + \varepsilon_m - \varepsilon_\alpha] - (mn|\alpha\beta) + 2(m\alpha|\beta n)$$

Ground-Doubles

$$\langle \Psi_0 | \mathbf{H}_{el} | {}^{1}\Psi_{\alpha\alpha}^{mm} \rangle = K_{m\alpha}$$

$$\langle \Psi_0 | \mathbf{H}_{el} | {}^{1}\Psi_{\alpha\alpha}^{mn} \rangle = 2^{1/2}(m\alpha|n\alpha)$$

$$\langle \Psi_0 | \mathbf{H}_{el} | {}^{1}\Psi_{\alpha\beta}^{mm} \rangle = 2^{1/2}(m\beta|m\alpha)$$

$$\langle \Psi_0 | \mathbf{H}_{el} | {}^{A}\Psi_{\alpha\beta}^{mn} \rangle = 3^{1/2}[(m\alpha|n\beta) - (m\beta|n\alpha)]$$

$$\langle \Psi_0 | \mathbf{H}_{el} | {}^{B}\Psi_{\alpha\beta}^{mn} \rangle = [(m\alpha|n\beta) + (m\beta|n\alpha)]$$

Doubles-Doubles

$$\langle {}^{1}\Psi_{\alpha\alpha}^{mm} | \mathbf{H}_{el} | {}^{1}\Psi_{\alpha\alpha}^{mm} \rangle = E_0^{HF} + 2(\varepsilon_m - \varepsilon_\alpha) + J_{\alpha\alpha} + J_{mm} - 4J_{m\alpha} + 2K_{m\alpha}$$

$$\langle {}^{1}\Psi_{\alpha\alpha}^{mn} | \mathbf{H}_{el} | {}^{1}\Psi_{\alpha\alpha}^{mn} \rangle = E_0^{HF} + (\varepsilon_m + \varepsilon_n - 2\varepsilon_\alpha) + J_{\alpha\alpha} + J_{mn} + K_{mn} - 2J_{m\alpha} - 2J_{n\alpha} + K_{m\alpha} + K_{n\alpha}$$

$$\langle {}^{1}\Psi_{\alpha\beta}^{mm} | \mathbf{H}_{el} | {}^{1}\Psi_{\alpha\beta}^{mm} \rangle = E_0^{HF} + (2\varepsilon_m - \varepsilon_\alpha - \varepsilon_\beta) + J_{mm} + J_{\alpha\beta} + K_{\alpha\beta} - 2J_{m\alpha} - 2J_{m\beta} + K_{m\alpha} + K_{m\beta}$$

$$\langle {}^{A}\Psi_{\alpha\beta}^{mn} | \mathbf{H}_{el} | {}^{A}\Psi_{\alpha\beta}^{mn} \rangle = E_0^{HF} + (\varepsilon_m + \varepsilon_n - \varepsilon_\alpha - \varepsilon_\beta) + J_{\alpha\beta} + J_{mn} - K_{\alpha\beta} - K_{mn} - (J_{m\alpha} + J_{m\beta} + J_{n\alpha} + J_{n\beta}) + \frac{3}{2}(K_{m\alpha} + K_{m\beta} + K_{n\alpha} + K_{n\beta})$$

$$\langle {}^{B}\Psi_{\alpha\beta}^{mn} | \mathbf{H}_{el} | {}^{B}\Psi_{\alpha\beta}^{mn} \rangle = E_0^{HF} + (\varepsilon_m + \varepsilon_n - \varepsilon_\alpha - \varepsilon_\beta) + J_{\alpha\beta} + J_{mn} + K_{\alpha\beta} + K_{mn} - (J_{m\alpha} + J_{m\beta} + J_{n\alpha} + J_{n\beta}) + \frac{1}{2}(K_{m\alpha} + K_{m\beta} + K_{n\alpha} + K_{n\beta})$$

$$\langle {}^{A}\Psi_{\alpha\beta}^{mn} | \mathbf{H}_{el} | {}^{B}\Psi_{\alpha\beta}^{mn} \rangle = \frac{\sqrt{3}}{2}(K_{m\alpha} + K_{n\beta} - K_{m\beta} - K_{n\alpha})$$

$$< \Psi_0|\Phi_0 > = 1. \quad (76.3)$$

Because of the eigenvalue relationship

$$\mathbf{H}_{el}|\Phi_0 > = E_0|\Phi_0 > \quad (76.4)$$

we get

$$(\mathbf{H}_{el} - E_0^{HF})\,|\Phi_0 > = E_{correl}|\Phi_0 > \quad (76.5)$$

so that the correlation energy may be obtained as

$$E_{correl} = < \Psi_0|\, \mathbf{H}_{el} - E_0^{HF}\, |\Phi_0 >. \quad (76.6)$$

If we expand the right hand side as in Eq. (76.1) and use Brillouin's theorem we get

$$E_{correl} = \sum_{a<b}\sum_{m<n} C_{ab}^{mn} <\Psi_0|\, \mathbf{H}_{el}\, |\Psi_{ab}^{mn}>. \quad (76.7)$$

The configuration $|\Psi_0 >$ cannot directly interact with triples, quadruples, etc. as the Hamiltonian $\mathbf{H}_{el}$ contains only one- and two-electron operators. Thus the ground state correlation energy is completely determined by the coefficients of the double excitations in the intermediately normalized ground state wave function.

The coefficients C_{ab}^{mn}'s are of course influenced by the occurrence of other types of excitations in $|\Phi_0 >$. To see this we can use Eq. (76.5) to write

$$< \Psi_a^m|\, \mathbf{H}_{el} - E_0^{HF}\, |\Phi_0 > = E_{correl} < \Psi_a^m|\Phi_0 > = C_a^m E_{correl}$$

which, using expansion (76.1), leads to the equality

$$\sum_{b,n} C_b^n <\Psi_a^m|\, \mathbf{H}_{el} - E_0^{HF}\, |\Psi_b^n> + \sum_{b<c}\sum_{n<p} C_{bc}^{np} <\Psi_a^m|\mathbf{H}|\Psi_{bc}^{np}>$$

$$+ \sum_{b<c}\sum_{n<p} C_{abc}^{mnp} <\Psi_a^m|\mathbf{H}|\Psi_{abc}^{mnp}> = C_a^m E_{correl}$$

etc. Similarly using $<\Psi_{ab}^{mn}|$, $<\Psi_{abc}^{mnp}|$, etc. a hierarchy of equations can be obtained which must be solved together to get all the coefficients.

Natural orbitals

For any arbitrary state the first order reduced density matrix is expressible in terms of the Hartree-Fock spin orbitals as

$$\rho^{(1)}(\mathbf{x}_1, \mathbf{x}_1') = \sum_{i,j} \psi_i(\mathbf{x}_1)\rho_{ij}^{(1)} \psi_j^*(\mathbf{x}_1') \quad (76.8)$$

such that one can define a matrix element

$$\rho_{ij}^{(1)} = \int d^4x_1 \int d^4x_1'\, \psi_i^*(\mathbf{x}_1)\rho^{(1)}(\mathbf{x}_1, \mathbf{x}_1')\, \psi_j(\mathbf{x}_1'). \quad (76.9)$$

It is easy to observe

$$\rho^{(1)}(\mathbf{x}_1', \mathbf{x}_1) = \rho^{(1)*}(\mathbf{x}_1, \mathbf{x}_1') \quad (76.10)$$

from which it is obvious that

$$\rho_{ji}^{(1)*} = \rho_{ij}^{(1)}, \tag{76.11}$$

that is, the density matrix $\rho^{(1)}$ is hermitian. It is thus possible to diagonalize $\rho^{(1)}$ by a unitary transformation of the form

$$\rho^{(1)} \rightarrow \mathbf{U}^{\dagger}\rho^{(1)}\mathbf{U} = \rho_{d}^{(1)}. \tag{76.12}$$

The same unitary transformation transforms the set of spin orbitals $\{\psi\}$ into the orthonormal set of spin orbitals $\{\eta\}$,

$$\eta_i(\mathbf{x}) = \sum_{s} \psi_s(\mathbf{x})U_{si}. \tag{76.13}$$

The transformed orbitals are called *natural (spin) orbitals* (NO) since they offer a diagonal density matrix. In terms of the NO's the reduced density matrix is given by

$$\rho^{(1)}(\mathbf{x}_1, \mathbf{x}_1') = \sum_{i} (\rho_{d}^{(1)})_{ii}\, \eta_i(\mathbf{x})\, \eta_i^*(\mathbf{x}_1'). \tag{76.14}$$

The CI expansion rapidly converges when natural orbitals are employed. A given accuracy needs fewer configurations that are built from the natural orbitals than the configurations formed from any other orthonormal basis. In practice one may form the natural orbitals from a small-scale CI calculation such as the SDCI calculation. These orbitals are then used in a large-scale CI treatment like the FCI.

Example 55: CI for helium atom

To make the matters clear we now discuss the configuration interaction treatment of the helium atom. Since there are only two electrons, the full CI is in general identical with SDCI. For simplicity, consider only the RHF orbitals 1s and 2s. The configurations involved are

$|\Psi_0\rangle \equiv |\Psi_{1s,1s}\rangle$ (Term ^1S)

$|^1\Psi_{1s}^{2s}\rangle$ (Single)

$|^1\Psi_{1s,1s}^{2s,2s}\rangle$ (Double)

and we have only DCI. The DCI equation is

$$\begin{pmatrix} (H_{el})_{00} & (H_{el})_{02} \\ (H_{el})_{20} & (H_{el})_{22} \end{pmatrix} \begin{pmatrix} C_0 \\ C_{1s1s}^{2s2s} \end{pmatrix} = E \begin{pmatrix} C_0 \\ C_{1s1s}^{2s2s} \end{pmatrix} \tag{76.15}$$

where

$$\begin{aligned} (H_{el})_{00} &= 2\varepsilon_{1s} - J_{1s,1s} \\ (H_{el})_{02} &= (H_{el})_{20} = K_{1s,2s} \\ (H_{el})_{22} &= 2\varepsilon_{2s} + J_{2s,2s} - 4\, J_{1s,2s} + 2K_{1s,2s}. \end{aligned} \tag{76.16}$$

When the Hamiltonian matrix is diagonalized, the diagonal elements are found to be

$$E_{\pm} = \frac{1}{2}[(H_{el})_{00}+(H_{el})_{22}] \pm \frac{1}{2}[\{(H_{el})_{22}-(H_{el})_{00}\}^2 + 4(H_{el})_{02}{}^2]^{1/2}. \tag{76.17}$$

The "minus" combination corresponds to the ground state energy. The correlation energy is calculated as

$$E_{correl} = E_{-} - (H_{el})_{00} = \frac{1}{2}[(H_{el})_{22}-(H_{el})_{00}] - \frac{1}{2}[\{(H_{el})_{22}-(H_{el})_{00}\}^2 + 4(H_{el})_{02}^2]^{1/2}$$

$$\approx -\frac{(H_{el})_{02}^2}{[(H_{el})_{22} - (H_{el})_{00}]}. \tag{76.18}$$

The approximate result can be obtained also from second order MBPT.

The intermediately normalized ground state can be written from Eq. (76.1) and the correlation energy expression (76.7):

$$|\Phi_0\rangle = |\Psi_0\rangle + \frac{E_{correl}}{(H_{el})_{20}}|{}^1\Psi_{1s,1s}^{2s,2s}\rangle$$

$$\approx |\Psi_0\rangle - \frac{(H_{el})_{02}}{[(H_{el})_{22} - (H_{el})_{00}]}|{}^1\Psi_{1s,1s}^{2s,2s}\rangle. \tag{76.19}$$

This wave function can be easily normalized. The approximate result corresponds to the first order correction to wave function (second order MBPT).

Texts Suggested for Further Reading

W. J. Hehre, L. Radom, P. V. R. Schleyer and J. A. Pople, "Ab Initio Molecular Orbital Theory" (J. Wiley, New York, 1986).
J. A. Pople and D. L. Beveridge, "Approximate Molecular Orbital Theory" (McGraw-Hill, New York, 1970).

PROBLEMS

1. Explicitly derive the results in (72.13) by applying $\boldsymbol{L}^2$ and $\boldsymbol{S}^2$ operators on the determinants in (72.12).
2. Write the Pople-Nesbet determinant for carbon atom. Derive expressions for the UHF total energy and orbital energies.
3. Find the natural orbitals for He atom from the approximate ground state given in Eq. (76.19).

CHAPTER 13

EXPLICIT TREATMENT OF MOLECULES

There is no fundamental difference between the quantum mechanical treatment of the atomic electrons and that of the molecular electrons under Born-Oppenheimer approximation. The electronic theory of atomic structure is only easier to grasp. This is so because of the existence of the angular symmetries in an atom. The methodologies applied to the investigation of the atomic structure are equally applicable to the investigation of the movement of electrons in a molecule. Thus one may carry out a Hartree-Fock calculation to determine the (approximate) molecular orbitals and then investigate the correlation effects by the CI method or by MBPT. Calculations of the molecular integrals are much more difficult. Therefore, it would be useful to examine extremely simple cases like the H_2^+ molecule ion and the H_2 molecule at first. It would also be an extremely valuable exercise to follow the simple valence-bond calculation explicitly for the H_2 molecule. Since the GTO basis is very useful for *ab-initio* molecular calculations, we would briefly examine how the different molecular integrals are calculated over this basis. In the final lecture on this topic, we would discuss a few semiempirical theories which are based partly on the calculation of some of the molecular integrals and in part on an empirical formulation for the rest of these integrals.

Lecture 77: Separation and Solution of the Wave Equation for a Diatomic Species

Consider diagram 2.3(iii). In elliptical coordinate system, coordinates of the electron is written as (μ, ν, φ) where

$$\mu = \frac{r_A + r_B}{R}, \tag{77.1}$$

$$\nu = \frac{r_A - r_B}{R},$$

and the angle φ is the same as that defined for the spherical polar coordinate system. The molecular axis is taken as the z-axis. See Figure 13.1. The Laplacian operator is given in terms of the elliptical coordinates by

$$\nabla^2 = \frac{4}{R^2(\mu^2-\nu^2)}\left[\frac{\partial}{\partial\mu}\{(\mu^2-1)\frac{\partial}{\partial\mu}\} + \frac{\partial}{\partial\nu}\{(1-\nu^2)\frac{\partial}{\partial\nu}\} + \frac{(\mu^2-\nu^2)}{(\mu^2-1)(1-\nu^2)}\frac{\partial^2}{\partial\varphi^2}\right]. \tag{77.2}$$

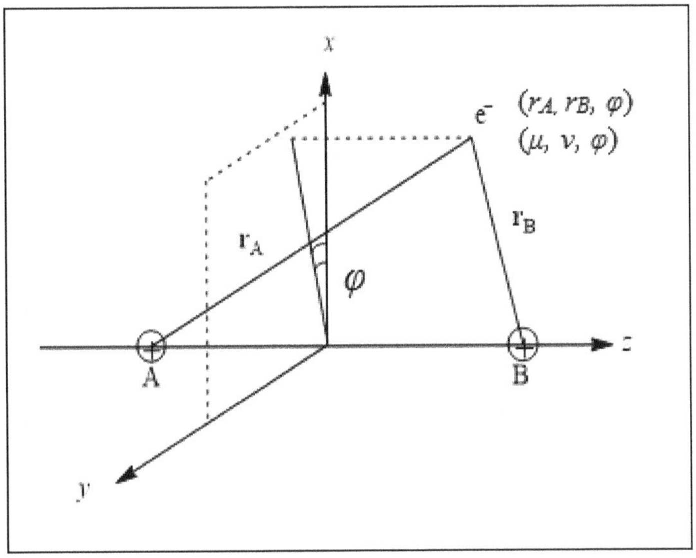

Figure 13.1 Elliptical coordinates of the electron in H_2^+ molecule-ion.

The volume element is written as

$$dv = \frac{R^3}{8}(\mu^2 - \nu^2)\, d\mu\, d\nu\, d\varphi. \tag{77.3}$$

Burrau (1927) pointed out that under Born-Oppenheimer approximation, the wave equation for the hydrogen molecule ion

$$-\frac{1}{2}\nabla^2\psi - \left(\frac{1}{r_A}+\frac{1}{r_B}\right)\psi + \frac{1}{R}\,\psi = E\psi \tag{77.4}$$

is separable in elliptical coordinate system. This can be shown in the following way.

In the elliptical coordinate system the wave equation reads as

$$\frac{\partial}{\partial\mu}\{(\mu^2-1)\frac{\partial\psi}{\partial\mu}\}+\frac{\partial}{\partial\nu}\{(1-\nu^2)\frac{\partial\psi}{\partial\nu}\}+\{\frac{1}{\mu^2-1}+\frac{1}{1-\nu^2}\}\frac{\partial^2\psi}{\partial\varphi^2}$$
$$+ [-(\mu^2-\nu^2)\varepsilon+2\mu R]\psi = 0. \tag{77.5}$$

where

$$\varepsilon = -\frac{R^2}{2}(E-\frac{1}{R}). \tag{77.6}$$

On replacing $\psi(\mu, \nu, \varphi)$ by the product

$$\psi(\mu, \nu, \varphi) = M(\mu)N(\nu)\,\Phi(\varphi) \tag{77.7}$$

the wave equation can be separated into three differential equations,

$$\frac{d^2\Phi}{d\varphi^2} = -\lambda^2\Phi;$$
$$\frac{d}{d\nu}[(1-\nu^2)\frac{dN}{d\nu}] + (\varepsilon\nu^2 - \frac{\lambda^2}{1-\nu^2} - \tau)\, N = 0; \tag{77.8}$$
$$\frac{d}{d\mu}[(\mu^2-1)\frac{dM}{d\mu}] + (-\varepsilon\mu^2 + 2R\mu - \frac{\lambda^2}{\mu^2-1} + \tau)\, M = 0.$$

The solution of the φ-equation yields the familiar result

$$\Phi(\varphi) = \frac{1}{\sqrt{2\pi}} e^{i\lambda\varphi}; \ \lambda = 0, \pm 1, \pm 2, \ldots \tag{77.9}$$

Solutions of the equations involving the variables ν and μ depend upon finding a satisfactory relation between τ and ε. Burrau (1927) carried out a numerical integration of the ν- and μ- equations and found approximate solutions for M and N. Hylleraas (1931) and Jaffé (1934) performed more accurate treatments. We avoid discussing these numerical procedures here in this lecture, but take note of the salient features of the solutions. The most important observation is that ε depends on $|\lambda|$, the absolute magnitude of the component of the orbital angular momentum along the molecular axis. Note that for the united atom λ has the possible values $-l$, $-(l-1)$, ... , $(l-1)$, l. The names of the molecular orbitals are related to the absolute magnitude of λ. The molecular obritals are called σ (Greek for s) for $\lambda = 0$, π (Greek for p) for $|\lambda| = 1$, δ (Greek for d) for $|\lambda| = 2$, etc. The states π, δ, etc. are all doubly degenerate corresponding to the two values $\lambda = \pm|\lambda|$ (like ±1, ±2, etc.).

Qualitative Molecular Orbitals

Understanding comes with the calculation of QMO orbital energy. We proceed to obtain a few integrals which are useful for a minimal basis STO treatment of H_2^+-like molecule-ions and H_2-like molecules. The general 1s STO is written as

$$\phi_{1s}(\mathbf{r}) = \frac{\zeta^{3/2}}{\pi^{1/2}} e^{-\zeta r} \tag{77.10}$$

which leads to the following electronic Hamiltonian matrix elements for the electron in H_2^+-like molecule-ions

$$(h_{el})_{11} = (h_{el})_{22} = -\frac{\zeta^2}{2} + \left(2 - \frac{\zeta}{Z}\right) J$$

and (77.11)

$$(h_{el})_{12} = (h_{el})_{21} = -\frac{\zeta^2}{2} S + \left(2 - \frac{\zeta}{Z}\right) K.$$

For $\zeta = Z$, these elements reduce to those given by expressions (35.8). The orbital energies for $\psi_+ \equiv [2(1+S)]^{-1/2} [\phi_{1s}(\mathbf{r}_A) + \phi_{1s}(\mathbf{r}_B)]$ and $\psi_- \equiv [2(1-S)]^{-1/2} [\phi_{1s}(\mathbf{r}_A) - \phi_{1s}(\mathbf{r}_B)]$ are now given by

$$E_+^{el} = -\frac{\zeta^2}{2} + \left(2 - \frac{\zeta}{Z}\right) \frac{J+K}{1+S}$$

And (77.12)

$$E_-^{el} = -\frac{\zeta^2}{2} + \left(2 - \frac{\zeta}{Z}\right) \frac{J-K}{1-S}$$

which reduce to Equations (35.10) for $\zeta = Z$. These expressions involve the integrals S, J and K which can be evaluated by using the elliptical coordinate system.

The overlap integral S turns out as

$$S=\frac{\zeta^3}{\pi}\int dv\, e^{-\zeta(\mathbf{r}_A+\mathbf{r}_B)}=\frac{1}{\pi}\int dv\, \mathrm{e}^{-(\,|\mathbf{r}+\mathbf{D}/2|+|\mathbf{r}-\mathbf{D}/2|\,)} \tag{77.13}$$

where $\mathbf{D}=\zeta\mathbf{R}$. On introduction of the elliptical coordinates the overlap integral can be written as

$$S=\frac{\mathrm{D}^3}{8\pi}\int_1^\infty d\mu\, e^{-\mathrm{D}\mu}\int_{-1}^{1} dv(\mu^2-v^2)\int_0^{2\pi} d\varphi=\frac{\mathrm{D}^3}{2}\int_1^\infty d\mu\,\mu^2 e^{-\mathrm{D}\mu}-\frac{\mathrm{D}^3}{6}\int_1^\infty d\mu\, e^{-\mathrm{D}\mu}$$

$$=e^{-\mathrm{D}}\left(1+\mathrm{D}+\frac{\mathrm{D}^2}{3}\right). \tag{77.14}$$

The integral J is [see equations (35.9)]

$$J=-\frac{Z\zeta^3}{\pi}\int dv\,\frac{1}{r_B}\,e^{-2\zeta\mathbf{r}_A}.$$

It is calculated as

$$J=-\frac{Z\zeta}{\pi}\int dv\,\frac{1}{|\mathbf{r}-\mathbf{D}/2|}\,e^{-2\zeta\,|\mathbf{r}+\mathbf{D}/2|}=-\frac{Z\zeta}{\pi}\frac{2}{\mathrm{D}}\int dv\,\frac{1}{\mu-v}\,e^{-\mathrm{D}(\mu+v)}$$

$$=-Z\zeta\frac{\mathrm{D}^2}{2}\left[\int_1^\infty d\mu\,\mu e^{-\mathrm{D}\mu}\int_{-1}^{1} dv\, e^{-\mathrm{D}v}+\int_1^\infty d\mu\, e^{-\mathrm{D}\mu}\int_{-1}^{1} dv\, v e^{-\mathrm{D}v}\right]$$

$$=-\frac{Z}{R}\left[1-(1+\mathrm{D})e^{-2\mathrm{D}}\right]. \tag{77.15}$$

Proceeding in a similar way we find for the integral K

$$K=-\frac{Z\zeta^3}{\pi}\int dv\,\frac{1}{r_A}\,e^{-\zeta(\mathbf{r}_A+\mathbf{r}_B)}$$

$$=-\zeta Z(1+\mathrm{D})\,e^{-\mathrm{D}}. \tag{77.16}$$

These integrals have been used to draw Figure 6.16 and 6.17.

Limits of bonding

In the limit of united atoms, $\zeta\to 2Z$. As we also have $\mathrm{D}\to 0$, we get $S\to 1$, $J\to -2Z^2$ and $K\to -2Z^2$ so that $(\mathrm{H}_{el})_{11}\to -2Z^2$ and $(\mathrm{H}_{el})_{12}\to -2Z^2$. Thus in this limit

$$\psi_+\to\left(\frac{8Z^3}{\pi}\right)^{1/2} e^{-2Zr},\quad E_+^{el}\to -2Z^2; \tag{77.17}$$

$$\psi_-\to 0,\quad E_-^{el}\to 0.$$

In separated atom limit, $\zeta\to Z$ and as $\mathrm{D}\to\infty$, we have $S\to 0$, $J\to 0$ and $K\to 0$. Thus $(\mathrm{H}_{el})_{11}\to -\frac{Z^2}{2}$ and $(\mathrm{H}_{el})_{12}\to 0$ such that

$$E_+^{el}\to -\frac{Z^2}{2};\qquad E_-^{el}\to -\frac{Z^2}{2};$$

$$\frac{1}{\sqrt{2}}(\psi_+ + \psi_-) \rightarrow \left(\frac{Z^3}{\pi}\right)^{1/2} e^{-Zr_A} = \phi_{1s}(\mathbf{r}_A) \equiv \phi_A; \tag{77.18}$$

$$\frac{1}{\sqrt{2}}(\psi_+ - \psi_-) \rightarrow \left(\frac{Z^3}{\pi}\right)^{1/2} e^{-Zr_B} = \phi_{1s}(\mathbf{r}_B) \equiv \phi_B.$$

Alternatively, if we keep the STO exponent fixed at Z, results would be grossly changed in both limits. The united atom limit results would be influenced as $E_+^{el} \rightarrow -\frac{3}{2}Z^2$ and $E_-^{el} \rightarrow \frac{1}{2}Z^2$ in atomic unit, but in separated atom limit both E_+^{el} and E_-^{el} equal the hydrogen-like value $-\frac{1}{2}Z^2$. The difference between the E_+^{el} value for the optimized exponent and that for a fixed value of the exponent ($\zeta = Z$) explains the lowering of the potential energy surface by the use of a flexible ζ in Figure 6.17.

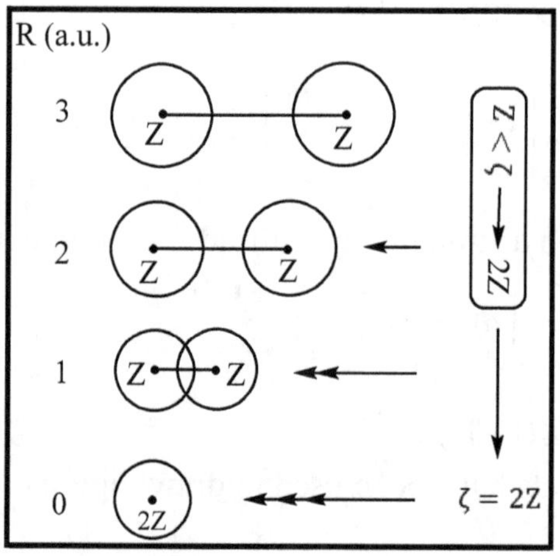

Figure 13.2 Variation of the STO exponent near united atom limit for H_2^+ molecule-ion.

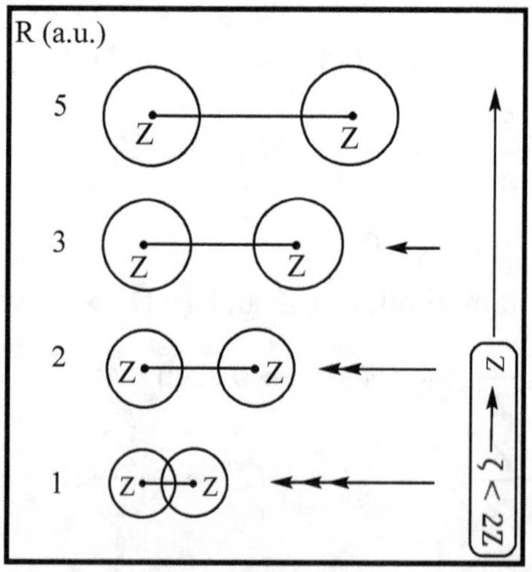

Figure 13.3 Variation of the STO exponent near separated atom limit for H_2^+ molecule-ion.

Figure 13.2 and Figure 13.3 show how the STO exponent varies in united atom limit and near separated atom limit, respectively. Each circle drawn in these diagrams represents the cross-section of a sphere of radius equal to the most probable radial distance r_{mp} (= $1/\zeta$) for the 1s electron in the corresponding atom. Near the united atom limit, $Z < \zeta \to 2Z$. Hence $1/Z > r_{mp} \to 1/2Z$. Near the separated atom limit, $2Z > \zeta \to Z$ so that $1/2Z > r_{mp} \to 1/Z$. Thus the values of the atomic r_{mp} are $1/2Z$ and $1/Z$, respectively, in these limits. The diagrams in Figure 13.2 and Figure 13.3 have been drawn with $Z = 1$, that is, for the H_2^+ molecule-ion.

Lecture 78: Hartree-Fock Calculation on H_2 Molecule

The minimal basis for H_2 leads to molecular orbitals $\psi 1s\sigma_g \equiv \psi_+$ and $\psi 1s\sigma_u^* \equiv \psi_-$ as a result of the molecular symmetry. Molecular orbitals of H_2^+ and those of H_2 are qualitatively of the same forms as discussed in Chapter 6, but for each internuclear distance R, the optimized exponents found for the two cases are different.

The electronic Hamiltonian for the H_2 molecule is

$$\mathbf{H}_{el} = \mathbf{h}(\mathbf{r}_1) + \mathbf{h}(\mathbf{r}_2) + \frac{1}{r_{12}} \tag{78.1}$$

where $\mathbf{h}$ is the electronic Hamiltonian for the H_2^+ system. The ground state configuration of the neutral molecule is

$$\begin{aligned} |1s\sigma g^2> &= 2^{-1/2} \| \, 1s\sigma g(\mathbf{r}_1)\alpha(s_1) \quad 1s\sigma g(\mathbf{r}_2)\beta(s_2) \, \| \\ &= \Psi(\mathbf{r}_1,\mathbf{r}_2)\, \mathrm{X}(s_1,s_2) \end{aligned} \tag{36.12}$$

where the spatial function $\Psi(\mathbf{r}_1,\mathbf{r}_2)$ is given by

$$\Psi(\mathbf{r}_1, \mathbf{r}_2) = 1s\sigma g(\mathbf{r}_1) 1s\sigma g(\mathbf{r}_2) \tag{36.13}$$

and $\mathrm{X}(s_1,s_2)$ is the singlet spin function of Eq. (36.14).

The corresponding Hartree-Fock total electronic energy is

$$E_{el} = 2E_+^{el} + J_{1s\sigma_g,1s\sigma_g} \tag{78.2}$$

where the direct coulomb repulsion integral can be expanded as

$$\begin{aligned} J_{1s\sigma_g,1s\sigma_g} = [2(1+S)]^{-1} \big[& J_{1s_A,1s_A} + J_{1s_B,1s_B} + 2K_{1s_A,1s_B} \\ & + 4(1s_A 1s_A | \, 1s_A 1s_B) \big]. \end{aligned} \tag{78.3}$$

The orbital energies are

$$\varepsilon_{1s\sigma g} = E_+^{el} + J_{1s\sigma g,1s\sigma g},$$

and (78.4)

$$\varepsilon_{1s\sigma u^*} = E_-^{el} + (2J_{1s\sigma g,1s\sigma u^*} - K_{1s\sigma g,1s\sigma u^*}).$$

Total energy can be found by adding the nuclear-nuclear repulsion energy to the total electronic energy. Explicit calculations were first carried out by Weinbaum (1933) who found the electronic dissociation energy as $De = 3.47$ eV and the equilibrium bond length as $R_{eq} = 0.73$ Å. The optimized exponent ζ at $R = R_{eq}$ was found to be equal to 1.193.

Lecture 79: Valence Bond Calculation on H_2 Molecule

Just after Burrau presented his discourse on the H_2^+ molecule-ion, Heitler and London put forward an elegant quantum mechanical treatment of the electronic structure of the H_2 molecule (1927). The approach laid down by Heitler and London represented a novel way of investigation of the electronic structure of molecules. Their work is, as stated by Pauling and Wilson in 1935, *the greatest single contribution* to the concept of valence since G. N. Lewis' work in 1916 describing the chemical bond as the sharing of a pair of electrons by two atoms. Following G. N. Lewis, Heitler and London chose the spatial (two-electron) wave functions

$$\Psi_1 = \phi_A(\mathbf{r}_1)\, \phi_B(\mathbf{r}_2)$$

and (79.1)

$$\Psi_2 = \phi_A(\mathbf{r}_2)\, \phi_B(\mathbf{r}_1)$$

as bases for the construction of a state function that may represent the covalent bond. The trial function is a general linear combination of Ψ_1 and Ψ_2:

$$\Psi = c_1\Psi_1 + c_2\Psi_2. \qquad (79.2)$$

The Ritz variation technique leads to the linear homogeneous equations

$$c_1((H_{el})_{11} - E) + c_2((H_{el})_{12} - Es) = 0$$

$$c_1((H_{el})_{21} - Es) + c_2((H_{el})_{22} - E) = 0 \qquad (79.3)$$

where s is the state overlap integral

$$s = <\Psi_1|\Psi_2> = |(\phi_A, \phi_B)|^2 \equiv |S|^2.$$

If the one-electron functions ϕ_A and ϕ_B are real then $s = S^2$ and $(H_{el})_{12} = (H_{el})_{21}$. When ϕ_A and ϕ_B are equivalent orbitals like the atom-centered orbitals $1s_A$ and $1s_B$ in the H_2 molecule, one also obtains $(H_{el})_{11} = (H_{el})_{22}$. The nontrivial solutions of the linear homogeneous equations in (79.3) are then associated with the energy values

$$E_S = \frac{(H_{el})_{11} + (H_{el})_{12}}{1+S^2},$$

and (79.4)

$$E_A = \frac{(H_{el})_{11} - (H_{el})_{12}}{1-S^2}.$$

The solution corresponding to the root E_S is found to be symmetric under the interchange of the spatial coordinates of the two electrons and the one associated with the root E_A turns out to be anti-symmetric,

$$\Psi_S = [2(1+S^2)]^{-1/2}\,[\phi_A(\mathbf{r}_1)\phi_B(\mathbf{r}_2) + \phi_A(\mathbf{r}_2)\phi_B(\mathbf{r}_1)],$$

$$\Psi_A = [2(1-S^2)]^{-1/2}\,[\phi_A(\mathbf{r}_1)\phi_B(\mathbf{r}_2) - \phi_A(\mathbf{r}_2)\phi_B(\mathbf{r}_1)]. \qquad (47.1)$$

Corresponding spin functions have been discussed earlier in Chapter 11. In the special case of the H_2 molecule the atomic functions ϕ_A and ϕ_B should be replaced by $1s_A$ and $1s_B$ orbitals respectively.

A numerical evaluation of the energies requires the explicit form of the Hamiltonian. The electronic Hamiltonian for H_2 is written as

$$\mathbf{H}_{el} = -\frac{1}{2}(\nabla_1^2 + \nabla_2^2) - \left(\frac{1}{r_{A1}} + \frac{1}{r_{B1}} + \frac{1}{r_{A2}} + \frac{1}{r_{B2}}\right) + \frac{1}{r_{12}}. \tag{79.5}$$

Using the basis function Ψ_1 given in Eq. (79.1) we get

$$(\mathrm{H}_{el})_{11} = \int dv_1 \int dv_2\, 1s_A(\mathbf{r}_1) 1s_B(\mathbf{r}_2) \left(2E_H - \frac{1}{r_{A2}} - \frac{1}{r_{B1}} + \frac{1}{r_{12}}\right) 1s_A(\mathbf{r}_1) 1s_B(\mathbf{r}_2)$$

$$= 2E_H + 2J + \mathbf{J}_{1s_A, 1s_B} \tag{79.6}$$

where E_H is the 1s energy of the hydrogen atom ($E_H = -0.5$ in atomic unit), J is the coulomb integral in Eq. (35.9) which is to be evaluated from Eq. (77.15) using $Z = 1$ and $\zeta = 1$, and $\mathbf{J}_{1s_A, 1s_B}$ is the direct Coulomb repulsion energy,

$$\mathbf{J}_{1s_A, 1s_B} = \int dv_1 \int dv_2\, 1s_A^2(\mathbf{r}_1) 1s_B^2(\mathbf{r}_2) \frac{1}{r_{12}}$$

$$= \frac{1}{R} - e^{-2R}\left(\frac{1}{R} + \frac{11}{8} + \frac{3}{4}R + \frac{1}{6}R^2\right). \tag{79.7}$$

The expression for $(\mathrm{H}_{el})_{12}$ can be found in a similar way. We obtain

$$(\mathrm{H}_{el})_{12} = \int dv_1 \int dv_2\, 1s_A(\mathbf{r}_1) 1s_B(\mathbf{r}_2) \left(2E_H - \frac{1}{r_{A1}} - \frac{1}{r_{B2}} + \frac{1}{r_{12}}\right) 1s_B(\mathbf{r}_1) 1s_A(\mathbf{r}_2)$$

$$= 2E_H S^2 + 2KS + \mathbf{K}_{1s_A, 1s_B} \tag{79.8}$$

where K is the resonance integral in Eq. (35.9) and it can be evaluated from Eq. (77.16) by using Z=1 and ζ=1. The exchange coulomb repulsion integral $\mathbf{K}_{1s_A, 1s_B}$ was explicitly calculated by Sugiura (1927):

$$\mathbf{K}_{1s_A, 1s_B} = \int dv_1 \int dv_2\, 1s_A(\mathbf{r}_1) 1s_B(\mathbf{r}_2) \frac{1}{r_{12}} 1s_B(\mathbf{r}_1) 1s_A(\mathbf{r}_2)$$

$$= \frac{1}{5}\Big[- e^{-2R}\left(-\frac{25}{8} + \frac{23}{4}R + 3R^2 + \frac{1}{3}R^3\right)$$

$$+ \frac{6}{R}\left\{S^2(\gamma + \log R) + S'^2 E_i(-4R) - 2SS' E_i(-2R)\right\}\Big]. \tag{79.9}$$

In the above γ is Euler's constant

$$\gamma = 0.5772... \tag{79.10}$$

and

$$S' = e^R\left(1 - R + \frac{1}{3}R^2\right). \tag{79.11}$$

The function E_i is known as the *integral logarithm:* (See Pauling and Wilson, 1935).

If we substitute Equations (79.6) and (79.8) in Equations (79.4) we get

$$E_S = 2E_H + \frac{2J + \mathbf{J}_{1s_A,1s_B} + 2KS + \mathbf{K}_{1s_A,1s_B}}{1+S^2}, \tag{79.12}$$

$$E_A = 2E_H + \frac{2J + \mathbf{J}_{1s_A,1s_B} - 2KS - \mathbf{K}_{1s_A,1s_B}}{1-S^2}.$$

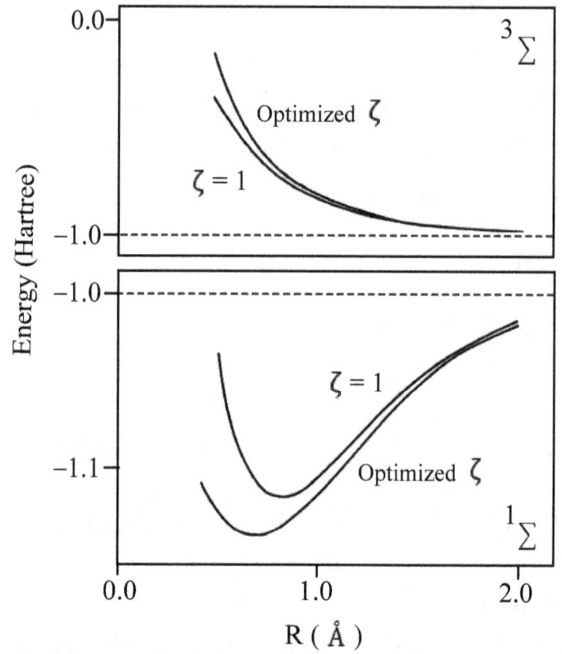

Figure 13.4. Bonding ($^1\Sigma$) and antibonding ($^3\Sigma$) energy curves computed for H_2 by the simple valence bond method. The effect of the optimization of exponents has been shown in each case.

The total energy for each valence-bond state is obtained by adding the internuclear repulsion energy to the electronic energy:

$$E_{total} = E_{S\text{ or }A} + \frac{1}{R}. \tag{79.13}$$

The energy curves are shown in Figure 13.4. As discussed earlier, the simple valence bond treatment gives an equilibrium bond length of 0.80Å which is in rough agreement with the experimental value 0.74 Å. The calculated dissociation energy 3.14 eV is better than that obtained from a simple Hartree-Fock treatment, but it is still a lot smaller than the experimental energy 4.72 eV. The calculated vibrational frequency is 4800 cm^{-1} that may be compared with the band spectral value of 4317.9 cm^{-1}.

The valence-bond calculation may be improved in several steps, namely, (i) by making a better choice of ϕ_A and ϕ_B, (ii) by considering many more covalent structures, and (iii) by bringing in ion-covalent resonance. A comparison of the results obtained from different procedures is given in Table 8.2 for the H_2 molecule, and the energy curves calculated by different methods are shown in Figure 13.5 and Figure 13.6.

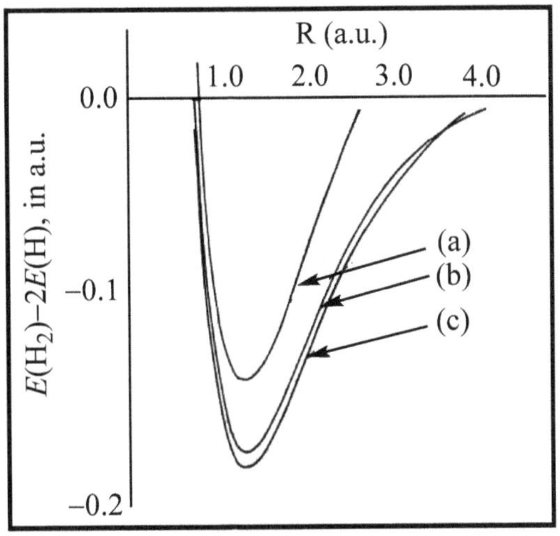

Figure 13.5 Comparison of the energy curves computed by different methods: (a) SCF vis-a-vis (b) MO Configuration Interaction and (c) 80-term series expansion.

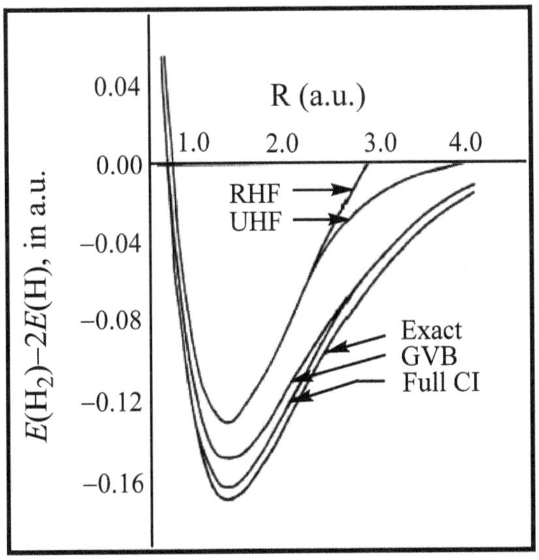

Figure 13.6 Comparison of the energy curves computed by different methods: SCF (both RHF and UHF) vis-a-vis GVB, FCI and the exact solution limit.

Lecture 80: Calculation of Molecular Integrals Over GTO Basis

After learning the initial treatment on the simplest molecular species, it would be fitting to know how the integrals required for the quantum mechanical treatment on a general molecule can be calculated. From Lecture 78 the reader might have guessed that the calculation of integrals over a STO basis is very complicated, and from Lecture 79 one may realize that analytical expressions like (79.9) must be ultimately evaluated by a numerical integration method or by making use of explicitly tabulated data. In either case it is not going to be an easy job. These are the reasons why the GTO basis is used almost invariably in all molecular calculations. In this lecture, the basis functions considered would be necessarily GTOs which are linear combinations or contractions of (atom-centered) Gaussian functions,

$$\chi_{A\mu}(\mathbf{r}) = \sum_{p=1}^{L_\mu} d_{p\mu}\, g_p(\alpha_{p\mu}, \mathbf{r}-\mathbf{R}_A). \qquad (74.3)$$

The one- and two-electron integrals over the basis functions are then simplified in terms of integrals over the Gaussian functions,

$$\langle \chi_{A\mu}|\mathbf{h}|\chi_{B\nu}\rangle = \sum_{p=1}^{L_\mu} \sum_{q=1}^{L_\nu} d_{p\mu}\, d_{q\nu}\, \langle g_p^A\, |\mathbf{h}|\, g_q^B\rangle,$$

and (80.1)

$$\langle \chi_{A\mu}(1)\chi_{C\sigma}C(2)|\frac{1}{r_{12}}|\chi_{B\nu}(1)\chi_{D\delta}(2)\rangle = \sum_{pqrs} d_{p\mu} d_{q\nu}\, d_{r\sigma} d_{s\delta}\, (g_p^A\, g_q^B\, |\, g_r^C\, g_s^D),$$

respectively. Furthermore,

$$\langle g_p^A\, |\mathbf{h}|\, g_q^B\rangle = -\frac{1}{2}\langle g_p^A |\nabla^2| g_q^B\rangle - \sum_C Z_C \langle g_p^A|\frac{1}{|\mathbf{r}-\mathbf{R}_C|}|g_q^B\rangle. \qquad (80.2)$$

Also, the overlap integral is given by

$$\langle \chi_{A\mu}|\chi_{B\nu}\rangle = \sum_{pq} d_{p\mu}\, d_{q\nu}\, \langle g_p^A | g_q^B\rangle. \qquad (80.3)$$

We will next see how to evaluate the Gaussian integrals $\langle g_p^A | g_q^B\rangle$, $\langle g_p^A|-\frac{1}{2}\nabla^2|g_q^B\rangle$, $\langle g_p^A |\frac{1}{r_C}| g_q^B\rangle$ and $(g_p^A\, g_q^B\, |\, g_r^C\, g_s^D)$.

In the Cartesian coordinates a normalized Gaussian function centered on atom A is written as

$$g^A = N^A\, x_A^{\,l}\, y_A^{\,m} z_A^{\,n}\, e^{-\alpha r_A^2},$$

$$N^A = \left\{\frac{2^{\,2(l+m+n)+3/2}\,\alpha^{\,l+m+n+3/2}}{(2l-1)!!(2m-1)!!(2n-1)!!\pi^{3/2}}\right\}^{1/2}, \qquad (80.4)$$

where $\mathbf{r}_A = \mathbf{r} - \mathbf{R}_A$ is the position vector with components x_A, y_A and z_A respectively along the three Cartesian axes, and l, m and n are quantum numbers having values zero or positive integers. The term $(2l-1)!!$ represents the product $1 \times 3 \times 5 \times \ldots \times (2l-1)$. The Gaussian integrals are calculated by using Eq. (74.2) and by making the observation

$$\mathbf{r}_A = (\mathbf{r} - \mathbf{R}_P) + (\mathbf{R}_P - \mathbf{R}_A) \equiv \mathbf{r}_P + \mathbf{R}_{PA} \qquad (80.5)$$

such that

$$x_A^{\,l_1}\, x_B^{\,l_2} = (x_P + X_{PA})^{l_1}\,(x_P + X_{PB})^{l_2} = \sum_{j=0}^{l_1+l_2} f_j(l_1, l_2, X_{PA}, X_{PB})\, x_P^{\,j}. \tag{80.6}$$

The new variable x_P is given by $x_P = x - X_P$ where $X_P = (\alpha_1 X_A + \alpha_2 X_B)/(\alpha_1+\alpha_2)$, and

$$f_j(l_1, l_2, X_{PA}, X_{PB}) = \sum_{k=0}^{l_1}\sum_{l=0}^{l_2}\binom{l_1}{k}\binom{l_2}{l} X_{PA}^{l_1-k} X_{PB}^{l_2-l}.$$

The following mathematical formulae are also used:

$$\int_{-\infty}^{\infty} dx\, x^{2n} e^{-\beta x^2} = \frac{(2n-1)!!}{2^{n+1}}\left(\frac{\pi}{\beta^{2n+1}}\right)^{1/2}, \tag{80.7}$$

$$\int_{-\infty}^{\infty} dx\, x^{2n+1} e^{-\beta x^2} = 0.$$

These relationships yield the Gaussian overlap and kinetic energy integrals in a straight-forward way. We get

$$\langle g_p^A | g_q^B \rangle \equiv \mathbf{S}(\alpha_1, l_1, m_1, n_1;\ \alpha_2, l_2, m_2, n_2)$$

$$= \int_{-\infty}^{\infty} dx \int_{-\infty}^{\infty} dy \int_{-\infty}^{\infty} dz\ g_p^A(\alpha_1, l_1, m_1, n_1)\, g_q^B(\alpha_2, l_2, m_2, n_2) = N_p^A\, N_q^B\, I_x I_y I_z \tag{80.8}$$

where, for example, integral I_x is written as

$$I_x = \int_{-\infty}^{\infty} dx\, x_A^{\,l_1}\, x_B^{\,l_2}\, e^{-\alpha_1 x_A^2 - \alpha_2 x_B^2}. \tag{80.9}$$

Further calculation reveals

$$I_x = \left(\frac{\pi}{\alpha_1+\alpha_2}\right)^{1/2} e^{-\alpha_1\alpha_2 X_{AB}^2/(\alpha_1+\alpha_2)} \sum_{i=0}^{[(l_1+l_2)/2]} \frac{(2i-1)!!}{[2(\alpha_1+\alpha_2)]^i} f_{2i}(l_1, l_2, X_{PA}, X_{PB})$$

with $[p/2]$ defined as $[p/2] = p/2$ if p is even, and $[p/2] = (p-1)/2$ if p is odd. Here we have used the fact that the product of two Gaussians is a constant times another Gaussian function,

$$e^{-\alpha_1 x_A^2 - \alpha_2 x_B^2} = e^{-(\alpha_1+\alpha_2)x_P^2}\, e^{-\alpha_1\alpha_2 X_{AB}^2/(\alpha_1+\alpha_2)}.$$

The last observation is also demonstrated in Figure 13.7 below.

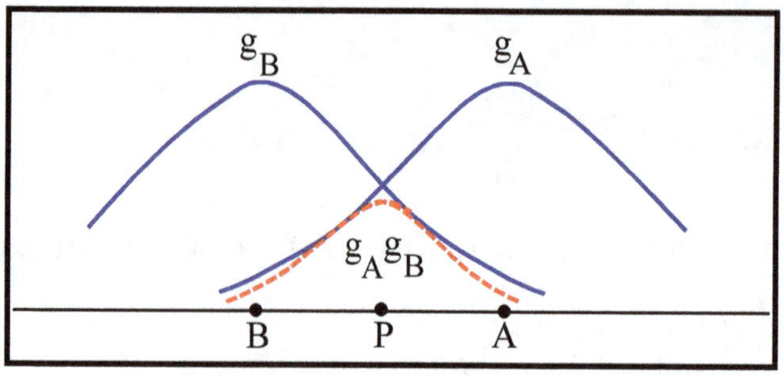

Figure 13.7 Illustrating that the product of two Gaussians is a constant times another Gaussian function.

We also obtain the kinetic energy integral

$$\langle g_p^A | -\frac{1}{2}\nabla^2 | g_q^B \rangle = N_p^A N_q^B \big[\alpha_2\{2(l_2+m_2+n_2)+3\}\,\mathbf{S}(\alpha_1,l_1,m_1,n_1;\alpha_2,l_2,m_2,n_2)$$
$$-2\alpha_2^2\,\{\mathbf{S}(\alpha_1,l_1,m_1,n_1;\alpha_2,l_2+2,m_2,n_2)$$
$$+\mathbf{S}(\alpha_1,l_1,m_1,n_1;\alpha_2,l_2,m_2+2,n_2)$$
$$+\mathbf{S}(\alpha_1,l_1,m_1,n_1;\alpha_2,l_2,m_2,n_2+2)\}$$
$$-\frac{1}{2}\{l_2(l_2-1)\,\mathbf{S}(\alpha_1,l_1,m_1,n_1;\alpha_2,l_2-2,m_2,n_2)$$
$$+m_2(m_2-1)\,\mathbf{S}(\alpha_1,l_1,m_1,n_1;\alpha_2,l_2,m_2-2,n_2)$$
$$+n_2(n_2-1)\,\mathbf{S}(\alpha_1,l_1,m_1,n_1;\alpha_2,l_2,m_2,n_2-2)\}\big] \qquad (80.10)$$

where $\mathbf{S}(\alpha_1,l_1,m_1,n_1;\alpha_2,l_2,m_2,n_2)$ is the overlap integral given by Eq. (80.8).

To calculate the nuclear attraction integral one makes use of the Singer transformation

$$\frac{1}{r} = \frac{1}{2\pi^2}\int \frac{d^3k}{k^2}\, e^{i\mathbf{k}\cdot\mathbf{r}} \qquad (80.11)$$

to write

$$\langle g_p^A | \frac{1}{r_C} | g_q^B \rangle = \frac{N_p^A N_q^B}{2\pi^2}\, e^{-\alpha_1\alpha_2 R_{AB}^2/(\alpha_1+\alpha_2)}$$
$$\times \sum_{i,j,k} f_i(l_1,l_2,X_{PA},X_{PB})\, f_j(m_1,m_2,Y_{PA},Y_{PB})\, f_k(n_1,n_2,Z_{PA},Z_{PB})$$

$$\times\int \frac{d^3k'}{k'^2}\, e^{i\mathbf{k}'\cdot\mathbf{p}} \int_{-\infty}^{\infty} dx\, x^i\, e^{ik'_x x-(\alpha_1+\alpha_2)x^2}$$

$$\times \int_{-\infty}^{\infty} dy\, y^j\, e^{ik'_y y-(\alpha_1+\alpha_2)y^2} \int_{-\infty}^{\infty} dz\, z^k\, e^{ik'_z z-(\alpha_1+\alpha_2)z^2} \qquad (80.12)$$

where $\mathbf{p}=\mathbf{P}-\mathbf{C}$. O-Ohata and coworkers 1966) made use of the integration

$$\int_{-\infty}^{\infty} dx\, x^n\, e^{ik'_x x-\beta x^2} = i^n \sqrt{\frac{\pi}{\beta}} \left(\frac{1}{2\sqrt{\beta}}\right)^n H_n\left(\frac{k'_x}{2\sqrt{\beta}}\right) e^{-k'^2_x/4\beta} \qquad (80.13)$$

where $Hm(z)$ is the Hermite polynomial given by Rodrigues formula in Table 3.1,

$$Hm(z) = m! \sum_{i=0}^{[n/2]} \frac{(-1)^i (2z)^{n-2i}}{i!(m-2i)!}.$$

If one substitutes this expression and Eq. (80.13) in Eq. (80.12), uses the identity

$$e^{-a} = a \int_1^{\infty} dx\, e^{-ax} \qquad (80.14)$$

such that

$$e^{-ak^2} = 2ak^2 \int_0^1 dt\, t^{-3} e^{-ak^2/t^2}, \qquad (80.15)$$

and finally makes use of the definition

$$F_i(t) = \int_0^1 ds\, s^{2i}\, e^{-ts^2}, \qquad (80.16)$$

one obtains

$$\langle g_p^A | \frac{1}{r_C} | g_q^B \rangle = N_p^A N_q^B \frac{2\pi}{(\alpha_1+\alpha_2)} e^{-\alpha_1\alpha_2 R^2_{AB}/(\alpha_1+\alpha_2)}$$

$$\times \sum_{i,r,u} A_{i,r,u}(l_1,l_2,X_A,X_B,X_C,\alpha_1+\alpha_2) \times \sum_{j,s,v} A_{j,s,v}(m_1,m_2,Y_A,Y_B,Y_C,\alpha_1+\alpha_2)$$

$$\times \sum_{k,t,w} A_{k,t,w}(n_1,n_2,Z_A,Z_B,Z_C,\alpha_1+\alpha_2) \times F_\mu\left(\frac{R^2_{CP}}{4\varepsilon}\right) \qquad (80.17)$$

where

$$\varepsilon = \frac{1}{4(\alpha_1+\alpha_2)}, \qquad (80.18a)$$

$$\mu = (i-2r-u) + (j-2s-v) + (k-2t-w) \qquad (80.18b)$$

and

$A_{i,r,u}(l_1,l_2,X_A,X_B,X_C,\gamma)$

$$= (-1)^{i+u} f_i(l_1,l_2,X_{PA},X_{PB}) \frac{i!X_{CP}^{i-2r-2u}}{r!u!(i-2r-2u)!(4\gamma)^{r+u}}. \tag{80.18c}$$

Proceeding along similar lines one finds that the electron repulsion integral can be written as

$$(g_p^A g_q^B | g_r^C g_s^D) = N_p^A N_q^B N_r^C N_s^D \times \frac{2\pi^2}{\gamma_1\gamma_2} \exp\left[-\frac{\alpha_1\alpha_2 R_{AB}}{\gamma_1} - \frac{\alpha_3\alpha_4 R_{CD}}{\gamma_2}\right]$$

$$\times \sum_{i_1=0}^{l_1+l_2} \sum_{i_2=0}^{l_3+l_4} \sum_{r_1=0}^{[\frac{i_1}{2}]} \sum_{r_2=0}^{[\frac{i_2}{2}]} \sum_{u=0}^{[\frac{i_1+i_2}{2}-r_1-r_2]} B_{i_1,i_2,r_1,r_2,u}(l_1,l_2,X_A,X_B,X_P,\gamma_1;l_3,l_4,X_C,X_D,X_Q,\gamma_2)$$

$$\times \sum_{j_1=0}^{m_1+m_2} \sum_{j_2=0}^{m_3+m_4} \sum_{s_1=0}^{[\frac{j_1}{2}]} \sum_{s_2=0}^{[\frac{j_2}{2}]} \sum_{v=0}^{[\frac{j_1+j_2}{2}-s_1-s_2]} B_{j_1,j_2,s_1,s_2,v}(m_1,m_2,Y_A,Y_B,Y_P,\gamma_1;m_3,m_4,Y_C,Y_D,Y_Q,\gamma_2)$$

$$\times \sum_{k_1=0}^{n_1+n_2} \sum_{k_2=0}^{n_3+n_4} \sum_{t_1=0}^{[\frac{k_1}{2}]} \sum_{t_2=0}^{[\frac{k_2}{2}]} \sum_{w=0}^{[\frac{k_1+k_2}{2}-t_1-t_2]} B_{k_1,k_2,t_1,t_2,w}(n_1,n_2,Z_A,Z_B,Z_P,\gamma_1;n_3,n_4,Z_C,Z_D,Z_Q,\gamma_2)$$

$$\times F_v(R_{PQ}^2/4\delta) \tag{80.19}$$

where $\gamma_1= \alpha_1+\alpha_2$, $\gamma_2= \alpha_3+\alpha_4$, and $\delta= (\gamma_1+\gamma_2)/4\gamma_1\gamma_2$. The position vector **Q** has similar relations to the vectors **C** and **D** as **P** has with **A** and **B**, and the terms B's are given by

$B_{i_1,i_2,r_1,r_2,u}(l_1,l_2,X_A,X_B,X_P,\gamma_1;l_3,l_4,X_C,X_D,X_Q,\gamma_2)$

$$= (-1)^{i_2+u} f_{i_1}(l_1,l_2,X_{PA},X_{PB}) f_{i_2}(l_3,l_4,X_{QC},X_{QD})$$

$$\times \frac{1}{(4\gamma_1)^{i_1-r_1}(4\gamma_2)^{i_2-r_2}\delta^{i_1+i_2-r_1-r_2}} \times \frac{i_1!i_2![i_1+i_2-2(r_1+r_2)]!}{(i_1-2r_1)!(i_2-2r_2)!r_1!r_2!}$$

$$\times \frac{1}{u![i_1+i_2-2(r_1+r_2)-2u]!} \times (PQ)_x^{i_1+i_2-2(r_1+r_2)-2u} \delta^u . \tag{80.20}$$

These forms for the Gaussian integrals were first calculated by O-Ohata and coworkers (1966).

Ab-initio calculations

The term *ab-initio* means "starting from the first principles." In an *ab-initio* methodology every quantity is calculated or evaluated from the first principles, that is, from the quantum mechanical postulates or axioms and theorems, without making any recourse to any assumption or any approximation not included within the framework of the involved theory, or to any experimental information in the form of numerical values or data. The theory involved may be an approximate one like the molecular electronic treatment under Born-Oppenheimer approximation.

Nevertheless, even in an *ab-initio* treatment, one may get hints from some qualitative or experimental information for the purpose of a quick yet efficient numerical evaluation of certain physicochemical properties. Knowledge of the experimental results in advance always guides one to make a pragmatic selection of the initial data or input for the *ab-initio* calculation, although the actual process of calculation never resorts to the use of any experimental data in any form. Thus if one defines a basis set, evaluates *all* the necessary integrals and performs a calculation like the Hartree or the Hartree-Fock calculation with or without a subsequent CI or any other many-body treatment, the calculation is *ab-initio*. The choice of the basis set generally depends on our knowledge of the atomic or molecular system on which the calculation is performed.

The maximum size of the molecule for an *ab-initio* treatment is mainly determined by the ability to compute and to process (while forming the Fock matrix) the large number of two-electron integrals. In a semi-empirical treatment some of the integrals discussed above are explicitly calculated using STO (or GTO) basis functions, some others are neglected, and yet some of the integrals are replaced by a set of numerical values determined from observed physical properties like ionization potential, electron affinity, dissociation energy of homonuclear diatomic species, etc.

An *ab-initio* calculation normally requires an intense computational effort. A semi-empirical calculation would require a vastly reduced effort for a molecule of the same size. Whereas *ab-initio* calculations can be performed either for all electrons or only for the valence electrons, semi-empirical calculations are carried out strictly for the valence electrons. The latter types of calculation are designed to get semi-quantitative chemical information at a small cost of time and effort, and it is well known that the valence electrons dictate most of the chemistry in ground state.

Lecture 81: Semi-Empirical SCF Theories

A semi-empirical method is normally intended for the studies of moderately large molecules for which the *ab-initio* calculations would require an unrealistically large computing effort. A semi-empirical method must have a few desirable characteristics. The approximations involved in a semi-empirical theory must not overlook the primary physical interactions which determine the equilibrium geometry of the molecule. The approximate wave function should be found from a general treatment that does not require an empirically derived knowledge such as the covalent or the ionic nature of a bond. The calculated results must be easy to interpret so that a qualitative discussion of the structure of the molecule and its reactivity in terms of the distribution of electronic charges on atoms may become possible.

The π-electron formalism of Pople (1953) was the earliest semi-empirical theory. The σ electrons are also chemically active. Thus truly successful semi-empirical theories like CNDO, INDO, MINDO, etc. were developed later. These are general enough to account for all types of valence electrons on an equal footing. The valence-electron treatment in these semi-empirical theories is based on the assumption that the nucleus and the "core" elctrons in each atom together form an atomic core that can be represented as a point charge Z_C such that the Hamiltonian for one valence electron can be written as

$$\mathbf{h}^{\text{core}} = -\frac{1}{2}\nabla^2 - \sum_{n}^{nuclei} \frac{Z_{Cn}}{r_n}. \tag{81.1}$$

This operator is normally called the *core hamiltonian.*

ZDO

The most important approximation in the semi-empirical treatments is the *zero differential overlap* (ZDO) approximation. This is equivalent to putting $\phi_\mu(\mathbf{r})\phi_\nu(\mathbf{r}) = 0$ everywhere when ϕ_μ and ϕ_ν are two different atomic orbitals of different atoms or even the same atom. The ZDO approximation was introduced by Parr (1952). It implies that

$$(\mu\nu|\lambda\sigma) = (\mu\mu|\lambda\lambda)\delta_{\mu,\nu}\delta_{\lambda,\sigma} = \gamma_{\mu\lambda}\ \delta_{\mu,\nu}\delta_{\lambda,\sigma} \tag{81.2}$$

and of course it satisfies the weaker condition

$$S_{\mu\nu} = \int dv\ \phi_\mu(\mathbf{r})\phi_\nu(\mathbf{r}) = \delta_{\mu,\nu}. \tag{81.3}$$

The core Hamiltonian integral is written as

$$h_{\mu\nu}^{\text{core}} = \int dv\ \phi_\mu(\mathbf{r})\ \mathbf{h}^{\text{core}}\ \phi_\nu(\mathbf{r}). \tag{81.4}$$

It involves overlap distribution, but this is not always neglected; instead, it is approximated in terms of suitable atomic and bond parameters.

Under the ZDO approximations the Hartree-Fock-Roothaan equation (73.3) for a closed-shell species reduces to

$$\sum_{\nu} F_{\mu\nu}c_{\nu i} = \varepsilon_i c_{\mu i} \tag{81.5}$$

while the elements of the Fock matrix as given by Eq. (73.13) appears as

$$F_{\mu\mu} = h_{\mu\mu}^{\text{core}} - \frac{1}{2}P_{\mu\mu}(\mu\mu|\mu\mu) + \sum_{\lambda} P_{\lambda\lambda}(\mu\mu|\lambda\lambda)$$

and (81.6)

$$F_{\mu\nu} = h_{\mu\nu}^{\text{core}} - \frac{1}{2}P_{\mu\nu}(\mu\mu|\nu\nu) \qquad \text{(for } \mu \neq \nu\text{)}.$$

These results find applications in many of the semi-empirical methods.

Pi-electron calculations

In Pople's π-electron theory, the core hamiltonian is written as

$$\mathbf{h}^{\text{core}} = -\frac{1}{2}\nabla^2 + \sum_{s} V_s \tag{81.7}$$

where V_s is the potential energy of a π electron in the field of the σ-bonded framework around the sth atom. If Z_s represents the number of π electrons with atom s then

$$\alpha_r = <\phi_r \mid \mathbf{h}^{core} \mid\phi_r> = W_r - \sum_{s(\neq r)}^{atoms} Z_s\gamma_{rs} \tag{81.8}$$

where W_r is given by

$$W_r = <\phi_r \mid -\frac{1}{2}\nabla^2 + V_r \mid\phi_r> \tag{81.9}$$

and the matrix element of V_s is approximated by

$$<\phi_r \mid V_s \mid\phi_r> = - Z_s <\phi_r \mid \frac{1}{|\mathbf{r}-\mathbf{R}_s|} \mid\phi_r>.$$

It would be easy if one writes

$$<\phi_t| \frac{1}{|\mathbf{r}-\mathbf{R}_s|} |\phi_t> = \gamma_{st} \tag{81.10}$$

so that the two-electron integrals and the nuclear attraction integrals can be treated at par with each other. The resonance integral is formulated as

$$\beta_{rs} = <\phi_r \mid \mathbf{h}^{core} \mid\phi_s> \approx <\phi_r \mid -\frac{1}{2}\nabla^2 + V_r + V_s \mid\phi_r> \tag{81.11}$$

where integrals of type $<\phi_r|V_t|\phi_s>$ with r, s and t all different have been neglected by the ZDO approximation.

Therefore, the elements of the π-electron Fock matrix are given by

$$\alpha^F_{rr} = W_r + \frac{1}{2} P_{rr}\gamma_{rr} + \sum_{s(\neq r)}^{atom} (P_{ss}-Z_s)\gamma_{rs}\,,$$

and (81.12)

$$\beta^F_{rs} = \beta_{rs} - \frac{1}{2} P_{rs}\gamma_{rs}\,.$$

The difference from a simple Hückel calculation is manifest. In fact Hückel's π-electron theory is based on the assumption of an average value of α^F_{rr} and a similarly averaged value for β^F_{rs}. The resonance integral is limited to the nearest-neighbor interaction in Hückel theory as it decreases very rapidly when $|\mathbf{R}_r - \mathbf{R}_s|$ increases. Actually the Hückel parameter β is often found from a comparison of energy differences such as the calculated and thermo-chemically measured values of resonance energy of benzene. For homo-nuclear conjugated organic molecules the Hückel parameter α can be chosen in different ways, but the accepted norm is to equate it to W_r, that is, to the negative of the atomic valence orbital ionization potential. For hetero-nuclear systems, different α values for different types of atoms enter into the energy expressions and so sufficient care must be taken for the choice of these parameters.

Invariance under transformations

Investigations on the electronic structure of molecules are normally carried out by using the LCAO-MO scheme, that is, by expressing the molecular orbitals as orthonormal linear combinations of atomic orbitals. The valence orbitals of each atom are linearly transformed among themselves as

$$\phi_\nu \rightarrow \phi_\nu' = \sum_\mu \phi_\mu T_{\mu\nu} \tag{81.13}$$

such that the transformed (hybrid) orbitals remain orthogonal to each other. Correspondingly, each molecular orbital can be written as a linear combination of the transformed orbitals. This is established as follows. The LCAO-MO scheme is given by

$$\psi = \boldsymbol{\phi}\mathbf{C} \Leftrightarrow \psi_i = \sum_\mu \phi_\mu c_{\mu i}. \tag{81.14}$$

Since by Eq. (81.13) $\boldsymbol{\phi}' = \boldsymbol{\phi}\mathbf{T}$, we write

$$\boldsymbol{\phi} = \boldsymbol{\phi}'\mathbf{T}^{-1} \tag{81.15}$$

which can be used to obtain

$$\psi = \boldsymbol{\phi}'\mathbf{T}^{-1}\mathbf{C} = \boldsymbol{\phi}'\mathbf{C}' \quad \Leftrightarrow \quad \psi_i = \sum_\mu \phi_\mu' c_{\mu i}'. \tag{81.16}$$

This transformation has the effect that hybrid orbitals are formed from the valence orbitals of an atom.

The *ZDO approximation*, $\phi_\mu\phi_\nu = 0$ for $\mu \neq \nu$ at every point in space, remains unchanged as $\phi_\mu'\phi_\nu' = 0$ for $\mu \neq \nu$ when μ and ν are orbitals on two different atoms. That is, the neglect of the diatomic differential overlap remains valid under any linear transformation that transforms a set of orbitals belonging to a specific atom to another set of orbitals belonging to the same atom. We will see in the next lecture that the transformed monatomic differential overlap need not necessarily vanish. Under the transformation, the integrals necessary for the Hartree-Fock calculation transform as

$$\begin{aligned} S_{\mu\nu} &\rightarrow S'_{\mu\nu} = \sum_{\alpha,\beta} T_{\alpha\mu}T_{\beta\nu}S_{\alpha\beta}, \\ h^{\text{core}}_{\mu\nu} &\rightarrow h^{\text{core}'}_{\mu\nu} = \sum_{\alpha,\beta} T_{\alpha\mu}T_{\beta\nu}\, h^{\text{core}}_{\alpha\beta}, \\ (\mu\nu|\sigma\lambda) &\rightarrow (\mu'\nu'|\sigma'\lambda') = \sum_{\alpha,\beta,\gamma,\delta} T_{\alpha\mu}T_{\beta\nu}T_{\gamma\sigma}T_{\delta\lambda}\,(\alpha\beta|\gamma\delta). \end{aligned} \tag{81.17}$$

In an approximate LCAO-MO SCF calculation we require that the transformed integrals should remain unchanged from the original ones for any arbitrary transformation of the type discussed above. This was first pointed out Pople, Santry and Segal in 1965.

Lecture 82: ZDO-Based Methods

CNDO

Pople, Santry and Segal (1965), and Pople and Segal (1965) introduced the method of *complete neglect of differential overlap* (CNDO). In order to keep up the invariance under local rotational transformations the ZDO result (81.2) is made more approximate by writing

$(\mu\nu|\lambda\sigma) = \gamma_{AB}\delta_{\mu,\nu}\delta_{\lambda,\sigma}$, (ϕ_μ is on atom A and ϕ_λ is on atom B). (82.1)

See Figure 13.8. The parameter γ_{AB} depends only on the nature of atoms A and B and the distance between them, but it is taken to be independent of the explicit nature of the orbitals ϕ_μ and ϕ_λ. The charge distribution around each atom is assumed to be spherically symmetric, as in s states.

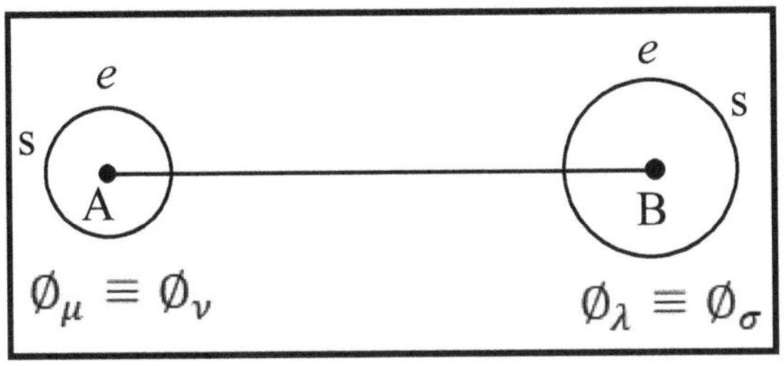

Figure 13.8 Invariance of two-center two-electron integrals under local rotational transformations.

The ZDO result (81.3) is rotationally invariant and it is retained in CNDO. The next logical step is to make the expression for the core Hamiltonian integrals invariant under the transformation. The core Hamiltonian is

$$\mathbf{h}^{core} = -\frac{1}{2}\nabla^2 + \sum_A V_A \tag{82.2}$$

where $V_A = -\frac{Z_A}{r_A}$, Z_A being the core charge of atom A. This gives rise to three types of core Hamiltonian integrals:

$$h^{core}_{\mu\mu} = U_{\mu\mu} + \sum_{B(\neq A)} <\phi_\mu|V_B|\phi_\mu> \qquad (\phi_\mu \text{ is on atom A})$$

$$h^{core}_{\mu\nu} = U_{\mu\nu} + \sum_{B(\neq A)} <\phi_\mu|V_B|\phi_\nu> \qquad (\text{Both } \phi_\mu \text{ and } \phi_\nu \text{ are on atom A})$$

$$h^{core}_{\mu\nu} = <\phi_\mu| -\frac{1}{2}\nabla^2 + V_A + V_B|\phi_\nu> + \sum_{C(\neq A,B)} <\phi_\mu|V_C|\phi_\nu>$$

(ϕ_μ is on atom A and ϕ_ν is on atom B)

where (82.3)

$$\mathbf{U} = -\frac{1}{2}\nabla^2 + V_A.$$

The rotational invariance is achieved in three steps. In the first step one writes

$$<\phi_\mu|V_B|\phi_\nu> = V_{AB}\,\delta_{\mu,\nu} \qquad (\text{Both } \phi_\mu \text{ and } \phi_\nu \text{ are on atom A}) \tag{82.4}$$

so that

$$h^{core}_{\mu\nu} = \delta_{\mu,\nu}\,[U_{\mu\mu} + \sum_{B(\neq A)} V_{AB}]. \tag{82.5}$$

The approximation (82.4) has been illustrated in Figure 13.9. In the second step one neglects the three-center integrals $<\phi_\mu|V_C|\phi_\nu>$. The resulting H_2^+–like integral

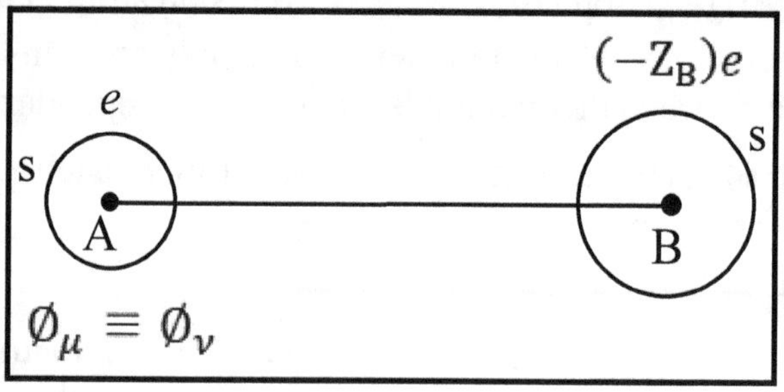

Figure 13.9 Illustrating approximation (82.4) that keeps the core hamiltonian matrix invariant under a local rotational transformation.

$$h_{\mu\nu}^{\text{core}} = \langle \phi_\mu | -\frac{1}{2}\nabla^2 + V_A + V_B | \phi_\nu \rangle$$

for ϕ_μ on atom A and ϕ_ν on atom B is approximated by

$$h_{\mu\nu}^{\text{core}} = \beta^0_{AB} S_{\mu\nu} \tag{82.6}$$

where β^0_{AB} is a proportionality constant.

If we use these approximations we find that the elements of the Fock matrix given by Eq. (81.6) appear in the vastly simplified forms

$$F_{\mu\mu} = U_{\mu\mu} + (P_{AA} - \frac{1}{2}P_{\mu\mu})\,\gamma_{AA} + \sum_{B(\neq A)} [-Q_B\gamma_{AB} + (Z_B\gamma_{AB} + V_{AB})]$$

and (82.7)

$$F_{\mu\nu} = \beta^0_{AB} S_{\mu\nu} - \frac{1}{2}P_{\mu\nu}\gamma_{AB} \qquad \text{(for } \mu \neq \nu\text{)}$$

where P_{AA} is the net Mulliken population on atom A and Q_B is the net charge on atom B,

$$P_{AA} = \sum_{\mu\in A} P_{\mu\mu}$$

and (82.8)

$$Q_B = Z_B - P_{BB}.$$

The Hartree-Fock total energy is found as

$$E_{\text{tot}} = \frac{1}{2}\sum_{\mu\nu} P_{\mu\nu}\,(h_{\mu\nu}^{\text{core}} + F_{\mu\nu}) + \sum_{A<B} Z_A Z_B R_{AB}^{-1}. \tag{82.9}$$

Parametrization

There are basically two systems of parametrization for the CNDO method of calculation, the older version being known as CNDO/1 and the improved one as CNDO/2 parameter schemes.

In the **CNDO/1** scheme the integrals V_{AB} and γ_{AB} are calculated by assuming that the charge distribution around each atom is spherically symmetric. The integrals are then expressed by employing the STO's which represent the valence s orbitals of the involved atoms:

$$V_{AB} = -Z_B \int dv_1 \; s_A^2(\mathbf{r}_1) \, r_{1B}^{-1}$$

and (82.10)

$$\gamma_{AB} = \int dv_1 \int dv_2 \; s_A^2(\mathbf{r}_1) s_B^2(\mathbf{r}_2) \, r_{12}^{-1}.$$

Compare these expressions with the diagrams in Figures 13.9 and 13.8. These integrals are explicitly evaluated in the CNDO procedure by employing the standard exponents available from Slater's rules.

The $U_{\mu\mu}$ terms are estimated from the expressions for the total energy of an atom (A) and that of its mono-positive ion (A^+), the neutral atomic configuration being $2s^{i}2p^{j}$:

$$E_A(2s^{i}2p^{j}) = iU_{2s,2s} + jU_{2p,2p} + \frac{1}{2}(i+j)(i+j-1)\gamma_{AA},$$

$$E_A(2s^{i-1}2p^{j}) = (i-1)U_{2s,2s} + jU_{2p,2p} + \frac{1}{2}(i+j-1)(i+j-2)\gamma_{AA}, \qquad (82.11)$$

$$E_A(2s^{i}2p^{j-1}) = iU_{2s,2s} + (j-1)U_{2p,2p} + \frac{1}{2}(i+j-1)(i+j-2)\gamma_{AA}.$$

These give the U matrix elements in terms of the orbital ionization potentials I_S and I_P,

$$U_{2s,2s} = -I_S - (i+j-1)\gamma_{AA} = -I_S - (Z_A-1)\gamma_{AA},$$

and (82.12)

$$U_{2p,2p} = -I_P - (i+j-1)\gamma_{AA} = -I_P - (Z_A-1)\gamma_{AA}.$$

The average ionization potentials necessary for the formation of the $U_{\mu\mu}$ parameters and the values of the resonance integrals β^0_{AA} used in the CNDO/1 scheme are given in Table 13.1.

The bonding parameter β^0_{AB} is assumed to be the average of bonding parameters for the corresponding homonuclear diatomic molecules,

$$\beta^0_{AB} = \frac{1}{2}(\beta^0_{AA} + \beta^0_{BB}). \qquad (82.13)$$

The overlap integrals $S_{\mu\nu}$ are explicitly calculated but they are only used to form the off- diagonal elements of the Fock matrix. The ZDO relationship

$$S_{\mu\nu} = \delta_{\mu,\nu} \qquad (81.3)$$

is still to be used in the Hartree-Fock-Roothaan equation.

Table 13.1 Average ionization potentials and resonance integral in electron Volt for the CNDO/1 parameter scheme for elements hydrogen ($Z = 1$) to fluorine ($Z = 9$).

Element	Atomic Number (Z)	Core Charge (Z_C)	Valence ionization potential s	Valence ionization potential P	β^0_{AA}
H	1	1	13.06		−9.
Li	3	1	5.39	3.54	−9.
Be	4	2	9.32	5.96	−13.
B	5	3	14.05	8.30	−17.
C	6	4	19.44	10.67	−21.
N	7	5	25.58	13.19	−25.
O	8	6	32.38	15.85	−31.
F	9	7	40.20	18.66	−39.

CNDO/2 modifications: The term $(Z_B\gamma_{AB}+V_{AB})$ in $F_{\mu\mu}$ in Eq. (82.7) represents the effect of the penetration of one electron that stays in the μth orbital of atom A into the shell of another atom B. When V_{AB} and γ_{AB} are calculated as different integrals as in equations (82.10), the penetration term leads to additional stability so much so that a non-negligible bond energy is found even when the bond order is zero! This problem is rectified by treating V_{AB} at par with γ_{AB} as it was done in Pople's π-electron theory. See Eq. (81.10). The expression for V_{AB} in Eq. (82.10) can be modified by writing

$$V_{AB} = -Z_B\gamma_{AB}. \tag{82.14}$$

This modification is illustrated by Figure 13.10. It constitutes the basis for the improved scheme CNDO/2. The diagonal elements of the Fock matrix are now written as

$$F_{\mu\mu} = U_{\mu\mu} + (P_{AA} - \frac{1}{2}P_{\mu\mu})\gamma_{AA} - \sum_{B(\neq A)} Q_B\gamma_{AB}. \tag{82.7a}$$

The off-diagonal element, $F_{\mu\nu}$ with $\mu \neq \nu$, remains unchanged from the second of the equations (82.7).

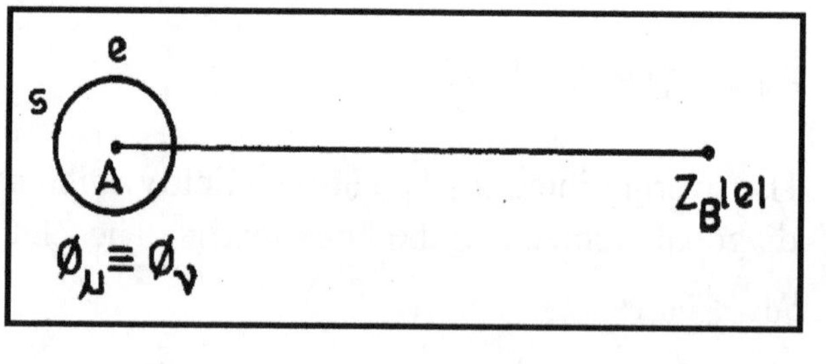

Figure 13.10 The Pariser-Parr-Pople type modification of the nuclear attraction integral in CNDO/2 scheme: this may also be viewed as the CNDO modification.

Another modification is made in CNDO/2. The CNDO/1 scheme defines the $U_{\mu\mu}$ matrix elements and conversely the ionization potential by Eq. (82.12). One can similarly express the electron affinity A_μ by

$$A_\mu = -U_{\mu\mu} - Z_A\gamma_{AA}. \tag{82.15}$$

Chemical properties are determined by the tendency of an atom to lose as well as gain one electron. If we consider ionization potential and electron affinity on an equal footing while forming the element $U_{\mu\mu}$, the treatment is expected to generate good values for thermochemical properties. Hence the parameter $U_{\mu\mu}$ is determined in the CNDO/2 scheme from the average of orbital ionization potential and orbital electron affinity, that is, from the orbital electronegativity. Orbital electronegativity values for different elements are given in Table 13.2.

Table 13.2 Orbital electronegativity and resonance integral for the CNDO/2 parameter scheme for elements hydrogen (Z=1) to chlorine (Z=17).

Element	Atomic Number (Z)	Core Charge (Z_C)	Orbital electronegativity s	p	d	β^0_{AA} (eV)
H	1	1	7.176			–9.
Li	3	1	3.106	1.258		–9.
Be	4	2	5.946	2.563		–13.
B	5	3	9.594	4.001		–17.
C	6	4	14.051	5.572		–21.
N	7	5	19.316	7.275		–25.
O	8	6	25.390	9.111		–31.
F	9	7	32.272	11.080		–39.
Na	11	1	2.804	1.302	0.150	–7.720
Mg	12	2	5.125	2.052	0.162	–9.447
Al	13	3	7.771	2.995	0.224	–11.301
Si	14	4	10.033	4.133	0.337	–13.065
P	15	5	14.033	5.464	0.500	–15.070
S	16	6	17.650	6.989	0.713	–18.150
Cl	17	7	21.591	8.708	0.977	–22.330

In the above we have come across the expressions for the elements of the Fock matrix for a closed-shell molecule. Similar expressions can be written down for the open-shell species. In passing we remark that the Pople-Nesbet open shell formalism is almost invariably adopted in different semiempirical methods.

The initial molecular orbital coefficients required for the SCF procedure are often found from a Hückel-type approach that involves the solution of the ZDO-level SCF equation (81.5) with the Fock matrix elements

$$F_{\mu\mu} = U_{\mu\mu},$$

$$F_{\mu\nu} = \beta^0_{AB} S_{\mu\nu} \qquad \text{(for } \mu \neq \nu). \qquad (82.16)$$

The CNDO/2 scheme has been further modified by Santry and Segal (1967) in order to include the third-row elements in the treatment. These authors suggested the rewriting of the resonance integral as

$$\beta^0_{AB} = \frac{1}{2} K (\beta^0_{AA} + \beta^0_{BB}). \qquad (82.17)$$

where K = 0.75. The quantity β^0_{AA} for a homonuclear diatomic of a third-row element A (belonging to the fourth row in the periodic table) can be estimated from the corresponding quantity β^0_{XX} for the second-row diatomic X_2 by

$$\beta^0_{AA} = \frac{U_{3s,3s}(A) + U_{3p,3p}(A)}{U_{2s,2s}(X) + U_{2p,2p}(X)} \beta^0_{XX}. \qquad (82.18)$$

Table 13.3 CNDO/2 parameters for the third-row elements, collected from various sources as discussed in the text.

Element	STO exponent		$U_{\mu\mu}$ in electron Volt			β^0_{AA} in electron Volt	
	s,p	d	s	p	d	sp	d
K	0.700	1.500					
Ca	0.825	1.783					
Sc	0.950	2.067	−3.657	−0.558	−3.793	−0.5	−15.0
Ti	1.075	2.283	−3.770	−0.690	−4.140	−2.0	−16.0
V	1.200	2.500	−3.822	−0.777	−4.475	−3.5	−16.5
Cr	1.325	2.650	−3.909	−0.876	−4.822	−5.5	−17.3
Mn	1.450	2.933	−3.983	−0.975	−5.157	−8.0	−18.0
Fe	1.575	3.150	−4.120	−1.062	−5.504	−12.0	−18.5
Co	1.700	3.367	−4.170	−1.160	−5.839	−17.5	−19.1
Ni	1.825	3.583	−4.306	−1.260	−6.182	−24.6	−19.5
Cu	1.950	3.733	−4.567	−1.347	−6.520	−35.0	−20.0
Zn	2.075	4.067					
Ga	~ 1.35	~ 1.35					
Ge	1.527	1.527	−11.435	−4.080	0.0	−10.0	−10.0
As	1.702	1.702	−13.335	−5.345	0.0	−13.0	−13.0
Se	1.878	1.878	−16.315	−7.100	0.0	−16.0	−16.0
Br	2.054	2.054	−19.630	−8.400	0.0	−22.0	−22.0

Optimized STO exponents for the elements of the first transition series have been calculated by Burns (1964). Clack, Hush and Yandle (1972) as well as Zerner and Gouterman (1966) have determined the corresponding $U_{\mu\mu}$ and β^0_{AA} parameters. Hase and Schweig (1973) have prepared similar parameters as well as the STO exponents for the elements with atomic numbers 32-35 (Ge, As, Se and Br). CNDO/2 parameters for the third-row elements, which have been collated from different sources, are listed here in the form of Table 13.3.

INDO

One of the most important CNDO approximations is given by Eq. (82.1) that takes into consideration only the direct repulsion between two atomic charge densities and neglects all other coulomb repulsion integrals. The exchange integrals, both one-center and two-center, are completely neglected in CNDO theory. Therefore, the Hartree-Fock total energies calculated by the CNDO method for different spin states but with the same electronic configuration are not different! Thus the ^3P, ^1D and ^1S states of the carbon atom, all arising from the valence electronic configuration $2s^22p^2$, are not energetically distinguished in CNDO. The CNDO method is also unable to lead to a net spin density in σ orbitals. This drawback gave rise to an improvement in terms of *intermediate neglect of differential overlap* (**INDO**) where the monatomic differential overlap is retained in one-center two-electron integrals (Pople, Beveridge and Dobosh, 1967). The retention of the monatomic exchange integrals results in a considerably improved treatment, though the theory is still not as accurate as the *ab-initio* formalism.

Since the only non-vanishing one-center, two-electron integrals are of type $(\mu\mu|\mu\mu)$, $(\mu\mu|\nu\nu)$ and $(\mu\nu|\mu\nu)$, the elements of the Fock matrix in the unrestricted INDO scheme are written as

$$F^{\alpha}_{\mu\mu} = U_{\mu\mu} + \sum_{\lambda \in A} [P_{\lambda\lambda}(\mu\mu|\lambda\lambda) - P^{\alpha}_{\lambda\lambda}(\mu\lambda|\mu\lambda)] + \sum_{B(\neq A)} (P_{BB} - Z_B)\gamma_{AB} \quad (\mu \text{ on A}),$$

$$F^{\alpha}_{\mu\nu} = (2P_{\mu\nu} - P^{\alpha}_{\mu\nu})(\mu\nu|\mu\nu) - P^{\alpha}_{\mu\nu}(\mu\mu|\nu\nu) \quad (\text{both } \mu \text{ and } \nu \text{ on atom A}), \qquad (82.19)$$

where $P^{\alpha}_{\mu\nu}$ is the (μ,ν) element of the density matrix for electrons with the α spin and $P_{\mu\nu}$ is the corresponding element of the full density matrix. The expression for $F^{\alpha}_{\mu\nu}$ with orbitals μ and ν belonging to different atoms remains unchanged from that in the CNDO/2 scheme. The restricted Fock matrix elements can be easily worked out from these expressions.

If we assume that *n*s and *n*p orbitals have the same radial parts, the non-vanishing integrals are found as

$$\begin{aligned}
(ss|ss) &= (ss|xx) = F^0 = \gamma_{AA}, \\
(sx|sx) &= \frac{1}{3} G^1, \\
(xy|xy) &= \frac{3}{25} F^2, \qquad (82.20)\\
(xx|xx) &= F^0 + \frac{4}{25} F^2, \\
(xx|yy) &= F^0 - \frac{2}{25} F^2,
\end{aligned}$$

and so on. In the above, F^0, G^1 and F^2 are two-electron integrals involving the radial parts of atomic orbitals (Slater, 1960). Integral F^0 can be explicitly calculated. A list of the integral values G^1 and F^2 for atoms of light elements is given in Table 13.4. The electronic energy of the average states of an element with the electronic configuration $ns^a\, np^b$ can be generally written as

$$E = aU_{2s,2s}+bU_{2p,2p}+\frac{1}{2}(a+b)(a+b-1)F^0-\frac{1}{6}abG^1-\frac{1}{25}b(b-1)F^2. \quad (82.21)$$

If one calculates I and A from this expression and similar ones for the ionic species, one can find the core Hamiltonian integrals $U_{\mu\mu}$'s in terms of the Slater-Condon parameters F^0, G^1 and F^2 and the core charge Z_C. These integrals are as follows:

For H: $U_{ss} = -\frac{1}{2}(\mathrm{I+A})_s - \frac{1}{2}\gamma_{HH}$

For Li: $U_{ss} = -\frac{1}{2}(\mathrm{I+A})_s - \frac{1}{2}F^0$

$U_{pp} = -\frac{1}{2}(\mathrm{I+A})_p - \frac{1}{2}F^0+\frac{1}{12}G^1$

For Be: $U_{ss} = -\frac{1}{2}(\mathrm{I+A})_s - \frac{3}{2}F^0+\frac{1}{12}G^1$

$U_{pp} = -\frac{1}{2}(\mathrm{I+A})_p - \frac{3}{2}F^0+\frac{1}{4}G^1$

B to F: $U_{ss} = -\frac{1}{2}(\mathrm{I+A})_s - (Z_C-\frac{1}{2})F^0+\frac{1}{6}(Z_C-\frac{3}{2})G^1$

$$U_{pp} = -\frac{1}{2}(\mathrm{I+A})_p - (Z_C-\frac{1}{2})F^0+\frac{1}{3}G^1+\frac{2}{25}(Z_C-\frac{5}{2})F^2 \quad (82.22)$$

Table 13.4. Empirical values of Slater-Condon parameters for the first-row elements.

	G^1		F^2	
Element	Rydberg	a.u.	Rydberg	a.u.
Li	0.184024	0.092012	0.09973	0.049865
Be	0.28140	0.14070	0.17825	0.089125
B	0.39853	0.199265	0.26082	0.13041
C	0.535416	0.267708	0.34744	0.17372
N	0.692058	0.346029	0.43811	0.219055
O	0.86846	0.43423	0.53283	0.266415
F	1.06461	0.532305	0.63160	0.31580

The INDO methodology can be successfully applied to the determination of very small energy changes that are normally associated with the solvation process, redox reactions and other biochemical reactions in solution. The latter processes often involve moderately large molecules as reactants and products. As the molecular size increases, the INDO calculation itself becomes more voluminous and intricate as the Hartree-Fock total energy needs to be optimized for an increasingly large number of bond lengths and bond angles.

This demands a meticulous work which, when done, pays a rich dividend. For example, the INDO calculation of the standard reduction potential of p-benzoquinone has shown (Mallik and Datta 1994) that if one is careful enough to optimize the molecular geometry in great detail, not only the calculated reduction potential turns out to be in good agreement with the experimentally determined potential but also the theoretical determination identifies the exact mechanism for the reductive process. Incidentally, the mechanism of the reduction of p-benzoquinone in water is known from polarographic studies and it cannot be guessed from an experiment that involves only a measurement of the reduction potential. Semi-empirical methodologies, especially CNDO and INDO, are now used to study aspects of protein dynamics, protein folding and electronic conduction through protein and DNA chains. These have also found applications to objects in nanoscale and solid state extensions, such as systems like carbon nanotubes, finite strips of graphene, and diatomic materials like BN.

Overview

It is often found that the results of a semi-empirical calculation are considerably better than the *ab-initio* results. For example, the semi-empirical ground state energy or binding energy may be much closer to the experimental one. The reason for this observation is that most of the semi-empirical theories have been generated with the supreme goal of reproducing the experimental results of a specific type, such as the thermo-chemical energy changes, from a brute-force numerical calculation. Because of the empiricism inherent in a semi-empirical theory, some of the correlation effects are already embedded in the treatment. The merit of a particular semi-empirical formalism should be appreciated only within this context. It would be grossly wrong to think that a semi-empirical theory yields all the results better than their *ab-initio* counterparts. The semi-empirical formalism is often based on such approximations and parametrizations that some of the properties like the energy of a reaction, equilibrium bond lengths and dipole moments are very nicely evaluated. But then the computed values of some other properties like the force constants, anharmonic constants, and spectroscopic transition energies would turn out to be quite inaccurate.

There are other semi-empirical methods such as (i) the *neglect of diatomic differential overlap* (**NDDO**) approximation constructed by Pople, Santry and Segal (1965) and by Sustmann and coworkers (1969), and (ii) the *medium intermediate neglect of differential overlap* (**MINDO**) methodology due to Dewar (1969, 1970, 1975). These methods are based on retaining some more of the multi-center two-electron integrals. Integrals of the type $(s_A p_A|s_B p_B)$ which represent the dipole-dipole interaction are not neglected in NDDO. The additional non-negligible two-electron integrals are either directly calculated or empirically chosen. Thus these schemes represent a higher level of precision than that expected for the INDO method, but one has to pay a higher price for the more precise calculation. The rise in price comes in the form of an increased computing effort.

A parameter scheme somewhat different from the usual one has been suggested for the CNDO treatment. This scheme has been based on spectroscopic data and it is used for a singles CI (SCI) treatment that follows a Hartree-Fock SCF calculation. The methodology is known as **CNDO/S**, the letter S standing for 'Spectroscopic'. The thrust applied in this method is towards the generation of good values for excitation energies and transition moments. The ground state properties are rather poorly generated by this method. The method has been moderately successful in the calculation of the peak frequency and the intensity of spectroscopic transitions that involve the excitation of a single electron from one of the occupied orbitals in the ground state, and highly successful in predicting the polarization of these absorptions. Zerner and his coworkers (1966, 1973, 1979, 1980, 1987) have developed an **INDO/S** treatment for an accurate computation of the spectroscopic characteristics of single excitations in general organic molecules, organometallic molecules, and inorganic complexes containing transition metal ions.

Texts Suggested for Further Reading

L. Pauling and E. B. Wilson, "Introduction to Quantum Mechanics", (McGraw-Hill, Tokyo, 1935).
J. A. Pople and D. L. Beveridge, "Approximate Molecular Orbital Theory", (McGraw-Hill, New York, 1970).
W. J. Hehre, L. Radom, P. v. R. Schleyer and J. A. Pople, "*Ab Initio* Molecular Orbital Theory", (Wiley, New York, 1986).

EXERCISES AND PROBLEMS

1. Write a program to compute the potential energy surface for the species H_2^+ according to the simple MO treatment.
2. Write a program to compute the potential energy surface for the species H_2 following the simple MO treatment;
3. Write a program to compute the potential energy surface for H_2 using the simple VB treatment.
4. Derive the semi-empirical expressions for the elements of the Fock matrix for an open shell atom and an open-shell molecule within the framework of the Pople-Nesbet HFSCF formalism. Try the examples of Lithium atom, Carbon atom, Diatomic oxygen molecule, methylene radical and methyl radical.

CHAPTER 14

QUANTUM MECHANICAL TUNNEL EFFECTS

Lecture 83: Tunnelling Through a Square Hill

Classically, when a particle of mass m collides with a one-dimensional square potential barrier V(x) as shown in Figure 14.1, it is reflected if its energy E is less than the barrier height V_0 and it continues unchanged in direction in case E is greater than V_0. The force at point $x = 0$ is $-V_0\delta(x)$ while it is $V_0\delta(x - a)$ at point $x = a$. When $E > V_0$, the linear momentum changes from p to p' at $x = 0$ and from p' to p at $x = a$, where $p' = [2m(E - V_0)]^{1/2}$. For $E < V_0$, the linear momentum changes from p to $-p$ at point $x = 0$ where $p = [2mE]^{1/2}$. The region $0 < x < a$ where a is the barrier height is classically forbidden for $E < V_0$. We shall see in a moment that the situation changes for a quantum mechanical particle: there is a finite probability of transmission to the other side of the barrier even when $E < V_0$.

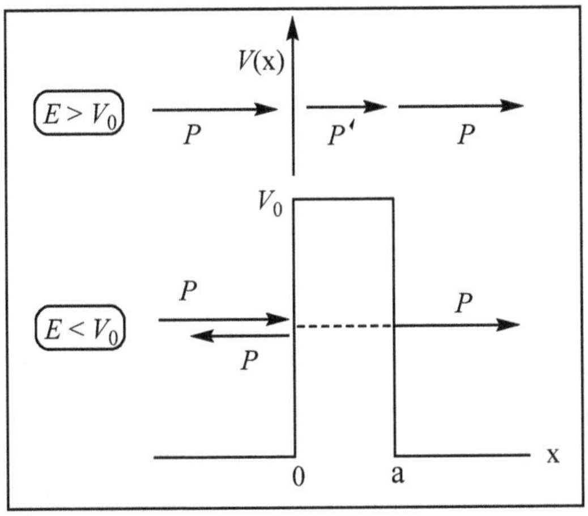

Figure 14.1 A one-dimensional square potential barrier of height V_0 and width a.

Consider the whole one-dimensional space divided into three regions. Compare Figure 14.1. Schrödinger equation for each region can be explicitly written as:

$$\begin{aligned}
&\text{Region I.} \quad && -\infty \le x \le 0 \quad && -\frac{\hbar^2}{2m}\frac{d^2\psi}{dx^2} = E\psi \\
&\text{Region II.} \quad && 0 \le x \le a \quad && -\frac{\hbar^2}{2m}\frac{d^2\psi}{dx^2} + V_0\psi = E\psi \qquad (83.1) \\
&\text{Region III.} \quad && a \le x \le \infty \quad && -\frac{\hbar^2}{2m}\frac{d^2\psi}{dx^2} = E\psi
\end{aligned}$$

The solution of the first and the last equations in (83.1) is $e^{\pm ikx}$ with momentum $\pm\hbar k$ along the x-axis where $k = (2mE)^{1/2}/\hbar$. The general solution in Region I is

$$\psi_{\rm I} = Ae^{ikx} + Be^{-ikx}. \tag{83.2}$$

The coefficient of the incident plane wave is A, and that of the reflected plane wave is B. The solution of the second equation in (83.1) is $e^{\pm ik'x}$ where $k' = [2m(E - V_0)]^{1/2}/\hbar$, so that the general solution in Region II is

$$\psi_{\rm II} = Ce^{ik'x} + De^{-ik'x}. \tag{83.3}$$

Notice that k' is real when $E > V_0$, and imaginary for $V_0 > E$. In the latter case we write $k' = -i\kappa$ such that κ is a real quantity given by $\kappa = [2m(V_0 - E)]^{1/2}/\hbar$, and the solution can be simplified as $\psi_{\rm II} = Ce^{\kappa x} + De^{-\kappa x}$. Finally, the general solution in Region III is like that in Region I,

$$\psi_{\rm III} = Fe^{ikx} + Ge^{-ikx}. \tag{83.4}$$

However, a particle incident from the negative side of the x-axis would be transmitted only, that is, $G = 0$ for the corresponding solution, though F will be non-zero unless the particle is fully reflected at any point $x \le$ a.

To know the ratio of the coefficients A, B, C, D and F, one imposes the conditions that the wave function ψ is well-behaved at points $x = 0$ and $x =$ a, that is, ψ is single-valued and continuous at these two points. Thus one requires

$$\psi_{\rm I}(x = 0) = \psi_{\rm II}(x = 0), \tag{83.5a}$$

$$\left(\frac{d\psi_{\rm I}}{dx}\right)_{x=0} = \left(\frac{d\psi_{\rm II}}{dx}\right)_{x=0}, \tag{83.5b}$$

$$\psi_{\rm II}(x = \mathrm{a}) = \psi_{\rm III}(x = \mathrm{a}), \tag{83.5c}$$

$$\left(\frac{d\psi_{\rm II}}{dx}\right)_{x=\mathrm{a}} = \left(\frac{d\psi_{\rm III}}{dx}\right)_{x=\mathrm{a}}. \tag{83.5d}$$

These conditions give

$$A + B = C + D, \tag{83.6a}$$

$$k(A - B) = k'(C - D), \tag{83.6b}$$

$$Ce^{ik'\mathrm{a}} + De^{-ik'\mathrm{a}} = Fe^{ik\mathrm{a}}, \tag{83.6c}$$

$$k'[Ce^{ik'\mathrm{a}} - De^{-ik'\mathrm{a}}] = kFe^{ik\mathrm{a}}. \tag{83.6d}$$

Using equations (83.6a) and (83.6b) one obtains

$$A = \frac{1}{2}\left(1 + \frac{k'}{k}\right)C + \frac{1}{2}\left(1 - \frac{k'}{k}\right)D, \tag{83.7a}$$

$$B = \frac{1}{2}\left(1 - \frac{k'}{k}\right)C + \frac{1}{2}\left(1 + \frac{k'}{k}\right)D. \tag{83.7b}$$

Similarly, using equations (83.6c) and (83.6d), one gets

$$C = \frac{1}{2}\left(1 + \frac{k}{k'}\right)e^{i(k - k')\mathrm{a}}F, \tag{83.8a}$$

$$D = \frac{1}{2}\left(1 - \frac{k}{k'}\right)e^{i(k + k')\mathrm{a}}F, \tag{83.8b}$$

If one inserts equations (83.8a) and (83.8b) in equations (83.7a) and (83.7b), one finds

$$A = \left[\cos k'a - i\frac{k^2 + k'^2}{2kk'}\sin k'a \right] e^{ika} F, \tag{83.9a}$$

$$B = -i\frac{k^2 - k'^2}{2kk'}\sin k'a\, e^{ika} F, \tag{83.9b}$$

so that the reflection and transmission coefficients, given by the ratios of the corresponding amplitudes with the incident amplitude, are

$$\frac{B}{A} = \frac{-i(k^2 - k'^2)\sin k'a}{2kk'\cos k'a - i(k^2 + k'^2)\sin k'a} \tag{83.10}$$

and

$$\frac{F}{A} = \frac{2kk'}{2kk'\cos k'a - i(k^2 + k'^2)\sin k'a} e^{-ika}. \tag{83.11}$$

The reflectance R and transmittance T are defined as the probabilities of reflection and transmission, respectively. For $E > V_0$, k' is real and

$$R = \left|\frac{B}{A}\right|^2 = \frac{(k^2 - k'^2)^2 \sin^2 k'a}{4k^2k'^2 + (k^2 - k'^2)^2 \sin^2 k'a} \tag{83.12}$$

and

$$T = \left|\frac{F}{A}\right|^2 = \frac{4k^2k'^2}{4k^2k'^2 + (k^2 - k'^2)^2 \sin^2 k'a}. \tag{83.13}$$

For $E < V_0$, k' is imaginary. So we replace k' by $-i\kappa$ to obtain

$$R = \frac{(k^2 + \kappa^2)^2 \sinh^2 \kappa a}{4k^2\kappa^2 + (k^2 + \kappa^2)^2 \sinh^2 \kappa a} \tag{83.14}$$

and

$$T = \frac{4k^2\kappa^2}{4k^2\kappa^2 + (k^2 + \kappa^2)^2 \sinh^2 \kappa a}. \tag{83.15}$$

In both cases,

$$R + T = 1, \tag{83.16}$$

that is, the particle is either reflected or transmitted.

For $E > V_0$ the reflectance R and transmittance T will be interpreted as the scattering probabilities in the backward and forward directions respectively.

For $E > V_0$ the transmittance T tends to zero as κa goes on increasing arbitrarily, that is, when $[2m(V_0 - E)]^{1/2}a/\hbar >> 1$. This happens in three situations: (i) when the mass of the particle becomes large, that is, for macroscopically large particles; (ii) when the barrier width a is macroscopically large; (iii) when the barrier V_0 is macroscopically high. In all the three cases, the unit of action $\hbar$ becomes extremely small in comparison to the numerator of the ratio. The classical limit is defined by $\hbar \to 0$.

For a particle of very small mass such as an electron, the probability to tunnel through a barrier that is only a few Angstrom in thickness (such as the one that exists due to the air gap between two conducting wire pieces tightly wound around each other), is quite high. This is a purely quantum mechanical effect that happens because the wave function in Region II becomes finite and represents a bound state oscillation. The tunnelling probability becomes negligibly small in the classical limit that is valid for the macroscopic world of larger mass and larger size or dimensions.

Let us take a closer look at the classical limit for $E = \hbar^2k^2/2m << V_0 = E + \hbar^2\kappa^2/2m$. The term $\frac{1}{4}(k^2+\kappa^2)^2 e^{2\kappa a}$ dominates in the denominator of Eq. (83.15) so that

$$T \simeq 16\frac{E}{V_0}\left(1-\frac{E}{V_0}\right)\times\exp\left[-\frac{2}{\hbar}[2m(V_0-E)]^{1/2}a\right]. \tag{83.17}$$

The transmittance is extremely sensitive to changes in the energy of the particle and the barrier width, and also to the particle mass.

Lecture 84: The WKB Approximation

The expression for transmittance in the classical limit also suggests an elegant way to calculate the probability of tunnelling through a potential barrier of arbitrary shape.

Our starting point is that the Schrödinger equation

$$i\hbar\frac{d\psi}{dt} = -\frac{\hbar^2}{2m}\nabla^2\psi + V(\mathbf{r})\psi \tag{5.6}$$

has a general solution of the form

$$\psi(\mathbf{r},t) = Ae^{iW(\mathbf{r},t)/\hbar} \tag{84.1}$$

such that $W(\mathbf{r},t)$ satisfies

$$\frac{\partial W}{\partial t} + \frac{1}{2m}\vec{\nabla}W\bullet\vec{\nabla}W + V - \frac{i\hbar}{2m}\nabla^2 W = 0\,. \tag{84.2}$$

In the classical limit ($\hbar \to 0$), this equation is the same as Hamilton's partial differential equation for the principal function $W(\mathbf{r},t)$,

$$\frac{\partial W}{\partial t} + H(\mathbf{r},\mathbf{p}) = 0 \tag{84.3}$$

where the linear momentum $\mathbf{p}$ is defined as the gradient of the principal function

$$\mathbf{p} = \vec{\nabla}W\,. \tag{84.4}$$

This shows that the possible trajectories are normal to the constant W surface. In other words, the trajectories in the classical limit are normal to the surface of constant phase of ψ.

Of special interest is the stationary state wave function

$$\psi(\mathbf{r},t) = u(\mathbf{r})e^{-iEt/\hbar} \tag{84.5}$$

for which W can be written as

$$W(\mathbf{r},t) = S(\mathbf{r}) - Et \tag{84.6}$$

such that

$$u(\mathbf{r}) = Ae^{iS(\mathbf{r})/\hbar}\,. \tag{84.7}$$

Then Eq. (84.2) converts into the form

$$\frac{1}{2m}\vec{\nabla}S\bullet\vec{\nabla}S - [E - V(\mathbf{r})] - \frac{i\hbar}{2m}\nabla^2 S = 0\,. \tag{84.8}$$

Wentzel (1926), Kramers (1926) and Brillouin (1926) solved this equation in the one-dimensional case by expanding $S(\mathbf{r})$ in a series involving powers of $\hbar$.

In one dimension, Eq. (84.8) turns out as

$$i\hbar\frac{d^2S}{dx^2}-\left(\frac{dS}{dx}\right)^2+\hbar^2k^2(x)=0 \qquad [E>\mathrm{V}(x)] \tag{84.9a}$$

where

$$k(x)=\frac{1}{\hbar}[2m\{E-\mathrm{V}(x)\}]^{1/2}, \tag{84.10a}$$

and

$$i\hbar\frac{d^2S}{dx^2}-\left(\frac{dS}{dx}\right)^2-\hbar^2\kappa^2(x)=0 \qquad [E<\mathrm{V}(x)] \tag{84.9b}$$

where

$$\kappa(x)=\frac{1}{\hbar}[2m\{\mathrm{V}(x)-E\}]^{1/2}. \tag{84.10b}$$

When the expansion

$$S(x)=S_0(x)+\hbar S_1(x)+\ldots \tag{84.11}$$

is incorporated in Eq. (84.9a) and Eq. (84.9b) and the total coefficient of each power of $\hbar$ is equated to zero, one obtains a series of equations like

$$\begin{aligned}&-S_0'^2+2m[E-\mathrm{V}(x)]=0,\\&iS_0''-2S_0'S_1'=0,\end{aligned} \tag{84.12}$$

etc. Thus one finds for $E>\mathrm{V}(x)$,

$$S_0(x)=\pm\hbar\int dx'\ k(x'),\qquad S_1(x)=\frac{i}{2}\ \ln k(x), \tag{84.13}$$

etc., and for $E<\mathrm{V}(x)$,

$$S_0(x)=\mp i\hbar\int dx'\ \kappa(x'),\qquad S_1(x)=\frac{i}{2}\ \ln \kappa(x), \tag{84.14}$$

etc. To this order of approximation the function $u(x)$ is given by

$$\begin{aligned}&u(x)=A'\ k^{-1/2}(x)\ e^{\pm i\int k(x)dx} \quad [\text{for } E>\mathrm{V}(x)],\\&u(x)=A''\ \kappa^{-1/2}(x)\ e^{\pm i\int \kappa(x)dx} \quad [\text{for } E<\mathrm{V}(x)].\end{aligned} \tag{84.15}$$

These are known as the WKB solutions.

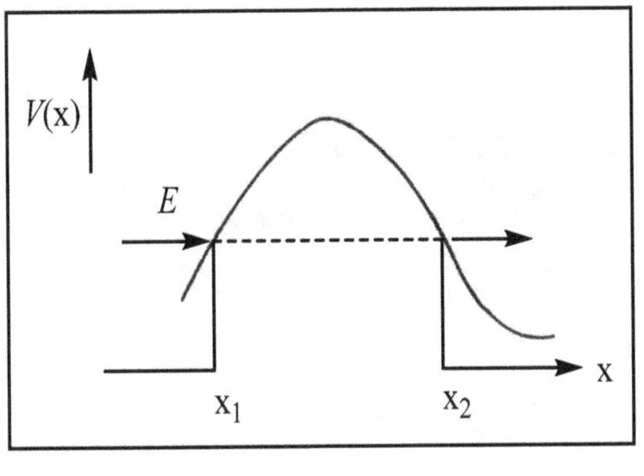

Figure 14.2 Tunnelling through a potential of an arbitrary shape: The WKB method becomes useful in such cases.

A calculation of the transmittance for tunnelling through a barrier requires a special consideration of the wave function at the classical turning points. The magnitude and slope of the solutions should match at each turning point. The transmittance can be written as

$$T=\frac{|u(x_2)|^2}{|u(x_1)|^2}=\frac{C_2}{C_1}\ \exp\left[-2\int_{x_1}^{x_2}\kappa(x)dx\right] \tag{84.16}$$

where C_1 is the ratio $|A'|^2/k$ at the point of entry x_1 and C_2 is a similar ratio $|A''|^2/\kappa$ at the point of exit x_2 for total energy E. See Figure 14.2. Compare this result with the classical limit of T in Eq. (83.17). The latter equation is valid for a square barrier. It is actually a special case of the WKB transmittance. Equation (84.16) is good through order $\hbar$.

The WKB method is applicable to the study of radioactive α decay in which case the coordinate x is identified with the radial distance as shown in Figure 14.3. The potential energy is then modified from the electrostatic (Coulombic) term by addition of the kinetic energy for angular motion:

$$\kappa(r)=\frac{1}{\hbar}\left[2m\left\{V(r)+\frac{\hbar^2 l(l+1)}{2mr^2}-E\right\}\right]^{1/2} \tag{84.17}$$

where m is the reduced mass and

$$V(r)=\frac{ZZ'e_0}{r}, \tag{84.18}$$

Z and Z' being the atomic number of the nucleus left and that of the escaping nucleus, respectively. For α decay, $Z'=2$ and $r_1 \approx R$, the nuclear radius. See Figure 14.3 where we have marked another effective radius r_2 at which the Coulomb potential equals the kinetic energy E ($= mv^2/2$) of the escaping body. For $l=0$, $r_2 = ZZ'e_0^2/E$. We have

$$\int_{r_1}^{r_2}\kappa(r)dr=\pi\frac{ZZ'e_0^2}{\hbar}\sqrt{\frac{m}{2E}}\left(1-\frac{2}{\pi}\sin^{-1}\gamma^{1/2}\right)-\frac{\sqrt{2mER}}{\hbar}(\gamma-1)^{1/2} \tag{84.19}$$

where γ is the ratio of the height of the Coulomb barrier to the kinetic energy E of the escaping particle,

$$\gamma=\frac{ZZ'e_0^2}{ER}. \tag{84.20}$$

Figure 14.3 Potential energy diagram for alpha decay: Coulombic repulsion is observed outside the effective radius of the nucleus, and an attractive nucleon-nucleon interaction inside explains binding energy.

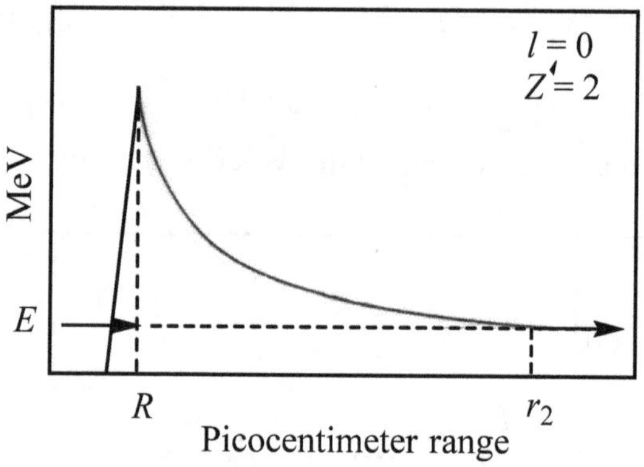

The tunnelling probability T linearly varies with the barrier penetration factor

$$P=\exp\left[-2\int_{r_1}^{r_2}\kappa(r)dr\right]. \tag{84.21}$$

If τ is the typical period of oscillation of the α particle within the radioactive nucleus of radius R, then τ/P is of the order of the lifetime for α decay. This treatment yields, for $R\sim10^{-12}$ cm and a typical period $\tau\sim10^{-21}$ s for α oscillation within nucleus, the correct Z and E dependence of the lifetime for α decay over an extremely wide range.

A chemical application of the WKB treatment is found in the study of electron and proton transfer reactions. A rather rapid proton transfer takes place when molecules with alcoholic or phenolic hydroxyl groups are dissolved in protonic solvents like water. Every chemistry student knows that the rapid transfer leads to a broadening of the proton magnetic resonance spectrum. The corresponding potential energy surface may be generated by quantum chemical calculations where, for simplicity, the proton may be considered to remain at various distances from the rest of the molecules. An explicit discussion of the treatment is beyond the scope of this text, but some of the aspects involved will be indicated in the next lecture.

Lecture 85: Electron and Proton Transfer Reactions

The Double Well

The double-well potential is a widely used model for the theoretical treatment of electron transfer, proton transfer, and other atomic transfer processes. The potential is generated when two bonding energy surfaces of the *same* symmetry cross each other as illustrated by Figure 14.4.

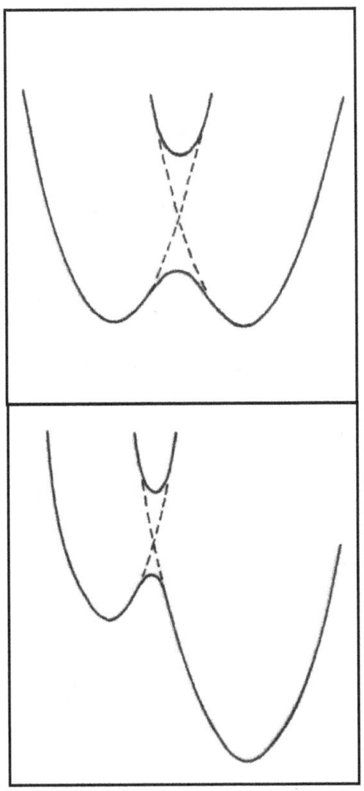

Figure 14.4 Symmetric and asymmetric double wells formed from the combination of two similar wells and two dissimilar ones, respectively, both the wells having the same symmetry features.

In principle, the quantized energy levels of a double well can be easily found from the energy levels of each individual well. From our discussion of the non-crossing rule in Lecture 20, we know that the mixing is strong when the two levels have approximately equal energy. Take $H_{11} = E_1$ as the energy corresponding to the wave function ψ_1 for the first well, and $H_{22} = E_2$ as the energy associated with the wave function ψ_2 for the second well such that $E_1 \sim E_2$. The double-well function is

$$\psi = c_1\psi_1 + c_2\psi_2 \tag{85.1}$$

which may be determined from the secular equation

$$\begin{vmatrix} E_1 - E & H_{12} - ES_{12} \\ H_{21} - ES_{21} & E_2 - E \end{vmatrix} = 0 \tag{85.2}$$

subject to the normalization constraint

$$|c_1|^2 + |c_2|^2 + (c_1 {}^* c_2 S_{12} + c_2 {}^* c_1 S_{21}) = 1. \tag{85.3}$$

Functions ψ_1 and ψ_2 are necessarily vibrational functions: the almost harmonic shape of each individual potential ensures it. Consider ψ_1 and ψ_2 as real functions of space. Then $H_{12}= H_{21}= \beta$ and $S_{12}= S_{21} = S$ are both real quantities, and the coefficients c_1 and c_2 are both real numbers. These considerations make the solution easier.

The overlap integral S is significant only above the barrier, for $E_1 \approx E_2 > V_0$. When E_1 ($\approx E_2$) is significantly less than V_0, the integral S is negligibly small. Thus we have two clear situations of interest.

Above the barrier, the splitting is large and the particle undergoes a large amplitude movement typical of an extensive delocalization. This is witnessed at an elevated temperature such that a substantially large fraction of molecules have energy in excess of the critical energy to overcome the barrier. The situation is more common than what you would normally think. It is encountered in all chemical reactions, and the said critical energy is often referred to as the activation energy. The reaction becomes irreversible when the vibrations in the second well are coupled with other forms of movement. The coupling leads to energy loss such that a part of the vibrational energy is converted into energy of other vibrational modes, another part transforms into translational energy in the medium (solvent or lattice), and yet another part changes into heat that dissipates to the environment. These steps are associated with a huge increase in entropy.

A more interesting situation develops when reactions occur at energies less than V_0. Below the barrier peak, $S \approx 0$ and tunnelling is the mechanism of transfer. We write

$$H_{12} = \frac{1}{2} h\nu_T \tag{85.4}$$

where ν_T is the tunnelling frequency. The secular equation for a symmetric double well yields

$$E = E_1 \pm \frac{1}{2} h\nu_T \Leftrightarrow c_1 = \pm c_2 = \frac{1}{\sqrt{2}}. \tag{85.5}$$

The double well splitting is $h\nu_T$, as shown in Figure 14.5. The solutions in (85.5) are for the stationary states Ψ_+ and Ψ_-.

How does it explain tunnelling? To visualize the phenomenon, start from the nonstationary state

$$\begin{aligned} \Phi(t) &= d_+\Psi_+(\mathrm{t}) + \mathrm{d}_-\Psi_-(\mathrm{t}) \\ &= \frac{1}{\sqrt{2}} e^{-iE_1 t/\hbar} \left[\left(d_+ e^{-i\pi\nu_T t} + d_- e^{i\pi\nu_T t} \right) \psi_1 + \left(d_+ e^{-i\pi\nu_T t} - d_- e^{i\pi\nu_T t} \right) \psi_2 \right]. \end{aligned} \tag{85.6}$$

Choose $d_+ = d_- = 1/2^{1/2}$ so that $\Phi(t = 0) = \psi_1$. Then $\Phi(t = 1/2\nu_T) = -i\,\psi_2$ and $|\Phi(t = 1/2\nu_T)|^2 = \psi_2^2$. Thus if the particle is localized in the first well at time $t = 0$, it is completely transferred to the second well at time $t = 1/2\nu_T$, and becomes localized again in the first well at time $t = 1/\nu_T$. Hence ν_T is the frequency of tunnelling. The transmission rate will be given by the ratio of the tunnelling frequency and the vibrational frequency of each well: $T = \nu_T/\nu_1$.

Proton Transfer

Although the discussion above is for a symmetric double well potential, a similar treatment is possible for an asymmetric double well. The splitting pattern for an asymmetric double well is shown in Figure 14.5. The formalism is applicable, for example, to the transfer of a proton from one molecule AH to another molecule B as

$$AH + B = A^- + HB^+ \qquad (85.7)$$

The potential wells in the vicinity of the two participating molecules need not be mirror images of each other, and the double well can be quite asymmetric. The tunnel effect is observed also for the transfer of other atoms.

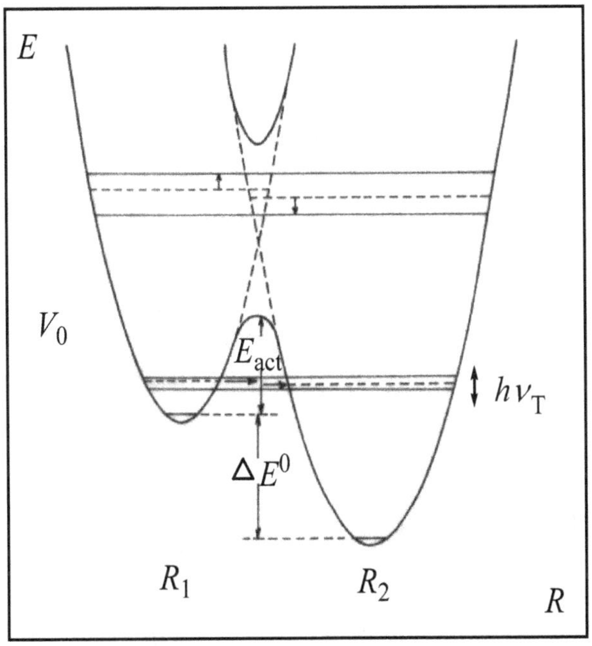

Figure 14.5 Tunnelling arising from the formation of delocalized states in an asymmetric double well.

Nonrigid Behavior

Fluxional molecules such as PF_5 are known to exhibit pseudorotations (Berry, 1960). The latter type of motion is different from a bodily rotation that involves the angular movement of the whole body of a molecule. During a pseudorotation, one part of the molecule rotates relative to the rest of the molecule. A common example is the almost free rotation of the methyl group, that is, the movement of the methyl hydrogen atoms, around the axis of the C—C single bond that holds the same group. The detailed mechanism for these pseudorotations may involve either a thermal mechanism to overcome the barrier associated with the process, or quantum mechanical tunnelling if the barrier for the interconversion of two configurations is suitably low and narrow (Dalton, 1966, 1971).

The geometric and electronic structure of XeF_6 had for a long time posed a problem for experimentalists as well as theoreticians. Various theoretical models led to conflicting predictions of the equilibrium geometry, while the experimental data were difficult to interpret in terms of any single structure. By considering XeF_6 as a nonrigid species like PF_5, Trindle et al. showed (1977) that the pseudorotations in Figure 14.6 can account for the difficulties encountered with this molecule. A simple double-well tunnelling could explain the typical temperature dependence of the infra-red and Raman spectra for this molecule. This work, along with the observation by Pitzer and Bernstein (1975) that the entropy data required low-energy movements in XeF_6 at frequencies not easily specified by the IR and Raman spectra, has been fairly conclusive about the fluxional nature of XeF_6.

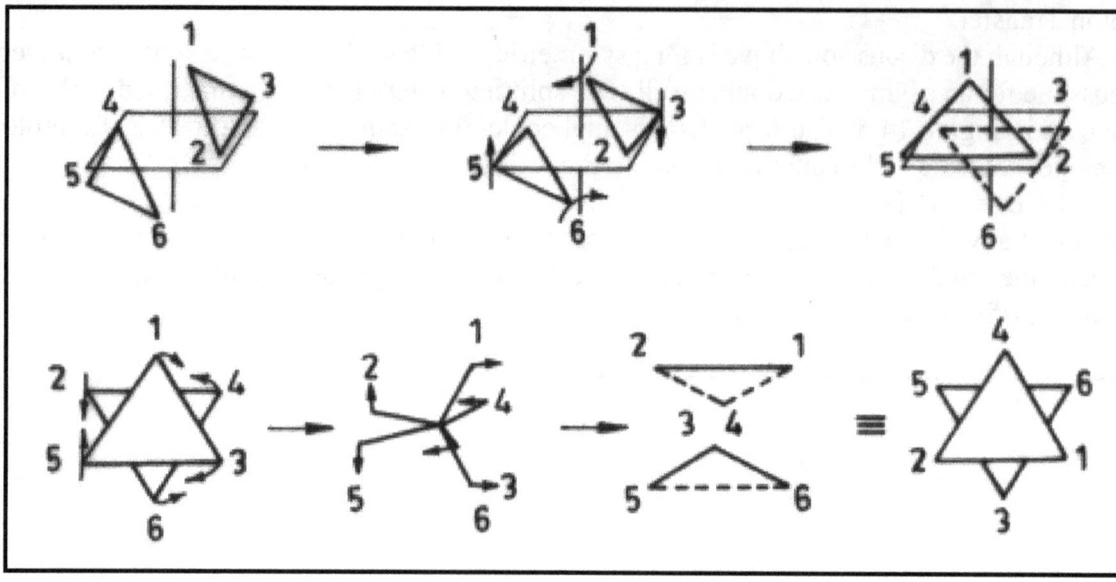

Figure 14.6 Pseudorotations in XeF_6. These motions are not bodily rotations, that is, the molecule does not rotate as a whole by these movements. The effect of these movements is to generate equivalent configurations. The pseudorotations physically appear in the form of low-frequency vibrations of large amplitudes. The low-energy motions shown here are required to explain the vibrationl spectra of XeF_6.

Electron Transfer

The double well potential may be used for studying electron transfer processes

$$D + A = D^+ + A^- \quad (85.8)$$

where D represents the donor molecule and A is the acceptor species. The first well may symbolize the potential of the complex D…A. Then the second one would represent the energy surface for $D^+ \ldots A^-$. The transfer may be adiabatic, or it may be nonadiabatic. In both cases the transfer involves no net change of the energy of the Donor-Acceptor system. This is illustrated in Figure 14.7.

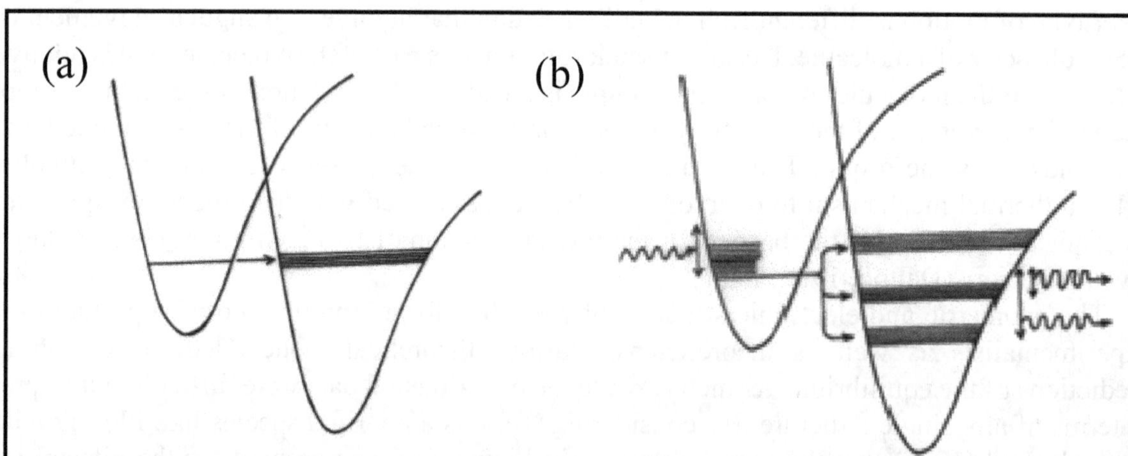

Figure 14.7 (a) Adiabatic electron transfer with vibrational structure shown for the group of final states; (b) Nonadiabatic electron transfer with energy supplied or excess energy dissipated as shown by incoming or outgoing wavy lines. This type of transfer is normally accompanied by a departure from Born-Oppenheimer approximation.

Adiabatic transfers involve DA and D^+A^- levels of equal energy. These levels may be electronic levels, but for an explicit theoretical consideration, the levels are at least electronic-vibrational in nature. The rate of transfer is obtained from the so-called Fermi Golden Rule that gives the probability of evolution of the group of final states from the initial state in unit time,

$$p = \frac{2\pi}{\hbar}\left|H_{DA,D^+A^-}\right|^2 \rho(E_{D^+A^-} = E_{DA}) \tag{85.9}$$

per molecule where ρ is the density of the group of final states at energy E_{D+A-}.. The equality $E_{D+A-} = E_{DA}$ specifies the resonance condition. The transfer rate P is normally derived from time-dependent perturbation theory that will be discussed in the next chapter. Nonadiabatic transfers involve vibronic levels of different energy. Obviously the energy imbalance is to be rectified, either in the form of a supply of energy or as loss of excess energy – a dissipative process.

Electron Transfer in Solution

For general considerations in the solid state or in the gaseous phase, electron transfer between two species requires them to be in close proximity of each other. The involved atoms or molecules as such remain bare. The situation, however, is quite different in solution as the donor and acceptor species, both as reactants and as products, are extensively solvated. Reactions of the types

$$\begin{aligned} &M^{n+}(s) + M^{(n-1)+}(s) \rightarrow M^{(n-1)+}(s) + M^{n+}(s) \\ &M^{n+}(aq) + M'^{\,p+}(aq) \rightarrow M^{(n-1)+}(aq) + M'^{\,(p+1)+}(aq) \\ &[ML_p]^{+m}(s) + [M'L'_q]^{+n}(s) \rightarrow [ML_p]^{+(m-1)}(s) + [M'L'_q]^{+(n+1)}(s) \end{aligned} \tag{85.10}$$

which take place in solution, are known to be electron transfer processes. For such reactions, a chemical mechanism such as the transfer of a charged species to a cation or an anion, hydrolysis, tautomeric change, etc., would require a free energy of activation much larger than what is observed.

It is known that istopic exchange reactions between ions that differ only in charge (the first reaction in 85.10) are slow when solvated simple cations are involved, but faster when the solvated ions are complex ions. Libby explained this phenomenon in terms of the normal decrease in electrostatic contribution to the free energy of activation while the distance between the ionic centers increases. So the thought was still about a thermal mechanism, conceptualized by van't Hoff and Arrhenius in the late nineteenth century.

As the two ions approach each other, the dielectric constant varies in the neighborhood until a dielectric saturation is achieved. For the Fe^{2+}/Fe^{3+} system in water, the dielectric saturation is achieved at a distance slightly less than 4.2 Å. Correspondingly, the entropy of activation of isotopic exchanges is large and negative. This was rationalized by Marcus, Zwolinsky and Eyring by postulating a reorganization of the solvent atmosphere around the donor-acceptor complex, in tune with the variation of dielectric constant.

The rather small value of the remaining activation free energy inferred from experiments indicates a physical process, and Weiss contended that the transfer of an electron is the most plausible mechanism of the physical type. From a crystal field arrangement of the solvent molecules, Hush estimated the electrostatic free energy of Fe^{x+}–$Fe^{(3-x)+}$ separated by 6 Å where $2 \le x \le 3$. A free energy of activation of 5 kcal mol^{-1} was estimated for electron tunneling. A good exposure to the subject has been given in "Chemical Kinetics" by K. J. Laidler, and qualitative calculations are in agreement with experimental numbers. Only in recent years it has become possible to confirm the electron transfer mechanism by carrying out explicit theoretical calculations.

A theory for the long-range electron transfer in solution was developed by Marcus (1956). It relies on a radical departure from the conventional treatment of chemical reactions in solvent phase by postulating the existence of two transition states that have equal free energy and are necessarily associated with the same non-equilibrium configuration of solvent molecules around the donor-acceptor complex. The actual transfer takes place between these two transition states, and it is fast enough for the solvent arrangement to undergo any substantial change. See Figure 14.8 for illustration, where the equilibrium solvent environment is indicated by (s) and the non-equilibrium distribution is represented by (s′), and k_2 and k_{-2} are the rate constants for electron transfers in the forward and backward directions. Marcus proposed a mechanism of five steps.

1. Approach of the solvated donor and acceptor species in thermal equilibrium towards each other to form complex DA with equilibrium solvent configuration around it: The Fe^{2+}-Fe^{3+} approach to 4.2 Å in water needs about 4.5 kcal mol^{-1}.
2. Solvent reorganization around the DA complex accompanying its activation to form the transition state (DA)*: The corresponding activation free energy is $E_1^{\#}$. For the Fe^{2+}/Fe^{3+} system in water, $E_1^{\#} = 6.5$ kcal mol^{-1}.
3. Electron transfer from the activated complex (DA)* to form another activated complex $(D^+A^-)^*$ with the same non-equilibrium arrangement of the neighboring solvent molecules: the k_2 process ($\Delta G^{\#} = 5$ kcal mol^{-1}). These three steps account for a total of 16 kcal mol^{-1} activation energy, compared to 16.8 kcal mol^{-1} found at 25°C.
4. Deactivation of $(D^+A^-)^*$ to form the complex D^+A^- with equilibrium-average solvent configuration around the latter: This step releases the deactivation energy $E_3^{\#}$.
5. Separation of D^+ and A^- from each other by association with solvent molecules and accompanied by thermal kicks

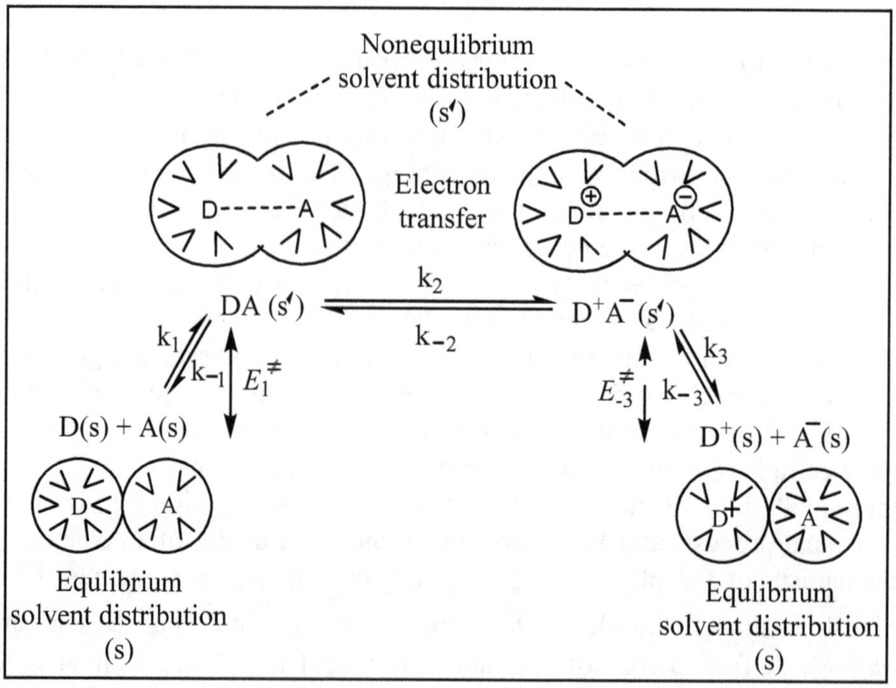

Figure 14.8 A physical representation of the Marcus-Hush scheme of long range electron transfer in solution.

The actual theoretical treatment is extremely detailed, but the final formulae worked out by Marcus are quite simple. Considering steady states for the two activated complexes one obtains the overall rate constant

$$k = \frac{k_1}{1+\left(1+\frac{k_{-2}}{k_3}\right)\frac{k_{-1}}{k_2}}. \tag{85.11}$$

For $k_2 >> k_{-1}$, $k \sim k1$. However, for $k_{-1} >> k_2$ and $k_3 >> k_{-2}$, the overall rate constant is

$$k = k_2\frac{k_1}{k_{-1}}. \tag{85.12}$$

The ratio k_1/k_{-1} has a well-documented statistical thermodynamical expression in transition state theory. The rate constant k_2 is obtained from a revised version of the Fermi Golden Rule that has the density of states replaced by a Franck-Condon factor, the latter expressed in terms of free energies:

$$k_2 = N_{Avo}\frac{2\pi}{\hbar}\left|H_{DA,D^+A^-}\right|^2 (4\pi\lambda RT)^{-1/2}\, e^{-(\Delta G^0+\lambda)^2/4\lambda RT}. \tag{85.13}$$

In the above ΔG^0 is the standard free energy change of reaction $DA \rightarrow D^+A^-$ and λ is the reorganization free energy. The latter is a sum of two parts: $\lambda = \lambda_0 + \lambda_{int}$. The first part λ_0 arises from the dielectric relaxation of the solvent, and Marcus used the double sphere model of reactions in solution and the Born free energy terms to find the solvent reorganization free energy

$$\lambda_0 = N_{\text{Avo}}(\Delta e)^2\left(\frac{1}{2a_1}+\frac{1}{2a_2}-\frac{1}{r}\right)\left[\frac{1}{D_{opt}}-\frac{1}{D_{stat}}\right] \tag{85.14}$$

where Δe is the charge transferred from one reactant to the other; a_1 and a_2 are the ionic radii of the two "spherical" reactants, and r is separation of their centers in the double sphere activated complex; D_{opt} and D_{stat} are the optical and static dielectric constants and N_{Avo} is the Avogadro number.

The second part of the reorganization energy is the internal or bond contribution approximated by

$$\lambda_{\text{int}} = \sum_j \frac{f_j^r f_j^p}{f_j^r + f_j^p}(\Delta q_j)^2 \tag{85.15}$$

Where f_j^r and f_j^p are the force constants of the *jth* normal mode of vibration in the reactant and product respectively, and Δq_j is the change in the equilibrium value of the normal coordinate.

The Marcus theory has become spectacularly successful in the study of long-range electron transfer kinetics in solution and also for bridge-mediated transfers. A review by Marcus and Sutin (1985) is a valuable document on this extraordinary subject. Investigation of the series of very fast electron transfer reactions that take place in the condensed phase of Chloroplast and in Bacteriochlorophyll has been a frontier area of research. The Marcus formulation has been proved to be very useful here. Explicitly done quantum chemical calculations generate a good fit with experimental kinetics. An example is given below where the theory involves a synthesis of the effects of the dielectric polarization with the influence of a molecular bridge to mediate the transfer.

It is known that in the first step of the so-called Z scheme of photosynthesis in green plants, the pigment molecule P700 (a Chlorophyll-a complex that has absorption peak at 700 nm) is excited by solar rays, and then it undergoes a charge separation that initiates a chain of electron transfer processes. During the course of the transfer, electron moves from one form of Plastoquinone Q_I to another form Q_{II} in reaction center in Chloroplast. In fact Q_{II} successively acquires two electrons:

$$Q_I^- + Q_{II} = Q_I + Q_{II}^-, \tag{85.16}$$

$$Q_I^- + Q_{II}^- = Q_I + Q_{II}^{2-} \tag{85.17}$$

The system involved is fairly large for a quantum mechanical calculation. A model chemical structure is shown in Figure 14.9. The molecules involved are not trivially small. There are two Plastoquinone molecules (with truncated side chains), each attached to a Histidine residue. The quinone Q_I is flanked by peptide chains on both sides, as illustrated by the two peptide fragments in Figure 14.9. One ferrous ion holds the two Histidine residues.

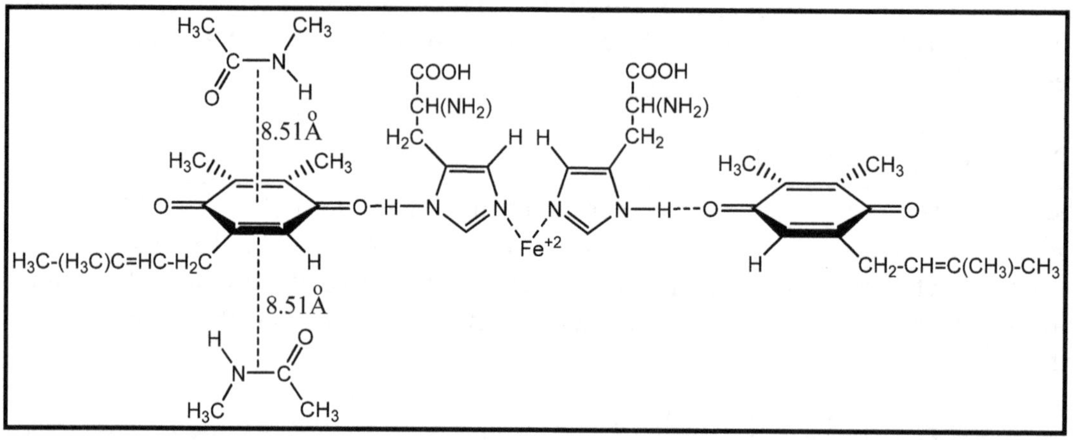

Figure 14.9 Chemical structure of a modelled QI-QII integrated system.

It is important to realize the complexity of the problem. The theoretical treatment involves not only the transfer of electrons but also medium polarization effects and thermal effects. The internal energy values were computed by INDO. In spite of the semi-empirical methodology for calculations and the drastic modelling of the reactive system, the $t_{1/2}$ estimated from the theory of Marcus [equations (85.12) – (85.15)] turned out to be in good agreement with the observed values, as evident from Table 14.1. The experimental $t_{1/2}$ has a spread of about ±20 μs.

Table 14.1 Calculated and observed $t_{1/2}$ for electron transfer from Q_I to Q_{II} in reaction centre of chloroplast.†

Histidine Binding	$t_{1/2}$ in μsec			
	$Q_I^- + Q_{II} = Q_I + Q_{II}^-$		$Q_I^- + Q_{II}^- = Q_I + Q_{II}^{2-}$	
	Calcd.	Obs.	Calcd.	Obs.
With Fe^{2+}	134	~100	181	~200
Without Fe^{2+}	247	~200	472	~400

† Datta and Mallik (1995).

Text Suggested for Further Reading

K. J. Laidler, "Chemical Kinetics", (Tata McGraw-Hill, New Delhi, 1991).

CHAPTER 15

TIME-DEPENDENT PERTURBATION THEORY

So far we have discussed mainly the stationary states for which the spatial dependence of the wave function is governed by the time-independent Schrödinger equation and the temporal dependence is strictly periodic. In fact, the time evolution of a general, that is, nonstationary state is physically more important. We have discussed in Chapter 2 that in quantum theory the outcome of an experiment is explained in terms of the probability of transition from a particular stationary state of the system (which represents the state before the experiment) to another stationary state of the system (which represents the state after the experiment). The state is not stationary during the experiment while the system is agitated by an externally applied perturbation. Thus our basic task is to solve the time-dependent Schrödinger equation

$$i\hbar\frac{d\Psi(t)}{dt}=\mathbf{H}(t)\Psi(t) \tag{5.6}$$

where $\mathbf{H}(t)$ is generally dependent on time. In the event $\mathbf{H}(t)$ is independent of time, it would be possible to separate the time dependence of $\Psi(t)$ from its dependence on the spatial coordinates when the wave function is an eigenfunction of $\mathbf{H}$. If $\Psi(t)$ is not an eigenfunction of $\mathbf{H}$ it would be possible to write it as a linear combination of stationary states and determine the time-dependent coefficients from the general wave equation. But most of the physical processes are more complicated in nature and the Hamiltonian varies with time during the course of an experiment.

Lecture 86: The Time-Dependent Formalism

The time-dependent Hamiltonian is assumed to consist of a time-independent part $\mathbf{H}^0$ and a time-dependent one $\mathbf{H}'(t)$ such that

$$\mathbf{H}=\mathbf{H}^0+\mathbf{H}'(t). \tag{86.1}$$

One further assumes that $\mathbf{H}'(t)$ is small enough to be treated as a perturbation. The perturbed wave function is expanded in the basis of stationary states of $\mathbf{H}^0$ as

$$\Psi(t)=\sum_n c_n(t)\Psi_n^0(t). \tag{86.2}$$

The method of the *variation of constants* is applied to determine the time-dependent expansion coefficients $c_n(t)$'s. This procedure must be carried out subject to the normalization of $\Psi(t)$, which requires

$$\sum_n |c_n(t)|^2 = 1 \tag{86.3}$$

as the stationary states are mutually orthogonal and normalized functions of the spatial coordinates of all the particles concerned.

The time-dependent Schrödinger equation is now written as

$$i\hbar\sum_n \dot{c}_n \Psi_n^0 + i\hbar\sum_n c_n \dot{\Psi}_n^0 = \mathbf{H}^0\sum_n c_n\Psi_n^0 + \mathbf{H}'(t)\sum_n c_n\Psi_n^0 . \tag{86.4}$$

The second term on the left hand side cancels the first term on the right hand side because Ψ_n^0's are the stationary eigenfunctions of $\mathbf{H}^0$. Hence

$$\sum_n \dot{c}_n \Psi_n^0 = -\frac{i}{\hbar}\sum_n c_n \mathbf{H}'(t)\Psi_n^0 . \tag{86.5}$$

Multiplying both sides by $\Psi_m^0\,*$, integrating over all space, and using the orthonormality of the stationary state wave functions, one obtains

$$\dot{c}_m = -\frac{i}{\hbar}\sum_n c_n < \Psi_m^0 \,|\, \mathbf{H}'(t) \,|\, \Psi_n^0 > . \tag{86.6}$$

We now explicitly write the stationary state wave functions

$$\Psi_s^0(\mathbf{r},t) = \psi_s(\mathbf{r})e^{-iE_s t/\hbar} \tag{86.7}$$

so that Eq. (86.6) can be formulated as

$$\dot{c}_m = -\frac{i}{\hbar}\sum_n c_n \, e^{i(E_m - E_n)t/\hbar} \, H'_{mn}(t) \tag{86.8}$$

where

$$H'_{mn}(t) = <\psi_m \,|\, \mathbf{H}'(t) \,|\, \psi_n > \equiv \int d^3r \, \psi_m{}^*(\mathbf{r})\mathbf{H}'(t)\psi_n(\mathbf{r}) . \tag{86.9}$$

Equation (86.8) needs to be solved subject to the normalization condition (86.3).

The general solution is

$$c_m(t) = c_m(0) - \frac{i}{\hbar}\sum_n \int_0^t d\tau \, e^{i(E_m - E_n)\tau/\hbar} H'_{mn}(\tau)c_n(\tau) . \tag{86.10}$$

This may be repeatedly expanded on the right hand side to obtain the infinite chain solution

$$\begin{aligned} c_m(t) = c_m(0) &- \frac{i}{\hbar}\sum_n c_n(0)\int_0^t d\tau \, e^{i(E_m - E_n)\tau/\hbar} H'_{mn}(\tau) \\ &+ \left(-\frac{i}{\hbar}\right)^2 \sum_n c_n(0)\sum_p c_p(0)\int_0^t d\tau \, e^{i(E_m - E_n)\tau/\hbar} H'_{mn}(\tau) \\ &\times \int_0^\tau d\tau' \, e^{i(E_n - E_p)\tau'/\hbar} H'_{np}(\tau') + \ldots \end{aligned} \tag{86.11}$$

Successive terms in this expansion are of order zero, one, two, etc. in perturbation $\mathbf{H}'$. These terms give rise to corrections to the wave function in successive orders.

At this juncture one makes the convenient assumption that the system remains in the specific stationary state with wave function Ψ_i^0 up to time $t = 0$ when the perturbation is switched on. The perturbation $\mathbf{H}'(t)$ is switched off at a later time $t = t'$. Since $\Psi(t{=}0) = \Psi_i^0(t{=}0)$, we get the initial condition

$$c_n(0) = \delta_{i,n} = 1 \quad (\text{for } n = i),$$
$$\qquad 0 \quad (\text{otherwise}). \tag{86.12}$$

When these values are inserted in Eq. (86.11), the expansion coefficients are obtained through first order in perturbation. Thus for $n = i$,

$$c_i(t') = 1 - \frac{i}{\hbar}\int_0^{t'} H'_{ii}(t)dt\,, \tag{86.13}$$

and for $n \neq i$,

$$c_n(t') = -\frac{i}{\hbar}\int_0^{t'} e^{i\omega_{ni}t} H'_{ni}(t)dt \tag{86.14}$$

where

$$\omega_{ni} = \frac{1}{\hbar}(E_n - E_i)\,. \tag{86.15}$$

The "first order" probability of finding the system in another stationary state Ψ_f^0 at time $t = t'$ is given by

$$P_{i\to f}(t') = |c_f(t')|^2 = \frac{1}{\hbar^2}\left|\int_0^{t'} e^{i\omega_{fi}t} H'_{fi}(t)dt\right|^2. \tag{86.16}$$

The rate at which this probability develops in time is called the transition probability per unit time. This is written as

$$p_{i\to f} = \frac{dP_{i\to f}(t')}{dt'}. \tag{86.17}$$

Often in the formalism t' will be taken to be sufficiently small so that $P_{i\to f}(t')$ becomes a *linear* function of t'. In such cases one obtains the transition rate as

$$p_{i\to f} = \frac{1}{t'}P_{i\to f}(t'). \tag{86.18}$$

This is the general solution of the time-dependent problem in quantum mechanics. The next step of solution depends on the precise functional appearance of the perturbation $\mathbf{H}'$.

Lecture 87: Constant Perturbation

A simple solution of the coefficient c_n and the corresponding transition probability are obtained when $\mathbf{H}'$ is explicitly independent of time. Then equations (86.13) and (86.14) yield

$$c_i(t') = 1 - \frac{it'}{\hbar}H'_{ii},$$

$$c_n(t') = \frac{1}{\hbar\omega_{ni}}\left(1 - e^{i\omega_{ni}t'}\right) H'_{ni} \quad (\text{for } n \neq i), \tag{87.1}$$

where the time-independent perturbation integrals are given by

$$H'_{mn} = \int d^3r\ \psi_m^*(\mathbf{r})\boldsymbol{H}'\psi_n(\mathbf{r}). \tag{87.2}$$

Thus we get the probability

$$P_{i\to f}(t') = \hbar^{-2}|H'_{fi}|^2 \frac{4\sin^2 \frac{1}{2}\omega_{fi}t'}{\omega_{fi}^2} = \hbar^{-2}|H'_{fi}|^2 t'^2\, g(\omega_{fi}, t') \tag{87.3}$$

Its magnitude depends on the function

$$g(\omega_{fi}, t') = \frac{4\sin^2 \frac{1}{2}\omega_{fi}t'}{\omega_{fi}^2 t'^2}. \tag{87.4}$$

This function is plotted against ω_{fi} in Figure 15.1. The sine function is bounded between -1 and $+1$, and its square is bounded between 0 and 1. Thus $g(\omega_{fi},t')$ remains positive except at points where the sine function vanishes. This causes an infinite number of peaks on two sides of the point $\omega_{fi} = 0$. The central peak is the most prominent one since the peak height is 1 whereas the height of any other peak is equal to $4/\omega_{fi}^2 t'^2$ that decreases very rapidly as ω_{fi} increases. The central peak also has a width of $4\pi/t'$ which is exactly double of the width of any other peak. However, the function $P_{i\to f}(t')$ is strictly discontinuous at the point $\omega_{fi} = 0$ as the function $g(\omega_{fi},t')$ is undefined there. The area under the central peak of $P_{i\to f}(t')$ increases as t'.

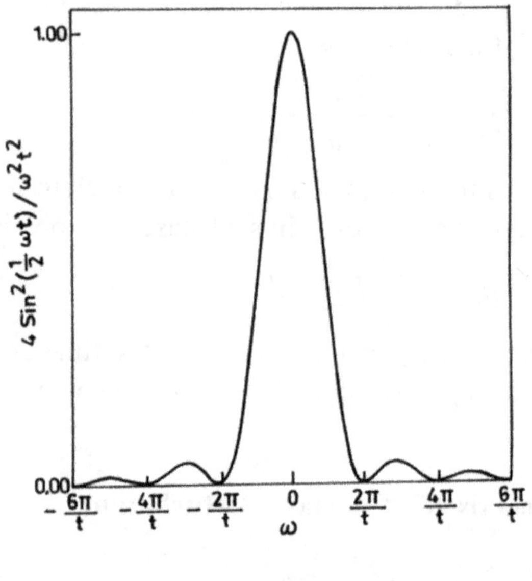

Figure 15.1 Function $g(\omega,t)$ plotted against ω. The zeroes occur at $\omega = 2\pi n/t$ where n is a nonzero integer.

Consider now the transition to take place from a discrete initial state to a group of final states associated with effectively continuous eigenvalues. Let the number of final states in the infinitesimally small interval between E_f and $E_f + dE$ be dN. The *density of states* is defined by

$$\rho(E_f) = \left(\frac{dN}{dE}\right)_{E = E_f} \tag{87.5}$$

The transition probability to the *group of final states* around $E = E_f$ is given by

$$\begin{aligned} P_{group}(t') &= \int_{-\infty}^{\infty} dE_f \, P_{i\to f}(t')\rho(E_f) \\ &= \frac{1}{\hbar}\int_{-\infty}^{\infty} d\omega_{fi}\, |H'_{fi}|^2 \frac{4\sin^2 \frac{1}{2}\omega_{fi}t'}{\omega_{fi}^2}\rho(E_f). \end{aligned} \tag{87.6}$$

Because of the nature of the function $g(\omega_{fi})$, this integral receives a predominant contribution from the transitions to the group of states in a very small region around $\omega_{fi} = 0$. Therefore, we approximate $P_{group}(t')$ by considering the density of states unchanged from its value at the midpoint $\omega_{fi} = 0$, that is, $E_f = E_i$:

$$\begin{aligned} P_{group}(t') &= \frac{|H'_{fi}|^2}{\hbar}\rho(E_f)\int_{-\infty}^{\infty} d\omega \frac{4\sin^2 \frac{1}{2}\omega t'}{\omega^2} \\ &= t'\frac{|H'_{fi}|^2}{\hbar}\rho(E_f)\int_{-\infty}^{\infty} dx \frac{4\sin^2 \frac{1}{2}x}{x^2} \end{aligned} \tag{87.7}$$

where *it is understood that* $E_f = E_i$, that is, the *resonance condition* holds. Note that in writing Eq. (87.7) we have also assumed that H'_{fi} does not vary much within the central peak, that is, t' is *large enough* to assign a vanishingly small width to each peak. Thus we have a curious pair of conditions to be imposed on t': (i) t' must be sufficiently small so that the first-order solutions in Eq. (87.1) hold, that is,

$$t'H'_{fi}/\hbar \ll 1; \tag{87.8}$$

and (ii) t' must be sufficiently large so that $|H'_{fi}|^2$ in Eq.(87.6) can be taken out of the integral as a nonvarying term, that is,

$$\sigma_\omega \approx 2\pi/t' \to 0. \tag{87.9}$$

This σ_ω corresponds to an energy uncertainty $\sigma_E \approx h/t'$ and a time uncertainty $\sigma_t \approx t'/4\pi$.

The integral in Eq. (87.7) can be calculated by complex integration method. The result is

$$\int_{-\infty}^{\infty} dx \frac{4\sin^2 \frac{1}{2}x}{x^2} = 2\pi. \tag{87.10}$$

Therefore,

$$P_{group}(t') = t'\frac{2\pi}{\hbar}|H'_{fi}|^2 \rho(E_f)|_{E_f = E_i} \tag{87.11}$$

which by Eq. (86.18) leads to the transition rate

$$p_{group} = \frac{2\pi}{\hbar}|H'_{fi}|^2 \rho(E_f)|_{E_f = E_i}. \qquad (87.12)$$

This expression is known as the *Fermi Golden Rule* (Fermi, 1950).

Auger Effect

The $2s^2$ configuration of helium is not stable enough. Its energy is greater than the energy of an isolated He^+ ion. So when a helium atom is excited from its ground state to an (impure) state resembling the $2s^2$ configuration or a configuration of still higher energy, the two electrons repel each other and eventually one is lost leaving aside a He^+ ion. This autoionization is known as the Auger effect. See Figure 15.2. By the by, this is a common feature of all atoms, molecules, clusters and solids, and the phenomenon is commonly known as the photoelectric effect.

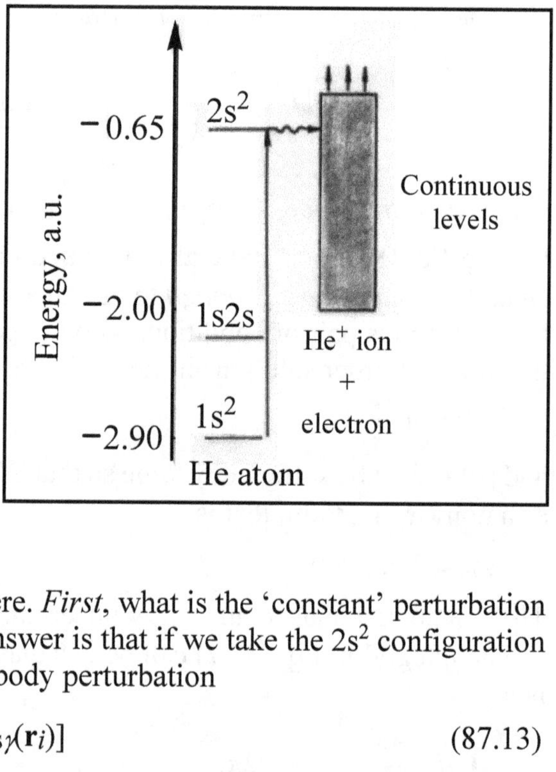

Figure 15.2 The Auger effect in helium: The atom autoionizes in the excited state.

Three theoretical points must be stressed here. *First*, what is the 'constant' perturbation involved in the autoionization process? The answer is that if we take the $2s^2$ configuration as the (albeit impure) excited state, the many-body perturbation

$$\mathbf{H}'_{mb} = \frac{1}{r_{12}} - \frac{1}{2}\sum_{i=1,2}\sum_{\gamma=\alpha,\beta} [\mathbf{J}_{2s\gamma}(\mathbf{r}_i) - \mathbf{K}_{2s\gamma}(\mathbf{r}_i)] \qquad (87.13)$$

serves as the constant perturbation. *Second*, why do we consider the bound states above the 1s2s level as impure, and not as eigenstates? Any calculation of the eigenstates must take into account the continuous levels in Figure 15.2. The perturbation $\mathbf{H}'_{mb}$ will strongly connect any configuration excited above the 1s2s energy level with an infinite member of continuous states. The coefficients of mixing of the continuous states can be calculated from Rayleigh-Schödinger perturbation theory. We have already seen from Eq. (42.20) that the first order correction to wave function contains terms with energy differences in the denominators, and these terms blow up as the energy differences approach the zero value. In other words, the bound state wave function will be intimately mixed with a large number of quasi-degenerate continuous state wave functions, and a bound state wave function with energy above the ground state energy of He^+ cannot be an eigenstate. *Third*, the case of any apparently bound eigenstate can also be considered in the light of the degenerate perturbation theory, or better, in terms of an extensive linear combination with the (quasi-degenerate) continuous states. The electronic Hamiltonian will connect the apparent eigenstate with an infinite member of continuous states, so much so that a Ritz variation treatment will lead to eigenvectors each with a vanishingly small contribution from the apparent eigenstate: thus the bound state wave function dissolves into the continuum of states, and figuratively speaking, it goes down the drain.

The decay rate for the $2s^2$ state of helium was calculated by Manson (1966). The calculated rate is 2.3×10^{14} s^{-1}. The Auger effect has since been found to occur with other atoms and with almost all molecules. The photoelectron spectrum of each two- or many-electron species almost invariably contains an unmistakable signature of the Auger effect.

Electron and Proton Transfer

The Golden rule (87.12) can be used to calculate the rate of electron and proton transfer processes.

In the case of proton transfers, the initial state represented by Ψ_i^0 can be taken as a vibrational state of one well, and the first state Ψ_f^0 can be viewed as a vibrational state of another well as depicted by Figure 14.5. The result (87.12) must be thermally averaged by taking into account the equilibrium population distribution (for simplicity, the Boltzmann distribution) in different initial states. The group of final states may be produced by coupling the vibrational levels of the second well with other molecular vibrations, rotations, or even translations. It only needs one to decide how extensively one would carry out the treatment. The thermal problem and the interaction among different kinds of motion generate such a vastly complex treatment that at different levels of theory one uses differently averaged values as parameters. A happy situation evolves, where with judicious ingenuity employed in the treatment almost everyone may claim to have a very good result. This statement is by no means a reflection on any shortcoming of theoretical formalism. It merely points out the triumph of quantum theory in explaining phenomena where electromagnetic interactions are involved.

The treatment of electron transfer is a more delicate affair as it should involve at least a consideration of the electronic-vibrational coupling. A drastic thermal modelling and a macroscopic averaging process leads to the substitution of the Franck-Condon factor for the density of states in Eq. (87.12), giving rise to the celebrated formula of Marcus, Eq. (85.13). Calculations performed by Marcus and his coworkers have established the quantitative accuracy of this formula. This again points out that theory is in a good state, that is, Quantum theory has clearly achieved the status of art in the explanation of properties based on the electromagnetic interactions.

Lecture 88: Harmonic Perturbation

We now take the perturbation $\mathbf{H}'(t)$ as a harmonic function of time,

$$\mathbf{H}'(\mathbf{r},t) = 2\mathrm{F}(\mathbf{r})\cos(\omega t+\delta)$$
$$= \mathrm{F}(\mathbf{r})\left[e^{i(\omega t+\delta)} + e^{-i(\omega t+\delta)}\right] \qquad (88.1)$$

Then the first-order coefficient $c_n(t')$ in Eq.(86.14) simplifies as

$$c_n(t'>0) = -\frac{i}{\hbar}F_{ni}\left[e^{i\delta}\int_0^{t'} e^{i(\omega+\omega_{ni})t}\,dt + e^{-i\delta}\int_0^{t'} e^{-i(\omega-\omega_{ni})t}\,dt\right]$$
$$= \frac{1}{\hbar}F_{ni}\left[\frac{1-e^{i(\omega+\omega_{ni})t'}}{\omega+\omega_{ni}}e^{i\delta} - \frac{1-e^{-i(\omega-\omega_{ni})t'}}{\omega-\omega_{ni}}e^{-i\delta}\right] \qquad (88.2)$$

where F_{ni} is the integral

$$F_{ni} = \int d^3r\ \psi_n^*(\mathbf{r})\mathrm{F}(\mathbf{r})\psi_i(\mathbf{r}). \qquad (88.3)$$

Suppose $\omega_{ni} > 0$, that is, $E_n > E_i$. The coefficient $c_n(t'>0)$ becomes large in absolute magnitude only when ω tends to ω_{ni}, and the second term in the square bracket in Eq. (88.2) predominates. For all other positive values of ω, each term in the square bracket remains small as the exponential function $e^{i\theta}$ is of modulus 1. Thus the only noticeable contribution to the coefficient is

$$c_n(t'>0) = -e^{-i\delta}\frac{F_{ni}}{\hbar}\left[\frac{1-e^{-i(\omega-\omega_{ni})t'}}{\omega-\omega_{ni}}\right]_{\omega\to\omega_{ni}}, \qquad (88.4)$$

the condition $\omega \to \omega_{ni}$ signifying *resonance*. The probability of transition $i \leftrightarrow f$ is

$$P_{i\to f}(t'>0) = \frac{4|F_{fi}|^2}{\hbar^2}\left[\frac{\sin\frac{1}{2}(\omega_{fi}-\omega)t'}{\omega_{fi}-\omega}\right]^2_{\omega\to\omega_{fi}} \qquad (88.5)$$

which is exactly of the form of Eq. (87.3) with $\mathbf{F}$ and $(\omega_{fi}-\omega)$ replacing $\mathbf{H}'$ and ω_{fi} respectively.

If we consider the perturbation to be monochromatic (of a specific angular frequency ω) and carry out a treatment analogous to that in Lecture 87, we will eventually find expressions similar to Eq.(87.11) for the probability of transition and the Fermi Golden rule (87.12) for the rate of transition to a group of final states. However, it is impossible to create a monochromatic perturbation. What appears to be monochromatic in a macroscopic scale is actually found to have a finite spread when viewed microscopically. Also, the final state is never discrete, but it is associated with the natural linewidth and collisional and other broadenings as encountered in ordinary atomic or molecular spectroscopy (Figure 15.3). Thus we require a different treatment that may start from Eq. (88.5). Nevertheless, there is enough similarity between the treatment in the case of the constant (in-built) perturbation and that for an externally applied perturbation varying harmonically with time.

Let us assign a frequency-dependent intensity to the harmonic perturbation, $I(\omega)$. The probability of transition i↔f due to the group of perturbations within the range of frequency $(0,\infty)$ is

$$P_{i\to f}^{group}(t') = \int_0^{\infty} d\omega\, I(\omega)\ |c(t'>0)|^2 \tag{88.6}$$

where $c_f(t'>0)$ is given by Eq. (88.2) for $f = n$. The noticeable nonvanishing c_f is only formed in the region around the resonance. Hence Eq. (88.5) is used to obtain

$$P_{i\to f}^{group}(t') = \frac{|F_{fi}|^2}{\hbar^2} \int_{\omega_{fi}-\delta\omega}^{\omega_{fi}+\delta\omega} \frac{4\sin^2\frac{1}{2}(\omega_{fi}-\omega)t'}{(\omega_{fi}-\omega)^2}\, I(\omega)\, d\omega. \tag{88.7}$$

Since $\delta\omega$ is sufficiently small, one can take $I(\omega_{fi})$ outside the integral and obtain

$$P_{i\to f}^{group}(t') = \frac{|F_{fi}|^2}{\hbar^2}\, I(\omega_{fi}) \int_{\omega_{fi}-\delta\omega}^{\omega_{fi}+\delta\omega} \frac{4\sin^2\frac{1}{2}(\omega_{fi}-\omega)t'}{(\omega_{fi}-\omega)^2}\, d\omega$$

$$\cong \frac{|F_{fi}|^2}{\hbar^2}\, I(\omega_{fi}) \int_{-\infty}^{+\infty} \frac{4\sin^2\frac{1}{2}\omega t'}{\omega^2}\, d\omega = 2\pi\, \frac{|F_{fi}|^2}{\hbar^2}\, I(\omega_{fi})\, t'. \tag{88.8}$$

The last step has been performed by using Eq. (87.10). The corresponding transition rate is given by

$$p_{i\to f}^{group}(t') = P_{i\to f}^{group}(t')/t' = 2\pi\, \frac{|F_{fi}|^2}{\hbar^2}\, I(\omega = \omega_{fi}) \tag{88.9}$$

where $\omega = \omega_{fi}$ stresses the resonance condition as discussed earlier.

Lecture 89: Semiclassical Theory of Interaction of Radiation with Matter

The classical interaction of a collection of charged species with the electromagnetic radiation is normally expanded in a series of increasingly higher moments as

$$\mathbf{H}' = -\boldsymbol{\mu}\bullet\boldsymbol{\mathcal{E}}(\omega) - \mathbf{m}\bullet\mathcal{H}(\omega) + \sum_{i,j=x,y,z} q_{ij}\ \mathcal{E}_i(\omega)\mathcal{E}_j(\omega) + \ldots \tag{89.1}$$

where $\boldsymbol{\mu}$ is the electric dipole moment, **m** the magnetic dipole moment and q_{ij} the (i,j) element of the electric quadrupole moment, etc., of the collection. This is known as the *multipolar expansion* of interaction energy. Note that

$$\boldsymbol{\mu} = \sum_i Q_i \mathbf{r}_i \tag{89.2}$$

where Q_i and $\mathbf{r}_i$ are the charge and the position vector of the *ith* species. Similarly,

$$\mathbf{m} = \sum_i \frac{Q_i \mathbf{J}_i}{2m_i c} \tag{89.3}$$

where $\mathbf{J}_i$ and m_i are the angular momentum and mass of the *ith* species, and c is the speed of light. The quadrupole moment is of order c^{-2}. Thus the strength of the interaction terms are approximately in the ratio 1: c^{-1}: c^{-2} etc., and the corresponding transition probabilities would be roughly in the ratio 1: c^{-2}: c^{-4} etc. The speed of light in vacuum c is equal to 137.036 in atomic unit (see Appendix I). Consequently the intensities of transitions caused by the electric dipolar term, the magnetic dipolar interaction and the electric quadrupolar mechanism are in the ratio 1: 10^{-3}–10^{-4}: 10^{-6}–10^{-8}, and similar ratios are found from experiment. This is why in normal treatment of interaction of radiation with matter one considers only the first term, the electric dipole mechanism, which contributes almost wholly to the intensities of ordinary spectroscopic transitions. Only in situations where the electric dipole transitions have zero probability one looks for contributions from the magnetic dipole or the electric quadrupole interaction. Examples of the latter kind are found in magnetic resonance spectroscopies like n.m.r. (nuclear magnetic resonance), e.s.r. (electron spin resonance), e.p.r. (electron paramagnetic resonance), etc.

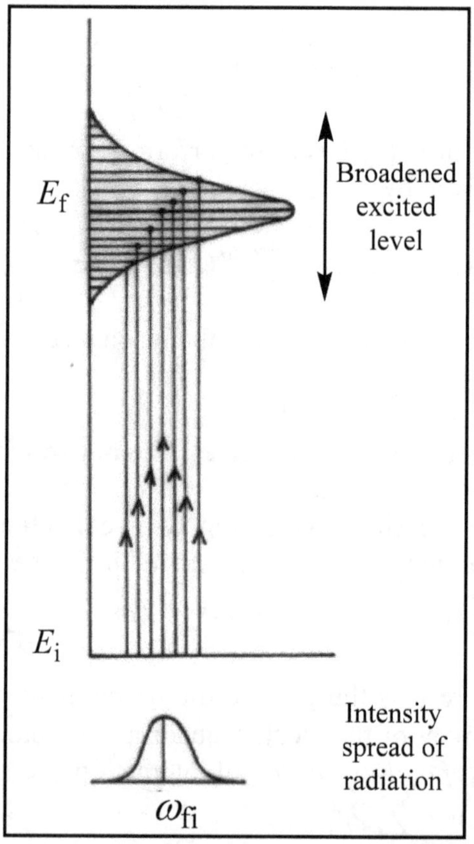

Figure 15.3 Spectroscopic transition between a discrete initial state and a final state that is usually broadened by natural uncertainty, collisions and other effects.

Let us take only the first term in the multipolar expansion (89.1) as perturbation,

$$\mathbf{H}'(t) = -\boldsymbol{\mu}\cdot E(\omega,t) \tag{89.4}$$

where the electric field of radiation is sinusoidal and the radiation is plane polarized such that

$$\boldsymbol{E}(\omega,t) = \boldsymbol{E}^0(\omega)\cos(\omega t+\delta), \tag{89.5}$$

δ being an arbitrary phase. The corresponding F(**r**) is

$$\mathrm{F}(\mathbf{r}) = -\frac{1}{2}\boldsymbol{\mu}\bullet\boldsymbol{E}^0(\omega) \tag{89.6}$$

which may be broken down into three contributions arising from the Cartesian components of the electric dipole vector and those of the electric field vector,

$$\mathrm{F}(\mathbf{r}) = -\frac{1}{2}\,[\mu_x E_x^0(\omega) + \mu_y E_y^0(\omega) + \mu_z E_z^0(\omega)\,]. \tag{89.7}$$

Thus the quantity $|F_{fi}|^2$ in Eq.(88.5) will be given by

$$|F_{fi}|^2 = F_{fi}\,F_{if} = \frac{1}{4}\sum_{\substack{\alpha,\beta\\=(x,y,z)}} (\mu_\alpha)_{fi}(\mu_\beta)_{if}\,E_\alpha^0(\omega)\,E_\beta^0(\omega)^*. \tag{89.8}$$

For an *isotropic* radiation with waves of *random phases* the contributions corresponding to the cross terms in the sum are averaged to zero *in the treatment*. Therefore,

$$|F_{fi}|^2 = \frac{1}{4}\sum_{\substack{\alpha\\=(x,y,z)}} |(\mu_\alpha)_{if}|^2|E_\alpha^0(\omega)|^2. \tag{89.9}$$

This result should be used in Eq. (88.5).

For the isotropic radiation with randomly varying phase of the superposing waves one also writes

$$I_x(\omega) = I_y(\omega) = I_z(\omega)$$

and (89.10)

$$|E_x(\omega)| = |E_y(\omega)| = |E_z(\omega)|$$

where the intensities are related to the electric fields by Eq. (54.9) as

$$I_\alpha(\omega) = \frac{c}{8\pi}|E_\alpha^0(\omega)|^2.$$

The intensity is the amount of radiant energy contained in *unit frequency* interval crossing unit area perpendicular to a specific axis per unit time, and is measured in units of erg per square centimeter per second per (cycle per second). The total intensity is given by

$$I(\omega) = I_x(\omega) + I_y(\omega) + I_z(\omega) = 3I_x(\omega). \tag{89.11}$$

Also, by Eq. (54.9),

$$I(\omega) = \frac{c}{8\pi}|\boldsymbol{E}^0(\omega)|^2 = c\rho(\omega) \tag{89.12}$$

where $\rho(\omega)$ is the energy density. Then we have, from Eq. (89.9),

$$|F_{fi}|^2 = \frac{2\pi}{3}|(\vec{\mu})_{if}|^2\rho(\omega) \tag{89.13}$$

where we have used the result

$$|(\vec{\mu})_{if}|^2 = (\vec{\mu})_{if}\bullet(\vec{\mu})_{fi} = |(\mu_x)_{if}|^2 + |(\mu_y)_{if}|^2 + |(\mu_z)_{if}|^2. \tag{89.14}$$

Integral $(\vec{\mu})_{if}$ is called the (electric) transition integral, and its square modulus is known as the (electric) dipole strength $D_{if} = |(\vec{\mu})_{if}|^2$. The probability of transition, found from Eq. (88.5), is proportional to the dipole strength:

$$P_{i\to f}(t'>0) = \frac{2\pi}{3\hbar^2}|(\vec{\mu})_{if}|^2 \left[\rho(\omega)\frac{4\sin^2\frac{1}{2}(\omega-\omega_{fi})t'}{(\omega-\omega_{fi})^2}\right]_{\omega\to\omega_{fi}}. \tag{89.15}$$

As the radiation is not monochromatic, one has to consider the integrated probability of transition. The latter will be given by

$$P_{i\to f}(t') = \frac{2\pi}{3\hbar^2}|(\vec{\mu})_{if}|^2 \int_0^\infty d\nu \, \frac{4\sin^2\frac{1}{2}(\omega-\omega_{fi})t'}{(\omega-\omega_{fi})^2}\rho(\omega). \tag{89.16}$$

Note that the integration is to be carried out here over the *circular frequency* ν. If we evaluate the integral following the method of the two earlier lectures we get

$$P_{i\to f}(t') = \frac{2\pi}{3\hbar^2}|(\vec{\mu})_{if}|^2 \rho(\omega_{fi})\, t' \tag{89.17}$$

with the transition rate

$$p_{i\to f} = \frac{2\pi}{3\hbar^2}|(\vec{\mu})_{if}|^2 \rho(\omega_{fi}). \tag{89.18}$$

For light propagating in a specific direction, say, along the positive direction of the *z*-axis, we obtain a somewhat different result. Consider the light to be polarized in the *x*-direction as indicated by Figure 15.4. We then have

$$p_{i\to f} = \frac{2\pi}{c\hbar^2}|(\mu_x)_{if}|^2 I_x(\omega_{fi}). \tag{89.19}$$

If we add similar contributions from the *y* and *z* polarizations to (89.19), we retrieve the transition rate in Eq. (89.18) for interaction with an isotropic radiation.

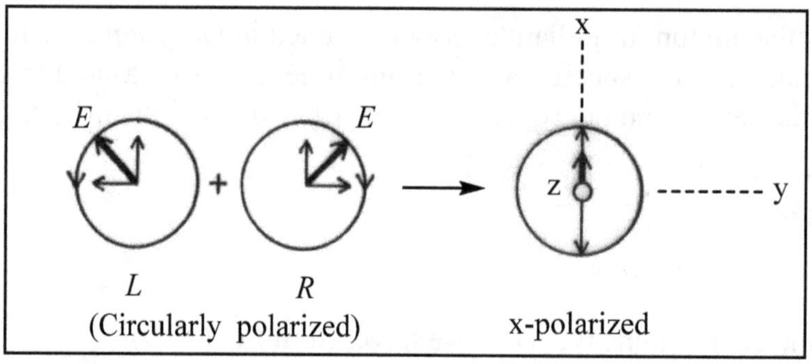

Figure 15.4 The *x*-polarized light formed from a linear superposition of two circularly polarized waves propagating in the *z*-direction.

Lecture 90: Einstein Coefficients

In Einstein's treatment, the rate of absorption of light which results in the transition of the system from state *i* to state *f* is proportional to the energy density $\rho(\omega_{fi})$,

$$p_{i\to f}^{induced} = B_{i\to f}\,\rho(\omega_{fi}) \qquad \text{(for } E_f > E_i\text{)} \tag{90.1}$$

where $B_{i\to f}$ is the proportionality constant. This quantity, $B_{i\to f}$, is called the *Einstein coefficient of induced absorption*. By comparing with the result (89.18) we find

$$B_{i\to f} = \frac{2\pi}{3\hbar^2}|(\vec{\mu})_{fi}|^2. \tag{90.2}$$

Similarly, in the presence of the radiation field, the rate of emission accompanying the transition of the atomic or molecular system from state f down to state i is written as

$$p_{f\to i}^{induced} = B_{f\to i}\,\rho(\omega_{fi}) \tag{90.3}$$

where the *Einstein coefficient of induced emission* is given by

$$B_{f\to i} = \frac{2\pi}{3\hbar^2}|(\vec{\mu})_{if}|^2. \tag{90.4}$$

Note that the coefficient of induced absorption equals that of induced emission.

$$B_{i\to f} = B_{f\to i}. \tag{90.5}$$

This is why spectroscopic transitions are denoted by $i\leftrightarrow f$.

Einstein argued that there will be emission even in the absence of the radiation field. To visualize the need for spontaneous emission we consider the radiation to be in thermal equilibrium with the atoms or molecules. For a macroscopically large ensemble, the number (N) of atoms or molecules in each state is given by the Boltzmann population distribution. If $E_f > E_i$, then $N_f < N_i$ and the rate of induced absorption, $N_i\,p_{i\to f}^{induced}$, exceeds the rate of induced emission, $N_f\,p_{f\to i}^{induced} = N_f\,p_{i\to f}^{induced}$. This leaves aside a nonzero net rate of induced absorption which must vanish in thermal equilibrium of the molecule-radiation system. Einstein proposed that the net rate of induced absorption is compensated by the rate of spontaneous emission $p_{f\to i}^{spontaneous} = A_{f\to i}$ that is independent of the radiation density. The coefficient $A_{f\to i}$ is the *Einstein Coefficient of spontaneous emission.* The thermal equilibrium is given by the *detailed balancing* condition

$$\boxed{N_i B_{i\to f}\,\rho(\omega_{fi}) = N_f[B_{f\to i}\,\rho(\omega_{fi}) + A_{f\to i}]} \tag{90.6}$$

which is schematically represented in Figure 15.5.

The coefficient $A_{f\to i}$ is to be determined from the detailed balancing condition (90.6). We get

$$A_{f\to i} = (N_i/N_f - 1)\,B_{i\to f}\,\rho(\omega_{fi}). \tag{90.7}$$

The Boltzmann distribution is

$$\frac{N_i}{N_f} = e^{\hbar\omega_{fi}/k_BT} \tag{90.8}$$

where k_B is Boltzmann's constant and T is the temperature in Kelvin scale. By using this equation and the expression for $B_{i\to f}$ from Eq. (90.2) we obtain

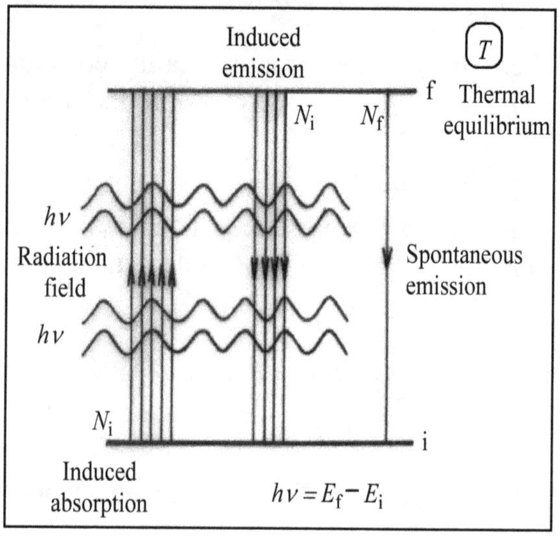

Figure 15.5 Schematic representation of the detailed balancing condition: Molecules in the ground state (*i*), those in the excited state (*f*), and the radiation density are all in thermal equilibrium at absolute temperature *T*.

$$A_{f\to i} = \frac{2\pi}{3\hbar^2}\left(e^{\hbar\omega_{fi}/k_BT} - 1\right) |(\vec{\mu})_{if}|^2 \rho(\omega_{fi}). \tag{90.9}$$

The energy density of the radiation in equilibrium with the system at temperature *T* is given by the *black body radiation law* of Planck,

$$\rho(\omega) = \frac{h\omega^3}{\pi^2 c^3}\left(e^{\hbar\omega_{fi}/k_BT} - 1\right)^{-1}. \tag{90.10}$$

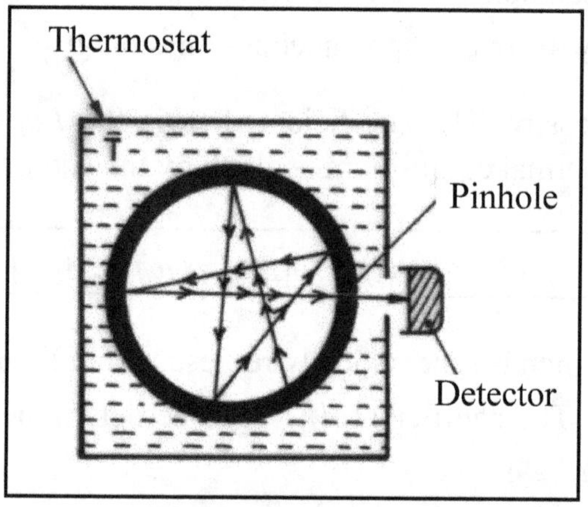

Figure 15.6 Experimental arrangement for the detection of black body radiation and recording the spectrum.

The experimental design for the detection of blackbody radiation is schematically represented in Figure 15.6. The corresponding spectrum of energy density per unit wavelength range is shown in Figure 15.7. If we insert the energy density from Eq. (90.10) into Eq. (90.9) we obtain the Einstein coefficient of spontaneous emission as

$$A_{f\to i} = (h\omega_{fi}^3/\pi^2 c^3)\, B_{f\to i} = (h\omega_{fi}^3/\pi^2 c^3)\, B_{i\to f}$$
$$= (4\omega_{fi}^3/3\hbar c^3)|\vec{\mu}_{fi}|^2. \tag{90.11}$$

In the semiclassical treatment spontaneous emission is postulated empirically, in order to account for the thermal equilibrium. In a purely quantum theoretical treatment where a photon represents a quantum of oscillation of the electromagnetic field, spontaneous emission arises from the form of interaction (electromagnetic coupling) itself. This is one of the great triumphs of quantum electrodynamics (QED).

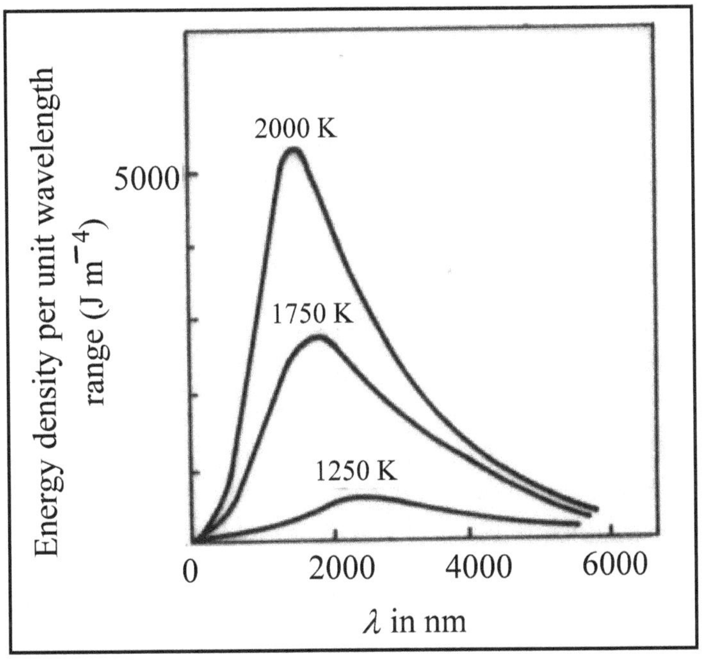

Figure 15.7 Spectrum of the radiation emitted by a black body at different temperature: This is in agreement with Planck's radiation law, Eq. (90.10).

Lecture 91: Basic Principles of Absorption Spectroscopy

Think of a very simple arrangement. Someone has taken a solution of a specific type of molecules in a suitable solvent. Let n_i and n_f be the concentrations in number per unit volume of molecules in initial (ground) and final (excited) states respectively in thermal equilibrium. Let this solution be agitated by a stream of light passing through unit cross-sectional area as shown in Figure 15.8. The system is no longer in thermal equilibrium with the new radiation, and there is a net induced absorption. The number of photons in the radiation that traverses an infinitesimal path length dp in unit time changes by the amount

$$dn_{photon} = -(n_i - n_f)\ p_{i\to f}^{induced}\ dp \tag{91.1}$$

that corresponds to the change in intensity

$$dI(\omega_{fi}) = \hbar\omega_{fi}\ dn_{photon}. \tag{91.2}$$

In writing Eq.(91.1) we have ignored the negligibly small contribution from spontaneous emission. Combining Eq. (91.1) with Eq. (91.2) and using Eq. (90.3) in the resulting expression one obtains

$$-\ dI(\omega_{fi})/dp = (n_i - n_f)\ B_{i\to f}\ \rho(\omega_{fi})\ \hbar\omega_{fi}. \tag{91.3}$$

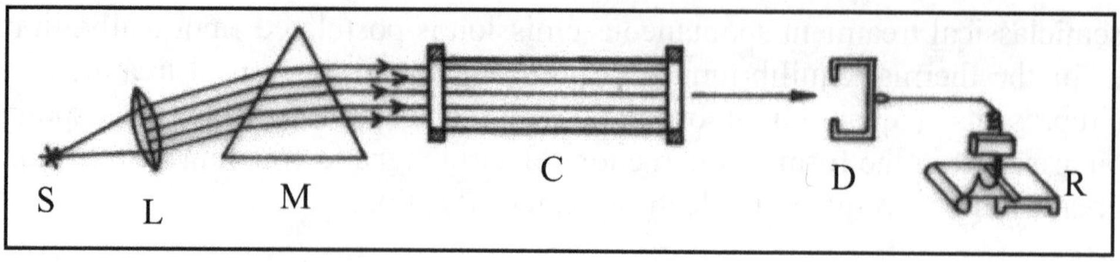

Figure 15.8 Experimental arrangement of absorption spectroscopy: S is the source of light, L is a lens to generate a collinear beam, M is a monochromator, and C is the sample cell of length p. The detector D detects the intensity of the transmitted light and the recorder R records the spectrum on chart paper.

We know that $\rho = I/c$, from which we get

$$-\frac{d \ln I}{dp} = \frac{1}{c}(n_i - n_f)\, B_{i\to f}\, \hbar\omega_{fi}. \tag{91.4}$$

For $\hbar\omega_{fi}/k_BT >> 1$, n_f is negligibly small compared to n_i and we write

$$-\frac{d \ln I}{dp} = \frac{1}{c} n_i\, B_{i\to f}\, \hbar\omega_{fi}. \tag{91.5}$$

One may replace n_i by the total number (n) of molecules in unit volume. Furthermore, if the total concentration is D in mol per liter, and the unit of length is centimeter, then $n = N_{Avo}D/1000$, N_{Avo} being the Avogadro number. These substitutions yield

$$-\frac{d \ln I}{dp} = \kappa D \tag{91.6}$$

where κ, the *absorption coefficient*, is given by

$$\kappa = (N_{Avo}/1000c)\, B_{i\to f}\, \hbar\omega_{fi} = (2\pi N_{Avo}/3000\,\hbar\, c)\, |\vec{\mu}_{if}|^2\, \omega_{fi}. \tag{91.7}$$

Equation (91.6) is in reality an integrated version of *Lambert-Beer* law that states that I, the intensity of the outgoing light, is related to the intensity of the incoming light I_0 by a simple exponential form

$$I = I_0 e^{-\alpha D p}, \tag{91.8}$$

where p is the path length (the length of the sample cell). The *optical density* is given by

$$O = \ln\left(\frac{I_0}{I}\right) = \alpha D p. \tag{91.9}$$

The optical density not only depends upon the concentration D and the width of the cell containing the solution, but also upon the frequency of the radiation. The (experimental) band intensity is given in terms of the (observed) integrated absorption coefficient,

$$A = \int d\nu\ \alpha(\nu). \tag{91.10}$$

Since theory predicts the total absorption for the transition and not that for a given frequency, we identify κ in Eq. (91.7) as the integrated absorption coefficient A.

The Lambert-Beer law embodies the basic principles of absorption spectroscopy. The experimental arrangement involves a source of light, a monochromator to make the light rays as monochromatic as possible, a cell containing the sample through which the light rays pass and a detector to detect the intensity of the outgoing light. Figure 15.8 shows a schematic arrangement of the experiment. The measurement of intensity needs to be corrected as a part of the light is absorbed by the cell and the solvent molecules contained in it. We take the transmitted intensity with only the solvent molecules in the cell as I_0, and the intensity detected when the sample molecules are also present and absorb another part of the radiation is I. The detector calculates the optical density and the recorder plots it as a function of the frequency of radiation. The resulting curve is called an *absorption spectrum* that has peaks around the resonance frequencies for absorptive transitions.

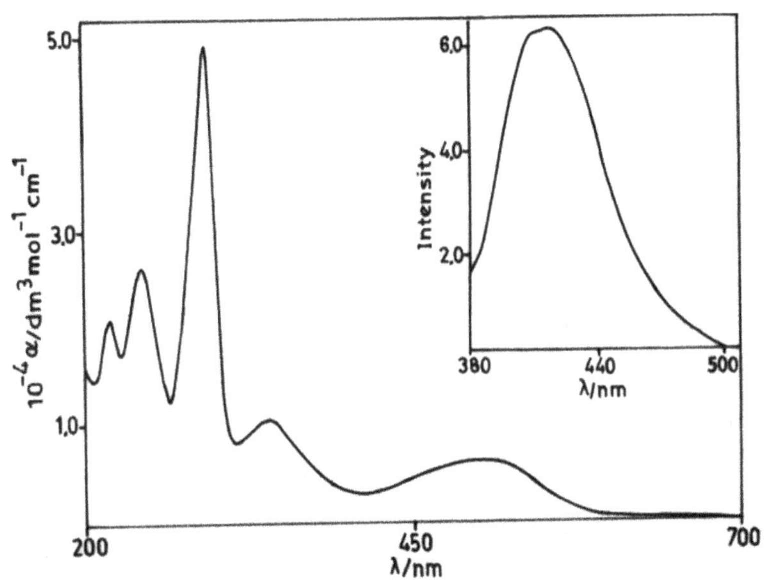

Figure 15.9 The UV-Visible spectrum of $[Ru(bipy)_2L]ClO_4$ in acetonitrile: Here bipy stands for 2,2′-bipyridine and the ligand L is pyridine-2-olate.

A typical uv-visible spectrum is shown in Figure 15.9. Note all the peaks. Each peak is assigned to a specific group of transitions. The spectrum is especially broad in the visible range, between 350 nm and 600 nm. The spectrum in this range can be resolved into a sum of several peaks of fairly low intensity. This is a standard feature of transition metal complexes. The broadening owes its origin to a reasonably strong vibronic coupling. The reason for the lower intensities is that the d↔d transitions are electric dipole forbidden. Though these transitions are allowed by magnetic dipole and electric quadrupole mechanisms, the intensities are too large for such processes to be responsible for the transitions. Actually the intensities are not too small because the ligand orbitals participate in the formation of metal-ligand bonds, and the transitions are no longer of the pure d↔d type, but they are somewhat like transitions between two molecular orbitals. We have already discussed this phenomenon while considering the theoretical aspects of ligand field theory.

Oscillator Strength

If a species of charge e, say, an electron, undergoes a simple harmonic oscillation in one dimension, the dipole moment operator is

$\mu = ex$ (91.11)

and the transition dipole is

$\mu_{if} = ex_{if}.$ (91.12)

In particular, $\mu_{10} = e(\hbar/2m\omega)^{1/2}$. A three-dimensional oscillator with equal classical vibrational frequencies for movement along three orthogonal axes is illustrated in Figure 15.10. This oscillator has three degenerate first excited levels. The integrated absorption coefficient for the 0→1 transition of the three-dimensional electronic oscillator is

$$A = \frac{\pi N_{Avo} e^2}{1000cm} \text{ sec}^{-1}\text{ cm}^{-1}\text{ mol}^{-1}\text{ liter,} \quad (91.13)$$

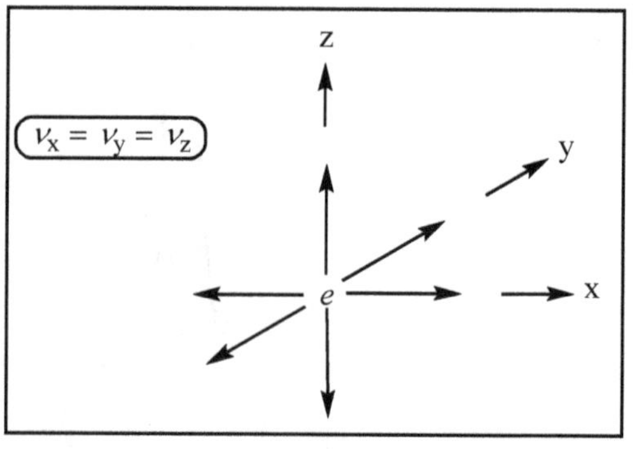

Figure 15.10 The isotropic three-dimensional electronic oscillator: The F centers in alkali halide crystals are examples of an oscillating trapped electron.

and in a different unit,

$$\bar{A} = \frac{\pi N_{Avo} e^2}{1000c^2 m} \text{ cm}^{-2}\text{ mol}^{-1}\text{ liter}$$
$$= 5.33 \times 10^8 \text{ cm}^{-2}\text{ mol}^{-1}\text{ liter} \quad (\text{for } e = \pm e_0). \quad (91.14)$$

The *oscillator strength* is defined by the ratio of the observed integrated absorption coefficient to the calculated $\bar{A}$ for the three-dimensional electronic oscillator:

$$f_{osc} = \left[\int \alpha(\bar{\nu})d\bar{\nu}\right]_{obs} / [\pi N_{Avo} e^2/1000c^2 m]$$
$$= 1.877 \times 10^{-9} \int \alpha(\bar{\nu})d\bar{\nu}. \quad (91.15)$$

The oscillator strength as the ratio of two similar quantities is dimensionless. Many electronic absorption bands have oscillator strengths near unity. When the oscillator strength is very small, the transition is considered to be forbidden. Experimental results are often quoted in terms of oscillator strength.

Lecture 92: Selection Rules

The spectrum corresponding to an absorption or an emission cannot be obtained when the probability of transition becomes zero. Since the transition probability depends on the dipole strength which is the square modulus of the transition integral, it becomes zero whenever the integral $\vec{\mu}_{if}$ vanishes. Often one finds that only one component of $\vec{\mu}_{if}$ is not zero, and the transition is said to be polarized along the same direction. From the symmetry of the molecule (see Appendix-III) one can determine which of the integrals $\vec{\mu}_{if}$ would vanish. The corresponding transitions are forbidden. For example, in the atomic case, $\vec{\mu}_{if} = 0$ whenever the change in the orbital angular momentum quantum number Δl is not equal to ± 1.

Selection rules were formulated empirically by experimental spectroscopists who observed that for isolated atoms, absorptions or emissions take place between states only when

$$\Delta l = \pm 1. \tag{92.1}$$

Thus one finds from theory a negative statement for selection rules that the transitions corresponding to

$$\Delta l \neq \pm 1 \tag{92.2}$$

do not occur as the corresponding transition probabilities are equal to zero. The negative way of reasoning forms the theoretical basis of selection rules. However, theory does not specify whether the transition would be allowed when $\Delta l = \pm 1$. The transition probability can still be zero because of some other symmetry. If the latter is known, the forbidden nature of the transition is clearly understood. If it is not known, the symmetry feature arises from certain hidden variable, such as spin. In any case one needs to explicitly calculate the transition rate in order to establish that the transition is indeed allowed by a certain mechanism.

Consider the integral

$$(\mu_x)_{if} = e\int dv\ \psi_i^* x \psi_f = e\int dx \int_{-\infty}^{+\infty} dy \int dz\ \psi_i^* x \psi_f. \tag{92.3}$$

This vanishes if the integrand $\psi_i^* x \psi_f$ is an odd function of x, and it *can be* nonvanishing only when the integrand is an even function. Even when the integrand is an even function of x, the integral over x can be zero. The integral is also equal to zero unless the product $\psi_i^* \psi_f$ is an even function of both y and z.

Suppose that the product $\psi_i^* \psi_f$ is an odd function of x but an even function of each of the variables y and z. Then $(\mu_x)_{if}$ can be nonzero, but $(\mu_y)_{if}$ and $(\mu_z)_{if}$ will definitely have zero values. The transition is then x-polarized. The y- and z- polarized electric dipole transitions would have zero probability to occur, that is, they would be forbidden.

In the language of Group Theory (see Appendix III), the transition $i \leftrightarrow f$ is allowed by electric dipole mechanism with x polarization when the triple product of representations $\Gamma_i \otimes \Gamma(x) \otimes \Gamma_f$ contains the totally symmetric representation. This implies that the direct product $\Gamma_i \otimes \Gamma_f$ must contain the representation $\Gamma(x)$ for the transition to occur. Theory only guarantees that the transition is forbidden. This has been discussed in the two earlier paragraphs.

When a transition is said to be forbidden we normally mean that it is forbidden by the electric dipole mechanism. An example of this sort may be provided by the d$\leftrightarrow$d transitions which are forbidden by the electric dipole mechanism. Transition metal complexes are in fact known to be weakly colored. They have excellent e.s.r. spectra. E.s.r. transitions, like the n.m.r. ones, take place by the magnetic dipole mechanism. UV-visible (electronic) spectra of complexes formed by elements of the inner transition series are more prominent. Lanthanide complexes are quite strongly colored. This property is attributed to the involvement of f and d orbitals as the initial and final single-particle states and a progressively pronounced participation of the ligand orbitals in forming the bonds in these complexes.

PROBLEMS

1. Derive Eq. (89.18) from Eq. (89.19).
2. Show that the Einstein coefficient of spontaneous emission for the electric dipole transition of an electron oscillating harmonically in one dimension is given by

$$A_{10} = \frac{2e^2\omega^2}{3mc^3} \, .$$

3. Calculate the rate of spontaneous emission of radiation of wavelength 4000 Å with unit oscillator strength.
4. Formulate the selection rules and the transition rates for electric dipole transitions (i) among vibrational levels and (ii) among rotational levels, and for magnetic dipole transition (iii) among atomic levels of the same l but different m.
5. Show that the selection rule for electronic transitions in a homonuclear diatomic molecule is g↔u.

CHAPTER 16

ASPECTS OF MANY-BODY PERTURBATION THEORY

The Hartree-Fock theory constitutes the *mean-field* picture, that is, the effective one-electron treatment, and so the Hartree-Fock level formalisms are known as treatments in the *one- electron approximation*. The separation of Hamiltonian given by

$$\mathbf{H} = \mathbf{H}_{HF} + \mathbf{H}' \qquad (75.4)$$

is the normally preferred starting point for the investigation of many-body effects. The Hartree-Fock Hamiltonian $\mathbf{H}_{HF}$ given by Eq. (75.3) usually takes the place of the zeroth order operator while the perturbation $\mathbf{H}'$ in Eq. (75.5) is referred to as the many-body perturbation. Other choices of the zero-order Hamiltonian is possible. For example the sum of the one-electron Hamiltonian operators may be taken as the zero-order operator while the whole of bare (instantaneous) interaction is considered as perturbation. But it is found that in the latter type treatments the convergence of the perturbation series becomes slow. The series expansion for energy is often oscillatory and at various orders shows large deviations from the converged value. The reader is referred to the texts suggested at the end of this Chapter for a critical discussion on the choice of perturbation.

Lecture 93: The Correlation Energy

The Rayleigh-Schrödinger perturbation theory (RSPT) may be applied to the study of many-body effects in an almost straight-forward way. We already know that the wave function $|\Psi\rangle$'s representing configurations are eigenfunctions of the Hartree-Fock Hamiltonian:

$$\mathbf{H}_{HF}\,|\Psi\rangle = E^{(0)}\,|\Psi\rangle \qquad (93.1)$$

where $E^{(0)}$ is the sum of the relevant orbital energies. In particular,

$$\mathbf{H}_{HF}\,|\Psi_0\rangle = E_0^{(0)}\,|\Psi_0\rangle \qquad (93.2)$$

where $|\Psi_0\rangle$ is the state vector representing the Hartree-Fock ground state. The eigenvalue $E^{(0)}$ (such as $E_0^{(0)}$) is merely the sum of the energies of the orbitals occupied in the particular configuration (such as the Hartree-Fock ground state configuration),

$$E^{(0)} = \sum_{\alpha=1}^{occ} \varepsilon_\alpha \, , \; E_0^{(0)} = \sum_{\alpha=1}^{N} \varepsilon_\alpha \tag{93.3}$$

Before proceeding any further we first discuss how one may evaluate the matrix elements of the perturbation $\mathbf{H}'$. For the sake of convenience we adopt the notations used in physics,

$$<ij|kl> \equiv (ik|jl) = \int d^4x_1 \int d^4x_2 \; \psi_i^*(x_1)\psi_j^*(x_2)\, r_{12}^{-1}\, \psi_k(x_1)\psi_l(x_2) \tag{93.4}$$

and

$$<ij||kl> = <ij|kl> - <ij|lk>. \tag{93.5}$$

Evaluation of the matrix elements of $\mathbf{H}'$requires the matrix elements of the instantaneous interaction as well as those of the Hartree-Fock interaction. Matrix elements of the instantaneous interaction are as follows:

$$<\Psi_0| \sum_{i<j}^{N} r_{ij}^{-1} |\Psi_0> = \frac{1}{2}\sum_{\alpha,\beta}^{N} <\alpha\beta||\alpha\beta>, \tag{93.6}$$

$$<\Psi_0| \sum_{i<j}^{N} r_{ij}^{-1} |\Psi_{\alpha\beta}^{mn}> = <\alpha\beta||mn>, \tag{93.7}$$

etc. Elements of the Hartree-Fock interaction matrix can be evaluated from the result

$$<i| \sum_{\alpha=1}^{N} (\mathbf{J}_\alpha - \mathbf{K}_\alpha)|j> = \sum_{\alpha=1}^{N} <i\alpha||j\alpha>. \tag{93.8}$$

When equations (93.6)-(93.8) are used together one obtains the $\mathbf{H}'$-integrals as well as the $\mathbf{H}$-integrals given in Table 12.5. The $\mathbf{H}'$ matrix elements are related to the Hamiltonian integrals as

$$\begin{aligned} <\Psi_P|\mathbf{H}'|\Psi_Q> &= <\Psi_P|\mathbf{H}|\Psi_Q> - <\Psi_P|\mathbf{H}_{HF}|\Psi_Q> \\ &= <\Psi_P|\mathbf{H}|\Psi_Q> - \delta_{P,Q} \sum_{i \in P} \varepsilon_i. \end{aligned} \tag{93.9}$$

Calculation of the integrals like $<\Psi_0|\mathbf{H}'|\Psi_\alpha^m>$ has not been shown in the above. By now the reader should be able to account for this type of integrals. Nevertheless, a clue for this omission is given in the following.

In order to obtain the first-order correction to the ground state energy one may subtract the expectation value of the Hartree-Fock potential

$$<\Psi_0| \sum_{\alpha=1}^{N} (\mathbf{J}_\alpha - \mathbf{K}_\alpha)|\Psi_0> = \sum_{\alpha,\beta}^{N} <\alpha\beta||\alpha\beta> \tag{93.10}$$

from the energy of instantaneous interaction in the Hartree-Fock ground state which is given in Eq.(93.6), and find

$$E_0^{(1)} = <\Psi_0|\mathbf{H}'|\Psi_0> = -\frac{1}{2}\sum_{\alpha,\beta}^{N} <\alpha\beta||\alpha\beta>. \tag{93.11}$$

The Hartree-Fock ground state energy appears as the sum of the zeroth order energy $E_0^{(0)}$ and the first order correction $E_0^{(1)}$,

$$E_0^{(0)} + E_0^{(1)} = \sum_{\alpha=1}^{N} \varepsilon_\alpha - \frac{1}{2}\sum_{\alpha,\beta}^{N} \langle\alpha\beta||\alpha\beta\rangle = E_0(\mathrm{HF}). \tag{93.12}$$

This is not surprising at all, since the Hartree-Fock ground state wave function is the zeroth-order wave function in the perturbative scheme and the zeroth-order function directly determines energy up to the first order in $\mathbf{H}'$.

The second order contribution to the ground state energy is obtained from the interaction of the ground state configuration with other configurations. Only the doubly excited intermediate states give non-vanishing integrals for interaction with $|\Psi_0\rangle$. This is a consequence of $\mathbf{H}$ or $\mathbf{H}'$ having only one- and two-electron operators and the fact that the singles do not directly interact with the Hartree-Fock ground state configuration (Brillouin theorem). Besides, only the instantaneous interaction $\sum_{i<j} r_{ij}^{-1}$ contributes, because the Hartree- Fock potential

$$\sum_{\alpha=1}^{N} (\mathbf{J}_\alpha - \mathbf{K}_\alpha),$$

is basically a one-electron operator and its integral over the ground state configuration and a doubly excited configuration is always equal to zero. Therefore, the second order energy correction from RSPT is

$$E_0^{(2)} = \sum_{\alpha<\beta}\sum_{m<n} \frac{|\langle\Psi_0|\sum_{i<j} r_{ij}^{-1}|\Psi_{\alpha\beta}^{mn}\rangle|^2}{E_0^{(0)} - E_{\alpha\beta}^{mn}} . \tag{93.13}$$

Use of Eq. (93.3) in the energy denominator and Eq. (93.7) in the numerator leads to the form

$$E_0^{(2)} = \sum_{\alpha<\beta}\sum_{m<n} \frac{|\langle\alpha\beta||mn\rangle|^2}{\varepsilon_\alpha + \varepsilon_\beta - \varepsilon_m - \varepsilon_n} \tag{93.14}$$

that can be written as a sum of first-order pair energies (first-order energy of two single excitations starting from two different spin orbitals that are occupied in the Hartree-Fock ground state configuration),

$$E_0^{(2)} = \sum_{\alpha<\beta} E_{\alpha\beta}^{pair} \tag{93.15}$$

where

$$E_{\alpha\beta}^{pair} = \sum_{m<n} \frac{|\langle\alpha\beta||mn\rangle|^2}{\varepsilon_\alpha + \varepsilon_\beta - \varepsilon_m - \varepsilon_n} . \tag{93.16}$$

The calculation of the third-order correction to the ground state energy is similar but quite tedious. If one pursues this task one finds

$$E_0^{(3)} = \frac{1}{8}\Bigg[\sum_{\alpha\beta\gamma\delta}\sum_{mn}\frac{<\alpha\beta\,||\,mn><mn\,||\,\gamma\delta><\gamma\delta\,||\,\alpha\beta>}{(\varepsilon_\alpha+\varepsilon_\beta-\varepsilon_m-\varepsilon_n)(\varepsilon_\gamma+\varepsilon_\delta-\varepsilon_m-\varepsilon_n)}$$

$$+\sum_{\alpha\beta}\sum_{mnpq}\frac{<\alpha\beta\,||\,mn><mn\,||\,pq><pq\,||\,\alpha\beta>}{(\varepsilon_\alpha+\varepsilon_\beta-\varepsilon_m-\varepsilon_n)(\varepsilon_\alpha+\varepsilon_\beta-\varepsilon_p-\varepsilon_q)} \tag{93.17}$$

$$+\sum_{\alpha\beta\gamma}\sum_{mnp}\frac{<\alpha\beta\,||\,mn><\gamma n\,||\,p\beta><mp\,||\,\alpha\gamma>}{(\varepsilon_\alpha+\varepsilon_\beta-\varepsilon_m-\varepsilon_n)(\varepsilon_\alpha+\varepsilon_\gamma-\varepsilon_m-\varepsilon_p)}.$$

More complicated expressions are obtained for the higher order energies.

The ground state correlation energy defined by Eq. (76.6) can be identified with the perturbation expansion

$$E_0^{correl} = E_0^{(2)} + E_0^{(3)} + \dots\ . \tag{93.18}$$

An important item to note is that the *nth*-order correction to wave function $|\Psi_0^{(n)}\rangle$ is sufficient to calculate up to the (2*n*+1)*st*-order correction to energy $E_0^{(2n+1)}$. Thus the first-order correction $|\Psi_0^{(1)}\rangle$ yields the correlation energy correctly through third order. The second-order correction $|\Psi_0^{(2)}\rangle$ requires a renormalization of the wave function through second order in $\mathbf{H}'$. This in turn requires renormalization corrections to be introduced to the calculation of physical quantities.

Another important point is that the ground state correlation energy correct through third order in $\mathbf{H}'$ is completely determined by the lowest-order, (that is, the first-order), *correlated ground state* wave function

$$|\Psi_g^1\rangle = |\Psi_0^{(0)}\rangle + \sum_{m<n}\frac{<mn\,||\,\alpha\beta>}{\varepsilon_\alpha+\varepsilon_\beta-\varepsilon_m-\varepsilon_n}|\Psi_{\alpha\beta}^{mn}\rangle. \tag{93.19}$$

This wave function has found widespread (semi-quantitative) applications in several types of many-body treatments such as Equations of motion methods for the calculation of spectroscopic transition frequencies, ionization energy, electron affinities, etc., Green's function analysis and kinematic analyses based on linear response and higher-order random phase approximations.

Lecture 94: The *N*-Dependence

The minimal-basis calculation on H_2 molecule gives two molecular orbitals, σ and σ^*, which correspond to the four spin molecular orbitals σ, $\bar{\sigma}$, σ^* and $\bar{\sigma}^*$. The Hartree-Fock ground state configuration is $|\sigma\bar{\sigma}>$, and it interacts with the only one available doubly excited configuration $|\sigma^*\bar{\sigma}^*>$. The interaction matrix element equals the exchange integral. To see it, write

$$<\sigma\bar{\sigma}|\mathbf{H}|\sigma^*\bar{\sigma}^*> = <\sigma\bar{\sigma}|\sigma^*\bar{\sigma}^*> = <\sigma\bar{\sigma}^*|\sigma^*\bar{\sigma}> \quad \text{(real spatial orbitals)}$$

$$= <\sigma\sigma^*|\sigma^*\sigma> \quad \text{(after spin integration)} \tag{94.1}$$

so that

$$<\sigma\bar{\sigma}|\mathbf{H}|\sigma^*\bar{\sigma}^*> = K_{\sigma\sigma^*} \equiv K. \tag{94.2}$$

The Hartree-Fock ground state energy is $E_0(\text{HF}) = 2\varepsilon_\sigma - J_{\sigma\sigma}$, and the doubly excited configuration is more energetic than the ground state configuration by the amount

$$\begin{aligned} 2\Delta &= <\sigma^*\bar{\sigma}^*|\mathbf{H}|\sigma^*\bar{\sigma}^*> - E_0(\text{HF}) \\ &= 2(\varepsilon_{\sigma^*} - \varepsilon_\sigma) + J_{\sigma\sigma} + J_{\sigma^*\sigma^*} - 4J_{\sigma\sigma^*} + 2K_{\sigma\sigma^*}. \end{aligned} \tag{94.3}$$

These matrix elements lead to the following DCI correlation energy for the monomer:

$$E_{correl}^{1-mer}(exact) = \Delta - (\Delta^2 + K^2)^{1/2}. \tag{94.4}$$

These results will be used in the immediate following to analyze the correlation energy in a polymer of *noninteracting* minimal-basis H_2 molecules.

Take the case of two noninteracting minimal-basis H_2 molecules with molecular orbitals (σ_1, σ_1^*) and (σ_2, σ_2^*) respectively. The single determinant

$$|\Psi_0> = |\sigma_1\bar{\sigma}_1\sigma_2\bar{\sigma}_2> \tag{94.5}$$

represents the Hartree-Fock ground state of the two-molecule system. Only two doubly excited configurations can mix with $|\Psi_0>$. These are $|\Psi_{1\bar{1}}^{1^*\bar{1}^*}> = |\sigma_1^*\bar{\sigma}_1^*\sigma_2\bar{\sigma}_2>$ and $|\Psi_{2\bar{2}}^{2^*\bar{2}^*}> = |\sigma_1\bar{\sigma}_1\sigma_2^*\bar{\sigma}_2^*>$. The following integrals are required while one formulates the DCI Hamiltonian matrix H_{DCI}:

$$< \Psi_0|\mathbf{H}|\Psi_0 > = 2E_0(\text{HF}),$$

$$< \Psi_{1\bar{1}}^{1^*\bar{1}^*}|\mathbf{H}|\Psi_{1\bar{1}}^{1^*\bar{1}^*} > = < \Psi_{2\bar{2}}^{2^*\bar{2}^*}|\mathbf{H}|\Psi_{2\bar{2}}^{2^*\bar{2}^*} > = 2E_0(\text{HF}) + 2\Delta,$$

$$< \Psi_0|\mathbf{H}|\Psi_{1\bar{1}}^{1^*\bar{1}^*} > = < \Psi_0|\mathbf{H}|\Psi_{2\bar{2}}^{2^*\bar{2}^*} > = K,$$

and (94.6)

$$< \Psi_{1\bar{1}}^{1^*\bar{1}^*}|\mathbf{H}|\Psi_{2\bar{2}}^{2^*\bar{2}^*} > = 0.$$

One obtains

$$H_{DCI} = \begin{bmatrix} 2E_0(HF) & K & K \\ K & 2[E_0(HF)+\Delta] & 0 \\ K & 0 & 2[E_0(HF)+\Delta] \end{bmatrix}. \tag{94.7}$$

The diagonalization of H_{DCI} yields the DCI correlation energy for the two-molecule system (the dimer). We get

$$E_{correl}^{2-mer}(\text{DCI}) = \Delta - (\Delta^2 + 2K^2)^{1/2}. \tag{94.8}$$

Thus from a DCI calculation one finds $E_{correl}^{2-mer} \neq 2E_{correl}^{1-mer}$. We say that the DCI calculation is not *size-consistent*. For N non-interacting molecules (an N-mer), we get

$$E_{correl}^{N-mer}(\text{DCI}) = \Delta - (\Delta^2 + NK^2)^{1/2}. \tag{94.9}$$

As $N \rightarrow \infty$, the DCI correlation energy behaves as

$$E_{correl}^{N-mer}(\text{DCI}) \approx -N^{1/2}K \tag{94.10}$$

such that the correlation energy per monomer vanishes:

$$\lim_{N\to\infty} \frac{1}{N} E_{correl}^{N-mer}\,(\mathrm{DCI}) = 0. \tag{94.11}$$

The exact correlation energy should behave as

$$E_{correl}^{N-mer}\,(\mathrm{DCI}) = N\,E_{correl}^{1-mer}\,(exact). \tag{94.12}$$

In fact, no truncated form of the CI treatment is found to be size-consistent. Only the full CI (FCI) is size-consistent. With the given basis set, FCI gives as exact solutions to the Schrödinger equation for a closed shell as possible. It is the only true gold standard method of calculation on open shells. However, FCI requires a very large number of configurations, and in most cases the calculation becomes too huge to be accomplished. The DQCI treatment (CI with doubly and quadruply excited configurations) yields moderately size-consistent results for small molecules with less than about 30 orbitals included in the single-particle basis set. The DQCI correlation energy per monomer certainly tends to zero as N becomes arbitrarily large.

Let us examine the same systems by MBPT. The DCI treatment may be compared with the first-order configuration mixing that yields the wave function correct through first order and energy correct through third order in perturbation. The first-order mixing may be seen as a systematic and prudent procedure. We would make use of the following monomer integrals:

$$<\sigma\bar{\sigma}|r_{12}^{-1}|\sigma\bar{\sigma}> \;=\; J_{\sigma\bar{\sigma}};$$

$$<\sigma\bar{\sigma}|2\mathbf{J}_\sigma(1)-\mathbf{K}_\sigma(1)+2\mathbf{J}_\sigma(2)-\mathbf{K}_\sigma(2)|\sigma\,\bar{\sigma}> = 2(J_{\sigma\sigma}+J_{\sigma\bar{\sigma}});$$

$$<\sigma^*\bar{\sigma}^*|r_{12}^{-1}|\sigma\bar{\sigma}> = K_{\sigma\sigma^*};$$

and (94.13)

$$<\sigma^*\bar{\sigma}^*|2\mathbf{J}_\sigma(1)-\mathbf{K}_\sigma(1)+2\mathbf{J}_\sigma(2)-\mathbf{K}_\sigma(2)|\sigma\bar{\sigma}> = 0.$$

The zeroth-order ground state vector for the monomer is given by

$$|\Psi_0^{(0)}(n{=}1)> = |\sigma\bar{\sigma}> \tag{94.14}$$

that leads to the energy values

$$E_0^{(0)}(1\text{-}mer) = 2\varepsilon_\sigma;$$

$$E_0^{(1)}(1\text{-}mer) = J_{\sigma\sigma} - 2J_{\sigma\sigma} = -\,J_{\sigma\sigma};$$

and (94.15)

$$E_0^{(0)}(1\text{-}mer) + E_0^{(1)}(1\text{-}mer) = E_0(\mathrm{HF};\,1\text{-}mer).$$

We also have

$$|\Psi_0^{(1)}(n{=}1)> = \frac{K_{\sigma\sigma^*}}{2(\varepsilon_\sigma - \varepsilon_{\sigma^*})}\,|\Psi_{\sigma\bar{\sigma}}^{\sigma^*\bar{\sigma}^*}(n{=}1)> \tag{94.16}$$

so that

$$E_0^{(2)}(1\text{-}mer) = \frac{K_{\sigma\sigma^*}^2}{2(\varepsilon_\sigma - \varepsilon_{\sigma^*})} \tag{94.17}$$

and

$$E_0^{(3)}(1\text{-}mer) = \sum_m{}' \sum_n{}' \frac{\langle \Psi_0^{(0)} | \mathbf{H}' | \Psi_m^{(0)} \rangle \langle \Psi_m^{(0)} | \mathbf{H}' | \Psi_n^{(0)} \rangle \langle \Psi_n^{(0)} | \mathbf{H}' | \Psi_0^{(0)} \rangle}{(E_0^{(0)} - E_m^{(0)})(E_0^{(0)} - E_n^{(0)})}$$

$$- E_0^{(1)}(1\text{-}mer) \sum_m{}' \frac{|\langle \Psi_0^{(0)} | \mathbf{H}' | \Psi_m^{(0)} \rangle|^2}{(E_0^{(0)} - E_m^{(0)})^2} \qquad (94.18)$$

$$= \frac{K_{\sigma\sigma^*}^2 (J_{\sigma\sigma} + J_{\sigma^*\sigma^*} - 4J_{\sigma\sigma^*} + 2K_{\sigma\sigma^*})}{4(\varepsilon_\sigma - \varepsilon_{\sigma^*})^2}$$

For N non-interacting minimal-basis H_2 molecules (the *N-mer*), the zeroth-order ground state is written as

$$|\Psi_0^{(0)}(n{=}N)\rangle = |\sigma_1 \bar{\sigma}_1 \ldots \sigma_N \bar{\sigma}_N\rangle, \qquad (94.19)$$

so that

$$E_0^{(0)}(N\text{-}mer) = 2N\varepsilon_\sigma = NE_0^{(0)}(1\text{-}mer)$$

and (94.20)

$$E_0^{(1)}(N\text{-}mer) = -NJ_{\sigma\sigma} = NE_0^{(1)}(1\text{-}mer).$$

Thus the size consistency is explicitly maintained so far. Moreover, the first-order correction to the *N-mer* ground state wave function is given by

$$|\Psi_0^{(1)}(n{=}N)\rangle = \sum_{i=1}^{N} \frac{K_{\sigma\sigma^*}}{2(\varepsilon_\sigma - \varepsilon_{\sigma^*})} |\Psi_{\sigma_i \bar{\sigma}_i}^{\sigma_i^* \bar{\sigma}_i^*}(n{=}N)\rangle \qquad (94.21)$$

which directly yields the second order energy correction

$$E_0^{(2)}(N\text{-}mer) = \sum_{i=1}^{N} \frac{K_{\sigma\sigma^*}}{2(\varepsilon_\sigma - \varepsilon_{\sigma^*})} \langle \Psi_0^{(0)}(n{=}N) | \mathbf{H}' | \Psi_{\sigma_i \bar{\sigma}_i}^{\sigma_i^* \bar{\sigma}_i^*}(n{=}N)\rangle$$

$$= N \frac{K_{\sigma\sigma^*}^2}{2(\varepsilon_\sigma - \varepsilon_{\sigma^*})} = NE_0^{(2)}(1\text{-}mer). \qquad (94.22)$$

Also, the two terms in $E_0^{(3)}(N\text{-}mer)$ are found to have different N-dependences yet preserving the correct N-dependence for $E_0^{(3)}$:

$$E_0^{(3)}(N\text{-}mer) = \left[-\frac{N^2 J_{\sigma\sigma} K_{\sigma\sigma^*}^2}{4(\varepsilon_\sigma - \varepsilon_{\sigma^*})^2} + \frac{NK_{\sigma\sigma^*}^2 (J_{\sigma\sigma} + J_{\sigma^*\sigma^*} - 4J_{\sigma\sigma^*} + 2K_{\sigma\sigma^*})}{4(\varepsilon_\sigma - \varepsilon_{\sigma^*})^2} \right]$$

$$+ \frac{N^2 J_{\sigma\sigma} K_{\sigma\sigma^*}^2}{4(\varepsilon_\sigma - \varepsilon_{\sigma^*})^2} \qquad (94.23)$$

$$= NE_0^{(3)}(1\text{-}mer)$$

Thus to these orders of perturbation, the RSPT is seen to yield a size-consistent, though approximate, correlation energy:

$$E_{correl}^{N-mer} = E_0^{(2)}(N\text{-}mer) + E_0^{(3)}(N\text{-}mer) + \ldots$$

$$= NE_0^{(2)}(1\text{-}mer) + NE_0^{(3)}(1\text{-}mer) + \ldots = NE_{correl}(1\text{-}mer). \qquad (94.24)$$

The Hartree-Fock total energy $E_0(\text{HF}) = E_0^{(0)} + E_0^{(1)}$ is of course size-consistent. Brueckner conjectured that the RSPT is size-consistent order by order.

Lecture 95: Diagrammatic Representation of RSPT

A simple way of representing the terms appearing in the higher-order energies was invented by Feynman who showed that a specific mathematical expression arising out of the perturbative treatment can be quantitatively described by a unique diagram. The first diagrammatic representation was introduced by Feynman in 1949 for the perturbation expansion of the *Green's functions* of the wave equation in *quantum electrodynamics* (QED).[1] Application of the diagrammatic technique rapidly became diversified to the evaluation of different physical quantities at different levels of theory, and powerful rules emerged for a systematic application of the diagrammatic principle in each case. Here we discuss the diagrammatic representation of the Rayleigh-Schrödinger perturbation expansion for total energy calculated in the framework of the strictly *nonrelativistic* theory based on the Schrödinger equation.

In the diagrammatic technique used here, the bra vector $<\Psi_0^{(0)}|$ and the ket vector $|\Psi_0^{(0)}>$ corresponding to the Hartree-Fock ground state configuration are represented by lines each with an arrow pointing down. Bras and kets for all excited configurations, $<\Psi_i^{(0)}|$ and $|\Psi_i^{(0)}>$ (for $i \neq 0$) are represented by lines with arrows pointing up. Two lines meet at a point. Hence the interaction $\mathbf{H'}$ is represented by a dot. An incoming line is a bra vector and an outgoing line stands for a ket vector. The matrix elements of $\mathbf{H'}$ can be represented by the very simple diagrams given in Eq. (95.1):

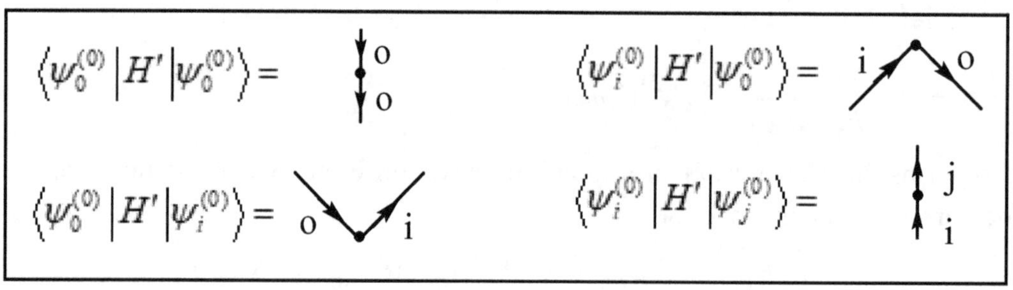

(95.1)

[1] These diagrams are in reality collections of world lines in 4-space (space-time). So there is always a correspondence between what happens to the particle and the diagram. Green's function is a measure of the probability amplitude for the particle to migrate from one space-time point to another. In a nonrelativistic theory, a similar interpretation can be given to the diagrams representing the perturbation corrections to the nonrelativistic Green's function. The energy diagrams can be rigorously derived from the diagrams for Green's functions. Therefore, each line may again be interpreted as a world line in space-time, or, depending upon the formulation of the particular diagrammatic treatment, as a world line in function space and time.

Each term in $E_0^{(n)}$ contains a product of n matrix elements of $\mathbf{H}'$. This permits $E_0^{(n)}$ to be systematically represented by diagrams that may be drawn according to three very simple rules.

<u>Rules for drawing energy diagrams:</u>

1) The n dots (interactions) are drawn in a vertical order.
   The vertical direction as such stands for the time axis.
2) All possible pairs of dots are connected by continuous lines shown by arrows.
   The whole diagram is comprehensively closed.
   Unlinked diagrams do not contribute.
3) All possible topologically distinct linked diagrams are drawn.
   Two diagrams are topologically distinct when there is no way to make them identical by distorting the connecting lines. That is, two distinct diagrams differ in the connectivities of the dots. Each distinct linked diagram corresponds to one type of terms in $E_0^{(n)}$ such that all the distinct diagrams together represent the whole *nth*-order correction. (95.2)

Figure 16.1 shows all topologically distinct linked diagrams up to fourth order.

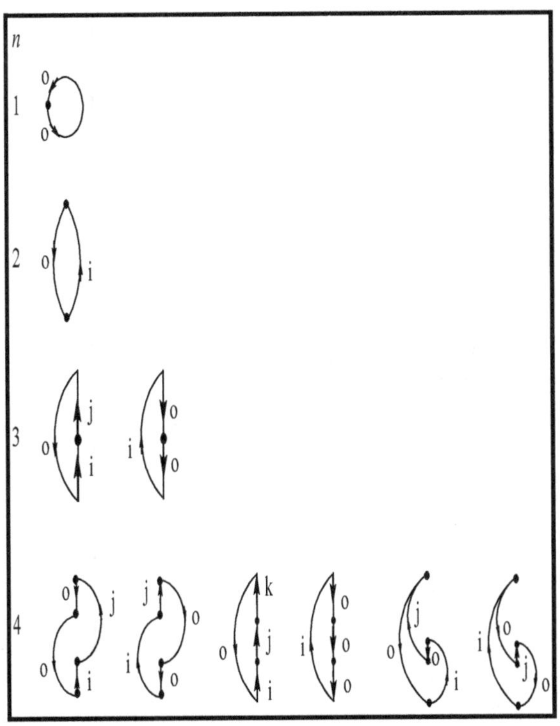

Figure 16.1 Topologically distinct, linked diagrams for first to fourth order corrections to ground state energy.

The diagrammatic technique is so useful because the rules for the evaluation of diagrams are exceedingly simple.

Rules for evaluating energy diagrams:

1) Each dot flanked by an incoming line and an outgoing line corresponds to a matrix element as given by Eq. (95.1). Remember that an incoming line stands for a bra vector and an outgoing line represents a ket vector. Matrix elements corresponding to all the *n* flanked dots in the diagram are to be multiplied together to form the numerator in the expression for the *nth*-order energy.
2) Each pair of adjacent dots contributes a factor like $(\sum E_0^{(0)} - \sum E_{\mathrm{i}}^{(0)})$ (95.3) to the denominator, the sums being over all the lines moving downwards and all those running upward which cross an imaginary horizontal line between the two dots.
3) Let the number of lines moving downward be *g* and the number of closed loops be *l*. Then the overall sign of the term is $(-1)^{g+l}$.
4) For each diagram an appropriate set of summations over the excited configurations (like sum over i, sum over j, etc.) must be included.

Diagrams in Figure 16.1 are easily evaluated by following these rules. This calculation is shown in Figure 16.2 for diagrams up to third order.

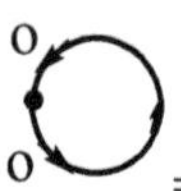

$$= <\Psi_0^{(0)}|\mathbf{H}'|\Psi_0^{(0)}> = E_0^{(1)}$$

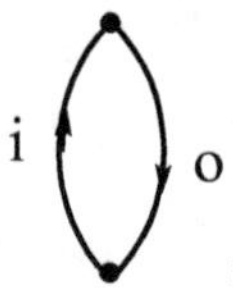

$$= \sum_i{}' \frac{|<\Psi_0^{(0)}|\mathbf{H}'|\Psi_i^{(0)}>|^2}{(E_0^{(0)} - E_i^{(0)})} = E_0^{(2)}$$

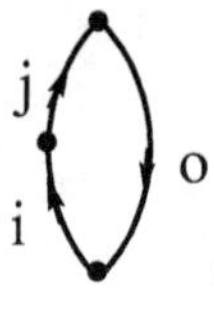

$$= \sum_i{}' \sum_j{}' \frac{<\Psi_0^{(0)}|\mathbf{H}'|\Psi_i^{(0)}><\Psi_i^{(0)}|\mathbf{H}'|\Psi_j^{(0)}><\Psi_j^{(0)}|\mathbf{H}'|\Psi_0^{(0)}>}{(E_0^{(0)} - E_i^{(0)})(E_0^{(0)} - E_j^{(0)})}$$

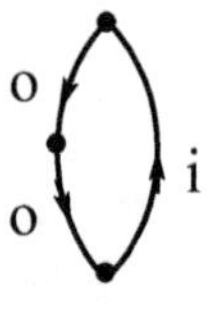

$$= -E_0^{(1)} \sum_i{}' \frac{|<\Psi_0^{(0)}|\mathbf{H}'|\Psi_i^{(0)}>|^2}{(E_0^{(0)} - E_i^{(0)})^2}$$

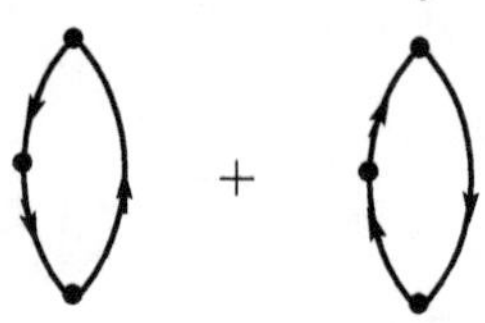

$$= E_0^{(3)}$$

Figure 16.2 Translation of the first-, second- and third- order diagrams in the previous figure.

The diagrams in Figure 16.2 represent contributions to the ground state energy. However, the technique is by no means limited to the ground state calculations. For calculating the contributions to the energy of any other state one has to only redefine the line moving downward as the vector for the specific zeroth-order state. Then the lines running upward represent all other zeroth-order states including the zeroth-order ground state. The summations are to be modified accordingly. So the technique described above can be extended to the calculation of $E_i^{(n)}$ for a general i without much difficulty.

Certain types of diagrams can be summed to infinite order. This is an extremely useful observation for it leads to a considerable reduction in the number of diagrams, and hence the number of terms involved at any specific order in the perturbation expansion is reduced by a great extent. It is also normally associated with a rich interpretation. The following example serves to illustrate these aspects of the infinite order summation.

Example 56: Summation to infinite order

Sum =

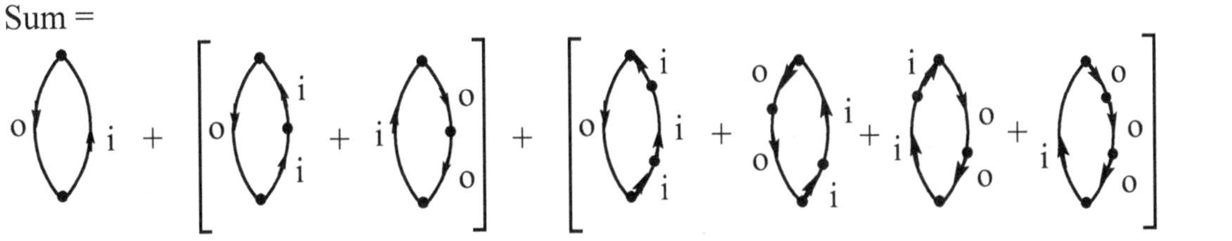

$+ \cdots$

(95.4)

First we show how to evaluate the sum. After translating the diagrams in the series we find

$$\text{Sum} = \sum_i{}' \frac{|\mathbf{H}'_{0i}|^2}{(E_0^{(0)} - E_i^{(0)})} + \sum_i{}' \frac{|\mathbf{H}'_{0i}|^2 (\mathbf{H}'_{ii} - \mathbf{H}'_{00})}{(E_0^{(0)} - E_i^{(0)})^2} + \sum_i{}' \frac{|\mathbf{H}'_{0i}|^2 (\mathbf{H}'_{ii} - \mathbf{H}'_{00})^2}{(E_0^{(0)} - E_i^{(0)})^3}$$

$$= \sum_i{}' \frac{|\mathbf{H}'_{0i}|^2}{(E_0^{(0)} - E_i^{(0)})}(1 + \theta + \theta^2 + \dots) \tag{95.5}$$

where

$$\theta = \frac{(\mathbf{H}'_{ii} - \mathbf{H}'_{00})}{(E_0^{(0)} - E_i^{(0)})}. \tag{95.6}$$

Further algebraic simplification yields

$$\text{Sum} = \sum_i{}' \frac{|\mathbf{H}'_{0i}|^2}{(E_0^{(0)} - E_i^{(0)})} \frac{1}{1-\theta}$$

$$= \sum_i{}' \frac{|\mathbf{H}'_{0i}|^2}{(E_0^{(0)} + \mathbf{H}'_{00}) - (E_i^{(0)} + \mathbf{H}'_{ii})}. \tag{95.7}$$

This is merely a second-order expression where each energy value appearing in the denominator has been shifted from the zeroth order to the first order. A similar shift occurs whenever the perturbed wave function requires normalization or it acquires an altered physical appearance when the perturbation is such that it causes the concerned particle to be coated by particles of other types. Basically the infinite order summation represents a renormalization process that accompanies the generation of a mean or an effective field, an effective mass, an effective charge, etc. For instance, such summations may be utilized to obtain perturbation corrections to the Hartree-Fock energy in a perturbed orbital approach.

Lecture 96: Goldstone Diagrams; The Linked-Cluster Theorem

Goldstone proposed a technique in which diagrams are drawn by one-electron state lines, more specifically, by hole and particle lines, with the instantaneous interaction r_{12}^{-1} being displayed by a dashed line. The Hartree-Fock ground state configuration is redefined as the vacuum state (the particle-hole vacuum), with an electron in an orbital (m, n, p, etc.) unoccupied in the Hartree-Fock ground state being called as a particle, and the state of an electron missing from an occupied orbital (α, β, γ, etc.) being interpreted as a hole state. When a particle is added to the (particle-hole) vacuum, a negatively charged ion is formed. If an electron escapes from the particle-hole vacuum, a positively charged hole corresponding to the positive ion is left behind. The particle lines move upward (forward in time) whereas the hole lines move downward. An electron moving backward in time looks like a positron moving forward. Similarly the absence of an electron moving upward appears as a hole moving downward.

In Goldstone diagrams a hole ket $|\alpha\rangle$ and a particle ket $|m\rangle$ are represented by out-going lines running downward and upward, respectively. Similarly a hole bra $\langle\alpha|$ and a particle bra $\langle m|$ are shown by incoming lines. The incoming lines stand for the annihilation of hole and particles whereas the outgoing lines indicate their creation in the redefined vacuum.

The particle-hole description and the use of the instantaneous electron-electron interaction results in the separation of matrix elements into parts representing the direct and exchange integrals of the r_{12}^{-1} operator. Goldstone diagrams for the first two matrix elements in Eq. (95.1) are

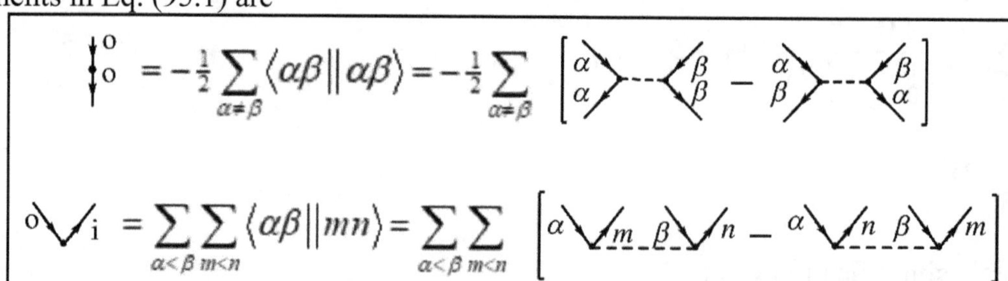

(96.1)

etc. The reader is urged to draw similar diagrams for the last two matrix elements in Eq. (95.1). Use caution in drawing the diagrams for the last (the fourth) matrix element. For a non-vanishing contribution, the states $|\Psi_i^{(0)}\rangle$ and $|\Psi_j^{(0)}\rangle$ must differ from each other either by a single excitation or by a double excitation. Also remember that the interaction dot in Eq. (95.1) is the many-body perturbation **H′**. The Goldstone diagrams corresponding to the diagrams in Figure 16.1 are given in Figure 16.3.

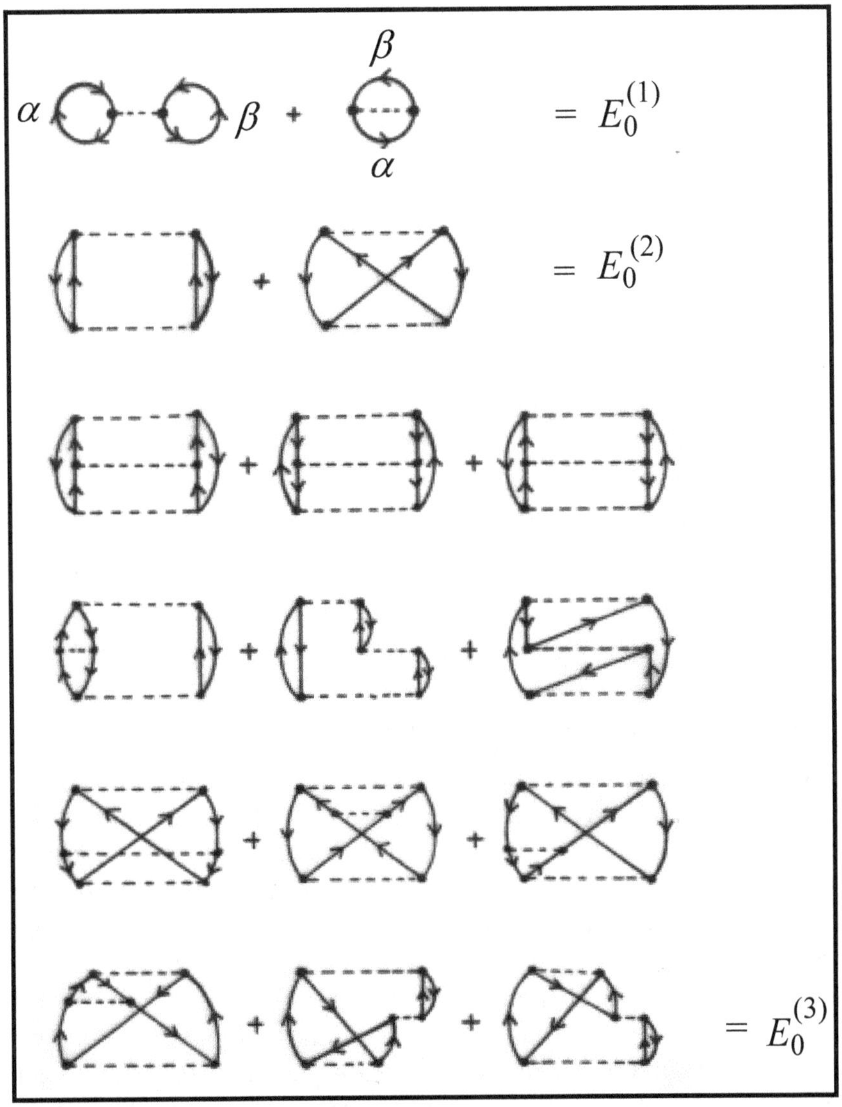

Figure 16.3 Goldstone diagrams for the perturbation expansion of the ground state energy through third order in interaction.

The following systematics are used to evaluate Goldstone diagrams.

Rules for evaluating Goldstone diagrams:

1) Each interaction line flanked by an incoming hole/particle line and an outgoing hole/particle line at each of the two ends contributes the multiplicative factor

$$< \text{incoming}_1\ \text{incoming}_2\ |\ r_{12}^{-1}|\ \text{outgoing}_1\ \text{outgoing}_2 >$$

to the numerator of each term.

2) There is a contribution of $\left(\sum \varepsilon_{\text{hole}} - \sum \varepsilon_{\text{particle}}\right)$ from each pair of adjacent interaction to the denominator. The sums run over the labels of all holes and particles which cross an imaginary horizontal line between the two interaction lines. (96.2)

3) There is an overall sign of $(-1)^{h+l}$ where h is the number of hole lines and l is the number of closed loops, and a sum over all relevant hole and particle indices. To avoid double-counting, check for a mirror plane perpendicular to the plane of the paper. Such a symmetry plane causes each distinct evaluation to occur twice. So its existence requires the multiplicative factor $\frac{1}{2}$.

4) The sum over spin orbitals of a closed-shell system equals 2^l times the sum over spatial orbitals.

We have already mentioned that Brueckner conjectured an order by order size-consistency for RSPT. The reason is that terms proportional to N^2 are represented by disjoint (unlinked) diagrams (diagrams consisting of more than one piece). Using a time-dependent formulation Goldstone showed that the unlinked diagrams never appear in the final results for the *nth*-order energy. Thus we get

Theorem 24. The linked cluster Theorem: For any $n > 0$, $E_0^{(n)}$ can be completely described by linked diagrams.
The theorem is a topological representation of size-consistency of RSPT.

Many-body techniques are best treated in the language of second quantization where the vacuum state is redefined, and the creation and destruction of both particles and holes are explicitly considered. The second quantization is in fact the realm of quantum field theory (QFT) where the electron field is quantized, and the adoption of the particle-hole picture leads to additional simplifications and a rich interpretation. An explicit formalism in QFT may take different appearances. Thus there are different QFT-based approaches to the investigation of many-body systems. To name a few, there are RSPT, other forms of MBPT, Green's function techniques, Equation of Motion methods, coupled-cluster analysis, pair formulations, kinematic formulations, etc. A detailed discussion of these approaches forms the subject matter of an advanced course in quantum chemistry. The theory becomes more sophisticated and the actual treatments more complicated by the introduction of the quantization of light. The QFT extended in this way is known as quantum electrodynamics (QED), the nonrelativistic version. The proper QED is presented in a relativistic framework. QED is the precise form of quantum theory. It is used in the study of the electromagnetic interaction, and it represents the greatest success of the idea of a quantum. It explains almost all effects that are purely electromagnetic in nature. It has successfully explained the results of sophisticated experiments involving particles and antiparticles as well as finer aspects of spectroscopy and scattering, and sometimes even predicted phenomena which have been verified later. It is the key to the understanding of a whole new world full of different particles and exotic interactions.

Texts Suggested for Further Reading

A. L. Fetter and J. D. Walecka, "Quantum Theory of Many-Particle Systems", (McGraw-Hill, New York, 1971).

P. Jørgensen and J. Simons, "Second Quantization Based Methods in Quantum Chemistry", (Academic Press, New York, 1981).

N. H. March, W. H. Young and S. Sampanthar, "The Many-Body Problem in Quantum Mechanics", (Cambridge University Press, Cambridge, 1967).

A. Szabo and N. S. Ostlund, "Modern Quantum Chemistry: Introduction to Advanced Electronic Structure Theory", (MacMillan, New York, 1982).

PROBLEMS

1. Take the (1s, 2s) basis for the helium atom. Calculate the correlation energy from the CI treatment and compare the result with the second-order correction to energy from RSPT.

2. Check the N-dependence of correlation energy for a collection of non-interacting (1s, 2s) helium atoms.

3. Construct a perturbative formalism where the unperturbed Hamiltonian is a sum of one-particle Hamiltonian operators

$$\mathbf{H}_0 = \sum_{i=1}^{N} \mathbf{h}(i),$$

and the perturbation is a sum of one-particle interaction operators

$$\mathbf{H}' = \sum_{i=1}^{N} \mathbf{v}(i).$$

Use your derivation to study the Hückel molecular orbitals of (a) linear and (b) cyclic polyenes.

APPENDIX I

FUNDAMENTAL PHYSICAL CONSTANTS AND PERIODIC TABLE

Table I.1 Fundamental physical constants.

Quantity	Symbol	Value	Units	Value in atomic units
Speed of light in vacuum	c	299 792 458	m s^{-1}	137.035 989
Permeability of vacuum	μ_0	$4\pi \times 10^{-7}$	N A^{-2}	$4\pi\alpha^2$
Permittivity of vacuum, $1/\mu_0 c^2$	ε_0	8.854 187 817	$10^{-12}\ \text{F m}^{-1}$	$1/4\pi$
Planck constant	h	6.626 075 5	$10^{-34}\ \text{J s}$	2π
in electron Volts, h/{e}		4.135 669 2	10^{-15} eV	
h / 2π	$\hbar$	1.054 572 66	$10^{-34}\ \text{J s}$	1
Elementary charge	e_0	1.602 177 33	10^{-19} C	1
Electron mass	m_e	9.109 389 7	10^{-31} kg	1
in a. m. u.		5.485 799 03	10^{-4} u	
in eV, $m_e c^2$/{e}		0.510 999 06	MeV	
Electron specific charge	$-e_0/m_e$	−1.758 819 62	$10^{11}\ \text{C kg}^{-1}$	−1
Compton wavelength, $h/m_e c$	λ_C	2.426 310 58	10^{-12} m	$2\pi\alpha$
Classical electron radius, $\alpha^2 a_0$	r_e	2.817 940 92	10^{-15} m	α^2
Bohr magneton, $e\hbar/2m_e$	μ_B	−9.274 015 4	$10^{-24}\ \text{J T}^{-1}$	−0.5
in electron Volts, μ_B/{e}		−5.788 382 63	$10^{-5}\ \text{eV T}^{-1}$	
in hertz, μ_B/h		−1.399 624 18	$10^{10}\ \text{Hz T}^{-1}$	
in wavenumbers, μ_B/hc		−46.686 437	$\text{m}^{-1}\ \text{T}^{-1}$	
in kelvins, μ_B/k_B		−0.671 709 9	K T^{-1}	
Proton mass	m_p	1.672 623 1	10^{-27} kg	1836.152 70
Proton-electron mass ratio	m_p/m_e	1836.152 701		1836.152 70
Nuclear magneton, $e\hbar/2m_p$	μ_N	5.050 786 6	$10^{-27}\ \text{J T}^{-1}$	$2.723\ 085\ 1 \times 10^{-4}$
in electron Volts, μ_N/{e}		3.152 451 66	$10^{-8}\ \text{eV T}^{-1}$	
in hertz, μ_N/h		7.622 591 4	Mhz T^{-1}	
in wavenumbers, μ_N/hc		2.542 622 81	$10^{-2}\ \text{m}^{-1}\ \text{T}^{-1}$	
in kelvins, μ_N/k_B		3.658 246	$10^{-4}\ \text{K T}^{-1}$	

Table I.1 Fundamental physical constants - continued.

Quantity	Symbol	Value	Units	Value in atomic units
Electron g factor, $2(1+a_e)$	g_e	2.002 319 304		2.002 319 3
Proton gyromagnetic ratio	γ_p	26752.212 8	10^4 s^{-1} T^{-1}	
	$\gamma_p/2\pi$	42.577 469	MHz T^{-1}	
Fine structure constant, $\mu_0 ce^2/2h$	α	7.297 353 08	10^{-3}	7.297 353 1 × 10^{-3}
Inverse fine structure constant	α^{-1}	137.035 989 5		137.035 989
Rydberg constant, $m_e c\alpha^2/2h$	R_∞	10973 731.534	m^{-1}	
in hertz, $R_\infty c$		3.289 841 949	10^{15} Hz	
in joules, $R_\infty hc$		2.179 874 1	10^{-18} J	0.5
in eV, $R_\infty hc/\{\mathbf{e}\}$		13.605 698 1	eV	
Bohr radius, $\alpha/4\pi R_\infty$	a_0	0.529 177 249	10^{-10} m	1
Hartree energy, $e^2/4\pi\varepsilon_0 a_0=2R_\infty hc$	E_h	4.359 748 2	10^{-18} J	1
in eV, E_h/e_0		27.211 396 1	eV	
Avogadro number	N_{Av}	6.022 136 7	10^{23} mol^{-1}	
Faraday constant, $N_{Av}e_0$	F	96 485.309	C mol^{-1}	N_{Av}
Molar gas constant	R	8.314 510	J mol^{-1} K^{-1}	3.166 829 × 10^{-6}
Boltzmann constant, R/N_{Av}	k_B	1.380 658	10^{-23} J K^{-1}	
in electron volts, k_B/e_0		8.617 385	10^{-5} eV K^{-1}	
in hertz, k_B/h		2.083 674	10^{10} Hz K^{-1}	
in wavenumbers, k_B/hc		69.503 87	m^{-1} K^{-1}	
Electron Volt, $(e_0/C)J=\{\mathbf{e}\}J$	eV	1.602 177 33	10^{-19} J	0.036 749 3
Unified atomic mass unit	u	1.660 540 2	10^{-27} kg	1822.888 53

$\pi = 3.141\ 592\ 653$

Table I.2 Factors for the conversion of energy from one unit to another.

Unit	kJ mol $^{-1}$	cm $^{-1}$	Hz	K	eV	Hartree (a.u.)
1 kJ /mol	1	83.593 46	2.506 069 $\times 10^{12}$	120.271 66	1.036 427 $\times 10^{-2}$	3.808 798 $\times 10^{-4}$
1 cm $^{-1}$	0.119 626	1	2.997 925 $\times 10^{10}$	1.438769	1.239 842 $\times 10^{-4}$	4.556 335 $\times 10^{-6}$
1 Hz	3.990 313 $\times 10^{-13}$	3.335 641 $\times 10^{-11}$	1	4.799 216 $\times 10^{-11}$	4.135 669 $\times 10^{-15}$	1.519 830 $\times 10^{-16}$
1 K	8.314 511 $\times 10^{-3}$	0.695 039	2.083 674 $\times 10^{10}$	1	8.617 385 $\times 10^{-5}$	3.166 829 $\times 10^{-6}$
1 eV	96.485 329	8 065.541	2.417 988 $\times 10^{14}$	11 604.45	1	3.674 931 $\times 10^{-2}$
1 a.u.	2.625 500 $\times 10^{3}$	219 474.63	6.579 684 $\times 10^{15}$	3.157 733 $\times 10^{5}$	27.211 396	1

Table I.3 Periodic table of the elements: Only atomic symbols have been given. Atomic numbers have been indicated in special cases. See text for the list of electronic configurations, atomic and ionic radii, ionization energies and electronegativities.

I A	II A	III B	IV B	V B	VI B	VII B	VIII B	I B	II B	III A	IV A	V A	VI A	VII A	VIII A
H															He
Li	Be									B	C	N	O	F	Ne
Na	Mg									Al	Si	P	S	Cl	Ar
K	Ca	Sc	Ti	V	Cr	Mn	Fe Co Ni	Cu	Zn	Ga	Ge	As	Se	Br	Kr
Rb	Sr	Y	Zr	Nb	Mo	Tc	Ru Rh Pd	Ag	Cd	In	Sn	Sb	Te	I	Xe
Cs	Ba	La •	Hf	Ta	W	Re	Os Ir Pt	Au	Hg	Tl	Pb	Bi	Po	At	Rn
Fr	Ra	Ac ••	Post Actinides •••												

Lanthanide Series •

58	59	60	61	62	63	64	65	66	67	68	69	70	71
Ce	Pr	Nd	Pm	Sm	Eu	Gd	Tb	Dy	Ho	Er	Tm	Yb	Lu

Actimide Series ••

90	91	92	93	94	95	96	97	98	99	100	101	102	103
Th	Pa	U	Np	Pu	Am	Cm	Bk	Cf	Es	Fm	Md	No	Lr

Post Actinides •••

IVB	VB	VIB	VIIB
104	105	106	107
Unq	Unp	Unh	Uns

APPENDIX II

USEFUL MATHEMATICAL RELATIONS

IIA. Derivative Operators

Cartesian coordinates

$V = \mathbf{i}V_x + \mathbf{j}V_y + \mathbf{k}V_z$ (vector in Cartesian coordinate system)

$\nabla = \mathbf{i}\frac{\partial}{\partial x} + \mathbf{j}\frac{\partial}{\partial y} + \mathbf{k}\frac{\partial}{\partial z}$ (derivative vector: gradient)

$\nabla^2 = \boldsymbol{\nabla}\bullet\boldsymbol{\nabla} = \frac{\partial^2}{\partial x^2} + \frac{\partial^2}{\partial y^2} + \frac{\partial^2}{\partial z^2}$ (Laplacian operator)

Spherical polar coordinates (Figure 3.6)

$x = r\sin\theta\cos\varphi$
$y = r\sin\theta\sin\varphi$
$z = r\cos\theta$

$\hat{\mathbf{r}} = \mathbf{i}\sin\theta\cos\varphi + \mathbf{j}\sin\theta\sin\varphi + \mathbf{k}\cos\theta$ (unit vector)
$\hat{\boldsymbol{\theta}} = \mathbf{i}\cos\theta\cos\varphi + \mathbf{j}\cos\theta\sin\varphi - \mathbf{k}\sin\theta$ (unit vector)
$\hat{\boldsymbol{\varphi}} = -\mathbf{i}\sin\varphi + \mathbf{j}\cos\varphi$ (unit vector)

$V = \hat{\mathbf{r}}V_r + \hat{\boldsymbol{\theta}}V_\theta + \hat{\boldsymbol{\varphi}}V_\varphi$ (vector in spherical polar coordinates)

$V_r = V\bullet\hat{\mathbf{r}};\quad V_\theta = V\bullet\hat{\boldsymbol{\theta}};\quad V_\varphi = V\bullet\hat{\boldsymbol{\varphi}}$ (vector components in SPC)

$\boldsymbol{\nabla} = \hat{\mathbf{r}}\frac{\partial}{\partial r} + \hat{\boldsymbol{\theta}}\frac{1}{r}\frac{\partial}{\partial\theta} + \hat{\boldsymbol{\varphi}}\frac{1}{r\sin\theta}\frac{\partial}{\partial\varphi}$ (gradient vector in SPC)

$$\nabla^2 = \boldsymbol{\nabla}\bullet\boldsymbol{\nabla} = \frac{1}{r^2}\frac{\partial}{\partial r}\left(r^2\frac{\partial}{\partial r}\right) + \frac{1}{r^2\sin\theta}\frac{\partial}{\partial\theta}\left(\sin\theta\frac{\partial}{\partial\theta}\right) + \frac{1}{r^2\sin^2\theta}\frac{\partial^2}{\partial\varphi^2}$$

(Laplacian in terms of SPC)

Orbital angular Momentum

Note the partial differentials

$$r=\sqrt{x^2+y^2+z^2} \Rightarrow \frac{\partial r}{\partial x}=\frac{x}{r}=\sin\theta\cos\varphi\,;\ \frac{\partial r}{\partial y}=\frac{y}{r}=\sin\theta\sin\varphi\,;\ \frac{\partial r}{\partial z}=\frac{z}{r}=\cos\theta\,.$$

$$\tan\theta=\frac{\sqrt{x^2+y^2}}{z} \Rightarrow \frac{\partial\theta}{\partial x}=\frac{\cos\theta\cos\varphi}{r};\ \frac{\partial\theta}{\partial y}=\frac{\cos\theta\sin\varphi}{r};\ \frac{\partial\theta}{\partial z}=-\frac{\sin\theta}{r}.$$

$$\tan\varphi=\frac{y}{x} \Rightarrow \frac{\partial\varphi}{\partial x}=-\frac{\sin\varphi}{r\sin\theta};\ \frac{\partial\varphi}{\partial y}=\frac{\cos\varphi}{r\sin\theta};\ \frac{\partial\varphi}{\partial z}=0\,.$$

Also, the partial derivatives with respect to x, y and z are expandable as

$$\frac{\partial}{\partial x}=\frac{\partial r}{\partial x}\frac{\partial}{\partial r}+\frac{\partial\theta}{\partial x}\frac{\partial}{\partial\theta}+\frac{\partial\varphi}{\partial x}\frac{\partial}{\partial\varphi},$$

$$\frac{\partial}{\partial y}=\frac{\partial r}{\partial y}\frac{\partial}{\partial r}+\frac{\partial\theta}{\partial y}\frac{\partial}{\partial\theta}+\frac{\partial\varphi}{\partial y}\frac{\partial}{\partial\varphi},$$

$$\frac{\partial}{\partial z}=\frac{\partial r}{\partial z}\frac{\partial}{\partial r}+\frac{\partial\theta}{\partial z}\frac{\partial}{\partial\theta}+\frac{\partial\varphi}{\partial z}\frac{\partial}{\partial\varphi}.$$

These expansions are to be used in the expressions for the Cartesian components of the orbital angular momentum

$$\begin{aligned} \mathbf{L}_x &= -i\hbar\left(y\frac{\partial}{\partial z}-z\frac{\partial}{\partial y}\right), \\ \mathbf{L}_y &= -i\hbar\left(z\frac{\partial}{\partial x}-x\frac{\partial}{\partial z}\right), \\ \mathbf{L}_z &= -i\hbar\left(x\frac{\partial}{\partial y}-y\frac{\partial}{\partial x}\right). \end{aligned} \tag{13.4}$$

The outcomes are expressions for the same components in terms of the angular variables,

$$\begin{aligned} \mathbf{L}_x &= i\hbar\left(\cot\theta\cos\varphi\frac{\partial}{\partial\varphi}+\sin\varphi\frac{\partial}{\partial\theta}\right); \\ \mathbf{L}_y &= i\hbar\left(\cot\theta\sin\varphi\frac{\partial}{\partial\varphi}-\cos\varphi\frac{\partial}{\partial\theta}\right); \\ \mathbf{L}_z &= -i\hbar\frac{\partial}{\partial\varphi}. \end{aligned} \tag{13.13}$$

IIB. Transformation of Classical Equations of Motion

Write

$$\frac{\partial L}{\partial \dot{q}_j} = \sum_i \frac{\partial L}{\partial \dot{x}_i}\frac{\partial \dot{x}_i}{\partial \dot{q}_j} = \sum_i \frac{\partial L}{\partial \dot{x}_i}\frac{\partial x_i}{\partial q_j}$$

where the last step results from Eq. (1.11a). Also,

$$\frac{\partial L}{\partial q_j} = \sum_i \frac{\partial L}{\partial x_i}\frac{\partial x_i}{\partial q_j}$$

If L has no explicit time dependence we get

$$\frac{d}{dt}\left(\frac{\partial L}{\partial \dot{q}_j}\right) - \frac{\partial L}{\partial q_j} = \sum_i \left[\frac{d}{dt}\left(\frac{\partial L}{\partial \dot{x}_i}\right) - \frac{\partial L}{\partial x_i}\right]\frac{\partial x_i}{\partial q_j} + \sum_i \frac{\partial L}{\partial \dot{x}_i}\frac{\partial \dot{x}_i}{\partial q_j}$$

where use has been made of Eq. (1.11b). As $\{\dot{x}\}$ are independent of {q}, we get Lagrange's equations in terms of the generalized coordinates

$$\frac{d}{dt}\left(\frac{\partial L}{\partial \dot{q}_j}\right) - \frac{\partial L}{\partial q_j} = 0 \tag{1.12}$$

in case Lagrange's equations in Cartesian coordinates (1.9) are valid. This leads to the great utility of the Lagarangian formulation.

IIC. Specific Complex Integrals

The function $\frac{e^{ix}}{x}$ has a simple pole at $x = 0$. The Cauchy principal value of integral $\int_{-\infty}^{\infty} \frac{e^{ix}}{x} dx$ is the sum of the two integrals along the real axis:

$$P\int_{-\infty}^{\infty} \frac{e^{ix}}{x} dx = \lim_{r\to 0+} \left[\int_{-\infty}^{-r} \frac{e^{ix}}{x} dx + \int_{r}^{\infty} \frac{e^{ix}}{x} dx\right].$$

To evaluate the Cauchy principal value, close the contour as shown in the figure below in the counter clockwise sense, and use the Residue theorem.

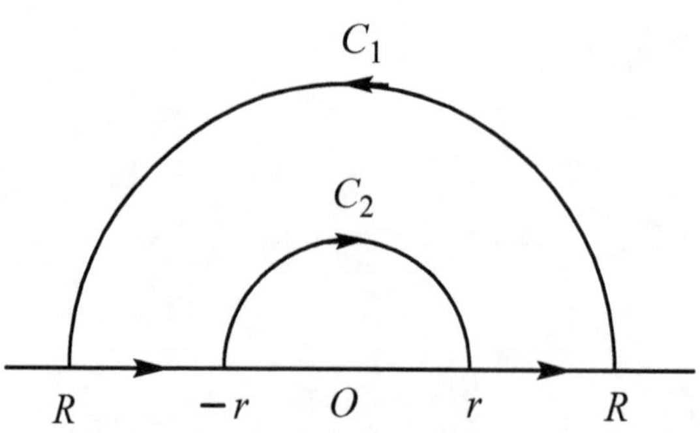

Theorem 25. The Residue theorem: $\oint \frac{e^{iz}}{z} dz = 2\pi i \times Residue$

Observe that this theorem implies, because the pole lies outside the closed path,

$$\lim_{R\to\infty} \int_{C_1} \frac{e^{iz}}{z} dz + P\int_{-\infty}^{\infty} \frac{e^{ix}}{x} dx + \lim_{r\to 0} \int_{C_2} \frac{e^{iz}}{z} dz = 0$$

where $Z = R\, e^{i\Phi}$ and $z = r\, e^{i\varphi}$ for contours C_1 and C_2, respectively.

The integral over the first contour C_1 is equal to 0 if we draw the horizontal line minutely above the x axis while letting R tend to ∞. Therefore,

$$P\int_{-\infty}^{\infty} \frac{e^{ix}}{x} dx = -\lim_{r\to 0} \int_{C_2} \frac{e^{iz}}{z} dz = i\pi.$$

Equating the imaginary parts on the two sides one obtains

$$P\int_{-\infty}^{\infty} \frac{\sin x}{x} dx = \pi.$$

As $x \to 0$, a series expansion of the integrand $\frac{\sin x}{x}$ becomes rapidly convergent, and it is observed to be not singular at $x = 0$. Therefore, one can actually write

$$\int_{-\infty}^{\infty} \frac{\sin x}{x} dx = \pi.$$

The same result may be visualized from the following steps which are quicker but requires an understanding of the Dirac delta function:

$$\int_{-\infty}^{\infty} \frac{\sin x}{x} dx = \int_{-\infty}^{\infty} \frac{\sin gx}{x} dx$$

$$= \lim_{g\to\infty} \int_{-\infty}^{\infty} \frac{\sin gx}{x} dx = \pi \int_{-\infty}^{\infty} \delta(x) dx = \pi.$$

The integral $\int_{-\infty}^{\infty} \frac{\sin^2 x}{x^2} dx$ may be evaluated in a similar way. We get

$$\int_{-\infty}^{\infty} \frac{\sin^2 x}{x^2} dx = \lim_{g\to\infty} \int_{-\infty}^{\infty} \frac{\sin x}{x} \frac{\sin gx}{x} dx$$

$$= \pi \int_{-\infty}^{\infty} \frac{\sin x}{x} \delta(x) dx$$

$$= \pi \int_{-\infty}^{\infty} \delta(x) dx = \pi.$$

This result has been used in Chapter 15.

APPENDIX III

SYMMETRY OF MOLECULES

IIIA. Importance of Symmetry

A molecule is said to be symmetric if a certain geometrical operation moves some or all of the atoms in it from their initial positions into the positions of another equilibrium configuration that is equivalent to the original configuration. A geometrical operation in the coordinate space can be described in terms of a linear transformation of Cartesian coordinates. The kinetic energy operator that involves ∇^2 remains invariant under a linear transformation, but the potential energy operator **V** changes with the geometrical operation. The symmetry operation **R** is a geometrical operation that leaves the potential energy **V** unchanged, that is,

$$\mathbf{RV} = \mathbf{VR}, \tag{A.1}$$

so that the Hamiltonian remains invariant. Otherwise the product configuration would not be equivalent to the initial one. Thus the symmetry of the Hamiltonian is the symmetry of the potential, and one may write

$$\mathbf{RH} = \mathbf{HR}. \tag{A.2}$$

This indicates that **H** and **R** can have common eigenfunctions. The nondegenerate eigenfunction of **H** must be an eigenfunction of **R**, and in the degenerate case the wave functions can be chosen in a specific linear combination form such that they also become eigenfunctions of **R**:

$$\mathbf{H}\psi_i = E_i\psi_i \Rightarrow \mathbf{R}\psi_i = r_i\psi_i. \tag{A.3}$$

This implies that the wave functions must be in some way symmetrically disposed in space.

Since the wave functions are symmetrical, it is possible to point out which of the integrals

$$\mathrm{O}_{ij} = \int dv \;\; \psi_i{}^*\mathbf{O}\psi_j \tag{A.4}$$

would be zero. The other integrals can have nonzero values. If it turns out that some of the other integrals are actually zero, there must be some hidden symmetry in the system.

In quantum theory, we always deal with expectation values and transition integrals. Hence the importance of symmetry can be easily realized.

Let the operator **O** be the Hamiltonian **H** of the system. Then the eigenvalue is the energy. When two or more eigenvalues are equal, the corresponding wave functions are called degenerate. The effect of a geometrical operation is to mix the degenerate functions. The degenerate wave functions must transform together under each symmetry operation, that is, they must change in a symmetrical way together and not individually. Thus degeneracy is best understood in terms of symmetry. Two states differing in symmetry can be only accidentally degenerate. In practice, however, an accidental degeneracy indicates the existence of a hidden symmetry. Although the discussion presented here is based on the Schrödinger wave functions (or the eigenfunctions of the Hamiltonian operator), it applies equally well to all other hermitean operators which commute with the Hamiltonian.

The mathematical theory that is used to quantitatively treat the features of symmetry is known as the theory of groups.

IIIB. Groups

A group ***G*** is a set of elements (E, A, B, C ...) which satisfy the four axioms:

i. The "product" of any two elements (including the self-product) must be an element of the set, that is, AB = C, while A, B and C all belong to ***G***, etc.
ii. One element is identity that commutes with all elements and leave them unchanged under "multiplication":
EA = A = AE for all elements A. The element E is called identity.
iii. The associative law of "multiplication" holds:
A (BC) = (AB) C etc.
iv. The inverse of every element is another element of the set:
Let AB = E. Then $A^{-1}AB = EB = B$ and $A^{-1}E = A^{-1}$ so that $B = A^{-1}$. One may similarly show $A = B^{-1}$. The element identity is its own inverse as EE = E.

The number of elements in a group is called the order (*h*) of the group (***G***).

In this connection we mention the important result that *the reciprocal of the product of elements is the product of the reciprocals in reverse order*:

$$(ABC...)^{-1} = ...C^{-1} B^{-1} A^{-1}. \tag{A.5}$$

This result can be easily established. Write AB...C = D. Using the associative law of multiplication we get

$$DC^{-1}...B^{-1}A^{-1} = AB...(CC^{-1})...B^{-1}A^{-1} = AB...E...B^{-1}A^{-1} = ... = ABEB^{-1}A^{-1}$$
$$= ABB^{-1}A^{-1} = AEA^{-1} = AA^{-1} = E.$$

Thus

$$D^{-1} = D^{-1}E = D^{-1} DC^{-1}...B^{-1}A^{-1} = EC^{-1}...B^{-1}A^{-1} = C^{-1}...B^{-1}A^{-1}$$

So far the exact nature of the "multiplication" has not been specified. It is understood to be a specific operation that combines the elements with each other in a certain way. This is clarified by the following examples.

IIIC. Examples of Group

Example 57: All real integers, positive and negative, and inclusive of zero, form a group of infinite order ($h = \infty$) under *arithmetic addition* that stands for "multiplication". Here the number zero takes the place of identity.

$$\otimes \equiv + \text{ (Summation)}$$
$$5 + 3 = 8 \text{ (Axiom 1)}$$
$$4 + 0 = 4 = 0 + 4 \quad \text{(axiom 2; identity is the number 0)}$$
$$-1 + (2 + 6) = 7 = (-1 + 2) + 6 \quad \text{(axiom 3)}$$
$$9 + (-9) = 0 = (-9) + 9 \quad \text{(axiom 4)}$$

See that 0 + 0 = 0: the identity is its own inverse.

Example 58: All real numbers, positive or negative, and excluding zero, form a group of infinite order ($h = \infty$) under *arithmetic multiplication* with 1 taking the place of identity.

$$\otimes \equiv \times \text{ (Multiplication)}$$
$$E \equiv 1$$

The number 0 is excluded as division by zero is undefined.[1]

Example 59: A finite group G_2 of order 2 can be represented by (E, A) with the multiplication table:

G_2	E	A
E	E	A
A	A	E

Example 60: A finite group G_3 of order 3 can be represented by (E, A, B) with only one multiplication table:

G_3	E	A	B
E	E	A	B
A	A	B	E
B	B	E	A

Example 61: There are two possible multiplication tables for the group G_4 of order 4:

G_4	E	A	B	C
E	E	A	B	C
A	A	B	C	E
B	B	C	E	A
C	C	E	A	B

G_4'	E	A	B	C
E	E	A	B	C
A	A	E	C	B
B	B	C	E	A
C	C	B	A	E

[1] Division of a number by zero is undefined in the following way. Division of a nonzero quantity x by zero is undefined as no solution can be found for x. If we write $\frac{x}{0} = y$ then x = y.0 and no number y can be found such that x ≠ 0. Division of zero by zero is undefined as there is every taker. For example, if $\frac{0}{0} = y$, then 0 = y.0 that is satisfied by every number y, rational or irrational, real or imaginary or even complex, including zero.

Examples 62-63: The groups G_5 and G_6 can have only one and only two multiplication tables, respectively. All these tables are collected under the heading Table A.1.

Table A.1 Multiplication tables for groups of order 5 and 6.

G_5	E	A	B	C	D
E	E	A	B	C	D
A	A	B	C	D	E
B	B	C	D	E	A
C	C	D	E	A	B
D	D	E	A	B	C

G_6	E	A	B	C	D	F
E	E	A	B	C	D	F
A	A	B	C	D	F	E
B	B	C	D	F	E	A
C	C	D	F	E	A	B
D	D	F	E	A	B	C
F	F	E	A	B	C	D

G_6'	E	A	B	C	D	F
E	E	A	B	C	D	F
A	A	E	D	F	B	C
B	B	F	E	D	C	A
C	C	D	F	E	A	B
D	D	C	A	B	F	E
F	F	B	C	A	E	D

We now state without proof the *rearrangement theorem* that may be used to construct the group multiplication table: *Each row and each column in the group multiplication table must contain every element once and only once.* Thus no two columns or rows can be identical.

The elements and the group operations may be chosen in different ways so as to obtain different groups like the groups of numbers, the groups of matrices, the groups of geometric operations, etc. If two different groups of the same order have similar multiplication tables, the elements of the two sets have a one-to-one correspondence and the two groups are said to be *isomorphic*. It is quite possible for two groups of different orders to have a many-to-one correspondence among their elements. These groups are homomorphic. Isomorphism is a special case of homomorphism.

Example 64: The set of elements ($X, X^2, X^3, \ldots, X^h \equiv E$) forms a *cyclic group* of order *h*. One may consult the following examples:

- G_2 (as $A^2 = E$);
- G_3 (as $A^2 = B$ and $A^3 = E$);
- G_4 (as $A^2 = B$, $A^3 = C$, $A^4 = E$);
- G_5 (as $A^2 = B$, $A^3 = C$, $A^4 = D$, $A^5 = E$); and
- G_6 (as $A^2 = B$, $A^3 = C$, $A^4 = D$, $A^5 = F$, $A^6 = E$).

A group is called abelian (in honor of Abel who invented Group theory) when every element commutes with all other elements, that is, $[A, B] = 0$ for all elements A and B belonging to $\boldsymbol{G}$.

Example 65: All cyclic groups are abelian.

IIID. Subgroups and Classes

A *subgroup* ***S*** is a group within a group ***G***. The order of the subgroup (*g*) must be an integral divisor of the order *h* of group ***G***. To give an example, the set of elements (E, D, F) is a subgroup of the group G_6'. See the multiplication table in Table A.1.

If X is an element of group ***G***, X^{-1} must be another element of the set (axiom 4). If A is still another element of ***G***, AX is also an element of the same group (by axiom 1) and thus

$$B = X^{-1}AX \tag{A.6}$$

must be an element of ***G*** (again by axiom 1). We say that B is the *similarity transform* of A. In turn, A is the similarity transform of B since Eq. (A.6) implies

$$A = XBX^{-1} \equiv Y^{-1}BY \tag{A.7}$$

where $Y = X^{-1}$ is an element of group ***G***. The elements A and B are said to be *conjugates* of each other. Each element of a group is conjugated with itself, since

$$A^{-1}AA = EA = A$$

as well as

$$E^{-1}AE = EAE = EA = A.$$

We have already shown that if B is conjugate of A, A is conjugate of B. Similarly one can prove that if A is conjugate of both B and C, the latter two elements are conjugates of each other.

A complete set of mutually conjugate elements form a *class* (*C*). For instance, the group G_6' has three classes in it:

(i) $X^{-1}EX = E$ (for X = E, A, B, C, D, E, F)
$C_1 = E$ (order 1)
The identity is self-conjugate, and it forms a class by itself.

(ii) $X^{-1}DX = D$ (for X = E, D, F); F (for X = A, B, C)
$X^{-1}FX = F$ (for X = E, D, F); D (for X = A, B, C)
$C_2 = (D, F)$ (order 2)

(iii) $X^{-1}AX = A$ (for X = E, A); B (for X = B, F); C (for X = C, D)
$C_3 = (A, B, C)$ (order 3)

The order of any class must be a factor of the order of the group.

The importance of a class may be described by a consideration of geometry. Let the elements be chosen as geometrical operations. *Geometrically equivalent operations together form a class*. This has important consequences which will be discussed in due course.

In an abelian group every element is its own conjugate, hence every element forms a class by itself.

IIIE. Symmetry Elements, Operations and Point Groups

A geometrical entity such as a point, a line or a plane with respect to which a geometrical operation can be carried out to bring the system into an equivalent configuration is called a *symmetry element*. The geometrical operation is known as a *symmetry operation*.

Molecular symmetry is specified by four types of symmetry elements – plane, center of inversion, proper axis and improper axis – and the corresponding operations. These elements, the corresponding operations and their symbols are collated in Table A.2. The operations form the so-called *point group* of the molecule. The point group can be found systematically by following a handful of relatively simple steps. These steps are discussed next.

The point group should not be confused with the *symmetry group* which consists of all distinct configurations obtainable from the permutation of the position of equivalent atoms. For example, the point group of ethane is D_{3d} in the staggered form and D_{3h} in the eclipsed conformation. The order of the point group is 12 in both cases. The symmetry group is of order $6! \times 2! = 1440$ corresponding to six equivalent hydrogen atoms and two equivalent carbon atoms.

Table A.2 Symmetry elements and operations for specifying molecular symmetry.

Elements	Symbol	Operations	Symbol
Plane	σ	Reflection in the plane	**σ**
Centre of inversion	i	Inversion through the center	**i**
Proper axis	C	Rotation about the axis	**C**
Improper axis	S	Rotation about the axis followed by reflection in a plane perpendicular to the axis	**S**

If one also combines the inversion operation with all the elements of the symmetry group one obtains what is generally known as the *full permutation inversion* (FPI) *group*. The symmetry group is a subgroup of the FPI group. Some of the permutations and permutation-inversions may be discounted on the energetic ground. A progressive elimination of the higher-energy permutations and permutation-inversions ultimately yields the point group. In a flexible molecule (fluxional molecule) some of the permutations and permutation-inversions outside the point group are actually operations of low energy. The latter can be taken into account to form the *non-rigid molecular* (NRM) *groups*. The NRM groups are necessarily larger than the point group though they may not be as large as the corresponding symmetry group or the FPI group. Normally the point group is a subgroup of the NRM group that is in turn a subgroup of the symmetry group or the FPI group.

Point groups arise when there is no translational symmetry in the system. The features of symmetry in a single molecule are described by the point group. There is translational symmetry in a crystal. So crystal structures are discussed in terms of the so-called *Space groups*.

IIIF. Determination of Point Groups

I. A molecule that has no symmetry element has the point group C_1 that consists of only the identity operation **E**. If the molecule has only a plane of symmetry, its point group is $\boldsymbol{C_s} \equiv$ (**E**,**σ**). If it has only a center of inversion, its point group is $\boldsymbol{C_i} \equiv$ (**E**, **i**).

II. If a rotation through the smallest nonzero angle $\theta = 2\pi/n$ around a specific axis brings about an equivalent equilibrium configuration, then the n-fold rotation restores the original configuration. This axis of rotation is called an n-fold rotational axis C_n and the n-fold rotation around this axis generates the identical configuration: $\mathbf{C_n^n = E}$. A molecule that has only one n-fold rotational axis of symmetry and no other symmetry element has the point group $\boldsymbol{C_n} \equiv (\mathbf{E = C_n^n, C_n, C_n^2, C_n^3 \ldots C_n^{n-1}})$.

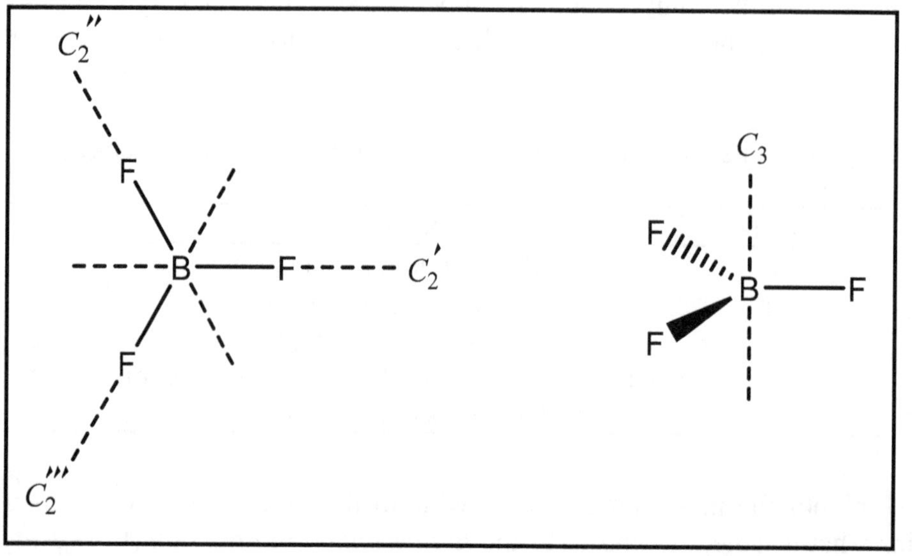

Figure A.1 Rotational axes of symmetry in BF_3: The C_3 axis is the principal axis. There are three C_2 axes each perpendicular to the principal axis. The proper rotational subgroup is $\boldsymbol{D_3}$.

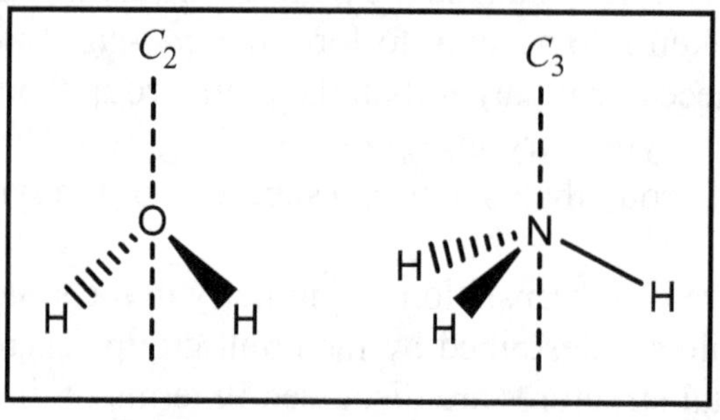

Figure A.2. Rotational axes of symmetry in water and in ammonia: The proper rotational subgroups are $\boldsymbol{C_2}$ and $\boldsymbol{C_3}$ respectively.

III. The $\boldsymbol{C_n}$ groups are *rotational groups*. They belong to the category of cyclic groups. In some cases the molecule has more than one rotational axis of symmetry. For example, the molecule BF_3 in Figure A.1 has one C_3 axis and three C_2 axes. The principal axis is the axis of the highest-fold symmetry. The C_3 axis is the principal axis in BF_3. When the molecule has an n-fold rotational axis as the principal axis of rotation and additional symmetry elements except another rotational axis, then the *proper rotational subgroup* is $\boldsymbol{C_n}$ and the point group is a higher group of this type. See Figure A.2. In this case the additional group elements are non-rotational operations other than the rotations around the principal axis.

IV. In case there are n number of C_2 axes of which each is perpendicular to the principal axis, the point group is $\boldsymbol{D_n}$. If there are more symmetry elements (as in BF_3 that has several planes of symmetry), the *proper rotational subgroup* is $\boldsymbol{D_n}$ and the point group is of the type $\boldsymbol{D_n}$?. The proper rotational subgroup for BF_3 is $\boldsymbol{D_3}$. See Figure A.1. Similarly the proper rotational subgroup is $\boldsymbol{C_2}$ for water, $\boldsymbol{C_3}$ for ammonia (Figure A.2), $\boldsymbol{D_3}$ for the ethane in both eclipsed and staggered forms (Figure A.3), $\boldsymbol{D_4}$ for $[Cu(CN)_4]^{-2}$, $\boldsymbol{D_5}$ for ferrocene and $\boldsymbol{D_6}$ for benzene (Figure A.4). *The proper rotational subgroup is representative of the bodily rotations of a species.*

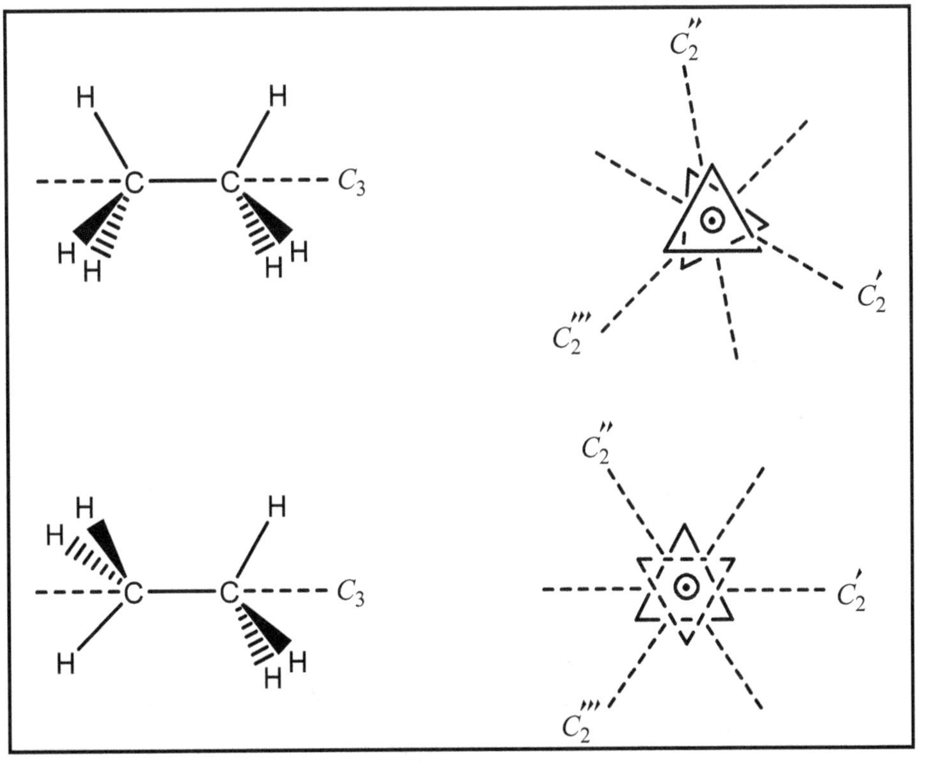

Figure A.3 Rotational axes of symmetry in eclipsed and staggered configurations of the ethane molecule: The proper rotational subgroup is D_3 in both cases.

V. A symmetry plane is called vertical (σ_v) when it contains the principal axis and horizontal (σ_h) when the principal axis is perpendicular to it.

If the order of the principal axis is even (n = 2p), some of the vertical planes go through the (2p) atoms forming the n corners of a regular polygon and some others pass through the midpoints between these corners. These two types of symmetry planes are distinguished by calling the former ones as diagonal planes (σ_d) because they connect the diagonally opposite atoms and simply naming the latter ones as vertical planes (σ_v).

For an odd n, the term diagonal plane is used when the point group is of type $\boldsymbol{D_n}$, in fact it is $\boldsymbol{D_{nd}}$, and the planes are called vertical when the proper rotational subgroup is only $\boldsymbol{C_n}$, the point group being $\boldsymbol{C_{nv}}$.

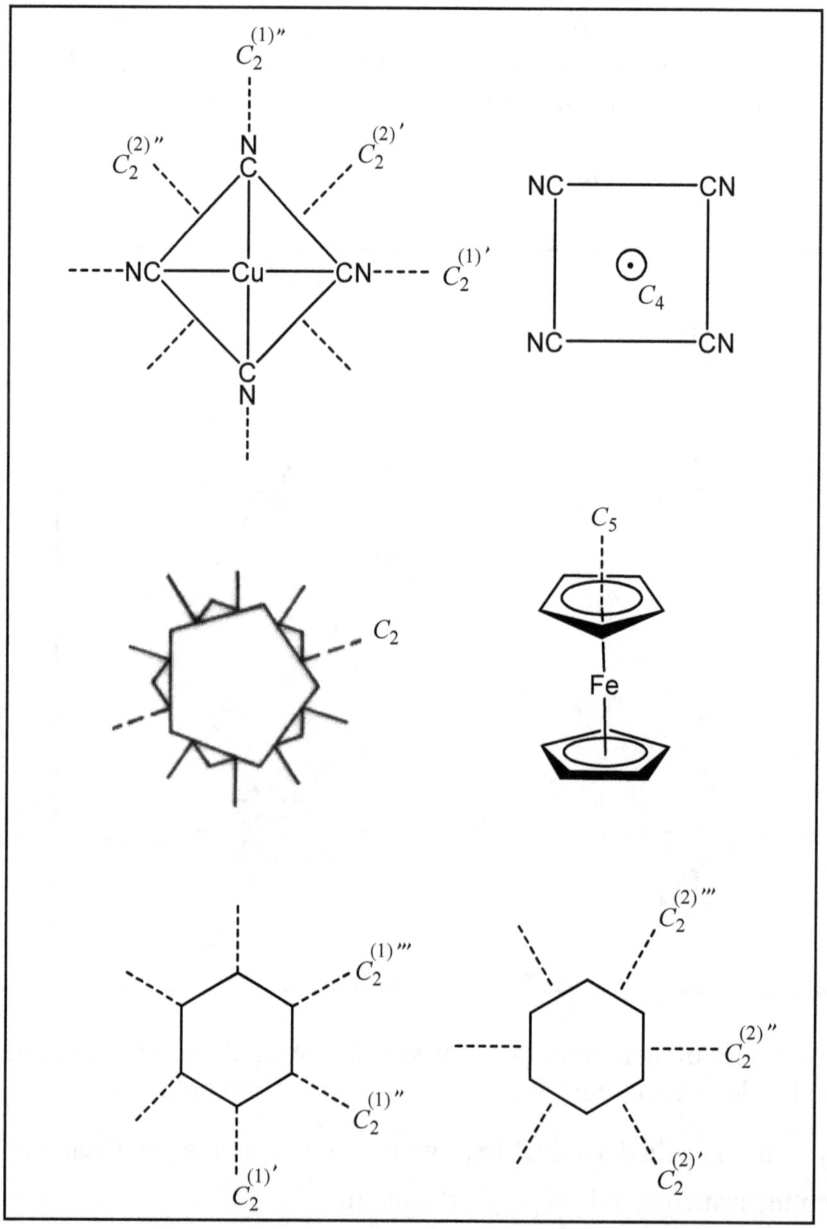

Figure A.4 Rotational axes of symmetry in $[Cu(CN)_4]^{-2}$, ferrocene and benzene.

VI. When the proper rotational subgroup is C_n and there are n vertical planes but no horizontal plane of symmetry, the point group is C_{nv}. The point group of water is C_{2v} and that of ammonia is C_{3v} as illustrated in Figure A.5. When the proper rotational subgroup is C_n and there is a horizontal plane of symmetry but no vertical planes, the point group is C_{nh}. Two examples of molecules belonging to the point group C_{3h} are given in Figure A.6. If there are n vertical planes as well as the horizontal plane, there necessarily exist n C_2 axes each perpendicular to the principal axis.

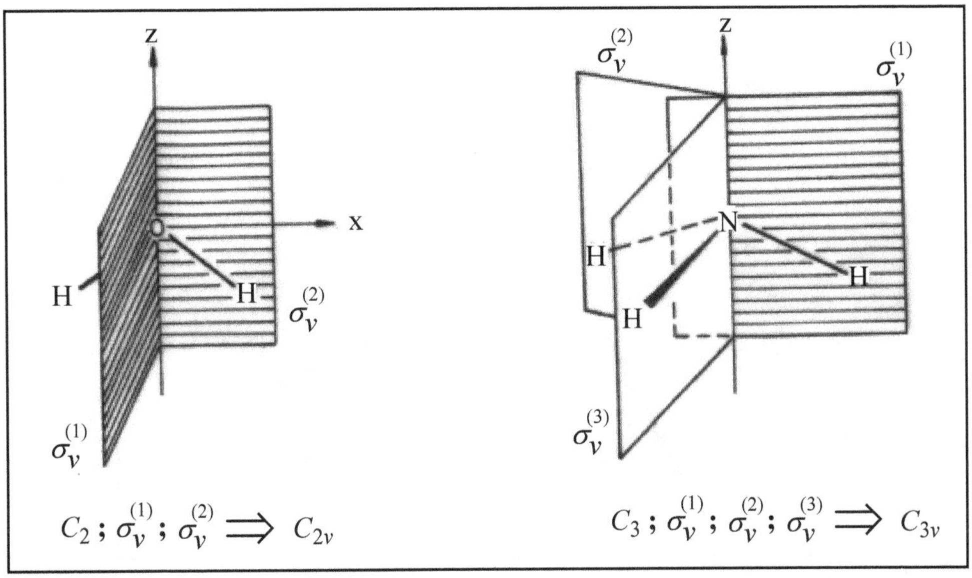

Figure A.5 Determination of the point groups of water and ammonia.

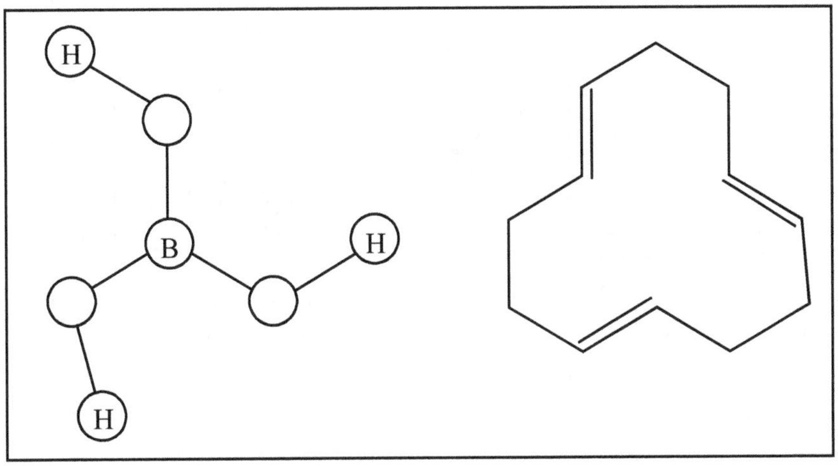

Figure A.6 Two examples of molecules belonging to the point group C_{3h}.

VII. When the proper rotational subgroup is $\boldsymbol{D_n}$ and there is a horizontal plane of symmetry, the point group is $\boldsymbol{D_{nh}}$. The existence of the horizontal plane and the C_2 axes each lying in the same plane ensures existence of vertical and diagonal planes. There is also a center of inversion when n is even. However, if there are vertical or diagonal planes but no horizontal plane of symmetry, the point group is $\boldsymbol{D_{nd}}$. In this case the molecule has a center of inversion. The point group of BF_3 is $\boldsymbol{D_{3h}}$. Similarly ethane in its eclipsed configuration has the point group $\boldsymbol{D_{3h}}$. These two cases have been illustrated in Figures A.7 and A.8 respectively. The staggered configuration of ethane corresponds to the point group $\boldsymbol{D_{3d}}$. See Figure A.9. Figures A.10, A.11 and A.12 show that the point groups for $[Cu(CN)_4]^{-2}$, ferrocene and benzene are $\boldsymbol{D_{4h}}$, $\boldsymbol{D_{5d}}$ and $\boldsymbol{D_{6h}}$ respectively.

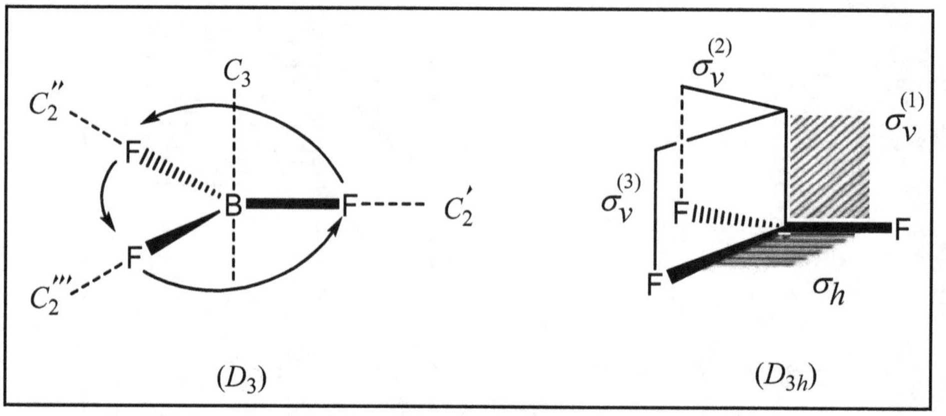

Figure A.7 Determination of the point group of BF_3.

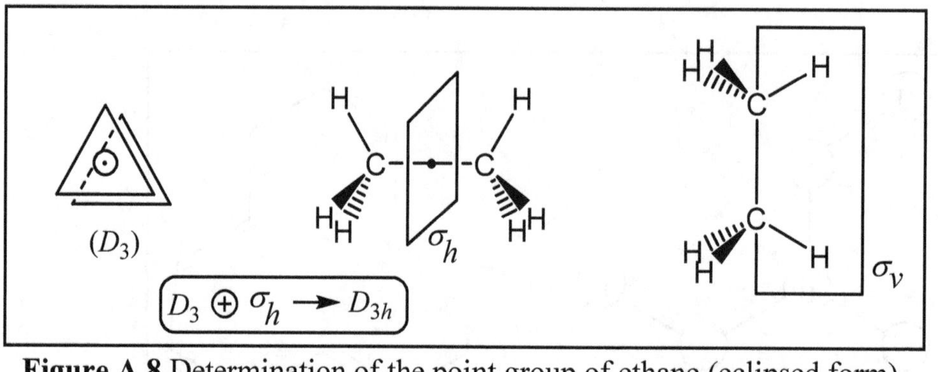

Figure A.8 Determination of the point group of ethane (eclipsed form).

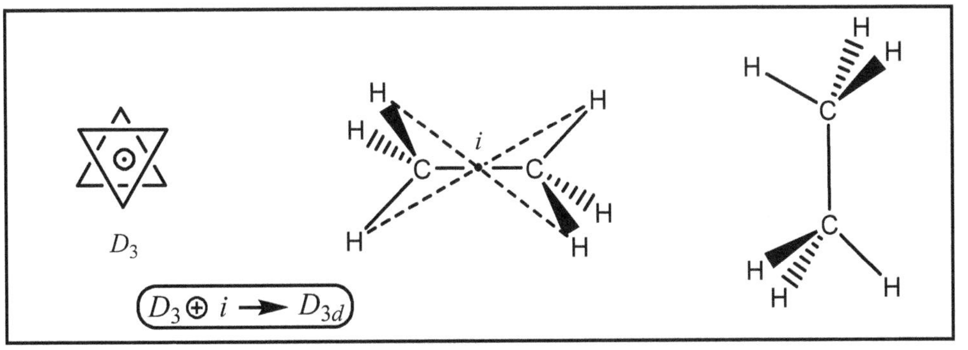

Figure A.9 Determination of the point group of ethane (staggered form).

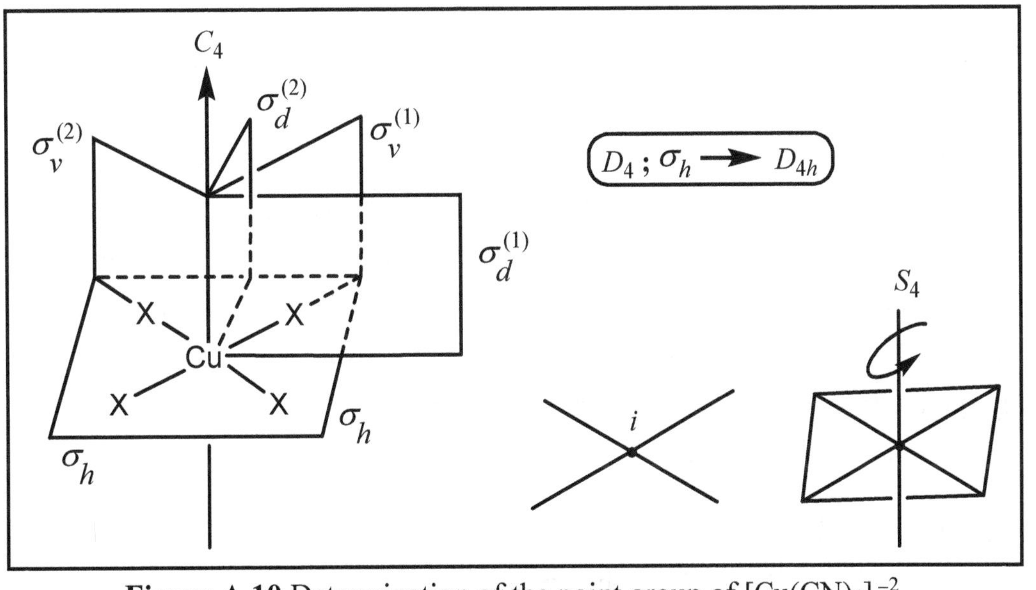

Figure A.10 Determination of the point group of $[Cu(CN)_4]^{-2}$.

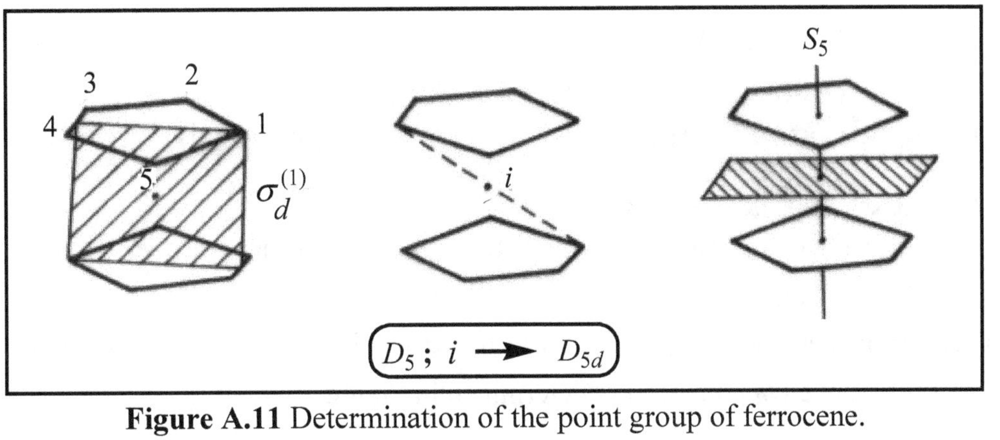

Figure A.11 Determination of the point group of ferrocene.

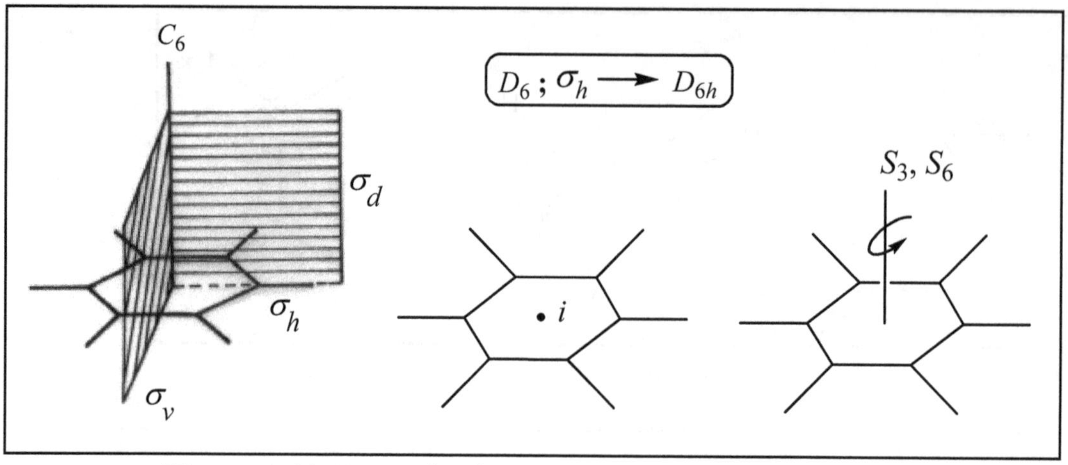

Figure A.12. Determination of the point group of benzene.

VIII. Sometimes the highest order axis is an improper axis of rotation (type S_{2n}). Note that $\mathbf{S_{2n}}^2 = \mathbf{C_n}$, $\mathbf{S_{2n}}^4 = \mathbf{C_n}^2$, etc., and $\mathbf{S_{2n}}^{2n} = \mathbf{C_n}^n = \mathbf{E}$. Furthermore, $\mathbf{S_{2n}}^n = \mathbf{i}$ if n is odd and $\mathbf{S_{2n}}^n = \mathbf{C_2}$ when n is even. In case the molecule possesses no other symmetry elements except S_{2n}, C_n, etc., including either C_2 or i, then the point group is $\boldsymbol{S_{2n}}$. Thus one obtains point groups like $\boldsymbol{S_4}$, $\boldsymbol{S_6}$, $\boldsymbol{S_8}$, etc. The point group $\boldsymbol{S_2}$ is identical with $\boldsymbol{C_i}$.

IX. Highly symmetric molecules have cubic groups as point groups.

Example 66: Cubic groups, such as

$\boldsymbol{T} = \{\mathbf{E}, 4(\mathbf{C_3},\mathbf{C_3}^2), 3\mathbf{C_2}\}$

$\boldsymbol{T_h} = T \otimes (\mathbf{E}, \mathbf{i}) = \{\mathbf{E}, 4(\mathbf{C_3},\mathbf{C_3}^2), 3\mathbf{C_2}, \mathbf{i}, 4(\mathbf{S_6},\mathbf{S_6}^5), 3\boldsymbol{\sigma}_\mathbf{h}\}$

$\boldsymbol{T_d} = \{\mathbf{E}, 4(\mathbf{C_3},\mathbf{C_3}^2), 3\mathbf{C_2}, 3(\mathbf{S_4},\mathbf{S_4}^3), 6\boldsymbol{\sigma}_\mathbf{d}\}$ (Tetrahedral)

$\boldsymbol{O} = \{\mathbf{E}, 3(\mathbf{C_4},\mathbf{C_4}^3), 3\mathbf{C_4}^2 \equiv 3\mathbf{C_2}, 4(\mathbf{C_3},\mathbf{C_3}^2), 6\mathbf{C_2}\}$

$\boldsymbol{O_h} = \boldsymbol{O} \otimes (\mathbf{E}, \boldsymbol{\sigma}_\mathbf{h}) = \{\mathbf{E}, 4(\mathbf{C_3},\mathbf{C_3}^2), 6\mathbf{C_2}, 3(\mathbf{C_4},\mathbf{C_4}^3), 3\mathbf{C_4}^2 \equiv 3\mathbf{C_2};$
$\mathbf{i}, 4(\mathbf{S_6}, \mathbf{S_6}^5), 6\boldsymbol{\sigma}_\mathbf{d}, 3(\mathbf{S_4}, \mathbf{S_4}^3), 3\boldsymbol{\sigma}_\mathbf{h}\}$ (Octahedral)

$\boldsymbol{I_h} = \{\mathbf{E}, 12(\mathbf{C_5},\mathbf{C_5}^2), 10(\mathbf{C_3},\mathbf{C_3}^2), 15\mathbf{C_2};$
$\mathbf{i}, 6(\mathbf{S_{10}},\mathbf{S_{10}}^9), 6(\mathbf{S_{10}}^3,\mathbf{S_{10}}^7), 10(\mathbf{S_6},\mathbf{S_6}^5), 15\boldsymbol{\sigma}_\mathbf{h}\}$ (Icosahedral).

The point group of methane is $\boldsymbol{T_d}$. This is illustrated in Figure A.13. Both SF_6 and $[CoCl_6]^{-4}$ complex ion have $\boldsymbol{O_h}$ point group, determined in Figure A.14.

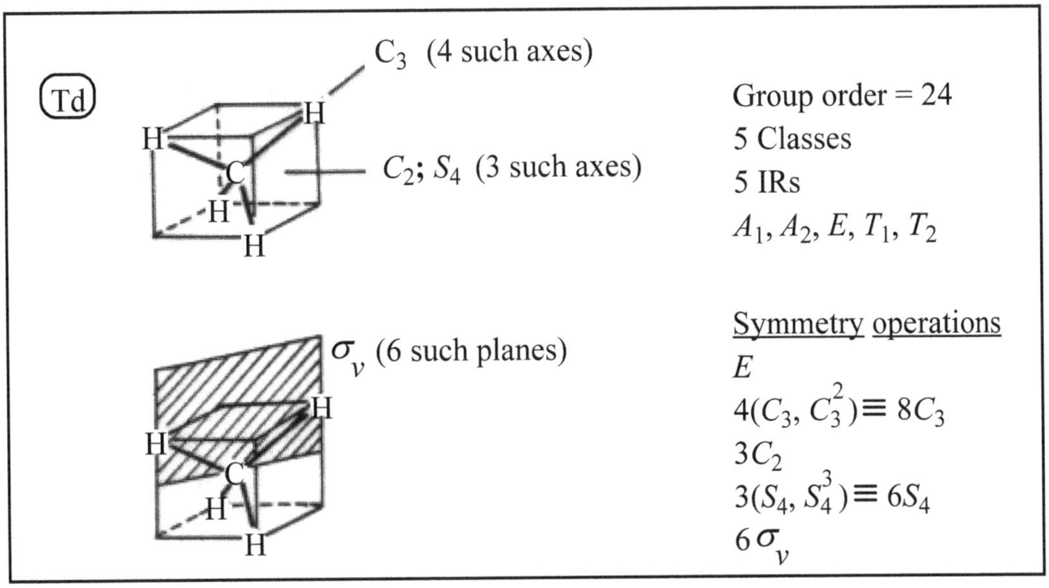

Figure A.13. The tetrahedral point group (***T_d***) for methane.

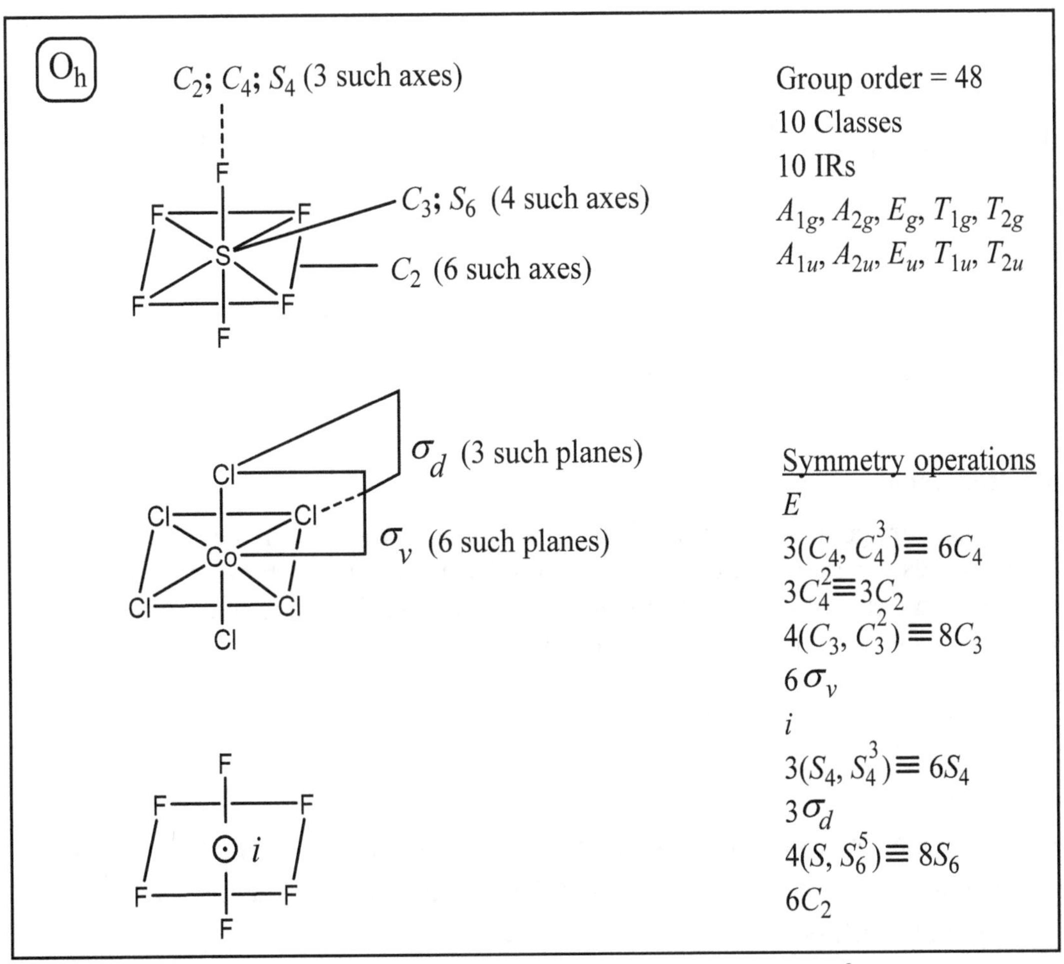

Figure A.14. Point group ***O_h*** for SF_6 and $[CoCl_6]^{-2}$.

X. Linear molecules are of exceptionally high symmetry. They have point groups of infinite order. Even an infinitesimally small rotation around the molecular axis generates an equivalent configuration. Thus the principal axis is C_∞ as $n = 2\pi/\theta \rightarrow \infty$ when $\theta \rightarrow 0$. When there is no inversion center such as in HCN, there are infinite number of vertical planes and the point group is $\boldsymbol{C_{\infty v}}$. A linear molecule like CO_2 has a center of inversion. It has an infinite number of C_2 axes each perpendicular to the principal axis. The proper rotational subgroup for such a molecule is $\boldsymbol{D_\infty}$. Since there is a horizontal plane of symmetry the point group is $\boldsymbol{D_{\infty h}}$. These two cases are illustrated by Figure A.15.

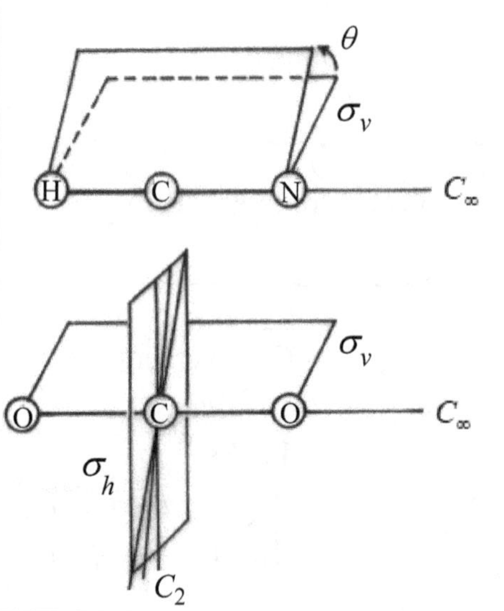

$$n\sigma_v$$

$$n = \lim_{\theta \to 0} \frac{2n}{\theta} = \infty$$

$$C_\infty \oplus \infty\sigma_v \Rightarrow C_{\infty v}$$

$$C_\infty \oplus \infty C_2 \perp C_\infty \Rightarrow D_\infty$$

$$D_\infty \oplus \sigma_h \Rightarrow D_{\infty h}$$

Figure A.15 Point groups of linear molecules.

IIIG. Matrix Representations

Symmetry operations are geometrical operations. The change in the coordinates of a given point under any of these operations can be described in terms of a linear transformation of coordinates, that is, in terms of a matrix transformation. Suppose that under the symmetry operation **R**, the point **P**(x,y,z) moves to the point **P′**(x′,y′,z′) in Figure A.16. We write

$$\mathbf{R}\begin{pmatrix} x \\ y \\ z \end{pmatrix} = \begin{pmatrix} x' \\ y' \\ z' \end{pmatrix} = \mathbb{R}\begin{pmatrix} x \\ y \\ z \end{pmatrix} \tag{A.8}$$

where $\mathbb{R}$ is a square matrix of rank 3. If we know the linear equations relating the primed coordinates to the unprimed ones explicitly, it will be possible to write the matrix $\mathbb{R}$. In this way one may find matrices like

$$\mathbf{E} = \begin{pmatrix} 1 & 0 & 0 \\ 0 & 1 & 0 \\ 0 & 0 & 1 \end{pmatrix}; \sigma_{xy} = \begin{pmatrix} 1 & 0 & 0 \\ 0 & 1 & 0 \\ 0 & 0 & -1 \end{pmatrix}; \mathbb{i} = \begin{pmatrix} -1 & 0 & 0 \\ 0 & -1 & 0 \\ 0 & 0 & -1 \end{pmatrix}; \text{etc.} \qquad \text{(A.9)}$$

These are *representation matrices*. The matrix representing an anticlockwise rotation through an angle θ around the *z*-axis is found to be

$$\mathbf{C}_{2\pi/\theta} = \begin{pmatrix} \cos\theta & -\sin\theta & 0 \\ \sin\theta & \cos\theta & 0 \\ 0 & 0 & 1 \end{pmatrix}. \qquad \text{(A.10)}$$

The multiplication table for the representation matrices is exactly parallel to that for the corresponding geometrical operations. There is a one-to-one correspondence. Thus if the geometrical operations form a group (the point group) the representation matrices form an isomorphic group. This fact gave rise to the idea of *group representations*.

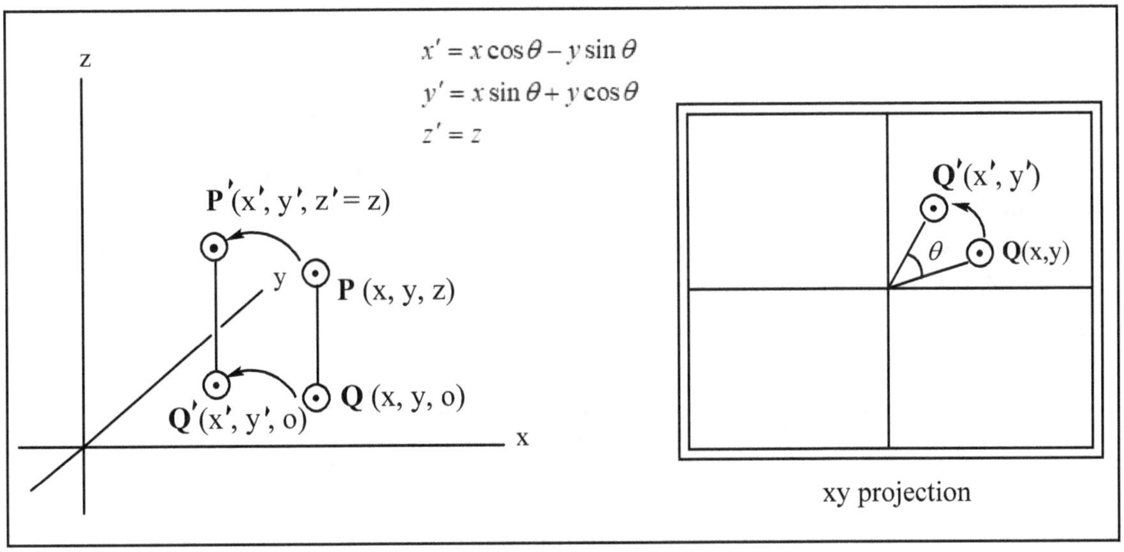

Figure A.16 Rotation in coordinate space: Relationships shown have been used to write the corresponding transformation matrix.

IIIH. Representations of Group

The group representation can be made clear with the help of two examples which we discuss in the following.

Consider the point group of water molecule, C_{2v}. See Figure A.5. The 3 × 3 matrices representing the point group operations are

$$\mathbb{E} = \begin{pmatrix} 1 & 0 & 0 \\ 0 & 1 & 0 \\ 0 & 0 & 1 \end{pmatrix}; \qquad \mathbb{C}_2 = \begin{pmatrix} -1 & 0 & 0 \\ 0 & -1 & 0 \\ 0 & 0 & 1 \end{pmatrix}; \tag{A.11}$$

$$\sigma_V(xz) = \begin{pmatrix} 1 & 0 & 0 \\ 0 & -1 & 0 \\ 0 & 0 & 1 \end{pmatrix}; \quad \sigma_V'(yz) = \begin{pmatrix} -1 & 0 & 0 \\ 0 & 1 & 0 \\ 0 & 0 & 1 \end{pmatrix}.$$

These obey the multiplication table that has elements in one-to-one correspondence with those in the group multiplication table:

$\boldsymbol{C_{2v}}$	$\mathbb{E}$	$\mathbb{C}_2$	σ_V	σ_V'
$\mathbb{E}$	$\mathbb{E}$	$\mathbb{C}_2$	σ_V	σ_V'
$\mathbb{C}_2$	$\mathbb{C}_2$	$\mathbb{E}$	σ_V'	σ_V
σ_V	σ_V	σ_V'	$\mathbb{E}$	$\mathbb{C}_2$
σ_V'	σ_V'	σ_V	$\mathbb{C}_2$	$\mathbb{E}$

⇔

$\boldsymbol{C_{2v}}$	**E**	**C$_2$**	$\boldsymbol{\sigma_V}$	$\boldsymbol{\sigma_V'}$
E	**E**	**C$_2$**	$\boldsymbol{\sigma_V}$	$\boldsymbol{\sigma_V'}$
C$_2$	**C$_2$**	**E**	$\boldsymbol{\sigma_V'}$	$\boldsymbol{\sigma_V}$
$\boldsymbol{\sigma_V}$	$\boldsymbol{\sigma_V}$	$\boldsymbol{\sigma_V'}$	**E**	**C$_2$**
$\boldsymbol{\sigma_V'}$	$\boldsymbol{\sigma_V'}$	$\boldsymbol{\sigma_V}$	**C$_2$**	**E**

See that all the four representations matrices in (A.11) are completely diagonal, that is, all of them are simultaneously diagonalizd into three 1×1 blocks. The group multiplication table is satisfied by the elements of each block which are, in the present case, 1×1 matrices or single numbers:

Element	$\mathbb{E}$	$\mathbb{C}_2$	σ_V	σ_V'	Effect	Name of Representation
(3,3)	1	1	1	1	transformation of z coordinate	A_1
(1,1)	1	-1	1	-1	transformation of x coordinate	B_1
(2,2)	1	-1	-1	1	transformation of y coordinate	B_2
A new set	1	1	-1	-1	transformation of $\boldsymbol{R}_z$ (rotation around the z axis) or $\boldsymbol{L}_z$	A_2

(A.12)

The last line is not visible from the blocks in the representation matrices shown in (A.11) but it can be seen if we write the representation matrices for rotation around the z axis or the angular momentum L_z or even for the product xy.

Each set of numbers shown in (A.12) follows the multiplication table for the point group $\boldsymbol{C_{2v}}$ and is said to form a *representation of the group*. Group representations can be reducible or irreducible. For instance, the matrices in (A.11) form a *reducible representation* as it can be broken down into a sum of representation B_1, B_2 and A_1 as shown in (A.12):

$$\Gamma = B_1 \oplus B_2 \oplus A_1 . \tag{A.13}$$

The representations in (A.12), namely, A_1, A_2, B_1 and B_2, involve the smallest block size $1 \otimes 1$. These are called *irreducible representations* as these cannot be diagonalized any further to form blocks of smaller (or reduced) size.

Not all irreducible representations are of dimension one. There are two-dimensional irreducible representations (general symbol: E), three-dimensional irreducible representations (symbol: T), etc. To see the two-dimensional irreducible representations, take the case of the simple molecule ammonia that has the point group $\boldsymbol{C_{3v}}$ with the multiplication table:

C_{3v}	E	C_3	C_3^2	σ_v'	σ_v''	σ_v'''
E	E	C_3	C_3^2	σ_v'	σ_v''	σ_v'''
C_3	C_3	C_3^2	E	σ_v'''	σ_v'	σ_v''
C_3^2	C_3^2	E	C_3	σ_v''	σ_v'''	σ_v'
σ_v'	σ_v'	σ_v''	σ_v'''	E	C_3	C_3^2
σ_v''	σ_v''	σ_v'''	σ_v'	C_3^2	E	C_3
σ_v'''	σ_v'''	σ_v'	σ_v''	C_3	C_3^2	E

Under these point group operations, the point $\mathbf{P}$(x,y,z) shifts to the location $\mathbf{P}'$(x′,y′,z′) and the corresponding representation matrices are written as

$$\mathbb{E} = \begin{pmatrix} 1 & 0 & 0 \\ 0 & 1 & 0 \\ 0 & 0 & 1 \end{pmatrix}; \quad \mathbb{C}_3 = \begin{pmatrix} -1/2 & -\sqrt{3}/2 & 0 \\ \sqrt{3}/2 & -1/2 & 0 \\ 0 & 0 & 1 \end{pmatrix}; \quad \mathbb{C}_3^2 = \begin{pmatrix} -1/2 & \sqrt{3}/2 & 0 \\ -\sqrt{3}/2 & -1/2 & 0 \\ 0 & 0 & 1 \end{pmatrix}; \tag{A.14}$$

$$\sigma_v' = \begin{pmatrix} 1 & 0 & 0 \\ 0 & -1 & 0 \\ 0 & 0 & 1 \end{pmatrix}; \quad \sigma_v'' = \begin{pmatrix} -1/2 & -\sqrt{3}/2 & 0 \\ -\sqrt{3}/2 & 1/2 & 0 \\ 0 & 0 & 1 \end{pmatrix}; \quad \sigma_v''' = \begin{pmatrix} -1/2 & \sqrt{3}/2 & 0 \\ \sqrt{3}/2 & 1/2 & 0 \\ 0 & 0 & 1 \end{pmatrix}.$$

Gathering the first 2×2 block from each representation matrix, one obtains a two-dimensional irreducible representation E. The next block of dimension one gives the one-dimensional *totally symmetric* irreducible representation A_1. Thus we have the following irreducible representations:

	E	C_3	C_3^2	σ_v'	σ_v''	σ_v'''
A_1:	1	1	1	1	1	1
E:	$\begin{pmatrix} 1 & 0 \\ 0 & 1 \end{pmatrix}$	$\begin{pmatrix} -1/2 & -\sqrt{3}/2 \\ \sqrt{3}/2 & -1/2 \end{pmatrix}$	$\begin{pmatrix} -1/2 & \sqrt{3}/2 \\ -\sqrt{3}/2 & -1/2 \end{pmatrix}$	$\begin{pmatrix} 1 & 0 \\ 0 & -1 \end{pmatrix}$	$\begin{pmatrix} -1/2 & -\sqrt{3}/2 \\ -\sqrt{3}/2 & 1/2 \end{pmatrix}$	$\begin{pmatrix} -1/2 & \sqrt{3}/2 \\ \sqrt{3}/2 & 1/2 \end{pmatrix}$

Another irreducible representation is possible:

A_2:	1	1	1	-1	-1	-1

Note that the matrices forming the irreducible representation E cannot be simultaneously diagonalized. For instance, if one diagonalizes the 2×2 $\mathbf{C_3}$ matrix, the transformed $\mathbf{E}$ matrix of the same rank is no longer diagonal. Also note that the reducible representation in (A.14) is decomposible as

$$\Gamma = \mathrm{E} \oplus \mathrm{A_1} \tag{A.15}$$

where the symbolism may be executed by extending the dimensionality of the respective matrices in appropriate ways while zeroes are taken for all additional elements.

IIII. The Great Orthogonality Theorem

Let us denote the matrices representing the operations $\mathbf{R_1}$, $\mathbf{R_2}$ ... $\mathbf{R_h}$ which form the *ith* irreducible representation as $\Gamma^{(i)}(\mathbf{R_1})$, $\Gamma^{(i)}(\mathbf{R_2})$, ... , $\Gamma^{(i)}(\mathbf{R_h})$. We will also write the (m, n) element of the matrix for $\mathbf{R_k}$ as $\Gamma^{(i)}_{\mathrm{mn}}(\mathbf{R_k})$. Let l_i be the dimension of the *ith* irreducible representation. The dimension and the elements of the matrices forming the *jth* irreducible representations are written in a similar way. Here we state without proof the so-called *Great Orthogonality theorem*,

Theorem 26. The Great Orthogonality Theorem

$$\frac{1}{h}\sum_{\mathbf{R}}\Gamma^{(i)}_{mn}(\mathbf{R})^*\Gamma^{(j)}_{pq}(\mathbf{R}) = \frac{1}{\sqrt{l_i l_j}}\delta_{ij}\delta_{mp}\delta_{nq}\,. \tag{A.16}$$

This theorem has been found to be extremely useful in the preparation and characterization of irreducible representations.

Matrices are changed by a similarity transformation, but the trace of a matrix remains invariant under such a transformation. The trace is hence called the *character* of the matrix. The character of the *ith* irreducible representation matrix for operation $\mathbf{R}$ is

$$\chi^{(i)}(\mathbf{R}) = \sum_{m=1}^{l_i}\Gamma^{(i)}_{mm}(\mathbf{R})\,. \tag{A.17}$$

A similarity transformation converts the operations belonging to the same class into one another. Therefore, operations belonging to the same class have the same character.

The group multiplication table is fully satisfied if we replace all the elements by 1. This action forms the totally symmetric irreducible representation that is necessarily a one-dimensional representation. We have already indicated that this representation is named $\mathrm{A_1}$. Characters of the matrices belonging to the totally symmetric representation are all equal to 1.

The Great Orthogonality theorem can be reduced to

Theorem 27. The Character Orthogonality Theorem

$$\frac{1}{h}\sum_{\mathbf{R}}\chi^{(i)}(\mathbf{R})^*\chi^{(j)}(\mathbf{R}) = \delta_{ij}\,. \tag{A.18}$$

Because $\chi^{(i)}(\mathbf{E}) = l_i$, we get

Theorem 28. The Dimensionality Theorem

$$\sum_i l_i^2 = h. \qquad \text{(A.19)}$$

The character orthogonality theorem (A.18) can be written in terms of characters for classes as

Theorem 29. The Class Orthogonality Theorem

$$\frac{1}{h}\sum_C n_C \chi^{(i)}(C)^* \chi^{(j)}(C) = \delta_{ij} \qquad \text{(A.20)}$$

where n_C is the number of elements in class C. A corollary of this theorem is quoted here without proof:

Corollary: Number of irreducible representations = Number of classes. (A.21)

For an extensive proof the interested reader may consult "Group Theory" by E. P. Wigner or "Group Theory and Quantum Mechanics" by M. Tinkham.

Another type of orthogonality relationship, called the *second orthogonality*, exists. This is written as

Theorem 30. The Second Orthogonality Theorem

$$\sum_i \chi^{(i)}(C_j)^* \chi^{(i)}(C_k) = \frac{h}{n_j}\delta_{jk} \qquad \text{(A.22)}$$

where the sum is over all the irreducible species.

These orthogonality relationships, namely, the dimensionality theorem, class orthogonality and the second orthogonality, along with the observation (A.21) can be used to set up the table of characters often by mere inspection.

IIIJ. Character Table

The $\boldsymbol{C_{2v}}$ point group has four elements: geometrical operations $\mathbf{E}$, $\mathbf{C_2}$, $\boldsymbol{\sigma_v}$ and $\boldsymbol{\sigma_v'}$. Each element forms a class by itself, that is, each is conjugated to itself. Thus there are four classes and by (A.21), four irreducible representations. Therefore,

$$l_1^2 + l_2^2 + l_3^2 + l_4^2 = h = 4 \text{ (by the dimensionality theorem).}$$

Since l_1, l_2, l_3 and l_4 must be positive integers, the only possibility is $l_1 = l_2 = l_3 = l_4 = 1$. Characters of the four irreducible representations A_1, A_2, B_1 and B_2 can be found by considering that A_1 is the totally symmetric representation with all characters equal to 1, the character of $\mathbf{E}$ is 1 in every irreducible species, the character orthogonality theorem stipulates that all other irreducible representations must have 1 as the character of two operations and –1 as the character of another two operations, that is, –1 is to be assigned for two of the three operations. The latter can be done in three unique ways and indeed we have three different irreducible representations that differ from A_1. These assignments are shown in (A.12) and they ensure the orthogonality of every two irreducible representations. The second orthogonality theorem may be used to confirm.

There are three classes in C_{3v} so that there are three irreducible species by (A.21). The dimensionality theorem gives

$$l_1^2 + l_2^2 + l_3^2 = h = 6.$$

This allows the possibility $l_1 = 1$, $l_2 = 1$ and $l_3 = 2$. There are two one-dimensional representations, A_1 and A_2, and one two-dimensional representation, E. Corresponding characters can be quickly found out from the character orthogonality relationships given by (A.20), with $n_1=1$, $n_2=2$ and $n_3=3$. These characters are revealed in Table A.3.

Table A.3 Character table for group C_{3v}.

C_{3v}	E	$2C_3$	$3\sigma_v$	Functions
A_1	1	1	1	$z;\ r;\ z^2;\ x^2+y^2;\ r^2$
A_2	1	1	-1	$xy;\ R_z$
E	2	-1	0	$(x, y);\ (R_x, R_y);\ (xz, yz)$

One may very easily check that the column orthogonality holds.

Each function shown on the right side of the character table transforms in a certain way under the group operations. The transformation matrices form the corresponding irreducible representation of the group. Traces or characters of these irreducible matrices have been shown along each row. Some functions such as x and y transform together under the operations of point group C_{3v}. They form the two-dimensional irreducible representation E in this specific case. Along with the function-variable z, they may form a three-dimensional irreducible representation in case of point groups of higher orders. The functions R_x, R_y and R_z are rotational operations: the corresponding quantum mechanical operators are $\boldsymbol{L}_x$, $\boldsymbol{L}_y$ and $\boldsymbol{L}_z$. The derivative operator $\frac{\partial}{\partial x}$ transforms as x and so on. Consequently, $\boldsymbol{L}_x$ transforms as yz, etc.

Atomic orbitals of the central atom (such as nigrogen in ammonia) are (i) dependent on the radial distance r and (ii) their orientation-dependent part (the angular wave function) is proportional to $\frac{x}{r}$, $\frac{y}{r}$ or $\frac{z}{r}$ (p_x, p_y or p_z orbital) or the ratio $\frac{xy}{r^2}$ (d_{xy} orbital), etc. The transformation matrices of these functions directly form the irreducible representation against which they are shown in character tables.

For molecular orbitals, one has to investigate the effect of applying the operations of the point group with more care as these operations generally convert the orbitals on one atom into the orbitals of another atom. Consider the molecule cyclobutene that has the point group $\boldsymbol{D_{4h}}$. Elements of $\boldsymbol{D_{4h}}$ are obtained by multiplying every element of $\boldsymbol{C_{4v}}$ with every element of $\boldsymbol{C_i}$, that is, **E** and **i**, (or every element of $\boldsymbol{C_s}$, that is, **E** and $\boldsymbol{\sigma_h}$). Results of the transformations of the p orbitals of the carbon atoms, under the operations of the point group C_{4v}, are given in Table A.4. With these results one may find how the π molecular orbitals are transformed. The corresponding characters are given in Table A.5. As every p_z orbital reverses its sign under the operation σ_h, the transformation properties of the molecular orbitals are as follows:

1. π_1 transforms as A_{1u};
2. π_2 and π_3 together transform as E_g;
3. π_4 transforms as B_{2u}.

These orbitals are usually denoted as a_{1u}, e_g and b_{2u} orbitals (with names in lower case letters).

Table A.4 Transformation of the p_z orbitals of carbon atoms in the cyclobutene molecule under the operations of point group C_{4v}, a subgroup of D_{4h}.

C_{4v}	E	C_4	C_4^3	C_2	σ_v'	σ_v''	σ_d'	σ_d''	i (⇒ D_{4h})
p_{z1}	p_{z1}	p_{z2}	p_{z4}	p_{z3}	p_{z2}	p_{z4}	p_{z1}	p_{z3}	$-p_{z3}$
p_{z2}	p_{z2}	p_{z3}	p_{z1}	p_{z4}	p_{z1}	p_{z3}	p_{z4}	p_{z2}	$-p_{z4}$
p_{z3}	p_{z3}	p_{z4}	p_{z2}	p_{z1}	p_{z4}	p_{z2}	p_{z3}	p_{z1}	$-p_{z1}$
p_{z4}	p_{z4}	p_{z1}	p_{z3}	p_{z2}	p_{z3}	p_{z1}	p_{z2}	p_{z4}	$-p_{z2}$

Table A.5 Characters for the transformation of π molecular orbitals under the operations of point group C_{4v}, a subgroup of D_{4h}.

C_{4v}	E	$2C_4$	C_2	$2\sigma_v$	$2\sigma_d$	I.R.	i (⇒ D_{4h})	I.R.
$\pi_1=\frac{1}{2}(p_{z1}+p_{z2}+p_{z3}+p_{z4})$	1	1	1	1	1	A_1	−1	A_{1u}
$\pi_2=\frac{1}{\sqrt{2}}(p_{z1}-p_{z3})$, $\pi_3=\frac{1}{\sqrt{2}}(p_{z2}-p_{z4})$	2	0	−2	0	0	E	1	E_g
$\pi_4=\frac{1}{2}(p_{z1}-p_{z2}+p_{z3}-p_{z4})$	1	−1	1	−1	1	B_2	−1	B_{2u}

IIIK. Symmetry-Adapted Linear Combination

To form the molecular orbitals we first form *symmetry-adapted linear combinations* (SALCs) or symmetry orbitals. Actually the systematic procedure for preparing the SALCs is to rely on the use of projection operators. A discussion of the method is skipped here. The interested reader may find a general description of the projection operator methodology in "Group Theory and its Application to the Quantum Mechanics of Atomic Spectra" by E. P. Wigner, "Group Theory and Quantum Mechanics" by M. Tinkham or "Chemical Applications of Group Theory" by F. A. Cotton.

In the previous example of cyclobutene, the π orbitals are examples of SALCs. Orbitals of the same symmetry interact with each other and they may have non-vanishing overlap integrals. In fact, *two functions belonging to different irreducible representations or to different rows of the same irreducible representations are always orthogonal to each other.* The final molecular orbitals are obtained as linear combinations of the SALC's (symmetry orbitals) belonging to the same irreducible species. This has been discussed in the text while considering the ligand field theory.

In cyclobutene the actual π_i orbital will be a linear combination of $2p\pi_i$, $3p\pi_i$, $4p\pi_i$, etc. Of course the most prominent contribution comes from $2p\pi_i$, but the contributions from other π orbitals of the same symmetry are non-vanishing. Hence the orbitals $2p\pi$'s are by no means molecular orbitals. They are in reality SALC's.

IIIL. Direct Product Representations

Let the functions $\psi_1 \ldots \psi_m$ and $\varphi_1 \ldots \varphi_n$ form the bases of two matrix representations, the m-dimensional $\Gamma^{(1)}$ and the n-dimensional $\Gamma^{(2)}$, reducible or irreducible. Then the product functions $\psi_i\varphi_j$ span a $(m \times n)$-dimensional matrix representation Γ. This representation, called the direct product and symbolized by

$$\Gamma(\mathbf{R}) = \Gamma^{(1)}(\mathbf{R}) \otimes \Gamma^{(2)}(\mathbf{R}), \tag{A.23}$$

is usually *reducible* to a linear combination of the irreducible species,

$$\Gamma(\mathbf{R}) = \sum_k a_{12k}\Gamma^{(k)}(\mathbf{R}). \tag{A.24}$$

Characters of the direct product representation are related accordingly,

$$\chi(\mathbf{R}) = \sum_k a_{12k}\chi^{(k)}(\mathbf{R}). \tag{A.25}$$

The coefficient a_{12k} can be found from the orthogonality theorem,

$$a_{12k} = \frac{1}{h}\sum_C n_C \chi^{(1)}(C)\chi^{(2)}(C)\chi^{(k)}(C)^*. \tag{A.26}$$

This result is very important, for it is used to determine the direct product representation as a linear combination of type (A.24). Important applications of this procedure are found in an analysis of molecular vibrations, and in different aspects of spectroscopy.

IIIM. Selection Rules

We have already discussed in the text that the selection rules in spectroscopy are based on the question whether the integrand in the transition integral is perfectly symmetric or not.

If it is antisymmetric in some way, the integral will be zero and the transition will turn out to be forbidden. In the language of group theory, the pertinent question is whether the direct product representing the integrand (the product of the initial state function, the final one and the transition moment) *contains* the totally symmetric representation A_1 (or A_{1g}) or not:

$$A_1 \text{ (or } A_{1g}) \in \Gamma(\psi_f) \otimes \Gamma(\text{moment}) \otimes \Gamma(\psi_i)\ ? \qquad \text{(A.27)}$$

The right hand side is in fact a triple product.

Another observation is that the totally symmetric representation always appears in the direct product of a specific representation with itself. If the specific representation is an irreducible representation, the direct product with another irreducible species does not contain the totally symmetric representation. Hence condition (A.27) is satisfied if and only if the direct product $\Gamma(\psi_f) \otimes \Gamma(\psi_i)$ contains Γ(moment).

Therefore, an electric dipole transition will be allowed with *x*, *y* or *z* polarization if the direct product of representations of the initial and final states is or contains the irreducible representation for transformation of the variables *x*, *y* or *z*. So we get the selection rule $\Delta v = \pm 1$ for IR transition, and $\Delta j = \pm 1$ for electron paramagnetic resonance.

Similarly, for a magnetic dipole transition that is involved in a magnetic resonance spectroscopy, the concerned direct product must be or must contain the irreducible representation according to which R_x, R_y or R_z transforms. This gives rise to rules like $\Delta m_s = \pm 1$ in electron spin resonance and $\Delta m_I = \pm 1$ in nuclear magnetic resonance spectroscopies.

Texts Suggested for Further Reading

B. Baumslag and B. Chandler, "Schaum's outline of Theory and Problems of Group Theory", (McGraw-Hill, New York, 1968).

F. A. Cotton, "Chemical Applications of Group Theory", (Wiley, New York, 1971).

I. Hargittai and M. Hargittai, "Symmetry through the Eyes of a Chemist", (VCH Verlags, Weinheim, 1986).

M. Tinkham, "Group Theory and Quantum Mechanics", (McGraw-Hill, New York, 1964).

E. P. Wigner, "Group Theory and its Application to the Quantum Mechanics of Atomic Spectra," (Academic, New York, 1959).

References

Chapter 1

1. Bohr N., *Phil. Mag.* **26**, 1 (1913).
2. Compton A. H., *Phys. Rev.* **21**, 483 (1923); *ibid.* **22**, 409 (1923).
3. Davisson C. J. and Germer L. H., *Phys. Rev.* **30**, 705 (1927).
4. De Broglie L., *Ann. Phys.* **3**, 22 (1925).
5. Debye P., *Ann. Phys.* **39**, 789 (1912).
6. Einstein A., *Ann. Phys.* **17**, 132 (1905).
7. Einstein A., *Ann. Phys.* **22**, 180 (1907).
8. Feynman R. P., *Phys. Rev.* **56**, 340 (1939).
9. Franck J. and Hertz G., *Verh. Dent. Phys. Ges.* **16**, 10 (1914).
10. Gibbs J. W., Scientific Papers, Vol. I (Longmans, Green and Co., New York, 1906).
11. Hellmann H., Einführung in die Quantenchemie, (Franz Denticke, Leipzig, 1937) p. 285.
12. Kekulé F. A., *Ann. Chem.* **104**, 129 (1857); *ibid.* **106**, 154 (1858).
13. Le bel J. A., *Bull. Soc. Chim. France [2]* **22**, 337 (1874).
14. Lennard Jones J. E., *Proc. Roy. Soc. (London)* **A106**, 463 (1924).
15. Lewis G. N., *J. Amer. Chem. Soc.* **38**, 762 (1916).
16. Planck M., *Ann. Phys.* **4**, 553 (1901).
17. Sidgwick N. V., *J. Chem. Soc.* **123**, 725 (1923); *Trans. Faraday Soc.* **19**, 469 (1923); The Electronic Theory of Valency, (Clarendon Press, Oxford, 1927).
18. Sommerfeld A., *Ann. Phys.* **51**, 1 (1916).
19. Stern O. and Gerlach W., *Z. Phys.* **7**, 249 (1921).
20. Thomson G. P., *Proc. Roy. Soc. (London)* **A117**, 600 (1928).
21. Van der Waals J. D., Die Kontinuität des gasförmingen und flüssigen Zustandes, (Barth, Leipzig, 1899-1900).
22. Van't Hoff J. H., *Arch. Neerland. Sci.* **9**, 445 (1874); *Bull. Soc. Chim. France [2]* **23**, 295 (1875).
23. Werner A., *Z. Anorg. Chem.* **3**, 267 (1893).
24. Wilson W., *Phil. Mag.* **29**, 795 (1915).

Chapter 2

1. Born M., *Z. Phys.* **37**, 863 (1926); *ibid.* **38**, 803 (1926); *Nature* **119**, 354 (1927).
2. Born M. and Oppenheimer J. R., *Ann. Phys.* **84**, 457 (1927).
3. Dirac P. A. M., The Principles of Quantum Mechanics, 4th ed., (Clarendon Press, Oxford, 1958).

4. Heisenberg W., *Z. Phys.* **39**, 499 (1926).
5. Schrödinger E., *Ann. Phys.* **79**, 361, 489 (1926); *ibid.* **81**, 109 (1926).

Chapter 3

1. Pauli W., *Z. Physik* **31**, 765 (1925).

Chapter 4

1. Eckart C., *Phys. Rev.* **36**, 878 (1930).
2. Hartree D. R., *Proc. Cambridge Phil. Soc.* **24**, 105 (1928); *Mem. Manchester Phil. Soc.* **77**, 91 (1932-1933).
3. Kellner G. W., *Z. Phys.* **44**, 91, 110 (1927).
4. Lord Rayleigh used this method in 1873 to calculate the vibrational frequencies of mechanical systems; see "Theory of Sound", 2nd rev. ed., Vol. 1, (Macmillan, London, 1937; reprinted by Dover, New York).
5. Ritz W., *J. Reine Angew. Math.* **135**, 1 (1908).
6. Y. Tanabe and S. Sugano, *J. Phys. Soc. Japan* **9**, 753, 766 (1954).

Chapter 5

1. Bose S. N., *Z. Phys.* **26**, 178 (1924).
2. Bush V. and Caldwell S. H., *Phys. Rev.* **38**, 1898 (1931).
3. Dirac P. A. M., *Proc. Roy. Soc.* **A112**, 661 (1926).
4. Einstein A., *Sitzber. Preuss. Akad. Wiss.* (1924) p. 261; *ibid.* (1925) p.3.
5. Fermi E., *Z. Phys.* **36**, 902 (1926).
6. Fermi E., *Atti Accad. Nazl. Lincei.* **6**, 602 (1927); *ibid.* **7**, 342, 726 (1928).
7. Fock V., *Z. Phys.* **61**, 126 (1930).
8. Froese-Fischer C., *Atomic Data* **4**, 302 (1972); *Atomic Data and Nuclear Data Tables* **12**, 87 (1973).
9. Hartree D. R., *Proc. Cambridge Phil. Soc.* **24**, 89 (1928).
10. Hohenberg P. and Kohn W., *Phys. Rev.* **78**, 29 (1964).
11. Hund F., *Z. Phys.* **33**, 345 (1925); *Z. Phys.* **73**, 24, 565 (1931); *Z. Phys.* **74**, 429 (1932).
12. Partridge H., *J. Chem. Phys.* **87**, 6643 (1987); *NASA Technical Memorandum* **101044** (NASA, Moffett Field, 1989).
13. Slater J. C., *Phys. Rev.* **34**, 1293 (1929); **36**, 57 (1930).
14. Stern O. and Gerlach W., *Z. Phys.* **7**, 249 (1921); *Ann. Phys.* **74**, 673 (1924).
15. Thomas L. H., *Proc. Cambridge Phil. Soc.* **23**, 542 (1927).
16. Tosi M. P. and Fumi F. G., *J. Phys. Chem. Solids* **25**, 31, 45 (1964).
17. Uhlenbeck G. E. and Goudsmit S., *Naturwissenschaften* **13**, 953 (1925).

Chapter 6

1. Albright T. A., Burdette J. K. and Whangbo M. H., Orbital Interactions in Chemistry, (Wiley, New York, 1985) Chapters 15-20.
2. Gimarc B. M., Molecular Structure and Bonding: The Qualitative Molecular Orbital Approach (Academic Press, New York, 1979) Chapter 7.
3. Hückel E., *Z. Physik* **60**, 423 (1930); **70**, 204 (1931).
4. Klessinger M. and McWeeny R., *J. Chem. Phys.* **42**, 3343 (1965).
5. Kolos W. and Roothaan C. C. J., *Rev. Mod. Phys.* **32**, 205, 219 (1960).

6. Mulliken R. S., *Rev. Mod. Phys.* **14**, 204 (1942).
7. Quantum Chemistry Program Exchange, Indiana University, Bloomington, Indiana.
8. Walsh A. D., *J. Chem. Soc.*, 2260, 2266, 2288, 2296, 2301, 2306, 2318, 2321, 2325 (1953).
9. Wolfsberg M. and Helmholz L., *J. Chem. Phys.* **20**, 837 (1952).

Chapter 7

1. Dirac P. A. M., *Proc. Roy. Soc. (London)* **A117**, 610 (1928).
2. Epstein P. S., *Phys. Rev.* **28**, 695 (1926).
3. Landé A., *Z. Phys.* **25**, 46 (1924).
4. Löwdin P. O., *J. Molec. Spectros.* **13**, 326 (1964); *J. Math. Phys.* **6**, 1341 (1965).
5. Raman C. V., *Ind. J. Phys.*, **March**, 24 (1928).
6. Schrödinger E., *Ann. Phys.* **80**, 437 (1926).
7. Zeeman P., *Phil. Mag.* **43**, 226 (1897).

Chapter 8

1. Darwent B. De B., Natl. Bur. Stans. (U.S.) Pub. NSRDS-NBS, No. **31** (1970).
2. Dewar M. J. S. and Dougherty R. C., "The PMO Theory of Organic Chemistry", (Plenum, New York, 1975).
3. Dibeler W. H., Walker J. A. and McCullouh K. E., *J. Chem. Phys.* **53**, 4414 (1970).
4. Evans M. G., *Trans. Faraday Soc.* **35**, 824 (1939).
5. Fisk G. A. and Kirtman B., *J. Chem. Phys.* **41**, 3516 (1964).
6. Gillespie R. J., "Molecular Geometry", (Van Nostrand, Princeton, 1972).
7. Gillespie R. J. and Nyholm R. S., *Q. Rev. Chem. Soc.* **11**, 339 (1957).
8. Heitler W. and London F., *Z. Phys.* **44**, 455 (1927).
9. Hinze J., Whitehead M. A. and Jaffé H. H., *J. Amer. Chem. Soc.* **85**, 148 (1963).
10. Hückel E., *Z. Physik* **60**, 423 (1930); **70**, 204 (1931).
11. Kolos W. and Roothaan C. C. J., *Rev. Mod. Phys.* **32**, 205, 219 (1960).
12. Kolos W. and Wolniewicz L., *J. Chem. Phys.* **41**, 3663 (1964).
13. McLean A. D., A. Weiss and M. Yoshimine, *Rev. Mod. Phys.* **32**, 211 (1960).
14. Mulliken R. S., *J. Chem. Phys.* **2**, 782 (1934); *J. Chem. Phys.* **3**, 573 (1935).
15. Pauling L., *J. Amer. Chem. Soc.* **53**, 1367 (1931); *Phys. Rev.* **54**, 899 (1938); *Proc. Roy. Soc.* **A196**, 343 (1949).
16. Pauling L., "The Nature of the Chemical Bond", 3rd. edn., (Cornell University Press, Ithaca, 1960).
17. Sugiura Y., *Z. Phys.* **45**, 484 (1927).
18. Wang S. C., *Phys. Rev.* **31**, 579 (1928).
19. Weinbaum S., *J. Chem. Phys.* **1**, 593 (1933).

Chapter 9

1. Dirac P. A. M., "The Principles of Quantum Mechanics", 4th. edn., (Oxford University Press, New York, 1958).
2. Heisenberg W., "The Physical Principles of the Quantum Theory", (University of Chicago Press, Chicago, 1930).
3. For a discussion on Hilbert Space also see: Powell J. L. and Craseman B., "Quantum Mechanics", (Narosa, New Delhi, 1988) p. 311.
4. Schrödinger E., *Ann. Phys.* **79**, 361, 489 (1926); **81**, 109 (1926).

Chapter 10

1. Fock V. A., "Fundamentals of Quantum Mechanics", (Mir Publishers, Moscow, 1978).

Chapter 11

1. Pauli W., *Z. Physik* **43**, 601 (1927).
2. Pauli W. and Weisskopf V., *Helv. Phys. Acta* **7**, 709 (1934).
3. Russell H. N. and Saunders F. A., *Astrophys. J.* **61**, 38 (1925).

Chapter 12

1. Boys S. F., *Proc. Roy. Soc. (London)* **A200**, 542 (1950).
2. Ditchfield R., Hehre W. J. and Pople J. A., *J. Chem. Phys.* **54**, 724 (1971).
3. Fock V., *Z. Phys.* **61**, 126 (1930).
4. Froese-Fischer C., *Atomic Data* **4**, 302 (1972); *Atomic Data and Nuclear Data Tables* **12**, 87 (1973).
5. Froese-Fischer C., "The Hartree-Fock Theory for Atoms", (J. Wiley, New York, 1977).
6. Hartree D. R., *Proc. Camb. Phil. Soc.* **24**, 89 (1928).
7. Hartree D. R. and Hartree W. R., *Proc. Roy. Soc.* **A166**, 450 (1938).
8. Hehre W. J., Ditchfield R. and Pople J. A., *J. Chem. Phys.* **56**, 2257 (1972).
9. Hehre W. J., Stewart R. F. and Pople J. A., *J. Chem. Phys.* **51**, 2657 (1969).
10. Krishnan R., Binkley J. S., Seeger R. and Pople J. A., *J. Chem. Phys.* **72**, 650 (1980).
11. Löwdin P. O., *J. Chem. Phys.* **18**, 365 (1950).
12. Mulliken R. S., *J. Chem. Phys.* **23**, 1833 (1955).
13. O-Ohata K., Taketa H. and Huzinaga S., *J. Phys. Soc. Jpn* **21**, 2306, 2313 (1966).
14. Pople J. A. and Nesbet R. K., *J. Chem. Phys.* **22**, 571 (1954).
15. Roothaan C. C. J., *Rev. Mod. Phys.* **32**, 179 (1960).
16. Thouless D. J., *Nucl. Phys.* **21**, 225 (1960).

Chapter 13

1. Burrau O., *Kgl. Danske Videnskab. Selskab.* **7**, 1 (1927).
2. Bacon A. D. and Zerner M. C., *Theoret. Chim. Acta* **53**, 21 (1979).
3. Baird N. C. and Dewar M. J. S., *J. Chem. Phys.* **50**, 1262 (1969).
4. Bingham R. C., Dewar M. J. S. and Lo D. H., *J. Amer. Chem. Soc.* **97**, 1285, 1294, 1302, 1307 (1975).
5. Burns G., *J. Chem. Phys.* **42**, 1521 (1964).
6. Clack D. W., Hush H. S. and Yandle J. R., *J. Chem. Phys.* **57**, 3503 (1972).
7. Dewar M. J. S., *Chem. Brit.* **11**, 97 (1975).
8. Dewar M. J. S., Haddon R. C., Li W.-K., Thiel W. and Weiner P. K., *J. Amer. Chem. Soc.* **97**, 4540 (1975).
9. Dewar M. J. S. and Haselbach E., *J. Amer. Chem. Soc.* **92**, 590 (1970).
10. Dewar M. J. S., Lo D. H. and Ramsden C. A., *J. Amer. Chem. Soc.* **97**, 1311 (1975).
11. Edwards W. D. and Zerner M. C., *Theoret. Chim. Acta* **72**, 347 (1987).
12. Gouterman M. and Zerner M. C., *Theoret. Chim. Acta* **4**, 44 (1966).
13. Hase H. L. and Schweig A., *Theoret. Chim. Acta* **31**, 215 (1973).
14. Heitler W. and London F., *Z. Physik* **44**, 455 (1927).
15. Hylleraas E., *Z. Physik* **71**, 739 (1931).

16. Jaffé G., *Z. Physik* **87**, 535 (1934).
17. Mallik B. and Datta S. N., *Int. J. Quantum Chem.* **52**, 629 (1994); Datta S. N. and Mallik B., *Int. J. Quantum Chem.* **53**, 37 (1995).
18. O-Ohata K., Taketa H. and Huzinaga S., *J. Phys. Soc. Japan* **21**, 2306 (1966).
19. Parr R. G., *J. Chem. Phys.* **20**, 239 (1952).
20. Pauling L., *Chem. Rev.* **5**, 185 (1928).
21. Pople J. A., *Trans. Faraday Soc.* **49**, 1375 (1953).
22. Pople J. A., Beveridge D. L. and Dobosh P. A., *J. Chem. Phys.* **47**, 2026 (1967).
23. Pople J. A., Santry D. P. and Segal G. A., *J. Chem. Phys.* **43**, S129 (1965).
24. Pople J. A. and Segal G. A., *J. Chem. Phys.* **43**, S136 (1965).
25. Ridley J. and Zerner M. C., *Theoret. Chim. Acta* **32**, 111 (1973).
26. Santry D. P. and Segal G. A., *J. Chem. Phys.* **47**, 158 (1967).
27. Slater J. C., *Phys. Rev.* **37**, 481 (1931); **38**, 1109 (1931).
28. Slater J. C., Quantum Theory of Atomic Strucure, Vol. **1** (McGraw-Hill, New York, 1960), pp. 339-342.
29. Sugiura Y., *Z. Phys.* **45**, 484 (1927).
30. Sustmann R., Williams J. E., Dewar M. J. S., Allen L. C. and Schleyer P. v. R., *J. Amer. Chem. Soc.* **91**, 5350 (1969).
31. Taketa H., Huzinaga S. and O-Ohata K., *J. Phys. Soc. Japan* **21**, 2313 (1966).
32. Weinbaum S., *J. Chem. Phys.* **1**, 317 (1933).
33. Zerner M. C., ZINDO Program Package (Quantum Theory Project, University of Florida, 1995).
34. Zerner M. C., Loew G. H., Kirchner R. F. and Mueller-Westerhoff U. T., *J. Amer. Chem. Soc.* **102**, 589 (1980).

Chapter 14

1. Berry R. S., *J. Chem. Phys.* **32**, 933 (1960).
2. Brillouin L., *Compt. Rend.* **183**, 24 (1926).
3. Dalton B. J., *Mol. Phys.* **11**, 265 (1966).
4. Dalton B. J., *J. Chem. Phys.* **54**, 4745 (1971).
5. Datta S. N. and Mallik B., *Int. J. Quantum Chem.* **61**, 865 (1997).
6. Kramers H. A., *Z. Physik* **39**, 828 (1926).
7. Marcus R. A., *J. Chem. Phys.* **24**, 966, 979 (1966)
8. Marcus R. A. and Sutin N., *Biochem. Biophys. Acta* **811**, 265 (1985).
9. For WKB see: Merzbacher E., "Quantum Mechanics", 2nd ed., Chap 7, (Wiley, New York, 1970) p. 116.
10. Pitzer K. S. and Bernstein L. S., *J. Chem. Phys.* **63**, 3849 (1975); Bernstein L. S. and Pitzer K. S., *J. Chem. Phys.* **62**, 2530 (1975).
11. Trindle C., Datta S. N. and Bouman T. D., *Int. J. Quantum Chem.* **11**, 627 (1977).
12. Wentzel G., *Z. Physik* **38**, 518 (1926).

Chapter 15

1. Condon E. U., *Phys. Rev.* **28**, 1182 (1926).
2. Einstein A., *Z. Phys.* **18**, 121 (1917).
3. Fermi E., Nuclear Physics, (University of Chicago Press, Chicago,1950).
4. Franck J., *Trans. Farad. Soc.* **21**, 536 (1925); *Z. Phys. Chem.* **120**, 144 (1926).
5. Manson S. T., *Phys. Rev.* **145**, 35 (1966).

Chapter 16

1. Bethe H. A. and Goldstone J., *Proc. Roy. Soc. (London)* **A238**, 551 (1957).
2. Brillouin L., Les Champs 'self-consistents' de Hartree et de Fock, (Hermann and Cie, Paris, 1934) p. 19.
3. Brueckner K. A., Levinson C. A. and Mahmoud H. M., *Phys. Rev.* **95**, 217 (1954); Brueckner K. A., *Phys. Rev.* **100**, 36 (1955); Brueckner K. A. and Gammel J. L., *Phys. Rev.* **109**, 1023, 1040 (1958); Brueckner K. A., The Many Body Problem, (Dunod, Paris, 1959).
4. Goldstone J., *Proc. Roy. Soc. (London)* **A239**, 267 (1957).

AUTHOR INDEX

SUBJECT INDEX

I

J

K

L

Q

R

www.ingramcontent.com/pod-product-compliance
Lightning Source LLC
LaVergne TN
LVHW081947200926
848484LV00009B/2517